NE능률 영어교과서

대한민국 고등학생 **10**명 중 **4.7**명이 보는 교과서

영어 고등 교과서 점유율 1위
(7차, 2007 개정, 2009 개정, 2015 개정)

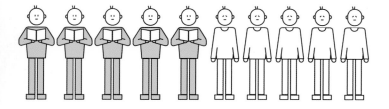

능률보카

그동안 판매된
능률VOCA 1,100만 부

대한민국 박스오피스
천만명을 넘은 영화
단 28개

리딩튜터

그동안 판매된
리딩튜터 1,800만 부
차곡차곡 쌓으면 18만 미터

에베레스트
20배 높이

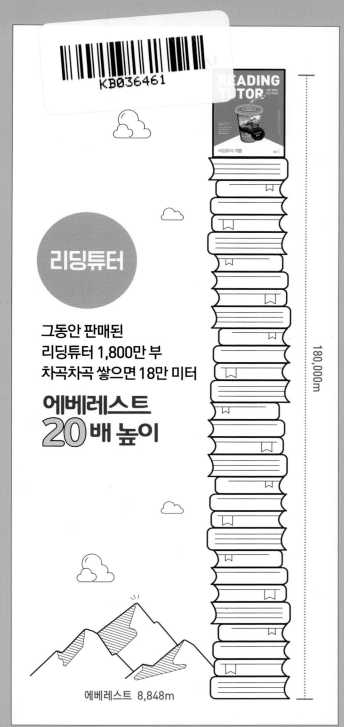

180,000m

에베레스트 8,848m

그래머존

그동안 판매된 400만 부의 그래머존을 바닥에 쭉 ~ 깔면
1000km 서울-부산 왕복가능

서울

부산

수능만만
영어듣기 35회

지은이	NE능률 영어교육연구소
선임연구원	신유승
연구원	이지영 권영주 박예지
원고작성	Danielle Josset Greg Bartz
영문교열	Patrick Ferraro August Niederhaus MyAn Le
디자인	민유화 김연주
내지 일러스트	윤병철
맥편집	김미진
영업	한기영 이경구 박인규 정철교 김남준 김남형 이우현
마케팅	박혜선 고유진 김여진

NE능률이
미래를
창조합니다.

건강한 배움의 고객가치를 제공하겠다는 꿈을 실현하기 위해
42년 동안 열심히 달려왔습니다.

앞으로도 끊임없는 연구와 노력을 통해
당연한 것을 멈추지 않고

고객, 기업, 직원 모두가 함께 성장하는 NE능률이 되겠습니다.

NE 능률

구성과 특징 ☁

하나. 100% 수능형 실전 모의고사

- 수능 개편안을 철저하게 분석·반영하여 문제 유형 및 배치를 실전과 동일하게 구성하였습니다. 또한 기출문제보다 난이도가 비슷하거나 살짝 더 어려우면서 출제 가능성이 높은 문제들로 편성하였습니다. 이를 통해 실제 수능에 대한 실전 감각을 기를 수 있을 것입니다.

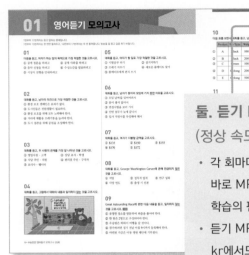

둘. 듣기 MP3 파일 무료 제공
(정상 속도와 빠른 속도)

- 각 회마다 삽입된 QR 코드를 스캔하면 바로 MP3를 청취할 수 있도록 하여 학습의 편의성과 효율성을 높였습니다.
- 듣기 MP3 파일을 www.nebooks.co.kr에서도 무료로 다운로드할 수 있습니다. 빠른 속도(1.2배속)로도 들어볼 수 있어 심화 듣기 학습도 가능합니다.

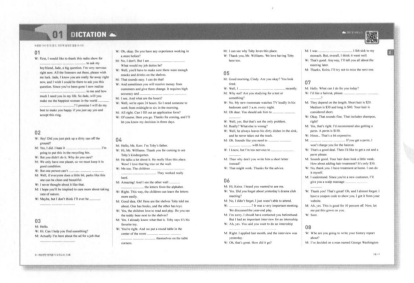

셋. DICTATION을 통한 듣기 집중 훈련

- 각 실전 모의고사 뒤에 DICTATION을 두어 문제의 단서가 되는 부분이나 핵심 어구를 한 번 더 듣고 써 볼 수 있도록 하였습니다. 복습뿐만 아니라 들을 때 주의해야 하는 내용이 어떤 것인지 다시 한 번 되짚어 봄으로써 학습 효과를 높일 수 있을 것입니다.

+ [부록] 수능듣기 MINI BOOK

PART 1 수능듣기 유형 탐구

- 수능 기출문제를 유형별로 정리하고, 최근의 출제 경향을 분석하여 해결 전략을 제시하였습니다.

PART 2 수능듣기 유형별 필수 어휘

- 수능에 출제된 필수 어휘들을 유형별로 정리하여 가지고 다니면서 암기할 수 있도록 하였습니다.

☁ 목차

01회 영어듣기 모의고사 ⋯⋯⋯⋯ p. 6

02회 영어듣기 모의고사 ⋯⋯⋯⋯ p. 12

03회 영어듣기 모의고사 ⋯⋯⋯⋯ p. 18

04회 영어듣기 모의고사 ⋯⋯⋯⋯ p. 24

05회 영어듣기 모의고사 ⋯⋯⋯⋯ p. 30

06회 영어듣기 모의고사 ⋯⋯⋯⋯ p. 36

07회 영어듣기 모의고사 ⋯⋯⋯⋯ p. 42

08회 영어듣기 모의고사 ⋯⋯⋯⋯ p. 48

09회 영어듣기 모의고사 ⋯⋯⋯⋯ p. 54

10회 영어듣기 모의고사 ⋯⋯⋯⋯ p. 60

11회 영어듣기 모의고사 ⋯⋯⋯⋯ p. 66

12회 영어듣기 모의고사 ⋯⋯⋯⋯ p. 72

13회 영어듣기 모의고사 ⋯⋯⋯⋯ p. 78

14회 영어듣기 모의고사 ⋯⋯⋯⋯ p. 84

15회 영어듣기 모의고사 ⋯⋯⋯⋯ p. 90

16회 영어듣기 모의고사 ⋯⋯⋯⋯ p. 96

17회 영어듣기 모의고사 ⋯⋯⋯⋯ p. 102

18회 영어듣기 모의고사 ⋯⋯⋯⋯ p. 108

19회 영어듣기 모의고사 ⋯⋯⋯⋯ p. 114

20회 영어듣기 모의고사 ⋯⋯⋯⋯ p. 120

21회 영어듣기 모의고사 ⋯⋯⋯⋯ p. 126

22회 영어듣기 모의고사 ⋯⋯⋯⋯ p. 132

23회 영어듣기 모의고사 ⋯⋯⋯⋯ p. 138

24회 영어듣기 모의고사 ⋯⋯⋯⋯ p. 144

25회 영어듣기 모의고사 ⋯⋯⋯⋯ p. 150

26회 영어듣기 모의고사 ⋯⋯⋯⋯ p. 156

27회 영어듣기 모의고사 ⋯⋯⋯⋯ p. 162

28회 영어듣기 모의고사 ⋯⋯⋯⋯ p. 168

29회 영어듣기 모의고사 ⋯⋯⋯⋯ p. 174

30회 영어듣기 모의고사 ⋯⋯⋯⋯ p. 180

31회 영어듣기 모의고사 ⋯⋯⋯⋯ p. 186

32회 영어듣기 모의고사 ⋯⋯⋯⋯ p. 192

33회 영어듣기 모의고사 ⋯⋯⋯⋯ p. 198

34회 영어듣기 모의고사 ⋯⋯⋯⋯ p. 204

35회 영어듣기 모의고사 ⋯⋯⋯⋯ p. 210

책속의 책 ⋯⋯⋯⋯⋯⋯ **정답 및 해설**

부록 ⋯⋯⋯⋯⋯⋯ **수능듣기 MINI BOOK**

수능
만만

영 어 듣 기
모 의 고 사

—

01회
-35회

1번부터 17번까지는 듣고 답하는 문제입니다.
1번부터 15번까지는 한 번만 들려주고, 16번부터 17번까지는 두 번 들려줍니다. 방송을 잘 듣고 답을 하기 바랍니다.

01

다음을 듣고, 여자가 하는 말의 목적으로 가장 적절한 것을 고르시오.
① 공개 청혼을 하려고
② 공개 사과를 하려고
③ 음악 신청을 하려고
④ 수상소감을 발표하려고
⑤ 시상식 진행을 안내하려고

02

대화를 듣고, 남자의 의견으로 가장 적절한 것을 고르시오.
① 환경 보호 캠페인은 효과가 없다.
② 도시인들은 전원생활이 필요하다.
③ 환경 보호를 위해 모두 노력해야 한다.
④ 거리에 재활용 쓰레기통을 늘려야 한다.
⑤ 도시 경관을 위해 공원을 조성해야 한다.

03

대화를 듣고, 두 사람의 관계를 가장 잘 나타낸 것을 고르시오.
① 영업사원 - 고객
② 상담 교사 - 학생
③ 식당 주인 - 직원
④ 편의점 주인 - 구직자
⑤ 요리사 - 웨이터

04

대화를 듣고, 그림에서 대화의 내용과 일치하지 않는 것을 고르시오.

05

대화를 듣고, 여자가 할 일로 가장 적절한 것을 고르시오.
① 시험공부 하기
② 설거지하기
③ 쓰레기 치우기
④ 새로운 룸메이트 찾기
⑤ 룸메이트에게 편지 쓰기

06

대화를 듣고, 남자가 동아리 모임에 가지 못한 이유를 고르시오.
① 모임 날짜를 잊어버려서
② 몸이 좋지 않아서
③ 면접시험을 보러 가서
④ 인턴 업무가 늦게 끝나서
⑤ 입사 지원서를 작성해야 해서

07

대화를 듣고, 여자가 지불할 금액을 고르시오.
① $135
② $150
③ $155
④ $170
⑤ $172

08

대화를 듣고, George Washington Carver에 관해 언급되지 않은 것을 고르시오.
① 직업
② 정치적 업적
③ 연구 성과
④ 사망 연도
⑤ 출생 시 신분

09

Great Astounding Race에 관한 다음 내용을 듣고, 일치하지 않는 것을 고르시오. 3점
① 유명한 장소를 방문하여 퍼즐을 풀어야 한다.
② 한 팀은 2명으로 구성되어야 한다.
③ 우승팀은 하와이 여행을 갈 것이다.
④ 참가하려면 경기 전날 아침 9시까지 등록해야 한다.
⑤ 마련된 기금은 아동 병원 재단에 기부된다.

10

다음 표를 보면서 대화를 듣고, 남자가 구입할 차를 고르시오.

	Product	Tea Type	Weight	Packaging	Availability
①	A	Black	100 g	Zip bag	Ships in several days
②	B	Black	200 g	Zip bag	In stock
③	C	Green	100 g	Tin	In stock
④	D	Oolong	200 g	Tin	Ships in several days
⑤	E	Oolong	500 g	Zip bag	In stock

11

대화를 듣고, 남자의 마지막 말에 대한 여자의 응답으로 가장 적절한 것을 고르시오.

① It was held last weekend.
② I won second prize at the fair.
③ I'm not sure yet. I haven't decided.
④ I have always wanted to go to that fair.
⑤ It starts on the 12th and lasts for a week.

12

대화를 듣고, 여자의 마지막 말에 대한 남자의 응답으로 가장 적절한 것을 고르시오.

① I already gave you your allowance this week.
② I told you not to borrow money from your friends.
③ Then you should be more careful with your money.
④ I want you to give my money back as soon as possible.
⑤ No way! I didn't give you permission to invite your friends home.

13

대화를 듣고, 여자의 마지막 말에 대한 남자의 응답으로 가장 적절한 것을 고르시오.

Man: _____

① I do like shaved heads personally.
② No, I don't like such people either.
③ I told you not to judge a man by his appearance.
④ Right. Everyone should look as normal as possible.
⑤ Yes. People should be respected for who they are.

14

대화를 듣고, 남자의 마지막 말에 대한 여자의 응답으로 가장 적절한 것을 고르시오. 3점

Woman: _____

① What type of phone did you get?
② Would you mind if I use your phone?
③ You are so lucky. I never win anything.
④ How can they deceive people like that?
⑤ I wish I could have a smartphone like yours.

15

다음 상황 설명을 듣고, Bob이 Gary에게 할 말로 가장 적절한 것을 고르시오. 3점

Bob: _____

① How ungrateful of you, Gary!
② Thank you for teaching me so much.
③ I hope you get your own car someday.
④ You'll have a fine career, Gary. Good luck.
⑤ All right. I'll introduce him to you as a mentor.

[16~17] 다음을 듣고, 물음에 답하시오.

16

남자가 하는 말의 주제로 가장 적절한 것은?

① the power of the pharaoh
② the process of making mummies
③ a symbol of life in ancient Egypt
④ the religions of ancient Egyptians
⑤ the afterlife that the ancient Egyptians imagined

17

언급된 전시품이 아닌 것은?

① wall   ② goddess   ③ artifact
④ mirror   ⑤ mummy

녹음을 다시 한 번 듣고, 빈칸에 알맞은 말을 쓰시오.

01

W: First, I would like to thank this radio show for _____ _____ _____ _____ to ask my boyfriend, Jade, a big question. I'm very nervous right now. All the listeners out there, please wish me luck. Jade, I know you are really far away right now, and I wish I could be there to ask you this question. Since you've been gone I now realize _____ _____ _____ _____ to me and how much I need you in my life. So Jade, will you make me the happiest woman in the world _____ _____ _____ _____? I promise I will do my best to make you happy if you just say yes and accept this ring.

02

W: Hey! Did you just pick up a dirty can off the ground?

M: Yes, I did. I hate it _____ _____ _____. I'm going to put this in the recycling bin.

W: But you didn't do it. Why do you care?

M: We only have one planet, so we must keep it in good condition.

W: But one person can't _____ _____ _____.

M: Well, if everyone does a little bit, parks like this one can be clean and beautiful.

W: I never thought about it like that.

M: I hope you'll be inspired to care more about taking care of nature.

W: Maybe, but I don't think I'll ever be _____ _____ _____ _____.

03

M: Hello.

W: Hi. Can I help you find something?

M: Actually I'm here about the ad for a job that _____ _____ _____.

W: Oh, okay. Do you have any experience working in a store before?

M: No, I don't. But I am _____ _____ _____. What would my job duties be?

W: Well, you'd have to make sure there were enough snacks and drinks on the shelves.

M: That sounds easy. I can do that!

W: And sometimes you will receive money from customers and give them change. It requires high accuracy and _____ _____ _____ _____.

M: I see. And what are the hours?

W: Well, we're open 24 hours. So I need someone to work from midnight to six in the morning.

M: All right. Can I fill out an application form?

W: Of course. Here you go. Thanks for coming, and I'll let you know my decision in three days.

04

M: Hello, Ms. Kerr. I'm Toby's father.

W: Hi, Mr. Williams. Thank you for coming to see Toby's kindergarten.

M: He talks a lot about it. He really likes this place. Wow! I love that big tree on the wall.

W: Me too. The children _____ _____ _____ _____ _____ _____. They worked really hard.

M: Amazing! And I see the other wall _____ _____ _____ the letters from the alphabet.

W: Right. This way, the children can learn the letters more easily.

M: Good idea. Oh! Here are the shelves Toby told me about. One has books, and the other has toys.

W: Yes, the children love to read and play. Do you see the teddy bear next to the shelves?

M: Yes. I already know what that is. Toby says it's his favorite toy.

W: You're right. And we put a round table in the center of the room _____ _____ _____ _____ _____ _____ themselves on the table corners.

M: I can see why Toby loves this place.

W: Thank you, Mr. Williams. We love having Toby here too.

05

M: Good morning, Cindy. Are you okay? You look tired.

W: Well, I _____ _____ _____ _____ recently.

M: Why not? Are you studying for a test or something?

W: No. My new roommate watches TV loudly in his bedroom until 3 a.m. every night.

M: Oh dear. You should ask him to _____ _____ _____.

W: Well, yes. But that's not the only problem.

M: Really? What else is wrong?

W: Well, he always leaves his dirty dishes in the sink, and he never takes out the trash.

M: Oh. Sounds like you need to _____ _____ _____ _____ with him.

W: I know, but I'm too nervous to _____ _____ _____.

M: Then why don't you write him a short letter instead?

W: That might work. Thanks for the advice.

06

M: Hi, Keira. I heard you wanted to see me.

W: Yes. Did you forget about yesterday's drama club meeting?

M: No, I didn't forget. I just wasn't able to attend.

W: _____ _____? It was a very important meeting. We discussed the year-end play.

M: I'm sorry. I should have contacted you beforehand. But I had an important interview for an internship.

W: Ah, yes. You said you want to do an internship _____ _____ _____ _____.

M: Right. I applied last month, and the interview was yesterday.

W: Oh, that's great. How did it go?

M: I was _____ _____ _____ I felt sick to my stomach. But, overall, I think it went well.

W: That's good. Anyway, I'll tell you all about the meeting later.

M: Thanks, Keira. I'll try not to miss the next one.

07

M: Hello. What can I do for you today?

W: I'd like a haircut, please. _____ _____ _____ _____?

M: They depend on the length. Short hair is $20. Medium is $30 and long is $40. Your hair is considered short.

W: Okay. That sounds fine. That includes shampoo, right?

M: Yes, that's right. I'd recommend also getting a perm. A perm is $150.

W: Hmm... That's a bit expensive.

M: _____ _____ _____. If you get a perm, I won't charge you for the haircut.

W: That's a good deal. Then I'd like to get a cut and a perm please.

M: Sounds good. Your hair does look a little weak. How about adding hair treatment? It's only $30.

W: No, thank you. I have treatment at home. I can do it myself.

M: I understand. Since you're a new customer, I'll give you a scalp massage _____ _____ _____.

W: Thank you! That's great! Oh, and I almost forgot. I have a coupon code to show you. I got it from your website.

M: Ah, yes. This is good for 10 percent off. Now, let me put this gown on you.

W: Sure.

08

W: Who are you going to write your history report about?

M: I've decided on a man named George Washington

Carver.

W: Wasn't he the first president of the United States?

M: That was George Washington. George Washington Carver was a scientist and inventor.

W: Oh, really? What did he invent?

M: He found ways to help poor farmers. For example, he showed them _____ _____ _____ the soil on their farms.

W: How did he do that?

M: He _____ _____ _____ _____ peanuts and soybeans instead of always growing cotton.

W: Peanuts and soybeans? Why was that helpful?

M: They add nitrogen to the soil, which _____ _____ _____.

W: I see. Is he still alive?

M: No, he passed away in 1943. But here's the most impressive thing: He was an African American who was born a slave.

W: You're kidding. That must have made it very difficult for him to succeed.

M: I'm sure it did. But he didn't _____ _____ _____ _____.

09

M: Do you like the thrill of competition? Do you enjoy solving puzzles? Then you will definitely want to _____ _____ this year's Great Astounding Race! In this event, you will solve puzzles as you visit famous landmarks around the city. You need to participate as a team, and each team should be made up of two people. The team that finishes all of the challenges in the shortest amount of time will _____ _____ _____ _____ _____. The race will be on October 5, and it will begin at Central Plaza at 11 a.m. You can register on the day of the race starting at 9 a.m. The registration fee is $20 per team, and all of the funds raised _____ _____ _____ _____ the Children's Hospital Foundation.

10

W: Is there anything I can help you find?

M: I'm looking for some tea _____ _____ _____ as a gift for my mother.

W: How nice! What type of tea does she usually drink?

M: She likes all types of tea. Unfortunately, she has a green tea allergy.

W: Oh, _____ _____ _____.

M: Yes. But I'd like to get her something nice. What sizes do you sell?

W: We have three sizes — 100 g, 200 g, and 500 g.

M: 500 g is too much. Either of the smaller sizes is okay, though.

W: All right. Would you prefer the tea in a tin or a zip bag?

M: A zip bag. It will be easier to carry in my luggage. I'm just _____ _____ _____.

W: I see. Then how about this one? It's not in stock now, but it ships in just a few days.

M: Actually, I'm leaving tomorrow, so I can't wait. I need something that's in stock now.

W: In that case, I recommend this one.

M: Great! I'll take it.

11

M: What's in that big box you're carrying?

W: This is my project for the science fair. It's still unfinished though.

M: I see. When does _____ _____ _____ _____?

W: (It starts on the 12th and lasts for a week.)

12

W: Dad, can you _____ _____ _____?

M: I don't know. I already give you $20 every week.

W: But that's not enough to eat with my friends after school.

M: (Then you should be more careful with your money.)

13

W: We shouldn't _____ _____ _____ by its cover.

M: What are you trying to say, Minyeong?

W: Some teachers believe that students who wear miniskirts or earrings are troublemakers. Do you agree, Minsu?

M: Absolutely not! Everyone has the _____ _____ _____ _____ as they wish.

W: That's what I believe as well.

M: But I must admit, I have treated people differently for having purple hair or a shaved head.

W: Why?

M: I felt that they were just out to cause problems, so I avoided them.

W: We need to be _____ _____ _____ who look different from us.

M: (Yes. People should be respected for who they are.)

14

W: Jake, where are you _____ _____ to?

M: Oh, hi. I'm going to the mobile communication service center.

W: Is there a problem with your phone?

M: Not at all. I got a phone call this afternoon from the telephone company saying that I've won a new smartphone from _____ _____ _____.

W: Wow, that's great! Then, are you going to pick it up now?

M: No, no. I told them my name and address to _____ that I was the winner. After that, I did some research and found out that the whole thing was a scam.

W: What? What kind of scam do you mean?

M: If I use the phone, _____ _____ _____ spend at least ₩30,000 to ₩40,000 a month on phone calls and text messages. If I don't, then I have to pay for the phone. It's not a prize at all!

W: (How can they deceive people like that?)

15

M: Gary wanted to race cars his entire life. One day he met Bob, a famous Formula One race car driver. Gary _____ Bob to teach him the art of race car driving. Bob decided that he liked the look of Gary and took him _____ _____ _____. Soon Gary was driving like a pro and was able to _____ _____ several times around the race track. Even Bob was impressed. Bob went to his manager and told him about Gary. Bob convinced his manager and financial backers to fund a car for Gary. When Bob told Gary he would be getting his own car, he was _____ _____ _____. He told Bob he would always remember his lessons and consider Bob his mentor. In this situation, what would Bob most likely say to Gary?

Bob: (You'll have a fine career, Gary. Good luck.)

16~17

M: Welcome to our special _____ _____ _____ _____. The wall you see here came from the tomb of a pharaoh. If you look here, you can see a symbol that was very common in ancient Egyptian culture. It is called an "ankh". You'll notice there are many just on this wall and on many of the artifacts in the exhibition. The ankh looks like a cross, but unlike Christian crosses, it has a loop _____ _____ _____ _____ it. The ankh is a symbol of life. Here, you see a goddess of the afterlife holding it because the Egyptians believed that she could bring the pharaoh _____ _____ _____. Now if you follow me over here, you will see a mirror that is shaped like an ankh. Actually, the word for *mirror* in Egyptian was *ankh*. Now, _____ _____ _____ walk around the exhibit and look at the numerous ankhs.

1번부터 17번까지는 듣고 답하는 문제입니다.

1번부터 15번까지는 한 번만 들려주고, 16번부터 17번까지는 두 번 들려줍니다. 방송을 잘 듣고 답을 하기 바랍니다.

01

다음을 듣고, 여자가 하는 말의 목적으로 가장 적절한 것을 고르시오.

① 행사 참여를 부탁하려고
② 동아리 행사를 안내하려고
③ 방과 후 활동을 권장하려고
④ 재정적인 후원을 요청하려고
⑤ 행사기금 마련 방법을 바꿔보려고

02

대화를 듣고, 남자의 의견으로 가장 적절한 것을 고르시오.

① 우주여행 비용을 낮춰야 한다.
② 우주 연구는 생존을 위해 필요하다.
③ 지구와 소행성 충돌 가능성은 희박하다.
④ 우주 연구에 대한 정부의 지원이 필요하다.
⑤ 과학 기술의 발달이 행복한 삶을 보장하지는 않는다.

03

대화를 듣고, 두 사람의 관계를 가장 잘 나타낸 것을 고르시오.

① 교사 – 학부모
② 의사 – 환자
③ 학부모 – 학생
④ 교사 – 학생
⑤ 경찰관 – 교사

04

대화를 듣고, 그림에서 대화의 내용과 일치하지 않는 것을 고르시오.

05

대화를 듣고, 여자가 남자에게 부탁한 일로 가장 적절한 것을 고르시오.

① to save her place in line
② to switch tickets with her
③ to recommend a good play
④ to let her borrow his cell phone
⑤ to give directions to the restroom

06

대화를 듣고, 남자가 기뻐할 수 없는 이유를 고르시오.

① 선수로 뛰지 못해서
② 축구 시합에서 져서
③ 동료가 부상을 당해서
④ 컨디션이 좋지 않아서
⑤ 친구한테 병문안을 가야 해서

07

다음을 듣고, 남자가 지불할 금액을 고르시오. 3점

① $99
② $110
③ $126
④ $140
⑤ $198

08

대화를 듣고, Ferdinand Magellan에 관해 언급되지 않은 것을 고르시오.

① 출생연도
② 국적
③ 항해 목적
④ 사망연도
⑤ 사망 원인

09

Breakwater Resort's activity에 관한 다음 내용을 듣고, 일치하지 않는 것을 고르시오.

① 내일 오전에 배드민턴 경기가 개최된다.
② 고래 관찰 여행은 11시에 시작한다.
③ 내일 오후에 전통 무용 공연이 마당에서 열린다.
④ 수영장에서 어린이를 위한 행사가 진행된다.
⑤ 특별 저녁 뷔페가 반값에 제공된다.

10

다음 표를 보면서 대화를 듣고, 여자가 구입할 선풍기를 고르시오.

Cool Star Electric Fans

	Model	Number of Speeds	Safety Certificate	Remote Control	Warranty
①	A	2	X	O	1 year
②	B	3	X	X	1 year
③	C	3	O	O	2 years
④	D	4	O	X	3 years
⑤	E	4	O	O	3 years

11

대화를 듣고, 여자의 마지막 말에 대한 남자의 응답으로 가장 적절한 것을 고르시오.

① Maybe I can show it to you again.
② Oh no, I'm late for my dance class!
③ My friend told me she wanted to watch it.
④ I hope I can go to a concert with you sometime.
⑤ Next time, let's go see something more interesting.

12

대화를 듣고, 남자의 마지막 말에 대한 여자의 응답으로 가장 적절한 것을 고르시오.

① No, I've never seen that fashion designer.
② Yes, I wish you could appear on the runway.
③ Yes. I'm not particularly interested in fashion.
④ Yes, but I heard it could be hard to work in that industry.
⑤ Congratulations! I knew that you could be a fashion designer.

13

대화를 듣고, 남자의 마지막 말에 대한 여자의 응답으로 가장 적절한 것을 고르시오.

Woman: _____

① Sorry, I'm on a diet.
② Can you also make me one?
③ Thank you. It's my favorite food.
④ Good. It'll be ready in 10 minutes.
⑤ I'd rather have a snack than eat nothing.

14

대화를 듣고, 여자의 마지막 말에 대한 남자의 응답으로 가장 적절한 것을 고르시오. 3점

Man: _____

① Speaking English is important in business.
② You're doing a lot to improve your English.
③ Well, no wonder you didn't improve much.
④ I was nervous while speaking to that stranger.
⑤ There's nothing better than lying around watching movies.

15

다음 상황 설명을 듣고, 직원이 Karen에게 할 말로 가장 적절한 것을 고르시오. 3점

Employee: _____

① We can't give you another discount.
② Please keep your ticket with you from now on.
③ You should have reserved your tickets in advance.
④ Don't worry. We'll be able to find out who stole your ticket.
⑤ Thanks for offering, but I can just give him a call after the movie.

[16~17] 다음을 듣고, 물음에 답하시오.

16

여자가 하는 말의 주제로 가장 적절한 것은? 3점

① why comets are on fire
② the components of a comet
③ how comets affect the Earth
④ the relationship between comets and the sun
⑤ the difference between comets and burning rocks

17

언급된 물질이 아닌 것은?

① dust　　② water　　③ ice
④ gas　　⑤ metal

녹음을 다시 한 번 듣고, 빈칸에 알맞은 말을 쓰시오.

01

W: Hi, everyone. This is the time of the year when our club _____ _____ to pay for our upcoming events. Traditionally, we have tried to find some creative ways to ask our family members to donate some money to our cause. Although this has been successful in the past, I don't think it's desirable for us to depend on our parents to financially support this club. We are old enough to _____ _____ _____ _____ _____. Most of us are working some kind of part-time job after school or on the weekends, aren't we? So why don't we try something novel and think of a new way to earn money ourselves for our club's budget, rather than _____ _____ _____?

02

M: Did you hear that James Cameron is supporting research in space?

W: Yes, but I don't see the value in it.

M: Really? What if we _____ _____ _____ _____ headed straight for Earth?

W: That doesn't seem very likely.

M: It's happened before, and scientists say it could happen again. But right now, we don't have the technology to save ourselves.

W: That sounds very scary. But if scientists are already _____ _____ _____ _____, they're probably preparing for it.

M: The way to prepare for it is by doing space research!

W: Yeah, I see now. I guess there is _____ _____ _____ _____ after all.

M: That's right!

03

W: Hello, Mr. Jones. Thank you for coming today. We need to talk about Bob.

M: Hello, Ms. Peters. Yes, I agree.

W: I'm very concerned about him. He started out the year with a good attitude.

M: Yes, he enjoyed your classes very much.

W: But he _____ _____ _____ _____ dramatically throughout the year.

M: My wife and I have noticed that his grades have gone down, and we're _____ _____ it.

W: Well, he hasn't been doing his homework, and he's been late 25 times already.

M: I see. I wasn't aware of those things. I will talk to him about that.

W: I am more concerned about his wellbeing than about his performance though. Is there anything going on at home?

M: _____ _____ _____, Bob has had a tough year. Both of his grandparents passed away.

W: That explains a lot. I am sorry _____ _____ _____. But now that I know, I can try to find a better way to help Bob.

M: I appreciate your concern.

04

M: Do you remember Ms. Brown, the new employee? She is working at the check-in counter.

W: Is she the lady _____ _____ _____?

M: Yes. Let's observe how she's doing.

W: Okay. Oh, the man who is checking in seems to be visiting Hawaii.

M: I think so, too. He's wearing a short-sleeved shirt and has sunglasses on his head. He'll _____ _____ _____ in Hawaii.

W: But I wonder where that boy waiting in the line is going.

M: Yes, he's wearing a sweater. He might be going somewhere cold. _____ _____ _____ at the back is too.

W: She is even carrying her coat over her arm.

M: Right. Oh, it's already ten to ten. We have a

meeting at 10 o'clock. Let's hurry.

05

W: Excuse me, sir. Would you mind if I _____ _____ _____ _____?

M: Not at all. We've been waiting in this line so long. I suppose you have things you need to do.

W: Yes. Actually, I need to _____ _____ _____ _____.

M: Oh, I see.

W: Could you _____ _____ _____ while I go over by the restroom and use my phone?

M: Certainly, but don't take too long. The line is moving slowly now, but it may speed up.

W: Yes. I would hate to miss my chance to get tickets.

M: That would be terrible. I've seen this play twice already, and I love it. You don't want to miss it.

W: That's what I heard. Anyway, thanks for _____ _____ _____.

M: It's my pleasure.

06

W: Hi, Kevin. You had a soccer game yesterday, didn't you?

M: Yes. We _____ _____ _____ _____ in the championship game.

W: Really? So, what was the final score?

M: We defeated them in overtime, four to three.

W: That's great. But you don't seem to be very excited about it.

M: Well, our goalkeeper _____ _____ toward the end of the game.

W: Oh, I see. I hope it wasn't anything serious.

M: He hit his head pretty hard on the goalpost. He's still in the hospital.

W: That's terrible. I guess soccer can be a rough sport.

M: Yes. Anyway, I really _____ _____ _____. I'm going to visit him later.

W: That's a good idea. I'm sure he will appreciate that.

07

W: Hi! Welcome to the My Town Fitness Center!

M: Hi. I'm thinking about joining. Can you tell me how much _____ _____ _____ costs?

W: Well, there's a two-day plan for $90 per month and a three-day plan for $120 per month.

M: Does that mean I can work out either two or three days per week?

W: Exactly. And there's also an unlimited plan for $200.

M: Three days a week would be enough for me. And do you have work-out clothes _____ _____ _____?

W: Yes, but that's an extra $20 per month.

M: That's fine. It's more convenient than doing laundry three times a week.

W: I feel the same way. Oh, are you a resident of this neighborhood?

M: Yes, I am. I live a couple of blocks away. Would I get a discount?

W: You would! Local residents get 10% off the total cost.

M: Excellent! _____ _____ _____. I'll sign up for a one-month membership today.

08

W: Who is the man in that painting?

M: That's Ferdinand Magellan, a famous explorer born in 1480.

W: Oh, really? His name sounds Spanish.

M: Actually, he was Portuguese. However, he was supported by the king of Spain.

W: Why is he so famous?

M: He _____ _____ _____ _____ to sail around the world. Actually, he wanted to go to

Indonesia to trade for spices. But he died before he could.

W: What happened?

M: He was killed _____ _____ _____ in the Philippines.

W: That's too bad. Did the expedition turn back?

M: No. The remaining crew went to Indonesia and got the spices they wanted. They finally returned home in 1522, three years after the start of the expedition.

W: That sounds impressive.

M: Yes, and that's why he _____ _____ _____ a great explorer.

09

M: Can I have your attention, please? I'd like to take a moment to announce the Breakwater Resort's activity schedule for tomorrow. We'll start the day with a family badminton tournament on the beach at 10 a.m. An hour later, our _____ _____ _____ will be leaving from the dock. It's an experience you really shouldn't miss. In the afternoon, we'll be having _____ _____ _____ _____ in the courtyard at 3, followed by exciting games for children in the main pool at 4:30. Finally we'll be having a special cocktail hour before dinner, with our bartenders serving up special drinks _____ _____ _____. It promises to be an exciting day, so I suggest you all get a good night's sleep.

10

W: I've decided to buy one of these Cool Star fans, but I don't know which one to order.

M: They all look similar. How are they different?

W: Well, _____ _____ _____, they have a different number of speeds. See, this one only has two.

M: Two speeds? Slow and fast?

W: Exactly. I need a fan with at least three speeds.

And some of them have a safety certificate.

M: Does that mean they're safer?

W: Yes, it does. I wouldn't buy a fan without a safety certificate.

M: I see. What about a remote control?

W: I definitely need one so I can turn the fan off _____ _____ _____ _____ _____.

M: Do you want one with a warranty too?

W: Yes. _____ _____, _____ _____. Electric fans often break down.

M: Well, then I guess this is the one you should get.

11

W: So what did you think about the show?

M: I really liked the dancers and the music, but _____ _____ _____ _____.

W: I'm surprised to hear that.

M: (Next time, let's go see something more interesting.)

12

M: What do you think would be a great job to have someday?

W: I want to _____ _____ _____ _____.

M: Wow! That would be amazing.

W: (Yes, but I heard it could be hard to work in that industry.)

13

W: Honey, I'm home. Where are you?

M: I'm in the bedroom.

W: I'm sorry I'm late. Have you eaten dinner?

M: No, I'm not hungry. I _____ _____ _____.

W: That doesn't mean you should skip dinner. You know that I hate it when you skip dinner.

M: Seriously, I'm not hungry.

W: Well, I'm going to _____ _____ _____ to eat.

What do you want?

M: Honey, I'm fine. I'll eat something later if I'm hungry.

W: What do you mean later? It's 9 o'clock already.

M: All right. Maybe I'll just _____ _____ _____ _____.

W: (Good. It'll be ready in 10 minutes.)

14

W: Wow, Jim! I'm so impressed!

M: What do you mean?

W: Well, you spoke Korean _____ _____ when that woman asked you for directions to the National Folk Museum.

M: Oh... Thanks. I guess it's because I've been practicing every day.

W: Really? How do you study?

M: I just practice speaking along to a radio program for learning Korean, and I always _____ _____ _____ with Koreans.

W: I wish I had your confidence. I get so embarrassed whenever I speak English to strangers.

M: But you studied English in America for six months, didn't you?

W: Yes, but I was _____ _____ _____ _____ _____ class. I mostly stayed in my dormitory and watched Korean movies.

M: (Well, no wonder you didn't improve much.)

15

M: Karen and her boyfriend go to the theater to watch a film. After they find their seats and sit down, Karen wants to _____ _____ _____ _____ at the snack counter before the movie starts. When she goes, however, she _____ _____ _____ on her seat. When she tries to re-enter the theater, she is stopped at the door by an employee. He asks to see her ticket, but she doesn't have it with her. The employee suggests calling her boyfriend, but her cell phone is in her jacket, which she also

left at her seat. Karen says she can go outside and _____ _____ _____ to call him, but the movie is starting and the employee decides to trust her. In this situation, what would the employee most likely say to her?

Employee: (Please keep your ticket with you from now on.)

16~17

W: Today, we are going to learn about comets. When people think of comets, they often think of giant burning rocks flying through space, but this is a very inaccurate image. _____ _____ _____ comets not on fire; they aren't even rocks! Comets _____ _____ _____ _____ dust, water, ice, and frozen gases. There might be some rocks in a comet, but they definitely do not make up a majority. So comets, far from burning, are actually extremely cold. They are sort of like dirty snowballs. Now _____ _____ _____ what the long trail behind a comet is. Well, solar radiation from the sun causes the gases in the comet to vaporize. The solar winds then carry the gases away from the comet, thus creating a tail. So a comet's tail is made of vaporized gases and dust. I hope that after this short introduction, you will _____ _____ _____ _____ comets as burning pieces of rock!

1번부터 17번까지는 듣고 답하는 문제입니다.
1번부터 15번까지는 한 번만 들려주고, 16번부터 17번까지는 두 번 들려줍니다. 방송을 잘 듣고 답을 하기 바랍니다.

01

다음을 듣고, 남자가 하는 말의 목적으로 가장 적절한 것을 고르시오.
① 대설 주의보에 대해 경고하려고
② 건물의 공사 현황을 보고하려고
③ 제설 작업에 참여를 요청하려고
④ 기상 악화로 인한 휴업을 알리려고
⑤ 건물의 미흡한 보수 관리에 대해 사과하려고

02

대화를 듣고, 여자의 의견으로 가장 적절한 것을 고르시오.
① 생선류가 다이어트에 효과적이다.
② 정기적인 성인병 검진을 받아야 한다.
③ 지방 함유량이 높은 음식은 건강에 해롭다.
④ 지방 함유량이 적은 음식은 영양소가 부족하다.
⑤ 고열량 음식이 반드시 살찌게 하는 것은 아니다.

03

대화를 듣고, 두 사람의 관계를 가장 잘 나타낸 것을 고르시오.
① 매니저 - 직원   ② 사장 - 직원
③ 면접관 - 지원자   ④ 검사관 - 공장 관리자
⑤ 회계사 - 공장 직원

04

대화를 듣고, 그림에서 대화의 내용과 일치하지 않는 것을 고르시오.

05

대화를 듣고, 남자가 할 일로 가장 적절한 것을 고르시오.
① 시 짓기   ② 음식을 요리해주기
③ 초대장 보내기   ④ 쇼핑 목록 작성하기
⑤ 요리법 공유하기

06

대화를 듣고, 여자가 은행을 방문한 이유를 고르시오.
① 환전하려고
② 돈을 이체하려고
③ 공과금을 내려고
④ 계좌를 새로 개설하려고
⑤ 새로운 은행 상품에 가입하려고

07

대화를 듣고, 여자가 지불할 금액을 고르시오. 3점
① $3   ② $6   ③ $9   ④ $15   ⑤ $24

08

대화를 듣고, science club에 관해 언급되지 않은 것을 고르시오.
① 회원 수   ② 모임 장소   ③ 담당 교사
④ 활동 내용   ⑤ 가입 요건

09

Westchester Mushroom Festival에 관한 다음 내용을 듣고, 일치하지 않는 것을 고르시오.
① 3일간 열리는 행사이다.
② 식용 야생 버섯에 관한 정보를 알려주는 게 취지이다.
③ 지역 전문가들의 발표회가 있다.
④ 12세 이하 어린이들의 입장료는 인당 4달러이다.
⑤ 마지막 날에 경품 행사가 있다.

10

다음 표를 보면서 대화를 듣고, 남자가 수강할 강좌를 고르시오.

Schedule of English Classes

	Class	Subject	Instructor	Days	Time
①	A	Listening Skills	Johnson	Mon./Wed.	6 p.m. - 7 p.m.
②	B	Listening Skills	Johnson	Tue./Thur.	8 p.m. - 9 p.m.
③	C	Reading Skills	Kim	Tue./Thur.	7 p.m. - 8 p.m.
④	D	Reading Skills	Garcia	Thur./Fri.	8 a.m. - 9 a.m.
⑤	E	Writing Skills	Garcia	Mon./Fri.	9 a.m. - 10 a.m.

11

대화를 듣고, 남자의 마지막 말에 대한 여자의 응답으로 가장 적절한 것을 고르시오.

① I need something to drink.
② I'll have some orange juice.
③ Sure. I'll take a cup of coffee.
④ I'm going to sleep well tonight.
⑤ This is my favorite coffee shop.

12

대화를 듣고, 여자의 마지막 말에 대한 남자의 응답으로 가장 적절한 것을 고르시오.

① You do not need to go there now.
② How about going climbing this Saturday?
③ These climbing boots are not comfortable.
④ You can buy the equipment at a sports store.
⑤ You need outdoor clothing, boots, and a walking stick.

13

대화를 듣고, 남자의 마지막 말에 대한 여자의 응답으로 가장 적절한 것을 고르시오. 3점

Woman: _____

① You really need to spend more time on this.
② You're our company's most dedicated trainee.
③ Sorry, the computer class has already finished.
④ All right. I'll tell Ms. Smith to wait a bit longer.
⑤ Sure. We don't have to be anywhere until 3.

14

대화를 듣고, 여자의 마지막 말에 대한 남자의 응답으로 가장 적절한 것을 고르시오. 3점

Man: _____

① Mr. Cho will be happy to hear that.
② I guess he must have made a mistake.
③ I'll tell him that you solved the problem.
④ She must have thought he'd be back soon.
⑤ Then I'll send her to your desk right away.

15

다음 상황 설명을 듣고, Elaine이 Jessica에게 할 말로 가장 적절한 것을 고르시오.

Elaine: _____

① We should have gone there together.
② Hopefully, it will clear up after lunch.
③ I'm sorry I didn't ask you to come along.
④ We still have time to ride the roller coaster.
⑤ That must have been such a disappointment.

[16~17] 다음을 듣고, 물음에 답하시오.

16

여자가 하는 말의 주제로 가장 적절한 것은?

① different uses of drones
② the history of the airplane
③ how the first drone was invented
④ useful features of modern aircraft
⑤ differences between drones and airplanes

17

언급된 분야가 <u>아닌</u> 것은?

① the military　② journalism　③ delivery
④ architecture　⑤ agriculture

녹음을 다시 한 번 듣고, 빈칸에 알맞은 말을 쓰시오.

01

M: Can I have the attention of all building residents, please? As you know, yesterday's big storm brought with it a huge amount of snowfall. Under normal circumstances, the building's maintenance staff takes care of _____ _____ _____ _____ as part of their daily routine. Unfortunately, this blizzard caught us unaware and we have been unable to clear the building's driveway and walkways in a timely manner. Therefore we are requesting the help of any residents who _____ _____ _____ _____ this morning. As many schools and businesses have closed, we hope that some of you are available and willing to spare 30 minutes to _____ _____ _____ _____. We'd sincerely appreciate any help we can get.

02

M: It's time for lunch. I feel like having a cheeseburger and French fries.

W: How about getting something a little less fatty? That kind of food just _____ _____ _____ _____ without giving you many nutrients.

M: I know, but it tastes good.

W: If you keep eating foods high in fat because they taste good, you might _____ _____ like high blood pressure.

M: Hmm... I guess you have a point there.

W: We can get something _____ _____ _____ that still tastes good. How about sushi?

M: All right. From now on, I'll consider what you said when choosing food.

03

W: We're from the Ministry of Labor. We were sent here for a routine safety check.

M: All right. Do you need to see the facilities?

W: Yes, but can you answer a few questions first?

M: Sure. Go ahead.

W: Does this factory _____ _____ _____ _____?

M: Yes, it does.

W: And have you seen any accidents on the job?

M: Not for five years. Before that, yes.

W: That was before the new management took over?

M: Yes, it was.

W: Do you think your workers _____ _____ _____?

M: Yes. We receive few complaints. They'd like to _____ _____ _____, of course.

W: Yes, of course. Thank you.

04

W: Excuse me. Can I use the baskets with two handles?

M: Sure. Here you go. Is there anything special you're looking for? We have _____ _____ _____ _____.

W: Oh, really? How much are these bananas?

M: They're $10 for a basket. There are four bunches in a basket.

W: Wow, that's not a bad price. What about the apples?

M: Well, they are normally $10 for one box, but they are _____ _____ today. So, you can get one for $9.

W: All right. The watermelons _____ _____!

M: All three of them came in fresh this morning.

W: I'd love to get one, but it would be too heavy for me to carry.

M: Don't worry. We can deliver it.

W: Oh, that banner says "Free Delivery"! I'll take two then.

05

M: Hey, Christine. Are you writing a poem or something?

W: No. I'm actually making a list of things to _____ _____ _____ _____.

M: Oh, yeah? What kind of party?

W: A surprise party for my boyfriend's birthday. I want to make it special.

M: What a good girlfriend! What do you have planned?

W: Well, I'm going to invite all his friends. And I also want to _____ _____ _____ _____.

M: Do you have anything in mind?

W: No, I don't. It has to be delicious, but I just don't have any great ideas.

M: Whenever I _____ _____ _____, I make this amazing taco salad I found online.

W: That sounds perfect. My boyfriend loves taco salad! Can you show me how to make it?

M: I can email you step-by-step instructions. Your boyfriend will love it.

W: I bet he will. Thanks!

06

M: Hello, Jane. I haven't seen you in a long time.

W: Hi, Tony. What are you doing here at the bank?

M: I _____ _____ to get these bills paid.

W: Oh, I see. The bank is really crowded today, isn't it?

M: Yeah. I heard this bank has a new savings plan available for a limited time only. So many people are coming to _____ _____ _____ _____.

W: Really? If I had known earlier, I would have opened one, too. But I don't have much time today because I'm going on vacation tomorrow.

M: Oh, where are you going?

W: Italy.

M: Wow, that sounds fantastic. I guess you came here to _____ _____ _____.

W: Yes. I hope I can do it before the bank closes!

M: Well, have a great time. And call me when you get back.

07

M: Hi. Can I help you with something?

W: Yes. I'd like to exchange these socks. I bought two packs of them yesterday.

M: All right. Is there a problem with them?

W: Well, I wanted socks that _____ _____ _____ _____ _____, but these are too short.

M: Ah, our knee socks are right here.

W: These are packs of six pairs. The packs I bought yesterday had four pairs each.

M: Yes, but these are the only ones we have. They cost _____ _____ _____.

W: Okay. Hmm... I guess having many socks is better than having too few. I'll get 12 pairs.

M: All right. But since the socks you bought yesterday cost a total of $24, you'll have to _____ _____ _____.

W: That's not a problem.

08

M: Michelle, somebody told me you're in the science club.

W: Yes, I am. Are you interested in joining?

M: I'm thinking about it, but I wanted to know more about it first.

W: Well, there are _____ _____ _____ _____. Right now, there are 11 of us.

M: I see. How often do you meet?

W: We meet _____ _____ _____ in the science lab. We mainly discuss science. But we have lots of special events too.

M: What kind of stuff do you do?

W: We go to science museums and take part in science fairs.

M: That sounds nice. So, can anyone join?

W: Sure! Just _____ _____ _____ you have at least a C in all your science classes.

M: I see. Well, thanks for answering my questions. I'll think more about it tonight.

W: All right. I hope you decide to join!

09

W: Welcome to the Westchester Mushroom Festival, a three-day event celebrating everyone's _____ _____ _____. This area, as you are probably aware, is famous for its large number of mushroom farms. But what you may not know is that there's a wealth of wild mushrooms that grow in our forests as well. This festival aims to educate people on what kind of wild mushrooms _____ _____ _____ and where they can be found. There will be several presentations by local experts, as well as a food area serving all kinds of delicious mushroom snacks. Admission is just $4, and kids 12 and under _____ _____ _____ _____ _____. And hold on to your tickets, as there will be _____ _____ _____ at the end of the day, with all kinds of mushroom-themed prizes.

10

M: Hello. I was just looking at this schedule, and I need some help.

W: Certainly, what can I do for you?

M: I'm thinking of _____ _____ _____ to improve a specific English skill.

W: Which English skill are you hoping to improve?

M: All of them! Writing is my first choice though, but I'll take _____ _____ _____ _____.

W: Mr. Kim isn't a native speaker, but his reading class is really popular with students. Are you free those nights?

M: I'm free every night except Wednesdays. I want a class with a native speaker, though.

W: Okay. Well, how about mornings?

M: Mornings are fine, as long as it's not too early. _____ _____ _____ I could get here before 9.

W: All right. Any other requirements?

M: Well, I'd like to be finished by 8 o'clock at night so I can meet my girlfriend.

W: In that case, it looks like this is the class for you.

11

M: Ellen, would you like to have a cup of coffee?

W: No thanks. I don't like coffee that much. It makes it hard for me to _____ _____ _____.

M: I see. What would you like to drink instead?

W: (I'll have some orange juice.)

12

W: Ethan, have you ever _____ _____?

M: Yes, I often hike at the mountains with my family. Why?

W: I'm planning to go this weekend. What equipment do I need?

M: (You need outdoor clothing, boots, and a walking stick.)

13

W: Excuse me, Dennis? Sorry to interrupt, but it's ten after two.

M: Is it already? I was just finishing up my final lesson of the class.

W: All right, but _____ _____ _____. The trainees are already late for their computer course.

M: I'm sorry. Give me 15 minutes, and then I'll send them on their way.

W: 15 minutes? I'm afraid that's too long.

M: Have the _____ _____ _____ already?

W: Yes, and if we wait too long, Ms. Smith, the computer instructor, will start without them.

M: I didn't realize I was holding up the whole training program!

W: Yes, unfortunately, we're running _____ _____
_____ _____.

M: In that case, how about 5 more minutes?

W: (All right. I'll tell Ms. Smith to wait a bit longer.)

14

[Phone rings.]

M: IT department, this is Vince.

W: Hi, Vince. This is Marie in the production department. Is Harry Cho there?

M: No. Unfortunately, Mr. Cho is _____ _____ _____ _____ all day today.

W: Oh, you're kidding! I really need his help right away.

M: I'm sorry to hear that. He was called away to one of our branches unexpectedly.

W: Well, maybe someone else there can help me.

M: Maybe. Can you describe the problem?

W: I can't print anything from my computer. Mr. Cho was able to _____ _____ _____ _____ last time this happened.

M: Hmm. If it's a problem with the connection, that's something Francine could handle.

W: Yes! I recall Mr. Cho saying something about _____ _____ _____.

M: (Then I'll send her to your desk right away.)

15

M: Elaine is eating lunch with her friend Jessica, who is complaining about the past weekend. It seems that Jessica and her boyfriend had decided to spend Saturday at an amusement park located outside of town. They got there early in the morning and each _____ _____ _____ _____, allowing them unlimited rides for the entire day. The passes were expensive, but they figured it would _____ _____ _____. However, about 20 minutes later it began to rain, and it didn't stop for the rest of the day. Because of this, they were _____ _____ _____ most of the rides. In this situation, what would Elaine most likely say to Jessica?

Elaine: (That must have been such a disappointment.)

16~17

W: Welcome to the seminar. My name is Professor Johnson, and I'd like to talk about aircraft. The first airplane flight _____ _____ _____ 1903. Now, more than 100 years later, unmanned aircraft are changing the world. They are known as drones, and they are controlled remotely. Drones were originally used by the military to safely observe enemy troops. Nowadays, however, they are being used in many different fields. Journalists, for example, use drones to _____ _____ _____ from high in the sky. This gives viewers a new, useful perspective. Also, some large stores are starting to use delivery drones. In the near future, customers may be able to order products online and have them delivered to their homes quickly. And, in the agriculture industry, farmers are already _____ _____ _____ _____ large fields of crops. If you think about it, the possibilities are endless.

1번부터 17번까지는 듣고 답하는 문제입니다.
1번부터 15번까지는 한 번만 들려주고, 16번부터 17번까지는 두 번 들려줍니다. 방송을 잘 듣고 답을 하기 바랍니다.

01
다음을 듣고, 여자가 하는 말의 목적으로 가장 적절한 것을 고르시오.
① 청중을 격려하려고
② 기부금을 요청하려고
③ 자신의 자서전을 소개하려고
④ 새로운 의학 기술을 알리려고
⑤ 수술비 마련 자선 행사를 하려고

02
대화를 듣고, 남자의 의견으로 가장 적절한 것을 고르시오.
① 초대된 파티에 선물을 가져가야 한다.
② 대인 관계에서는 배려가 가장 중요하다.
③ 초대된 행사에는 되도록 참여해야 한다.
④ 꽃장식에 따라 파티의 분위기도 바뀐다.
⑤ 성공적인 파티를 위해 서둘러 준비해야 한다.

03
대화를 듣고, 두 사람의 관계를 가장 잘 나타낸 것을 고르시오.
① 연예인 - 팬
② 우체국 직원 - 우체부
③ 고객 - 영업사원
④ 주민 - 택배 배달원
⑤ 여행객 - 여행 가이드

04
대화를 듣고, 그림에서 대화의 내용과 일치하지 않는 것을 고르시오.

05
대화를 듣고, 여자가 할 일로 가장 적절한 것을 고르시오.
① 여행 짐 싸기
② 커피 추출기 선물하기
③ 커피 추출기 고치기
④ 함께 원두 사러 가기
⑤ 커피 만드는 법 알려주기

06
대화를 듣고, 남자가 여자를 찾아간 이유를 고르시오.
① 시험 날짜를 조정하려고
② 시험 점수가 잘못 나와서
③ 수영 강습 시간을 바꾸려고
④ 경기에 대한 조언을 얻으려고
⑤ 특별 활동반을 바꾸고 싶어서

07
대화를 듣고, 남자가 지불할 금액을 고르시오. 3점
① $7.00
② $7.60
③ $8.00
④ $8.55
⑤ $9.00

08
대화를 듣고, 태양에 관해 언급되지 않은 것을 고르시오. 3점
① 현재 상태
② 생성 과정
③ 소멸 과정
④ 나이
⑤ 소멸 시기

09
rock concert에 관한 다음 내용을 듣고, 일치하지 않는 것을 고르시오.
① 지진 피해자들을 돕기 위한 자선 콘서트이다.
② 토요일 저녁 8시에 열린다.
③ 4개 학교에서 온 음악 밴드들이 참여한다.
④ 입장료는 10달러이다.
⑤ 구호품을 지참해야 입장할 수 있다.

10

다음 표를 보면서 대화를 듣고, 여자가 구입할 손목시계를 고르시오.

	Model	Shape	Strap	Display	Price
①	A	Oval	Metal	Analog	$300
②	B	Oval	Leather	Analog	$280
③	C	Square	Leather	Analog	$170
④	D	Square	Leather	Digital	$125
⑤	E	Triangle	Metal	Digital	$200

11

대화를 듣고, 여자의 마지막 말에 대한 남자의 응답으로 가장 적절한 것을 고르시오.

① He is the tallest guy on the court.
② Sorry, but I don't really like basketball.
③ Really? I'd love to join your basketball team.
④ He is looking for a book about basketball in the sports section.
⑤ I don't like that player since he never blocks any shots.

12

대화를 듣고, 남자의 마지막 말에 대한 여자의 응답으로 가장 적절한 것을 고르시오.

① Sure, I can help you tomorrow.
② No, I think your presentation was perfect.
③ No, I don't think I'll have time this afternoon.
④ Yes. Can you make some copies of the handouts?
⑤ I'd love to. But you'd better ask the science teacher.

13

대화를 듣고, 남자의 마지막 말에 대한 여자의 응답으로 가장 적절한 것을 고르시오.

Woman: _____

① Good luck at your meeting.
② That's kind of a bad time for me.
③ Well, I guess we'll just have to cancel.
④ I know. It could have happened to anyone.
⑤ Thanks. I know this is short notice, so I really appreciate it.

14

대화를 듣고, 여자의 마지막 말에 대한 남자의 응답으로 가장 적절한 것을 고르시오.

Man: _____

① Would you like to borrow my skateboard?
② You should have told me it was your birthday.
③ Just keep practicing, and you'll improve over time.
④ Can you teach me how to ride a skateboard instead?
⑤ You're doing great. Now try some more difficult tricks.

15

다음 상황 설명을 듣고, Tracy가 남자에게 할 말로 가장 적절한 것을 고르시오.

Tracy: _____

① That's too far for us to walk.
② Actually, I planned on sightseeing today.
③ I wish I could help you, but I lost my map.
④ Can you tell me how to get to Times Square?
⑤ Thanks for going out of your way to help me.

[16~17] 다음을 듣고, 물음에 답하시오.

16

남자가 하는 말의 주제로 가장 적절한 것은?

① diseases caused by food
② the most nutritional foods
③ foods that are endangered
④ the effects of climate change
⑤ foods that are vulnerable to bacteria

17

언급된 음식이 아닌 것은?

① honey   ② bananas   ③ coffee
④ salmon   ⑤ potatoes

녹음을 다시 한 번 듣고, 빈칸에 알맞은 말을 쓰시오.

01

W: You may think that you are going through the most difficult time in the world. But let me tell you about the pain _____ _____ _____ _____ to become who I am today. I was a healthy girl, but an accident took away my legs and paralyzed my arms. Still, I _____ _____ _____ _____ and, after much struggle, I relearned how to move my hands again. Then, I decided to help others and became a doctor. You can overcome your obstacles, just like me. _____ _____.

02

M: Don't forget that we're going to Vicky's house tomorrow.
W: Oh, that's right. She's having a dinner party to _____ _____ _____.
M: I think it's going to be a lot of fun. All of our friends will be there, and she's cooking a delicious dinner.
W: What time does it start?
M: It starts at 7 p.m., but I'd like to leave a little earlier than that to _____ _____ _____ _____ along the way.
W: What do you need flowers for?
M: Well, I think it would be rude to _____ _____ _____. I'd like to have something to give to our hostess.
W: That's very thoughtful of you.

03

W: Hello. Can I help you with something?
M: Yes, I _____ _____ _____ for Diane Peterson.
W: That's me. But I wasn't expecting anything. Can you tell me who it's from?
M: Let's see... It's from David Peterson in England.
W: Oh, that's my brother. What a nice surprise!

M: Can you _____ _____ _____, please?
W: Certainly. [Pause] By the way, do you have anything for the house next door?
M: Actually, I do. Why do you ask?
W: They're away on vacation for a few weeks and asked me to _____ _____ _____ _____ _____ any deliveries.
M: I see. Well, can you sign for this one as well?
W: Sure, I'd be glad to. [Pause] There you go.
M: Thanks a lot. Here it is.

04

M: Hey, Sarah. Look at this picture. My brother took it at the winter wonderland exhibition.
W: Wow! Everything in the picture is made of ice!
M: I like the ice car on the left. The door of the car is open _____ _____ _____ _____ _____.
W: Yes. Wouldn't it be fun to drive it?
M: Yeah. I wonder how fast it could go. Which of the ice sculptures is your favorite?
W: I love the one in the center. It looks like the Eiffel Tower, doesn't it?
M: Yes, it does. It's pretty impressive. And look at this harp on the right. It looks so delicate!
W: The strings on it are so thin. It looks like it could be played.
M: I'd like to try to play it. Behind it there is an ice swan as well.
W: Its wings are _____ _____ _____.
M: It looks like it's trying to fly. What is the sculpture behind the car?
W: It's a Christmas tree with a big star on top.
M: _____ _____ _____ _____!

05

M: Hi, Nancy. What are you doing?
W: Hi, Mike. I'm brewing some coffee. I got a new coffee machine _____ _____ _____

yesterday.

M: It smells wonderful. Where did you get the coffee beans?

W: I bought them at the market. They're from Indonesia.

M: Weren't they expensive?

W: Yes, but they're worth it. The coffee tastes really good.

M: I see. _____ _____ _____ _____ a coffee machine, too. Then I would make coffee every morning.

W: Actually, I was planning to sell my old one to a second-hand shop. If you want, I'll give it to you.

M: Really? I'll give you some money for it. I can't just take it.

W: You don't have to. Your birthday _____ _____ _____, anyway. Just take it as a present.

M: Oh, I can't tell you how much I appreciate it.

06

W: Hello, Albert. What can I do for you?

M: Hello, Ms. Harris. It's about the test on the 24th. I _____ _____ _____ that day.

W: Why not?

M: I'm on the swim team. We're going to a competition in Tokyo on that day.

W: Oh, I see. Are there any other swimmers in the class?

M: No, just me. The others are all in Mr. Teasedale's class, and he scheduled the test for a different day.

W: In that case, could you _____ _____ _____ before leaving for Tokyo?

M: Okay, I can do that. Would tomorrow be okay?

W: No, I'll have to _____ _____ _____ _____ for you since you're taking it earlier than the other students. How about Monday after class?

M: That would be fine.

W: And good luck in Tokyo.

M: Thanks.

07

M: Hi. I'd like some ice cream, please.

W: Sure! _____ _____ _____ _____ _____ today?

M: I'd like three cups of ice cream — one chocolate and two strawberry.

W: Okay. We have two sizes. A small scoop is $2, and a large scoop is $4.

M: I see. Then I'll take a large scoop of chocolate and two small scoops of strawberry.

W: All right. _____ _____ _____ _____?

M: No, that'll be all.

W: Are you sure? Don't you want some toppings? You can have sprinkles or syrup for $1.

M: That's fine. Oh, and I have a coupon for 5% off. Can I still use it?

W: _____ _____ _____. *[Pause]* Yes, it's still good until the end of the month.

M: Great. Then I'll use it now.

W: All right. Here's your ice cream.

08

W: I watched a scary movie last night. It was called *Dead Sun*.

M: That's the one about the sun _____ _____ _____ _____ _____, right?

W: Exactly. That couldn't really happen, could it?

M: Actually, it will happen. The sun is a star, and all stars die.

W: You're kidding!

M: Nope. Right now the sun is a very stable yellow star.

W: Well, that's good.

M: But eventually it will become a red giant star. It will _____ _____ _____ and hotter.

W: Oh, I see.

M: Over time, it will _____ _____ _____ _____ _____. And after that, it will die.

W: How long will that take to happen?

M: Well, the sun is about five billion years old. And it probably has about five billion years left. So we have plenty of time!

09

M: I'm sure you are all aware of last week's terrible earthquake in South America. To _____ _____ to help the survivors, our school will be hosting a special rock concert. It will be held at 8 p.m. on Saturday, March 8 in the auditorium. _____ _____ _____ _____ from four different local high schools. Admission is $10 and all the money will be donated directly to charities working in the region affected by the tragedy. People attending the concert are also encouraged _____ _____ _____ of canned food or used clothing. Tickets go on sale today and can be purchased at the administrative office.

10

M: Can I help you?

W: Yes, _____ _____ _____ a new watch.

M: Take a look at these. They're our most popular models.

W: Those are all very nice, except for the triangular one. It's too strange.

M: Yes, but it's popular with teenagers. What kind of strap do you prefer?

W: I'd like one with a leather strap.

M: I see. How about this one? It's very _____ _____ _____.

W: No, thank you. Actually, I don't like digital watches. They look cheap.

M: I understand. Then why don't you take a closer look at these two?

W: Wow! I really like them both. How much are they?

M: You can see their price tags right here.

W: Oh. That one is too much. I'll _____ _____ _____ _____ _____.

11

W: Hi James, what are you doing here?

M: My cousin _____ _____ for this school. I'm cheering for him.

W: I didn't know that. Which player is he?

M: (He is the tallest guy on the court.)

12

M: I heard you're giving a presentation in science class this afternoon.

W: Yes. That's why I am _____ _____ _____ right now.

M: Well, is there anything that I can do to help you?

W: (Yes. Can you make some copies of the handouts?)

13

[Phone rings.]

M: Hello?

W: Hello, Patrick. This is Seonhee.

M: Oh. Hi, Seonhee.

W: I'm sorry to call so late, but I have to cancel tomorrow's lesson.

M: Oh, I hope _____ _____ _____.

W: No, nothing is wrong. I just have a meeting tomorrow afternoon.

M: Okay, well, would you like to reschedule?

W: _____ _____ _____ _____, that would be great.

M: When would be a good time for you?

W: How about Friday afternoon at 4 o'clock?

M: _____ _____. Same time, same place.

W: (Thanks. I know this is short notice, so I really appreciate it.)

14

W: This is too difficult. _____ _____ _____!

M: What's wrong, Hillary?

W: I got this skateboard for my birthday, but I can't ride it.

M: Oh, you've never ridden a skateboard before?

W: No. At first it looked really fun, but it's too hard. Now I'm just frustrated.

M: I understand. It took me a long time to learn _____ _____ _____ _____.

W: Every time I try, I fall down.

M: You need to be patient. Just try to ride it in a straight line for a few days first.

W: That's boring. I want to do fun things, like jumps and tricks.

M: You can learn those later. First, you need to learn the basic skills.

W: I know. But I don't think I can. I can't even _____ _____ _____ on it now.

M: (Just keep practicing, and you'll improve over time.)

15

W: Tracy is visiting New York for the first time. She decides to _____ _____ _____ _____ all of the famous sights. Things start off well, but then she leaves her tourist map on the subway. Unable to find Times Square, she stops a passing man and _____ _____ _____ _____. He explains the best way to get there, but his directions are very complicated. He repeats them several times, but Tracy cannot understand. Finally, the man says that it is _____ _____ _____, and he will walk there with Tracy to make sure she doesn't get lost. In this situation, what would Tracy most likely say to the man?

Tracy: (Thanks for going out of your way to help me.)

16~17

M: Welcome to biology class. I have a question for you. What are some of your favorite things _____ _____ _____ _____? Well, you'd better enjoy them while you can because some of them might _____ _____ _____. For example, bee populations are decreasing all around the world. If bees die out, there will be no more honey _____ _____ _____ _____. Bananas are also facing danger. Banana trees are suffering from a deadly disease in some places. The disease attacks their roots and eventually kills the whole tree. Do you drink coffee? Well, coffee is a very sensitive crop. Global warming and other factors are making it more and more difficult to grow coffee. Rising temperatures are also threatening salmon. These fish are a favorite of seafood lovers. But they need cold water to survive, and our oceans are getting warmer. Hopefully, scientists and farmers will _____ _____ _____ to all of these problems soon.

1번부터 17번까지는 듣고 답하는 문제입니다.
1번부터 15번까지는 한 번만 들려주고, 16번부터 17번까지는 두 번 들려줍니다. 방송을 잘 듣고 답을 하기 바랍니다.

01

다음을 듣고, 남자가 하는 말의 목적으로 가장 적절한 것을 고르시오.
① 유괴 예방법을 소개하려고
② 안내소의 위치를 알리려고
③ 분실물 신고 절차를 설명하려고
④ 소지품 분실 주의를 요청하려고
⑤ 미아를 보호하고 있음을 알리려고

02

대화를 듣고, 여자의 의견으로 가장 적절한 것을 고르시오.
① 퀵서비스 비용은 너무 비싸다.
② 우체국 영업시간을 연장해야 한다.
③ 소포를 속달 서비스로 보내는 게 낫다.
④ 택배 기사의 근무 환경이 개선돼야 한다.
⑤ 우체국의 속달 서비스는 이용하기 편리하다.

03

대화를 듣고, 두 사람의 관계를 가장 잘 나타낸 것을 고르시오.
① 화가 - 기자
② 플로리스트 - 손님
③ 정원사 - 관리자
④ 파티플래너 - 고객
⑤ 보석 가게 주인 - 손님

04

대화를 듣고, 그림에서 대화의 내용과 일치하지 않는 것을 고르시오.

05

대화를 듣고, 두 사람이 할 일로 가장 적절한 것을 고르시오.
① 컴퓨터 게임을 하기
② 주말 계획 변경하기
③ 밤새워서 보고서 완성하기
④ 늦어도 과제를 제대로 하기
⑤ 과제 마감일을 연장해 달라고 요청하기

06

대화를 듣고, 남자가 고국으로 돌아온 이유를 고르시오.
① 향수병에 걸려서   ② 여행 경비가 부족해서
③ 여행 일정이 끝나서   ④ 영어를 가르치려고
⑤ 외국 문화에 적응하지 못해서

07

대화를 듣고, 여자가 지불한 금액을 고르시오.
① $300   ② $500   ③ $650
④ $850   ⑤ $900

08

대화를 듣고, Black Mountain National Park에 관해 언급되지 <u>않</u>은 것을 고르시오.
① 개장 시기   ② 입장료   ③ 위치
④ 등산로의 수   ⑤ 주변 관광지

09

school trip에 관한 다음 내용을 듣고, 일치하지 <u>않</u>는 것을 고르시오.
`3점`

① 졸업반 학생들을 대상으로 후보지 투표가 진행되었다.
② 작년에는 수도로 다녀왔다.
③ 작년 수학여행은 대성공이었다.
④ 스키 여행이 자연 체험 여행을 큰 표차로 앞질렀다.
⑤ 개표는 두 번에 걸쳐 진행되었다.

10

다음 표를 보면서 대화를 듣고, 두 사람이 참석할 워크숍을 고르시오.

17th Annual Citywide Education Conference

	Workshop	Topic	Time	Q&A
①	A	Elementary Education	9:00-11:00	O
②	B	Higher Education	9:30-11:00	O
③	C	Adult Education	11:00-12:00	O
④	D	Bilingual Education	11:00-12:30	X
⑤	E	Teacher Education	14:00-16:30	X

* Lunch will be served at 12:30 p.m.

11

대화를 듣고, 남자의 마지막 말에 대한 여자의 응답으로 가장 적절한 것을 고르시오.

① She is our new English teacher.
② She is one of my oldest friends.
③ She came to our class by accident.
④ She seems kind and warm-hearted.
⑤ She said her mother is from New York.

12

대화를 듣고, 여자의 마지막 말에 대한 남자의 응답으로 가장 적절한 것을 고르시오.

① I will tell her the news.
② She is not going to pass the exam.
③ Which song would be good for the audition?
④ We should get her autograph in advance.
⑤ When will she find out the result of her audition?

13

대화를 듣고, 남자의 마지막 말에 대한 여자의 응답으로 가장 적절한 것을 고르시오. 3점

Woman: _____

① Thank you for supporting our cause.
② I hope your daughter gets better soon.
③ You could always help us by joining our protest.
④ Any amount you can contribute will be appreciated.
⑤ The healthcare facilities in our town are outstanding.

14

대화를 듣고, 여자의 마지막 말에 대한 남자의 응답으로 가장 적절한 것을 고르시오. 3점

Man: _____

① Good idea. Let's go together.
② Yes. I drive more cautiously now.
③ I agree. That's why we need insurance.
④ I expect that paper on my desk tomorrow.
⑤ Don't worry about the paper. Go be with your father.

15

다음 상황 설명을 듣고, Sean의 어머니가 Sean에게 할 말로 가장 적절한 것을 고르시오.

Sean's mother: _____

① Congratulations! Let's go for a drive!
② Can you give me a ride to the supermarket?
③ I'll go with you because it's your first time driving.
④ Were you hurt? I don't care about the car as long as you're okay.
⑤ Don't worry. It was an old car and we're getting a new one anyway.

[16~17] 다음을 듣고, 물음에 답하시오.

16

남자가 하는 말의 주제로 가장 적절한 것은?

① the body language of dogs
② efficient dog training methods
③ common dog behavioral problems
④ proper responses to a dog attack
⑤ various games to play with dogs

17

언급된 부위가 <u>아닌</u> 것은?

① the tail   ② the eyes   ③ the ears
④ the teeth   ⑤ the fur

녹음을 다시 한 번 듣고, 빈칸에 알맞은 말을 쓰시오.

01

M: Ladies and gentlemen, may I have your attention please? There is a seven-year-old child here at the front desk. Would her parents kindly contact a salesclerk and then _____ _____ _____ to the front desk to pick her up? When you arrive, you will be asked to _____ _____ _____ _____ and give her full name before you are able to see her. This is simply a safety precaution to _____ _____. Thank you for your attention.

02

M: I'm going to the post office to send this package.

W: Why don't you just use a courier service?

M: Wouldn't that be _____ _____ _____ the post office?

W: It does cost more. However, they would send someone to _____ _____ _____. That would save you a trip.

M: I guess that would be convenient.

W: And the package would be delivered quickly. The post office would _____ _____ _____ _____ _____ _____.

M: But the post office has an express service.

W: Yes, but that costs more. So, there wouldn't be much of a difference in price if you're not sending it too far.

M: Okay, I'm convinced.

03

W: Hello! Can I help you with anything today?

M: Well, tomorrow is my girlfriend's birthday. I've already bought a pretty ring for her. Now I'd like to _____ _____ _____ _____.

W: That's wonderful. What kinds of flowers does she like?

M: I'm not sure. Maybe you can help me choose.

W: Sure. Tell me, how would you describe her?

M: Well, she's _____ _____ _____. She loves to dance.

W: How about the purple flowers over here? They look special.

M: They are pretty, but how about something more romantic?

W: Ah, I understand. Here are some lovely roses. Red is for romance; _____ _____ _____ _____ _____.

M: I'll take two dozen red roses, then.

W: I'm sure she'll love them.

M: Thank you. I really appreciate your help.

04

M: I think we should clean the storage room.

W: I agree. It's been a while. What shall we do first?

M: Well, there's a huge spider web _____ _____ _____. But I couldn't get it down because the ladder is too short.

W: All right. I think I can borrow a longer one from a neighbor.

M: Okay. The window is broken, too. We'll have to get it fixed.

W: Do we have any glass to _____ _____?

M: I ordered some. It might come this afternoon. Oh, I think we can start by _____ _____ _____ _____.

W: There are only three. Oh, what about the large chair we used last summer?

M: Oh, it's covered. We just need to wash the cover.

W: Good.

05

W: I can't believe this project is due tomorrow.

M: Yeah, I guess we shouldn't have played computer games all weekend.

W: I just didn't realize how much work it would involve.

M: I know. Now that I'm looking at it, I realize we

should have started this a while ago.

W: If we ___ ___ ___ tonight, we're not going to do a very good job.

M: Maybe we could ask for an extension on the deadline.

W: I doubt it. We have ___ ___.

M: That's true. What's the late penalty?

W: I think it's five points a day.

M: Hmm... Do you think it would be worth it to ___ ___ ___ late?

W: Well, we could stay up all night and get it in on time.

M: Yeah, but we would probably lose more than five points because it won't be very good.

W: You're right. We should ___ ___ ___ even if it means taking the penalty.

06

W: Jacob! Wow! I haven't seen you in years!

M: Hi, Ashley. I've been ___ ___. I just got back.

W: I heard that you were going to travel around Europe.

M: Well, that was my plan. But I needed to earn some money first.

W: ___ ___ ___ ___?

M: I got a job teaching English in Korea.

W: Wow! That sounds exciting. So, did you save enough to go to Europe?

M: I did. But I liked Korea so much that I stayed there for three years instead.

W: Three years! You must have really liked it there.

M: Yes. But I finally decided to come back home.

W: Why? Did something happen?

M: No, I just ___ ___, I guess.

07

M: Welcome back! How was the Caribbean cruise?

W: It was great. But I spent too much money.

M: Really? I thought you got a ___ ___.

W: Yes, the cruise itself was only $300, but I spent a lot of money on the different islands.

M: I guess ___ ___ ___ happen.

W: Yes, it took one day to get there, and then we stopped in the Bahamas for two days. I spent about $200 there.

M: Wow! Where else did you go?

W: We also went to Jamaica ___ ___ ___.

M: How much did you spend there?

W: About $150. Then we went to Barbados for two days, and I spent about $250 there.

M: I hope that was the last stop.

W: Yes, but then it took another day to get back.

08

M: Where should we ___ ___ this weekend?

W: Let's go to Black Mountain National Park.

M: I've never heard of it.

W: It just opened last year, so not many people know about it.

M: Oh, is it nearby?

W: It's just east of Silver City. It takes about three hours to drive there.

M: All right. Is it big?

W: It's not very big, but it has a pretty camping area along the river.

M: That sounds nice. How about hiking trails?

W: There are three different trails that ___ ___ ___. The view from the top is great.

M: Are there any tourist attractions around the park?

W: Yes, there are many historical sites ___ ___ ___ ___. We'll have plenty of stuff to do!

09

W: As you know, last week all of the seniors voted on where they would like to go for their school trip. The _____ _____ _____! Although last year's trip to the capital was _____ _____ _____, only 213 students voted to go there this year. As for the rest of the votes, they were split between a ski trip and a nature excursion to Eagle Rock. The vote was so close we had to _____ _____ _____. But pack up your ski gear, students, because this year we are off to Mt. Rocky Top.

10

W: The city's annual education conference is next month. You're going to go with me, right?

M: Of course! We go every year. But we should choose a workshop to attend soon.

W: I have the schedule with me now. _____ _____ _____ _____ _____?

M: Hmm... Well, I don't want to go to the adult education workshop again.

W: Yes, we went there last year. And I don't want to go to the one after lunch.

M: Oh, really? Why not?

W: I always _____ _____ _____ after I eat lunch. Don't you?

M: No. But I do get sleepy during long workshops. So let's choose one that's shorter than two hours.

W: _____ _____ _____ _____. How do you feel about question-and-answer sessions?

M: I think they're the most important part of a workshop.

W: I agree. So let's choose this workshop. It has a question-and-answer session.

M: That looks like a good choice!

11

M: Carolina, did you see the new student in our class?

W: The new girl? Yes, I _____ _____ _____ for

a while.

M: Did you? What is she like?

W: (She seems kind and warm-hearted.)

12

W: Did you hear that Leigh passed the audition?

M: Really? Well, I know that she's _____ _____ _____ and dancing.

W: I think so, too. Maybe she will become a TV star some day!

M: (We should get her autograph in advance.)

13

M: Excuse me, are you _____ _____?

W: No. We're raising money to build a new children's hospital here in town.

M: What a wonderful idea!

W: We think so. We already have the support of local businesses, politicians and many parents.

M: I can understand why. When my daughter was sick last year, we had to drive for almost an hour to the nearest hospital.

W: Many of us have had similar experiences. That's why we are doing our best to _____ _____ _____ _____.

M: So, how are you raising money?

W: Well, we're collecting donations from citizens like you and conducting various fundraising events at local schools and businesses.

M: Well, I'd like _____ _____ _____ _____, but I don't have very much money.

W: (Any amount you can contribute will be appreciated.)

14

W: Hello, Dr. Keating. Are you busy right now?

M: Not really. Come on in. What can I do for you?

W: Well, I'm afraid I won't be able to _____ _____

_____ _____ on time.

M: You know the rules. You lose two points a day for every day that it's late.

W: I understand. Unfortunately, my father is in the hospital.

M: I'm sorry to hear that. I hope it isn't anything serious.

W: Actually, he had a car accident and _____ _____ _____.

M: That's terrible. You must be very worried.

W: Yes, I am. He'll probably be in a hospital for a while.

M: I was in an accident several years ago. It _____ _____ _____ _____ _____.

W: I just hope he gets better. Anyway, I think I can finish my paper by next Wednesday.

M: (Don't worry about the paper. Go be with your father.)

15

W: Sean just _____ _____ _____ _____. His mother loaned him her car after giving him a long lecture about driving responsibly. Now Sean has been in an accident and put a dent in his mother's car. He is very worried about telling her because it's a new car. And even though the accident wasn't his fault, he _____ _____ because she trusted him with the car. He is certain that her _____ _____ _____ the accident and decides to tell her honestly what happened. He walks in the door and says, "Mom, I've got some bad news. I had a little accident, and there's a dent in the car." In this situation, what is his mother most likely to say to Sean?

Sean's mother: (Were you hurt? I don't care about the car as long as you're okay.)

16~17

M: Today, I'd like to talk about how to _____ _____ _____ _____. As you train your dog,

it is important to know how it feels. People often want to guess an animal's feelings, but guessing is not necessary with dogs. You can simply watch the way the dog is _____ _____ _____ _____ _____ to understand how it is feeling at that particular moment. For example, a happy dog tends to wag its tail. And if it wants to play, you'll notice that it bends its front legs and points its ears and tail up. On the other hand, if its tail is tucked between the legs and _____ _____ _____ _____, then the dog is probably afraid. Most importantly, you should stay away from a dog that is showing its teeth and has the fur on its back sticking up. This is a sign that the dog is angry.

1번부터 17번까지는 듣고 답하는 문제입니다.
1번부터 15번까지는 한 번만 들려주고, 16번부터 17번까지는 두 번 들려줍니다. 방송을 잘 듣고 답을 하기 바랍니다.

01

다음을 듣고, 여자가 하는 말의 목적으로 가장 적절한 것을 고르시오.
① 학교 축제를 홍보하려고
② 저소득층 지원의 확대를 촉구하려고
③ 각 지역의 고유한 문화를 홍보하려고
④ 봉사 활동에 참여할 것을 권장하려고
⑤ 거리의 쓰레기 문제의 심각성을 알리려고

02

대화를 듣고, 남자의 의견으로 가장 적절한 것을 고르시오.
① 시간을 효율적으로 사용해야 한다.
② 병원에 예약 시간을 확인해야 한다.
③ 병원의 접수 과정을 간소화해야 한다.
④ 정기적으로 치과 검진을 받아야 한다.
⑤ 예약 시간보다 일찍 병원에 도착하는 것이 좋다.

03

대화를 듣고, 두 사람의 관계를 가장 잘 나타낸 것을 고르시오.
① 컴퓨터 강사 - 학생   ② 컴퓨터 수리공 - 고객
③ 영업사원 - 손님   ④ 경비원 - 입주민
⑤ 인터넷 설치 기사 - 집주인

04

대화를 듣고, 그림에서 대화의 내용과 일치하지 <u>않는</u> 것을 고르시오.

05

대화를 듣고, 여자가 남자에게 부탁한 일로 가장 적절한 것을 고르시오.
① 차 수리하기   ② 세차하기   ③ 아기 돌보기
④ 양말 사기   ⑤ 빨래하기

06

대화를 듣고, 남자가 길을 잃은 이유를 고르시오.
① 지하철역이 보이지 않아서
② 이전한 박물관 위치를 몰라서
③ 여자의 말을 이해하지 못해서
④ 지하철 출구 번호가 바뀌어서
⑤ 여행 책자 종류가 너무 많아서

07

대화를 듣고, 여자가 지불할 금액을 고르시오. 3점
① $2.4   ② $4.8   ③ $7.2   ④ $10.5   ⑤ $16.0

08

대화를 듣고, 남자의 개에 관해 언급되지 <u>않은</u> 것을 고르시오.
① 털의 색상   ② 이름   ③ 산책 횟수
④ 좋아하는 장난감   ⑤ 수면 장소

09

도서관의 신규 정책에 관한 다음 내용을 듣고, 일치하지 <u>않는</u> 것을 고르시오. 3점
① 3층은 그룹 스터디를 위한 공간이다.
② 컴퓨터는 선착순으로 사용할 수 있다.
③ 주중에는 새벽 1시에 문을 닫는다.
④ 학생들은 한 번에 6권까지 대출할 수 있다.
⑤ 도서 반납 연체료가 두 배로 인상되었다.

10

다음 표를 보면서 대화를 듣고, 남자가 구매할 우쿨렐레를 고르시오.

	Product	Size	Wood Type	Condition	Price
①	A	Soprano	Laminate	Used	$65
②	B	Soprano	Laminate	New	$70
③	C	Tenor	Solid	Used	$80
④	D	Concert	Laminate	New	$90
⑤	E	Concert	Solid	Used	$110

11

대화를 듣고, 여자의 마지막 말에 대한 남자의 응답으로 가장 적절한 것을 고르시오.

① I see. Do you have your receipt?
② You should go see a doctor for that.
③ We are all sold out of sweaters now.
④ But I don't want to wear a sweater today.
⑤ You'd better get someone to massage your shoulder.

12

대화를 듣고, 남자의 마지막 말에 대한 여자의 응답으로 가장 적절한 것을 고르시오.

① The hospital wasn't too far from here.
② I was surprised to see you at the hospital.
③ I expect you to visit your teacher at the hospital.
④ My dad had a minor accident, but he's okay now.
⑤ My sister wanted me to drop by the youth hostel.

13

대화를 듣고, 여자의 마지막 말에 대한 남자의 응답으로 가장 적절한 것을 고르시오.

Man: _____

① All right. I think I can afford it.
② Okay, thank you for calling me.
③ Give my regards to Jan and Mark.
④ Well, it all depends on your budget.
⑤ It was a really expensive trip, wasn't it?

14

대화를 듣고, 남자의 마지막 말에 대한 여자의 응답으로 가장 적절한 것을 고르시오.

Woman: _____

① Me too. My room doesn't need to be painted.
② Never mind. I need to find something new to do.
③ Great! Do you want to hear me play the song now?
④ Don't worry. I'll help you finish everything on time.
⑤ Thanks. I really appreciate your concern and advice.

15

다음 상황 설명을 듣고, Blake가 Jane에게 할 말로 가장 적절한 것을 고르시오.

Blake: _____

① Would you mind if I had some pizza, too?
② Thanks for your help! I'll see you tomorrow!
③ Could you please talk quietly? I can't concentrate.
④ I'm not sure how to do this part. Could you show me?
⑤ It's not fair that I've done all the work while you've done nothing.

[16~17] 다음을 듣고, 물음에 답하시오.

16

여자가 하는 말의 목적으로 가장 적절한 것은?

① 새로 개장한 레스토랑을 홍보하려고
② 안전하지 않은 제품에 관해 공지하려고
③ 몸에 좋은 음식의 중요성을 역설하려고
④ 비위생적인 지역 식당에 관해 경고하려고
⑤ 아이들을 위한 적절한 장난감을 소개하려고

17

피해 증상으로 언급된 것은?

① 장염　　　② 식중독　　　③ 호흡곤란
④ 심장마비　　　⑤ 소화불량

녹음을 다시 한 번 듣고, 빈칸에 알맞은 말을 쓰시오.

01

W: Good afternoon, students. This Saturday is National Make a Difference Day. On this day in 2004, three million people cared enough about their communities to volunteer, _____ _____ _____ thousands of projects in hundreds of towns. The students at Chelsea High School have already signed up to paint two homes in Valley View for low-income families in order to help _____ _____ _____ _____. You can also spend time at an orphanage, or _____ _____ _____ _____ the streets. Pick something that interests you and do something good for your community. Your help really does make a difference.

02

W: We have to go to the dentist tomorrow.
M: Oh, I had forgotten about that. What time do we need to be there?
W: Our appointments are _____ _____ 11:40.
M: Then we should plan to get there around 11:20.
W: You think so? That's a lot of extra time.
M: Well, we're new patients. So we'll have to _____ _____ _____.
W: True.
M: Plus, if they finish up with one of their patients _____ _____ _____, they'll call us in early.
W: Oh, that would be great. It would leave us more time for our lunch appointment.
M: You're right. We should leave the house at 11 to get to Dr. Park's office on time.

03

[Phone rings.]
W: Hello. ACD Computer Service. How may I help you?

M: Well, I think my computer _____ _____ _____.
W: Why do you think that?
M: Well, I can turn it on, but when I hit the keys, nothing happens.
W: Have you _____ _____ _____?
M: Yes, and they all seem to be secure.
W: Does anything come up on your screen when you turn it on?
M: Just the desktop. I can see all the icons.
W: And when you _____ _____ _____ _____, does anything happen?
M: Not at all. That's the problem.
W: Okay, I think you'll have to bring the computer to the store.

04

W: Tony, are you ready to enjoy the play?
M: Of course! I've been _____ _____ _____ _____ Mozart for two months. It's a famous play.
W: Right. And the set looks amazing. The lights are so bright!
M: It looks like Mozart's house. There is a piano on the left. There is also a piece of paper with some writing on it.
W: Yes. It looks like Mozart is composing music. And some more paper is scattered around the piano.
M: Yeah. There are _____ _____ _____ _____ on the table. Most of them are empty.
W: He must have drunk them all. Unfortunately, Mozart became poor and unhealthy in his old age.
M: I see. Something is covering the stand on the right. What is it?
W: Mozart's wig. I think it _____ _____ _____ _____ _____ when he looks at that wig.
M: Oh, the lights are getting dark now! The play is about to begin!

05

M: Jane, I'm _____ _____ to the auto shop. I need to pick up some new windshield wipers for the car.

W: All right. But come here first and take a look at this.

M: Sure. What is it?

W: Mindy's socks _____ _____ _____ _____ anymore. And we just bought them!

M: That's unbelievable! Our little girl is growing so fast.

W: I know. Could you _____ _____ _____ _____ while you're out?

M: Sure. I'll go to the kid's clothing store just down the road from the auto shop.

W: Thanks. But make sure they have animals on them. Otherwise she won't wear them.

M: I know. Our little angel is _____ _____ when it comes to clothes.

W: I'll say. I think she gets it from me.

06

W: Hello. You look lost. Maybe I can help you.

M: Oh hello. I'm looking for the Folk Museum, but I can't see it anywhere.

W: I think you _____ _____ _____ _____.

M: According to my travel book, it's located right here outside Exit 7 of this subway station.

W: Really? May I see your book for a moment?

M: Sure.

W: [Pause] Ah, I see. This book is a few years old. The subway station _____ _____ a couple of years ago, and the exit numbers were changed.

M: Oh, I see. That's why I couldn't figure it out.

W: You need to go to Exit 9 now. It's about 100 meters _____ _____ _____ _____.

M: That's great. Thanks for the help.

07

M: Hello, ma'am. Can I see your parking slip, please?

W: Sure, here you go. _____ _____ _____ _____ _____ _____?

M: Well, the fee is $3 per hour, and you were here for four hours.

W: I bought a few things at the mall.

M: Could you show me your receipts? If you spend more than $100 at the mall, parking is free.

W: I only spent about $70. Can I _____ _____ _____ for that?

M: Actually, you get two hours of parking for free for spending more than $50. You still have to pay for the remaining hours.

W: Oh, I see. Wait, there is a discount for small cars, right? My car is small.

M: Let me see. [Pause] You're right. You can get 20% off.

W: Great. I might have to pay with a credit card because _____ _____ _____ _____ _____.

M: That's not a problem.

08

W: Why do you have white hair all over your coat?

M: It's my dog's actually. My parents gave me a dog for my birthday. I named him Chopin.

W: Oh, you've always wanted a dog. But why do you call him Chopin?

M: I decided to _____ _____ _____ my favorite musician.

W: That's funny. But it isn't easy taking care of a dog, is it?

M: No, it's not. I have to _____ _____ _____ _____ _____ three times a day.

W: I used to have a dog. He _____ _____ _____ _____ sometimes.

M: My dog's too big for that. He sleeps in a doghouse in our yard. [Pause] Oh! I have to go now.

W: Where are you going?

M: I'm going to the pet store. I should buy some snack for Chopin.

W: Can I come with you?

M: Sure, let's go.

09

M: The college library will be instituting some new policies this semester. First of all, the entire third floor is now _____ _____ _____ _____. Students looking to study alone should use the first and second floors. The computers have been moved to the basement and can no longer be reserved. Instead, they're available on a first come, first served basis. We've also _____ _____ _____. From now on, we will close at 1 a.m. during the week and 11 p.m. on weekends. Also, the number of items that can be checked out at one time has been increased from six to eight. However, _____ _____ have also been raised, from 50 cents to $1 a day. If you have any questions, please ask any available staff member.

10

M: Jennifer, you play the ukulele, don't you? Help me pick one out!

W: Sure. They come in three sizes — soprano, concert, and tenor.

M: What size is that big one over there?

W: That's a tenor. It's the biggest size.

M: Oh, okay. I don't want that size. It's too big for me to _____ _____ _____ around.

W: All right. Next you need to choose the material. There's solid wood and laminate wood.

M: _____ _____ _____?

W: Solid wood has a better sound, but laminate wood is stronger.

M: I'll take a laminate wood one. I don't want it to break _____ _____ _____ _____.

W: Okay. And this shop has both new and used ukuleles.

M: I don't want a used one. I want a new one.

W: Great. And finally, what's your budget?

M: It has to be $80 or less. That's all I can spend.

W: Then you should buy this one. And I'll give you a free lesson tonight!

11

W: Hello. I'd like to _____ _____ _____.

M: All right. Is there anything wrong with it?

W: Yes, there's a hole here near the shoulder.

M: (I see. Do you have your receipt?)

12

M: Jenny, where have you been? I've been waiting for a while.

W: I'm so sorry for being late. _____ _____ _____ at the hospital than I expected.

M: Oh, why did you have to go to the hospital?

W: (My dad had a minor accident, but he's okay now.)

13

M: Hi, Melissa. What are you doing?

W: I'm reading a travel guide, see?

M: Are you planning a trip?

W: Jan, Mark, and I are _____ _____ _____ through Europe this September. Do you want to come along?

M: Sure. That sounds like fun! Where would we be going?

W: Well, we hope to land in Greece, _____ _____ through Italy and across southern France, loop around Spain, and then fly out of Paris.

M: _____ _____ _____! How long would we be gone?

W: About five weeks, I think.

M: Have you already bought tickets for the flights?

W: Yes, they were about one million won _____.

M: (All right. I think I can afford it.)

14

M: What are you doing, Gina? I heard music.

W: Hi, Dad. I'm _____ _____ how to play a new song on the piano.

M: Oh. I thought you were going to paint your bedroom today. You've been talking about doing it for weeks.

W: Well, I started this morning, but I only finished two walls. Maybe I'll do the rest tomorrow.

M: Tomorrow? Why not today? Do you remember that sweater you were going to knit for your grandmother? You started last month, and you _____ _____ _____.

W: I'm sorry, Dad. Don't be angry at me. There are just so many things that I want to do.

M: I'm not angry. I think it's great that you have so many interests. However, it's important to _____ _____ _____ _____.

W: You're right. I'll finish painting my room now. I can learn this song later.

M: I think that's a good idea.

W: (Thanks. I really appreciate your concern and advice.)

15

M: For two weeks, Blake had been working on a group project with Jane, one of his classmates. He _____ _____ because he was the only one who had been working on the project. It was two days _____ _____ _____ _____, and Jane was eating pizza and watching TV while he worked on it alone. The project was on the floor unfinished. He looked at the pictures that had been collected and at the model that had been built. How would they get it all done by the day after tomorrow? He tried to _____ _____ his research paper in the noisy room. Suddenly, he became upset with her and he decided to _____ _____. In this situation, what did Blake most likely say to Jane?

Blake: (It's not fair that I've done all the work while you've done nothing.)

16~17

W: I'm Sarah Kim, with Channel Nine News. These days, many local parents have been bringing their children to Smile Burger, the popular fast-food restaurant. Although most children enjoy the hamburgers and French fries served there, _____ _____ _____ is the Fun Food sets. When kids order these sets, they receive a fun toy along with their meal. However, it has recently come to our attention that one of the new toys being offered is not safe for children _____ _____ _____ _____ _____. It's a talking train with small parts that can be removed. Yesterday, a local three-year-old put one of these parts in his mouth and accidentally swallowed it. This _____ _____ _____ _____, and he had to be treated at City Hospital. According to the manager of Smile Burger, this toy will no longer be offered. If you have already purchased a Fun Set _____ _____ _____, you should take it away from your child immediately. If you have any questions, you can call the Smile Burger customer service center.

1번부터 17번까지는 듣고 답하는 문제입니다.

1번부터 15번까지는 한 번만 들려주고, 16번부터 17번까지는 두 번 들려줍니다. 방송을 잘 듣고 답을 하기 바랍니다.

01

다음을 듣고, 남자가 하는 말의 목적으로 가장 적절한 것을 고르시오.
① 정책 변경을 알리려고
② 정기 회의를 연기하려고
③ 웹사이트를 소개하려고
④ 중고 물품을 광고하려고
⑤ 반상회 일정을 공지하려고

02

대화를 듣고, 여자의 의견으로 가장 적절한 것을 고르시오.
① 첫 해외 여행지로 남미가 좋다.
② 여행을 통해 견문을 넓힐 수 있다.
③ 여행 계획은 되도록 자세히 세워야 한다.
④ 어렸을 때 다양한 문화 체험을 해야 한다.
⑤ 해외여행 시 음식에 적응하는 것이 가장 어렵다.

03

대화를 듣고, 두 사람의 관계를 가장 잘 나타낸 것을 고르시오.
① 학생 - 교사
② 학생 - 도서관 사서
③ 세입자 - 임대주
④ 고객 - 은행원
⑤ 고객 - 서점 직원

04

대화를 듣고, 그림에서 대화의 내용과 일치하지 않는 것을 고르시오.

05

대화를 듣고, 남자가 할 일로 가장 적절한 것을 고르시오.
① 우편물 보내기
② 휴대전화 요금 내기
③ 출장 갈 준비하기
④ 여자에게 연락처 주기
⑤ 여자를 사무실에 데려다주기

06

대화를 듣고, 여자가 남자에게 고마워하는 이유를 고르시오.
① 함께 영화를 보러 가서
② 숙제를 도와줘서
③ 책을 빌려줘서
④ 과학 시험공부를 도와줘서
⑤ 숙제에 관해 알려줘서

07

대화를 듣고, 남자가 부담할 금액을 고르시오. 3점
① $1,000
② $1,250
③ $2,000
④ $2,250
⑤ $2,500

08

대화를 듣고, 시험에 관해 언급되지 않은 것을 고르시오.
① 문항 수
② 시험 일자
③ 합격자 수
④ 시험 장소
⑤ 성적 발표일

09

job opening에 관한 다음 내용을 듣고, 일치하지 않는 것을 고르시오. 3점
① 지원자는 2년 이상의 영업 경험이 필요하다.
② 지원자는 일어 회화 실력을 갖추어야 한다.
③ 연봉은 지원자의 경력에 따라 다르다.
④ 현재 재직 중인 직원은 고려 대상에서 제외된다.
⑤ 외부 신청자들은 3월 1일부터 지원할 수 있다.

10

다음 표를 보면서 대화를 듣고, 여자가 선택할 무선 스피커를 고르시오.

	Model	Price	Weight	Charging Hour	Shape
①	A	$190	1.2 kg	1.5 hr.	oval
②	B	$200	980 g	1.2 hr.	oval
③	C	$225	950 g	50 min.	oval
④	D	$240	800 g	45 min.	cube
⑤	E	$260	750 g	45 min.	cube

11

대화를 듣고, 남자의 마지막 말에 대한 여자의 응답으로 가장 적절한 것을 고르시오.

① You can do a lot of activities on this site.
② One of my friends recommended it to me.
③ First, you should sign up for this program.
④ Writing letters to them makes me feel good.
⑤ My friends went to another country to volunteer.

12

대화를 듣고, 여자의 마지막 말에 대한 남자의 응답으로 가장 적절한 것을 고르시오.

① I will play my best next time.
② The match was boring to watch.
③ The match will take place in the field.
④ I don't know. We should wait until the rain stops.
⑤ We should wash up and get ready for the game now.

13

대화를 듣고, 남자의 마지막 말에 대한 여자의 응답으로 가장 적절한 것을 고르시오. 3점

Woman: _____

① So I'm stuck with this ugly skirt?
② That's good to know, but it's your fault.
③ I guess I'll just have to exchange it then.
④ I'll come back later with my receipt then.
⑤ Thank you for making an exception in this case.

14

대화를 듣고, 여자의 마지막 말에 대한 남자의 응답으로 가장 적절한 것을 고르시오.

Man: _____

① Sadly, I can't come either.
② Yes, I'm sorry to hear that.
③ In that case, I can go after all.
④ Then I'll see you at the workshop!
⑤ But the holidays will be over by then.

15

다음 상황 설명을 듣고, Daniel이 Julie에게 할 말로 가장 적절한 것을 고르시오.

Daniel: _____

① I hope you feel better soon.
② Cheating is never a good idea.
③ I guess I just got lucky this time.
④ That's what I was worried about.
⑤ Do you mean that I studied for nothing?

[16~17] 다음을 듣고, 물음에 답하시오.

16

남자가 하는 말의 주제로 가장 적절한 것은?

① the most efficient learning method
② different learning styles of students
③ dealing with students' bad behavior
④ how to be your students' favorite teacher
⑤ common teacher mistakes in the classroom

17

언급된 학습 수단이 아닌 것은?

① graphics ② chants ③ quizzes
④ role-playing ⑤ discussions

녹음을 다시 한 번 듣고, 빈칸에 알맞은 말을 쓰시오.

01

M: Good afternoon, everyone. Before we start our monthly residents' meeting, I'd like to take a moment to remind you that the building management has set up _____ _____ _____ _____ for us. The website address is written on the cards that I just handed out. It's the best place to find up-to-the-minute _____ _____ _____ concerning the building. There's also a forum where you can connect with other neighbors. People are already using it to post personal news and photos, as well as to _____ _____ _____ for sale or to trade. So please take a look, and if you have any questions, just let me know.

02

M: How was your trip to South America?

W: I had a great time! I stayed for a few weeks, so I had a lot of time to _____ _____ _____.

M Did you eat any interesting foods?

W: Of course! I tried a few new dishes that were seasoned with special spices.

M: Was this your first time _____ _____?

W: No. I try to take international trips as often as possible. Every time I do, I learn something new.

M: It sounds like it really changes your view of the world.

W: It does. I guess experiencing another culture _____ _____ _____ on everything.

M: In that case, maybe I should plan a trip myself.

03

W: Hi. I need some help.

M: Certainly. What can I do for you?

W: I have _____ _____ _____ _____. Unfortunately, they're four days late.

M: I see. I hope you understand that you'll need to

_____ _____ _____ _____.

W: I understand. I was told that it's 50 cents per book per day.

M: I'm afraid so. That means your total fee is $6.

W: But I don't have that much cash on me. And I need to _____ _____ this DVD for a class.

M: Well, you can't check out any new items until you pay your late fee. Do you have a debit card?

W: Yes, I do. Can I use it here?

M: Sure. Just run it through this machine and type in your passcode.

W: Wow, that's really convenient. Thanks.

04

W: Peter, look at this picture. It is from the new zoo.

M: Oh! That zoo just opened, right?

W: Yes, and it has a lot of facilities and exhibits.

M: That picnic area on the left looks great. It has parasols, tables, and trees.

W: I agree. It would be the perfect place _____ _____ _____ _____.

M: The waterfall and lake in the center look nice too. Are there animals in the lake?

W: Yeah, two hippos are swimming in it. They look really happy.

M: _____ _____ _____ _____. And check out that funny popcorn stand on the right. It looks like a huge box of popcorn.

W: How clever! What are those two students doing?

M: They are petting some rabbits through a fence. They must be at the petting zoo.

W: The zoo looks like _____ _____ _____ _____ _____.

M: Yes, we should go there sometime soon.

W: Great idea!

05

[Phone rings.]

W: Hello?
M: Are you Tina Patterson?
W: Yes, this is Tina.
M: Hi. I _____ _____ _____ in the park the other day, and I think it belongs to you.
W: Oh, you found my purse! But how did you get my number?
M: There was an old cell phone bill in it with your number.
W: That's great. There were some precious photos in my purse that I would hate to lose.
M: Well, you can come by my office and _____ _____ _____ _____.
W: Unfortunately, I'm away on a business trip at the moment.
M: In that case, give me your address and I'll _____ _____ _____ _____ this afternoon.
W: That would be great. You've been very kind.
M: I'm just glad to be of help.

06

W: Hey, Justin. Do you want to go to the movie theater with me?
M: I'm sorry, but I have _____ _____ _____ to do.
W: Really? Is there an assignment we have to do?
M: Don't you remember? We have a science assignment _____ _____ _____.
W: Really? What was it?
M: It's a book report. We have to read a book about the ecosystem.
W: Oh, now I remember! I completely forgot about that. What should I do?
M: Hmm... You'd better start doing it right now. There are lots of pages to read.
W: Thanks for letting me know. Otherwise I _____ _____ _____ _____ tomorrow.
M: You're welcome.

07

M: Linda, do you have any suggestions for _____ _____ _____ _____?
W: You're looking for work? What do you need the money for?
M: I'm taking a study course in Spain this summer.
W: Oh, I took one of those courses last year. It costs about $2,000, right?
M: Actually, _____ _____ _____ is $2,500. But since I booked it through the school, I get a 10% discount.
W: So you're going to need a job that pays pretty well.
M: Well, actually my parents are paying $1,000. I just need to _____ _____ _____.
W: In that case, get a job waiting tables in a café. You can earn about $300 a month.
M: That would be perfect. Thanks for the advice.
W: No problem. Good luck finding a job.

08

M: Ms. Woods, thank you for applying for this job. We were very _____ _____ _____ _____.
W: I'm glad to hear that.
M: We'd like you to come in at 10 o'clock on June 15 and take a computer skills test.
W: All right. How long is the test?
M: Well, there are four different sections. Most people finish in about two hours.
W: How many people are taking it?
M: There will be 30 others. _____ _____ _____ will be selected for the job.
W: I see. So I should come back here to take the test?
M: Actually, you'll be taking it in the conference room _____ _____ _____ _____.
W: Great. When will the results be announced?
M: We will announce the results a week after the test.
W: Thank you for the information.

09

W: This message is to notify all employees of an upcoming job opening in the international sales department. With the retirement of Ann Beasley, we will be seeking ＿＿＿ ＿＿＿ ＿＿＿ ＿＿＿ for the Japanese market. All qualified applicants will have a college degree and at least ＿＿＿ ＿＿＿ ＿＿＿ ＿＿＿ ＿＿＿ in this or another company. They must also be able to speak Japanese at a conversational level. The salary for this position is between $60,000 and $75,000 per year, ＿＿＿ ＿＿＿ ＿＿＿. Current employees will have first consideration for this position, but if no one suitable is found, we will be accepting outside applications starting on March 1.

10

M: Hello. How can I help you?

W: Hi. Can you show me some wireless speakers, please?

M: Yes, of course. We have five models, and their prices ＿＿＿ ＿＿＿ ＿＿＿ ＿＿＿.

W: Okay. I don't think I can spend more than $250.

M: Then these are the ones ＿＿＿ ＿＿＿ ＿＿＿.

W: How much do they each weigh? I'm going to carry them around a lot, so I need something light, preferably under 1 kg.

M: Okay. Here are the models that are under 1 kg.

W: I'm wondering about how long they take to charge. I'd like the speakers to be fully charged in under an hour.

M: We have two models in your budget ＿＿＿ ＿＿＿ ＿＿＿. What about this oval shape model?

W: I don't know. I think the cube one is better. It looks very neat.

M: Great choice! Would you like to pay with cash or credit card?

W: I'll pay with cash.

11

M: Anna, what is this website about?

W: It's a homepage for charitable donations. I can send letters and ＿＿＿ ＿＿＿ ＿＿＿ in other countries.

M: That's great! How did you first find out about it?

W: (One of my friends recommended it to me.)

12

W: Alex, do you think we can play soccer today?

M: No, it's raining outside. It is hard to ＿＿＿ ＿＿＿ ＿＿＿ ＿＿＿.

W: You're right. So when should we have the match then?

M: (I don't know. We should wait until the rain stops.)

13

W: Hi. I bought this skirt here, but I'd like to return it.

M: All right. Is there something wrong with it?

W: It just ＿＿＿ ＿＿＿ ＿＿＿.

M: Okay. If you have your receipt, you can get a full refund or exchange it for another item.

W: Here it is. And I'll just take a refund.

M: Sure. *[Pause]* Oh, hold on. You purchased the skirt three weeks ago?

W: Yes, it will be three weeks tomorrow.

M: I'm sorry, but our return policy only ＿＿＿ ＿＿＿ ＿＿＿ ＿＿＿ after purchase.

W: Oh, does it? Can you ＿＿＿ ＿＿＿ ＿＿＿?

M: I'm sorry, but I can't do that. However, our exchange policy does last for six weeks.

W: (I guess I'll just have to exchange it then.)

14

W: Hi, Jacob. How's it going?

M: Not bad. I'm just finishing up my monthly financial reports.

W: Great. Listen, I just wanted to let you know that I'll be holding _____ _____ _____ at my house.

M: Oh, really? That sounds great. When is it?

W: It'll be on Saturday, December 20 starting at 2 p.m.

M: The 20th? Oh, I'm sorry, but I won't be able to make it.

W: Oh, that's too bad. Everyone from the department will be there.

M: I'd love to come, but I've got _____ _____ _____ _____ on that day.

W: Oh. Wait, do you mean the workshop for the new payroll system?

M: Yes, that's right.

W: I heard _____ _____ _____ until next month.

M: (In that case, I can go after all.)

15

W: Daniel is a high school student. As he has an important science test next week, he _____ _____ _____ _____ for it. On Monday morning, Daniel is exhausted but feeling confident about the test. However, he suddenly realizes that he has forgotten to _____ _____ _____ _____. Sitting in math class and waiting for the teacher to arrive, he desperately tries to _____ _____ _____ a good excuse. He can't think of one, so he turns to his friend Julie and explains the problem. Julie tells Daniel that the teacher is feeling ill and won't be coming to class today. In this situation, what would Daniel most likely say to Julie?

Daniel: (I guess I just got lucky this time.)

16~17

M: Hello, teachers. In today's seminar, I'd like to talk about your students. You've probably already noticed that they're all different. But these differences _____ _____ _____ _____ and personalities. Each student learns in a unique way. Some of them rely on their eyes. They can quickly understand graphics, such as charts and pictures. Others prefer to use their ears. They learn better when they hear the material. Chanting is a good technique for teaching this type of learner. Of course, there are some kids who learn best _____ _____ _____ _____. They should be given lots of quizzes that require them to write the answers. And there are also kids who need to use their bodies. To help them learn more efficiently, give them _____ _____ _____ to move around the classroom. Role-playing is a fun and active way for these students to learn. So if you want to be a great teacher, _____ _____ _____ _____ _____.

1번부터 17번까지는 듣고 답하는 문제입니다.
1번부터 15번까지는 한 번만 들려주고, 16번부터 17번까지는 두 번 들려줍니다. 방송을 잘 듣고 답을 하기 바랍니다.

01

다음을 듣고, 여자가 하는 말의 목적으로 가장 적절한 것을 고르시오.
① 주문을 변경하려고
② 배송일을 늦추려고
③ 주문 수량을 확인하려고
④ 프린터 수리를 맡기려고
⑤ 실수에 대해 불평하려고

02

대화를 듣고, 두 사람이 하는 말의 주제로 가장 적절한 것을 고르시오.
① 전기 요금을 줄이는 법
② 대체 에너지 개발의 어려움
③ 에너지를 절약해야 하는 이유
④ 에너지 효율이 높은 전자 제품
⑤ 에너지 소비와 환경 오염의 관계

03

대화를 듣고, 두 사람의 관계를 가장 잘 나타낸 것을 고르시오.
① 남편 - 아내
② 점원 - 손님
③ 친구 - 친구
④ 세입자 - 집주인
⑤ 부동산 중개인 - 집주인

04

대화를 듣고, 그림에서 대화의 내용과 일치하지 않는 것을 고르시오.

05

대화를 듣고, 남자가 여자에게 부탁한 일로 가장 적절한 것을 고르시오.
① to set up some chairs
② to recruit volunteers
③ to design some shirts
④ to sell festival tickets
⑤ to go to a film festival

06

대화를 듣고, 여자가 돌고래 관광을 갈 수 없는 이유를 고르시오.
① 뱃멀미를 해서
② 부모님이 허락하지 않아서
③ 웅변대회 연습을 해야 해서
④ 연극 연습을 해야 해서
⑤ 현장 학습을 가야 해서

07

대화를 듣고, 남자가 지불할 금액을 고르시오. 3점
① $20   ② $22   ③ $24   ④ $26   ⑤ $28

08

대화를 듣고, 여자의 가게에 관해 언급되지 않은 것을 고르시오.
① 판매 물품
② 위치
③ 휴무일
④ 할인 품목
⑤ 할인 폭

09

Introduction to Computers 수업에 관한 다음 내용을 듣고, 일치하지 않는 것을 고르시오.
① 일주일에 두 번 진행될 것이다.
② 교재는 학생들에게 무료로 제공된다.
③ 컴퓨터의 업무적인 측면을 다룰 것이다.
④ 과제는 많지 않을 것이다.
⑤ 복습을 위해 컴퓨터실을 이용할 수 있다.

10

다음 표를 보면서 대화를 듣고, 여자가 구입할 휴대전화를 고르시오.

	Phone	ROM Storage	Mega-pixels	Price	Customer Rating
①	A	16 GB	13	$550	★★★★
②	B	32 GB	13	$600	★★★
③	C	32 GB	21	$850	★★★
④	D	64 GB	16	$950	★★★★
⑤	E	128 GB	16	$1,050	★★★★

11

대화를 듣고, 여자의 마지막 말에 대한 남자의 응답으로 가장 적절한 것을 고르시오.

① No, Mom will come here next week.
② I told you I already mopped the floor.
③ That's a good idea. Let's clean it together.
④ Thank you for washing the dishes for me.
⑤ Knock first! Mom might be in the bathroom.

12

대화를 듣고, 남자의 마지막 말에 대한 여자의 응답으로 가장 적절한 것을 고르시오.

① She must be really excited.
② I think she took her puppy to the vet.
③ She never told me about her problem.
④ Wow. She must be so good with plants.
⑤ Well, I've seen him, but he wasn't cute at all.

13

대화를 듣고, 남자의 마지막 말에 대한 여자의 응답으로 가장 적절한 것을 고르시오. 3점

Woman: _____

① I'll keep that in mind whenever I use it.
② If your symptoms continue, see a doctor.
③ Bring your phone to a repair shop to be fixed.
④ We've been shopping for a new microwave oven.
⑤ I didn't know plants absorb electromagnetic waves.

14

대화를 듣고, 여자의 마지막 말에 대한 남자의 응답으로 가장 적절한 것을 고르시오.

Man: _____

① I'll get him right now.
② I'm afraid you can't. He quit.
③ This is much easier than I thought.
④ I'm sure he'll understand. Good luck!
⑤ When will I be able to speak to Mr. Douglas?

15

다음 상황 설명을 듣고, Kathy가 직원에게 할 말로 가장 적절한 것을 고르시오. 3점

Kathy: _____

① Sorry, but this steak is too expensive.
② Could you explain how I should cook this?
③ These roses are all withered. I want to get a refund.
④ This is expired. I'd like to exchange it for a different one.
⑤ I don't like red wine. Can you show me the white wine, please?

[16~17] 다음을 듣고, 물음에 답하시오.

16

남자가 하는 말의 주제로 가장 적절한 것은?

① how to apply to future jobs
② the least safe jobs in the future
③ training robots to replace humans
④ new jobs that require special skills
⑤ tasks that machines could never do

17

언급된 직업이 아닌 것은?

① telemarketers
② tax preparers
③ legal assistants
④ fast-food cooks
⑤ healthcare workers

녹음을 다시 한 번 듣고, 빈칸에 알맞은 말을 쓰시오.

01

W: Hello, this is Laura Blaine with Delstar Industries. I guess your office has already closed for the day. I'm calling _____ _____ _____ _____ _____ for two dozen cartons of laser printer paper that I placed with you yesterday morning. I _____ _____ _____ about the amount of paper, and it turns out that we have more paper in our supply room than we thought we did. If the order hasn't been shipped yet, _____ _____ _____ _____ to one dozen cartons instead of two? If there's any problem with this request, please give me a call first thing in the morning. My number is 862–8088, extension 43. Thanks.

02

M: Wow, another expensive utility bill! I can't believe how much I'm spending on energy every month.

W: There are a lot of things you can do to make it lower.

M: Like what?

W: First of all, you can _____ _____ _____ _____ ones that are more efficient.

M: I don't think I can afford to buy new appliances right now.

W: Okay, then how about using a fan instead of _____ _____ _____ _____? Fans use a lot less electricity.

M: I guess I could try that.

W: And, when it's really hot outside, keep your curtains closed. The sunlight shining in will only make the room hotter.

M: I've never thought about that before. Do you really think that will help me save money on energy?

W: Definitely. Sometimes small changes can _____ _____ _____ _____.

03

M: I saw this apartment last week. It's quite nice.

W: It seems large.

M: That's what I thought. And look at the closets — they're huge!

W: That will be good. We have _____ _____ _____.

M: Right. We do need the space.

W: Definitely. Especially with a baby on the way.

M: The agent said that the washing machine was _____ _____.

W: That's important.

M: And if we move here you'll live near your parents. We'll see them more often.

W: Yes, that's _____ _____. This might be the place for us.

M: Great! I'll tell the agent. She's waiting downstairs.

W: Wait. Let's take some time to think about this.

M: Okay. I'll tell her we'll call her tomorrow.

04

M: I had a great yard sale. Here, let me show you some of the pictures.

W: Okay. Oh, the mirror on the ground _____ _____ _____ me.

M: The square one? That used to be in our living room.

W: That's right. Oh, the woman with the baby seems to like that dress. Did she buy it?

M: Yes, her daughter will look good in that little dress with the ribbon.

W: Oh, is the _____ _____ _____ your son's?

M: Yes. He's too big for it now.

W: Oh, there's a backpack for two dollars. _____ _____ _____.

M: Yeah. It was sold quickly.

W: I see. So did you sell everything that you put out?

M: Everything except what's in the flower print box

under the table.

05

W: Wow, it looks like you guys are really busy.

M: Yes. We're working on the local independent film festival.

W: You're really going to do that? I thought it was just an idea.

M: Well, it was an idea, but it's becoming reality. We're holding it during the first week of May.

W: That's really great. Do you _____ _____ _____?

M: Of course we do. Would you like to help us?

W: Sure. Just tell me what I can do.

M: Let me think. *[Pause]* Hey! _____ _____ _____ _____ for the school fair, didn't you?

W: Yes, I did. They were pretty popular.

M: Right. Well... Do you think you could _____ _____ _____ for our film festival?

W: I'd love to. It sure beats selling tickets or setting up chairs.

M: Great! I really appreciate your help.

06

M: _____ _____ _____ this weekend!

W: Why is that? What's going on this weekend?

M: I'm going dolphin watching with my friends.

W: Wow! I'm jealous. I went dolphin watching with my family last summer, but we didn't see any dolphins.

M: You don't need to be jealous. You can join us this weekend.

W: I wish I could, but I can't _____ _____ _____.

M: Why? Won't your parents give you permission?

W: My parents wouldn't mind, but I don't have time.

M: Is it because you are going to _____ _____ _____ a speech contest next month?

W: No, but the school play is next week, so this weekend will be our last chance to rehearse.

M: Oh, I see. Well, I'll take some pictures for you!

W: Cool. I hope you see some dolphins. Have fun on the trip!

07

M: Hello. I'd like to _____ _____ _____ for my mother. How much are your roses?

W: They're $20 a dozen.

M: Is that $10 for half a dozen?

W: No, $12 for half a dozen.

M: Okay, that's fine. Six roses, please — red ones.

W: Okay. What else?

M: A dozen carnations... No, half a dozen as well.

W: Certainly. That's $10 for the carnations.

M: What else would go with that?

W: Baby's breath — a kind of tiny white flowers. Very pretty and only $2 more.

M: Well, maybe not. Just _____ _____ a couple more roses please.

W: Extra roses are $2 each. But I really recommend the baby's breath; it _____ the roses and carnations very well.

M: Okay, forget the extra roses and I'll take the baby's breath then.

W: Great choice!

08

M: Hi, Allison! What have you been up to?

W: Hi, Mike! I've opened my own store.

M: That's great. What kind of store?

W: It's a jewelry store. I sell _____ _____, bracelets, rings, and hair accessories.

M: Really? Do you make them yourself?

W: No, I buy them from local artists I know.

M: I'd love to come check it out sometime.

W: Great. It's in the Townville Mall on Fifth Avenue. Come by on Wednesday, we're _____ _____

_____.

M: I think I will. Will there be big discounts?

W: Yes! From 9 a.m. to 1 p.m., all the necklaces and rings in the store _____ _____ _____ _____.

M: I'll be sure to get there early. And maybe you and I can have lunch together.

W: Sounds good. See you soon.

09

M: Hello everyone and welcome to the class. This is, of course, Introduction to Computers, and we will be meeting twice a week at 8 p.m. You're all required to _____ _____ _____ for this course. It's listed on your syllabus and is available at most bookstores. Now this course will focus on _____ _____ _____ _____ _____. It has been designed specifically for adults who are returning to the workplace and need to develop a basic understanding of computers. I understand you're busy people, so there won't be much homework. However, I expect you to practice what you learn in class. If you don't have a computer of your own, you can _____ _____ _____ _____ _____ _____.

10

M: Hello, ma'am. What are you looking for today?

W: I want a new smartphone _____ _____ _____ _____ _____.

M: I see. How much storage do you need?

W: My current phone has 16 GB, and that's not enough.

M: All right. How about the camera? Is 13 megapixels okay?

W: No, it isn't. I like to _____ _____ _____ _____. It has to be at least 16 megapixels.

M: That's not a problem. What about the price?

W: I can't spend more than $1,000.

M: In that case, these two phones fit your needs. The one on the left is cheaper.

W: Oh, okay. It looks pretty nice.

M: However, the one on the right has a higher customer rating.

W: Really? Then I'll take the one on the right. Quality is _____ _____ _____ _____.

11

W: Mom will come home tomorrow. We have to clean the kitchen.

M: Okay. I'll wash all the dishes. Can you _____ _____ _____?

W: Sure. I'll go get the cleaning supplies. We should clean the bathroom, too.

M: (That's a good idea. Let's clean it together.)

12

M: Did you know that Suji _____ _____ _____ _____?

W: No, but I know she loves animals. What does it look like?

M: It's small and white with brown spots. He's so cute!

W: (She must be really excited.)

13

M: Hey, Molly. I read an interesting article this morning.

W: Oh, really? Tell me about it.

M: It said we are constantly _____ _____ _____ electromagnetic waves.

W: How is that possible?

M: Many of the things that we use in our daily lives, such as cell phones and microwaves, produce them.

W: That sounds dangerous!

M: It's not harmful in small amounts, but over time, regular exposure could _____ _____ _____

_____ our bodies and even lead to cancer.

W: That's terrible. How can I protect myself?

M: Well, if you keep certain kinds of plants in your apartment, they will absorb some of the electromagnetic waves.

W: All right. Is there anything else?

M: Yes. When you talk on your cell phone, _____ _____ _____ _____ your face.

W: (I'll keep that in mind whenever I use it.)

14

M: Hello, I'd like to _____ _____ _____ _____.

W: Well, Mr. Douglas, our loan officer, is not here today, but I can try to help you. What kind of loan were you looking for?

M: For a new home.

W: Okay. Is this your first home?

M: No, second.

W: I see. I can tell you in advance that you might not _____ _____ _____ some special programs.

M: That's okay. I didn't expect to be.

W: I can _____ _____ _____ _____ to look over about the loans we offer.

M: Good. How difficult will it be to get one?

W: I'm afraid I can't exactly answer that. Mr. Douglas can let you know more about it.

M: (When will I be able to speak to Mr. Douglas?)

15

W: Kathy received a call informing her that her husband _____ _____. To celebrate, she decided to cook a special dinner with wine and steak. So she went to the supermarket and bought some red wine and steaks. She even _____ _____ _____ _____ and bought a bouquet of roses. But when she was about to start cooking, she realized that _____ _____ _____ on the steak was the day before. Feeling irritated, she went back to the supermarket. In this situation, what is Kathy most likely to say to the clerk?

Kathy: (This is expired. I'd like to exchange it for a different one.)

16~17

M: Good morning, everyone. I'm Scott Fernandez from the Community Career Center. Are you worried about robots _____ _____ _____ _____? You might be _____ _____ _____ _____. I'm here today to talk about some of the jobs that are likely to disappear soon. First, telemarketers are expected to be out of work. Computers are already able to _____ _____ _____ _____. Tax preparers are also at risk. The routine calculation work that they do can be done easily and efficiently by machines. The same is true for legal assistants. Machines can do many of their repetitive tasks quickly. In addition, people should no longer expect to work as fast-food cooks. More and more of the jobs in the fast food industry are becoming automated. Are there any other jobs that _____ _____ _____ _____ by technology over the next 20 years? Take a few minutes to think about it.

1번부터 17번까지는 듣고 답하는 문제입니다.
1번부터 15번까지는 한 번만 들려주고, 16번부터 17번까지는 두 번 들려줍니다. 방송을 잘 듣고 답을 하기 바랍니다.

01

다음을 듣고, 여자가 하는 말의 목적으로 가장 적절한 것을 고르시오.
① 졸업식 행사를 소개하려고
② 졸업 무도회 참석을 요청하려고
③ 졸업 무도회 일정 변경을 알리려고
④ 졸업식 장소 변경에 대해 사과하려고
⑤ 졸업 무도회에서의 금지 사항을 공지하려고

02

대화를 듣고, 남자의 의견으로 가장 적절한 것을 고르시오.
① 고전 문학 작품을 읽어야 한다.
② 좋은 성적은 수업 태도가 결정짓는다.
③ 성적보다는 배움에 초점을 맞춰야 한다.
④ 학생들은 다양한 과외 활동 경험이 필요하다.
⑤ 모든 학교 시험의 평가가 정확한 것은 아니다.

03

대화를 듣고, 두 사람의 관계를 가장 잘 나타낸 것을 고르시오.
① 비서 - 임원   ② 학생 - 교사   ③ 조교 - 교수
④ 승객 - 기장   ⑤ 학생 - 조교

04

대화를 듣고, 그림에서 대화의 내용과 일치하지 않는 것을 고르시오.

05

대화를 듣고, 여자가 할 일로 가장 적절한 것을 고르시오.
① 공원에 가기   ② TV 시청하기
③ 영화 보러 가기   ④ 비디오 빌려 보기
⑤ 집에 가서 책 읽기

06

대화를 듣고, 남자가 실망한 이유를 고르시오.
① 그리스어 시험에서 낙제해서
② 그리스어 수업이 폐강되어서
③ 그리스 여행 계획이 취소되어서
④ 친구와 함께 수업을 듣지 못해서
⑤ 수업 시간에 선생님께 야단을 맞아서

07

대화를 듣고, 여자가 지불할 금액을 고르시오. 3점
① $60   ② $70   ③ $80
④ $90   ⑤ $100

08

대화를 듣고, 자동차 운전에 관해 언급되지 않은 것을 고르시오.
① 안전벨트 매기   ② 백미러 조정하기
③ 후진 기어 넣기   ④ 시동 걸기
⑤ 연료계 확인하기

09

다큐멘터리 관람에 관한 다음 내용을 듣고, 일치하지 않는 것을 고르시오. 3점
① 관람 시 휴대전화의 전원을 꺼야 한다.
② 관람을 방해하는 학생들은 퇴실 조치된다.
③ 일부 학생들이 복도에서 기부금을 받을 것이다.
④ 기부금 일부는 암 치료법을 개발하는 데 사용된다.
⑤ 기부금의 액수에는 제한이 없다.

10

다음 표를 보면서 대화를 듣고, 남자가 구매할 기차표를 고르시오.

Train to Toronto: Available Tickets

	Train	Time	Transfers	Available Seats	Price
①	A	7:00	0	Aisle	$50
②	B	8:00	0	Window	$40
③	C	10:00	2	Aisle and window	$30
④	D	11:00	1	Window	$35
⑤	E	13:00	0	Aisle and window	$45

11

대화를 듣고, 여자의 마지막 말에 대한 남자의 응답으로 가장 적절한 것을 고르시오.

① I'm sure those will make great gifts.
② Personally, I almost never wear neckties.
③ Hairstyle trends change so quickly nowadays.
④ I don't think that matches the suit I'm wearing.
⑤ Well, you'd better choose the one with the color you like.

12

대화를 듣고, 남자의 마지막 말에 대한 여자의 응답으로 가장 적절한 것을 고르시오.

① I haven't seen your tent anywhere.
② Do you think it's safe to camp here?
③ I've never seen a TV show about fishing.
④ I saw some on sale at a department store.
⑤ My son would also like to look at the fish in the aquarium.

13

대화를 듣고, 남자의 마지막 말에 대한 여자의 응답으로 가장 적절한 것을 고르시오.

Woman: _____

① You should wear a sweater.
② Okay, I'll call the Smiths then.
③ Don't forget to take out the trash.
④ He doesn't remember putting it there.
⑤ We really need to clean out the garage.

14

대화를 듣고, 여자의 마지막 말에 대한 남자의 응답으로 가장 적절한 것을 고르시오.

Man: _____

① I don't think you should go there.
② Yes, I'm really looking forward to it.
③ Okay. Thank you for inviting me along.
④ Not really. I don't really like that sort of thing.
⑤ I did. I would really recommend seeing them if you get the chance.

15

다음 상황 설명을 듣고, Sam의 아버지가 Sam에게 할 말로 가장 적절한 것을 고르시오.

Sam's father: Sam, _____

① I told you that you couldn't go out!
② are you hungry? Your dinner is on the table.
③ all the noise your friends made woke me up.
④ I'm so angry with you for failing your exams.
⑤ where are your friends? Aren't they going to sleep here?

[16~17] 다음을 듣고, 물음에 답하시오.

16

남자가 하는 말의 주제로 가장 적절한 것은?

① misuse of valuable foreign coins
② the best cleaning methods for coins
③ various ways of using leftover coins
④ the popularity of collecting foreign coins
⑤ different types of coins from around the world

17

언급된 활용 방안이 <u>아닌</u> 것은?

① souvenirs   ② online sales   ③ cleaners
④ magnets   ⑤ donations

녹음을 다시 한 번 듣고, 빈칸에 알맞은 말을 쓰시오.

01

W: Good day, students. As you all know, tonight is prom night. I would like to wish you all a most _____ _____ _____ _____. On that note, we would like to remind everyone that _____ _____ will not be permitted at the prom, and anyone who appears to be intoxicated will not be permitted to enter. Prom night should be one of _____ _____ _____ _____ of high school, and alcohol could turn this special night into the most unfortunate one. Thank you, and have a wonderful evening.

02

W: I'm never going to finish *Jane Eyre* by Thursday.
M: So finish it when you can.
W: But the test is Thursday!
M: I know. But if you _____ _____ _____ _____, you're not going to get much out of it.
W: But I might do better on the test.
M: Is that really the point?
W: Yeah! I want a good grade.
M: You care too much about your grades. The point of education is to learn. _____ _____ _____ _____ if you never learn anything?
W: The grades I get now will affect my opportunities in the future.
M: Look... _____ _____ _____ and less on the exams. Good grades will follow.

03

W: Excuse me, _____ _____ _____ you could help me with something. I need to take a biology class. Does Dr. Griffin teach any in September?
M: Yes, he is teaching biology next semester. Do you want to sign up for it?
W: Yes, I have to. I am planning on graduating in December.
M: Oh, no... Unfortunately, that class is already full. There is a waiting list though. If someone were to drop the course, you could take their place.
W: Yes, please _____ _____ _____ _____. How many people are on it?
M: Let me check. *[Pause]* Oh dear, there are already 14 people on the list. I will still _____ _____ _____ _____.
W: It sounds unlikely that I'll get in them. That's a problem because this class is a requirement for me to graduate.
M: Yes, I understand. Dr. Griffin has made exceptions in the past though.
W: What do you mean?
M: If there are special circumstances, he sometimes allows one more person to join the class.
W: Oh, well then I will try to _____ _____ _____ _____.
M: Okay. I'll set up an appointment for you then.

04

W: Do you want some ice cream?
M: Oh, I can see the store over there.
W: They have a giant ice cream cone on their sign.
M: Oh, there are people at all of the tables already. Where should we sit?
W: Look! There is a bench _____ _____ _____ _____. Let's sit there.
M: Okay. What do you want to have?
W: Hmm. I'll have a scoop of vanilla ice cream. How about you?
M: I'll have an ice cream cone like the one that boy is eating.
W: You mean the boy _____ _____ _____?
M: Yes. I think his ice cream looks yummy.
W: All right. The guy with the mustache is _____ _____. He looks strong, and I bet he can scoop ice cream better than anyone else.

05

M: What are you doing after work?

W: _____ _____. If it's not raining, I think I might go to the park.

M: By the looks of those clouds, I'd say you're _____ _____ _____.

W: Yeah, I think you're right. I guess I'll probably just go home and read.

M: Well, some of us are going to the movies. If you're interested, come along.

W: Really? What movie is playing?

M: *The Ring*. It's a scary movie.

W: Oh, I don't like it.

M: Really? How about a comedy movie? I think some other people are going to see it.

W: _____ _____ _____, I will come along. I can read at home anytime.

06

W: Hi, Nick. You look down. What's the matter?

M: I'm _____ _____ _____ _____ my schedule this semester.

W: Oh really? Why is that?

M: Well, I registered for Greek, but now it's been canceled.

W: I see. That's a shame. What happened?

M: I guess not many students _____ _____ _____ _____ _____.

W: Hey, maybe you can register for German instead. I'm going to take that this semester.

M: But I'm planning to visit Greece next summer, and I really wanted to be able to speak some of the language.

W: Why don't you _____ _____ _____ _____ _____ and study the language when you have some free time?

M: That's not a bad idea. I guess I'll have to do that.

07

W: Hello. I'd like to order a Marple High School yearbook.

M: Hello. _____ _____ _____ _____. The general yearbook is $50. And the yearbook by grade is $60.

W: How is the yearbook by grade different from the general yearbook?

M: The general yearbook includes photos of all the students. The yearbook by grade only has the students in your grade.

W: Great. I would like the one for my grade only. I was in tenth grade.

M: Okay. Do you want digital files of the pictures too?

W: How many pictures would I get? _____ _____ _____ _____ _____?

M: You can get 50 digital files for $10. And you can select the ones you want.

W: Hmm... That's nice. I'll take 150 pictures. Here is my credit card.

M: Okay. Please _____ _____ _____ _____ to request your yearbook and photos.

08

W: Okay, I know this is your first time driving a car. There are a few things you need to do before we go.

M: Right. I'm very nervous.

W: Don't be. The first thing you need to do is _____ _____ _____ _____ _____.

M: Oh, yes. I almost forgot.

W: Good. Now if you're comfortable, why don't you _____ _____ _____ _____?

M: Okay. That's good.

W: Now you're ready to start the car. Have you got your foot on the brake?

M: Yes.

W: Okay, go ahead. *[Pause]* Now, if there's nothing behind you, you can put the car _____ _____ and ease your foot off the brake.

M: Okay, we're in reverse... Uh-oh.

W: What's wrong?

M: It doesn't move. I think we're almost out of gas.

09

M: Welcome to Norris Auditorium. Before the documentary begins we would like to make a few announcements. Please _____ _____ _____ _____ _____ so that students around you can enjoy the documentary. _____ _____ _____ during the documentary will not be tolerated, and those students will be asked to leave by security personnel. Today we also have some students walking up and down the aisle collecting money for the Jenny Fund. The Jenny Fund will use all of its collected donations to _____ _____ _____ _____ for cancer. Any donation you can make will be greatly appreciated. Thank you for your attention and please enjoy the show.

10

M: Hello. I'd like _____ _____ _____ _____ _____ for tomorrow.

W: Certainly, sir. What time would you like to leave?

M: My schedule is flexible, but I'd like to leave some time before noon.

W: All right. Some of our trains go directly to Toronto, but others _____ _____ _____ _____ _____.

M: Transferring once would be okay, but I don't want to transfer twice.

W: And do you have a seat preference?

M: Yes, I would like a window seat. I enjoy watching the scenery go past.

W: In that case, we have two trains you can take. But the earlier one is a bit more expensive.

M: Oh, really? Why is that?

W: _____ _____ _____ _____ _____, so it's popular with commuters.

M: I see. Then I'll take a ticket for the cheaper train.

W: All right. *[Pause]* Here you are, sir.

11

W: Hey, which of these ties do you think would be a better gift?

M: I like that one on the left more. The one on the right _____ _____ _____ _____.

W: Really? I kind of thought the opposite.

M: (Well, you'd better choose the one with the color you like.)

12

M: I can't wait for our camping trip.

W: Me neither. I've never been camping. Can we go fishing there?

M: Yes, but I need to _____ _____ _____ _____ _____.

W: (I saw some on sale at a department store.)

13

W: Let's go camping in the mountains this weekend.

M: What a great idea! It's going to be sunny and clear over the weekend.

W: Yeah, I heard that this weekend is going to be _____ _____ _____. Let's get ready. Do you know where the tent is?

M: Oh, I think the Smiths borrowed it and haven't _____ _____ yet.

W: I thought we picked it up when we went to their barbecue two weeks ago.

M: Hmm, I don't remember. *[Pause]* Oh, now I remember! _____ _____ _____, but we forgot.

W: Why don't you go look in the garage? I'll get the sleeping bags and the camera. *[Pause]* Did you find it?

M: No, it's not there.

W: (Okay, I'll call the Smiths then.)

14

W: How was the concert?

M: It was great. They're a really great band.

W: I didn't like their last album.

M: Yeah, I like their earlier stuff _____ _____ _____.

W: Did they play much of their old stuff?

M: Yes, they played a little bit of everything. They had a lot of energy too.

W: Who did you go with?

M: Some old friends from school. We _____ _____ _____ that band.

W: Had you ever seen them before?

M: No, that's why it was so exciting.

W: Well, it sounds like you _____ _____ _____ _____ _____.

M: (I did. I would really recommend seeing them if you get the chance.)

15

W: Yesterday, Sam asked his father _____ _____ _____ _____ to a late night movie with his friends. His father didn't think it was a good idea, and told Sam that he should stay home and get ready for his exams. Later, after his father went to bed, Sam quietly opened the door and _____ _____ _____. He and his friends had a good time at the movie theater. They ate some popcorn and played some games after the movie. When Sam arrived home, he _____ _____ _____ _____ his father awake and waiting in the living room. In this situation, what did his father most likely say to him?

Sam's father: Sam, (I told you that you couldn't go out!)

16~17

M: These days, many people travel around the world and come back home with _____ _____. It's difficult to spend every coin before you leave, and exchange counters rarely accept them. Even though coins are heavy and dirty, you probably don't want to _____ _____ _____. Here are some useful tips on _____ _____ _____ _____ your leftover coins. First, try to use up as many coins as you can at the airport. For example, you can buy some small souvenirs for friends in a gift shop. Second, sell them online. You won't get their full value, but it's better than nothing. Third, you can stick small magnets to them and use them to decorate your refrigerator. Or you can donate them to a charity. Many charities will accept foreign coins and send them to the people who need them most. This is an easy way to help others! So from now on, _____ _____ _____ use your leftover foreign coins wisely.

1번부터 17번까지는 듣고 답하는 문제입니다.
1번부터 15번까지는 한 번만 들려주고, 16번부터 17번까지는 두 번 들려줍니다. 방송을 잘 듣고 답을 하기 바랍니다.

01

다음을 듣고, 여자가 하는 말의 목적으로 가장 적절한 것을 고르시오.
① 관객을 모으려고
② 연극부원을 모집하려고
③ 연극이 취소됨을 공지하려고
④ 상영 예정인 공연을 비판하려고
⑤ 행사 진행의 미숙함을 사과하려고

02

대화를 듣고, 두 사람이 하는 말의 주제로 가장 적절한 것을 고르시오.
① 불교의 상징물
② 연꽃 꽃말의 유래
③ 힌두교와 불교의 차이
④ 연꽃이 지닌 다양한 의미
⑤ 고대 문화에서 연꽃의 의미

03

대화를 듣고, 두 사람의 관계를 가장 잘 나타낸 것을 고르시오.
① 의사 - 간호사
② 간호사 - 환자
③ 의사 - 환자
④ 약사 - 손님
⑤ 서점 직원 - 손님

04

대화를 듣고, 그림에서 대화의 내용과 일치하지 않는 것을 고르시오.

05

대화를 듣고, 남자가 여자에게 부탁한 일로 가장 적절한 것을 고르시오.
① to take part in a play
② to charge the battery
③ to hire a photographer
④ to take a picture of him
⑤ to take the camera with her

06

대화를 듣고, 여자가 연극에 참여할 수 없는 이유를 고르시오.
① 축구 연습을 해야 해서
② 연극이 마음에 들지 않아서
③ 사람들 앞에 서는 게 두려워서
④ 남자가 연극에 참여하지 않아서
⑤ 대사를 다 외울 수 없을 것 같아서

07

대화를 듣고, 남자가 지불할 금액을 고르시오.
① $4.00
② $4.25
③ $4.50
④ $4.75
⑤ $5.00

08

대화를 듣고, 에메랄드에 관해 언급되지 않은 것을 고르시오.
① 색상
② 단단한 정도
③ 생산지
④ 가치
⑤ 식별법

09

summer book fair에 관한 다음 내용을 듣고, 일치하지 않는 것을 고르시오. 3점
① 날씨가 안 좋으면 도서관 로비에서 열린다.
② 6시간 동안 개최된다.
③ 마련된 기금은 교과서 구입에 사용된다.
④ 자원봉사할 학생을 모집 중이다.
⑤ 음식과 음료가 무료로 제공된다.

10

다음 표를 보면서 대화를 듣고, 두 사람이 참석할 축제를 고르시오.

	Festival	Type of Music	Period	Place	Price per Day
①	A	Rock	Aug. 5-6	Seoul	$40
②	B	Rock	Aug. 12-13	Seoul	$60
③	C	Jazz	Aug. 26-27	Busan	$35
④	D	Jazz	Sep. 9-10	Seoul	$50
⑤	E	Classical	Oct. 14-15	Busan	$15

11

대화를 듣고, 여자의 마지막 말에 대한 남자의 응답으로 가장 적절한 것을 고르시오.

① Okay, take this one.
② You probably did fine.
③ No, that's not my pencil.
④ I'll study for the test later.
⑤ Thank you for lending it to me.

12

대화를 듣고, 남자의 마지막 말에 대한 여자의 응답으로 가장 적절한 것을 고르시오.

① Oh, let me see your arm.
② Abby would never say that.
③ We should go visit her tomorrow.
④ No, I don't like watching the news.
⑤ I'm glad that nothing bad happened.

13

대화를 듣고, 여자의 마지막 말에 대한 남자의 응답으로 가장 적절한 것을 고르시오.

Man: _____

① Let's dance with Theo.
② Let's take a dance class.
③ I'd like some more shrimp.
④ Let's step outside for some fresh air.
⑤ Show me the way to the dance floor.

14

대화를 듣고, 남자의 마지막 말에 대한 여자의 응답으로 가장 적절한 것을 고르시오. 3점

Woman: _____

① You talked me into it.
② No, I can't go. I need to study.
③ I need to babysit my little sister.
④ Yes, I'll put the book on the table.
⑤ Yes, I just need to grab my wallet and we're off.

15

다음 상황 설명을 듣고, Stephanie가 John에게 할 말로 가장 적절한 것을 고르시오. 3점

Stephanie: _____

① I think the report looks better in color.
② Let's have our meeting in the café instead.
③ I'm sorry, I must have asked for the wrong report.
④ Unfortunately, the president won't be at the meeting.
⑤ Please run to the internet café and print out the report.

[16~17] 다음을 듣고, 물음에 답하시오.

16

여자가 하는 말의 주제로 가장 적절한 것은?

① the origin of flamenco
② the history of minor tribes in Spain
③ the popularity of flamenco in Spain
④ the many representations of flamenco
⑤ the variety of Spanish traditional dances

17

언급된 민족이 아닌 것은?

① the Romani
② Muslims
③ South Americans
④ Jewish people
⑤ the Spanish

녹음을 다시 한 번 듣고, 빈칸에 알맞은 말을 쓰시오.

01

W: Want to see a musical but don't know where or when? The students and staff of Whitman High School have ＿＿＿＿ ＿＿＿＿ a musical, *Cinderella's Glass Slipper*. Students have been working since March — learning the music, memorizing lines, and attending rehearsals — all while ＿＿＿＿ ＿＿＿＿ ＿＿＿＿ their homework and other activities. This play is a remake of the classic Cinderella story with some comical twists. So come see *Cinderella's Glass Slipper* on the ＿＿＿＿ ＿＿＿＿ ＿＿＿＿ of May at 7 p.m. in the auditorium. Admission is $5 at the door. This is family-friendly entertainment, so bring along the children!

02

W: What do you have there, Eric?
M: It's a lotus. I got it at the flower shop on the corner.
W: I thought ＿＿＿＿ ＿＿＿＿ ＿＿＿＿ ＿＿＿＿.
   The lotus is not only beautiful but also meaningful. Did you know that this flower ＿＿＿＿ ＿＿＿＿ ＿＿＿＿ several ancient cultures?
M: Of course! I read that in ancient Egypt it was a symbol of rebirth.
W: It is also significant in Buddhism.
M: Oh, really?
W: Yes. In that religion, it symbolizes purity and perfection.
M: That's similar to its meaning in Hinduism.
W: So, in Hinduism they use it to represent purity?
M: Yes, but that's not all. The Hindus also ＿＿＿＿ ＿＿＿＿ ＿＿＿＿ ＿＿＿＿ ＿＿＿＿ and youth. Who knew that one flower could be important to so many different people?

03

W: What is the problem today?
M: I haven't been feeling myself. And I ＿＿＿＿ ＿＿＿＿ ＿＿＿＿ ＿＿＿＿.
W: I see. Have you been taking anything?
M: Just aspirin. And something else the pharmacist recommended.
W: Is it helping at all?
M: Not really. The sore throat might even be getting worse.
W: You may have an infection. Let's ＿＿＿＿ ＿＿＿＿ ＿＿＿＿.
M: I don't think I have a fever.
W: Well, it's a little high. The doctor will probably need to do more tests.
M: Okay. Is it serious?
W: I can't really say. She will be in to see you shortly.
M: Thank you.
W: Here are some magazines there that you can look at ＿＿＿＿ ＿＿＿＿ ＿＿＿＿.
M: Thanks.

04

W: David, I'm so happy to be at the beach. Isn't it wonderful?
M: Yes, it's terrific. ＿＿＿＿ ＿＿＿＿ ＿＿＿＿ ＿＿＿＿.
W: Let's put our towel near those tall palm trees. They make me feel like I'm on a tropical island.
M: Good idea! There is a hammock hanging from them.
W: Yes. It looks so peaceful. Hey! What are those two people doing in the ocean?
M: They are paddle boarding. It's a fun sport.
W: Oh! Do you think ＿＿＿＿ ＿＿＿＿ ＿＿＿＿ ＿＿＿＿?
M: Sure. We can take lessons at the surf shop near here.
W: Cool! Let's go there together after eating lunch.
M: Okay. Look over there. A boy and girl are making

W: Yeah, next to them, some women are _____ _____ _____. What an exciting game!

M: I'm so glad we came here today.

W: Me too! Anyway, let's take out our picnic lunch!

05

W: Wow! It's your big night! You must be excited!

M: I'm really nervous, actually.

W: Oh, no! Why?

M: I've never played in front of so many people.

W: Don't worry. We'll be _____ _____ _____. It'll be great.

M: Could you do me a favor?

W: Of course.

M: Could you hold my camera? I'm afraid to leave it backstage.

W: Yeah, no problem. I'll _____ _____ _____.

M: No need to do that. Some photography student has offered to do that for free.

W: You could have a few more.

M: I know, but the battery is dead.

W: Oh, okay. Well, don't worry. I'll _____ _____ _____ _____ it.

M: Thanks. See you later.

06

M: Hey, Nara, did you hear about the school play? It's going to be a performance of *The Lost Princess*.

W: Oh, really? That's my favorite play!

M: I like it, too. Do you think you will try out?

W: No, I won't.

M: Why not? I bet _____ _____ _____ in the leading role.

W: The princess has a lot of lines, though.

M: You would have plenty of time to practice and memorize them.

W: That's not the problem. I am afraid of performing in front of all those people.

M: _____ _____ _____! You can do it.

W: I don't think so. How about you? Will you try out?

M: I won't, either. I wouldn't have enough time because of soccer practice.

W: It looks like neither one of us will be in the play. Let's just go see it together when they _____ _____ _____.

07

M: Can you believe it? I only have $5.

W: We'd better eat at the student cafeteria then.

M: Okay. _____ _____ _____ _____ here?

W: Well, the fish is $3. Plus $1 for a salad. You have enough.

M: Yes, fish sounds good. But _____ _____ _____ _____. I'm thirsty, though.

W: Drink water.

M: No, I want tea. It's only 50 cents.

W: Okay.

M: And I'll _____ _____ _____ too!

W: That's not very healthy!

M: I don't care. How much are they?

W: $1.25.

M: Good. I have enough.

08

W: It's my birthday next month, Todd.

M: You were born in May? That means your birthstone is the emerald.

W: I didn't know that. What do emeralds look like?

M: They're similar to diamonds, but they're deep green in color.

W: Oh yes, I've seen those before. Are they _____ _____ _____ _____?

M: No, they're not. They're strong, but they break more easily than diamonds.

W: I see. Where are they found?

M: They're _____ _____ _____ _____ _____,

but mostly South America and Africa.

W: Are they valuable?

M: Yes. An emerald _____ _____ _____ can be worth more than a diamond.

W: Wow, I want one.

M: In the US, emeralds are the traditional gift for the 55th wedding anniversary. So you'll have to wait!

09

M: Goldberg Elementary School will be holding a summer book fair on Sunday, August 3. It will take place on the sidewalk in front of the downtown library. _____ _____ _____ bad weather, however, it will be moved into the library lobby. The fair starts at 10 a.m. and will last until 4 p.m. Books for adults and _____ _____ _____ _____ will be on sale at special discounted prices. The money raised at the fair will be used to purchase new textbooks for the school. All Goldberg students _____ _____ _____ _____. We are also looking for parents who are willing to volunteer their time selling books. There will be free food and drinks and a puppet show performance by a local theater group, too!

10

M: Music festival season is starting soon. I downloaded a schedule. We should go to one this year.

W: Great! _____ _____ _____ _____ do you like?

M: I like rock and jazz. I also like classical music, but a classical festival would be too boring.

W: I agree. Which month is best?

M: August and October are both okay. Unfortunately, we have exams in September.

W: That's right. Let's go to one of the festivals in Busan. It's a beautiful city.

M: I love Busan, but it's too far. Let's pick one in Seoul.

W: All right. How much can you spend on a ticket?

M: _____ _____ _____ _____. I saved up a lot of money from my part-time job.

W: That's great. But I can only afford $50 per day.

M: Then let's go to this one.

W: _____ _____! I'll order a pair of tickets online today.

11

W: Hey Tim, can you _____ _____ _____ _____?

M: Shh! We're not supposed to talk during the test.

W: I know, but my pencil broke. I didn't know what else to do.

M: (Okay, take this one.)

12

M: I'm sorry Chloe, but I have some bad news.

W: Oh no, what happened?

M: It seems that your friend Abby _____ _____ _____ _____ _____.

W: (We should go visit her tomorrow.)

13

W: Are you enjoying the party?

M: Very much. Thank you for inviting me.

W: I couldn't have come without you.

M: I don't know many people here, though.

W: Well, you _____ _____ _____, Vicky and Theo. I work with Theo.

M: I know.

W: And most of these other people work with either Theo or Vicky. I don't know all of them.

M: Where does Vicky work?

W: Some advertising company, I think. How's the food?

M: _____ _____ _____ on all that shrimp. I

couldn't possibly eat any more.

W: Well, it's _____ _____ _____ then.

M: Not yet, I'm too full. A little later.

W: What would you like to do then?

M: (Let's step outside for some fresh air.)

14

M: Hey! Are you still studying? It's Friday night.

W: I _____ _____ _____ _____ _____ for my English class.

M: That can wait. Let's go out.

W: I have to read *Huckleberry Finn* by Monday. See _____ _____ _____ _____.

M: Don't worry. Everybody knows that story.

W: But we have a test. It's important.

M: _____ _____ _____ _____ _____... You know what they say.

W: Maybe you're right. I have all weekend.

M: Listen. Lily and Joe are going to be over at the café. Let's go meet them.

W: _____ _____. But it will have to be a little later. Maybe after 30 minutes?

M: Good. I knew you'd want to.

W: (You talked me into it.)

15

M: Stephanie is _____ _____ _____ a meeting with her company's president. Yesterday, she asked her coworker John to print out a copy of an important sales report. But when he brings it, she realizes that he printed out the wrong version. She asks him to print out the correct version, but he tells her that the printer _____ _____ _____. She knows that there's another printer on the second floor, but it only prints in black and white. She needs the report in color. Then Stephanie remembers that the internet café across the street has a color printer. The meeting starts _____ _____ _____ _____ _____, and she really needs the correct report. In this situation, what would Stephanie most likely say to John?

Stephanie: (Please run to the internet café and print out the report.)

16~17

W: Good evening and thank you for coming to tonight's flamenco performance. Before the show begins, I'd like to give you some background information about this passionate art. Flamenco _____ _____ _____ the mountains of Spain, where the Romani, Muslims, and Jewish people had fled to avoid the Spanish government. Indeed, at the time, the Spanish government rejected these minorities and _____ _____ _____. Thus, the three cultures blended to give birth to flamenco. Once the Spanish government became more accepting, the people of the mountains moved back to the cities, where they shared flamenco. There, the Spanish contributed their own ideas to the art. Today, flamenco is performed in various venues, but it is also often seen on the streets of Spain. These spontaneous events are often the best. Tonight, you will _____ _____ _____ passionate dancers and musicians perform this old — yet still highly popular — art. Now let the show begin!

11 영어듣기 모의고사

1번부터 17번까지는 듣고 답하는 문제입니다.
1번부터 15번까지는 한 번만 들려주고, 16번부터 17번까지는 두 번 들려줍니다. 방송을 잘 듣고 답을 하기 바랍니다.

01
다음을 듣고, 남자가 하는 말의 목적으로 가장 적절한 것을 고르시오.
① 영업시간 종료를 알리려고
② 신설 체육관을 광고하려고
③ 매장의 임시 폐점을 알리려고
④ 주차 요금 인상을 안내하려고
⑤ 오늘의 할인 품목을 소개하려고

02
대화를 듣고, 여자의 의견으로 가장 적절한 것을 고르시오.
① 시험 준비 기간이 너무 짧다.
② 과외는 성적 향상에 도움이 된다.
③ 생물학 수업은 진도가 너무 빠르다.
④ 도서관에서 공부하면 집중이 잘 된다.
⑤ 친구와 함께 공부하는 것이 효과적이다.

03
대화를 듣고, 두 사람의 관계를 가장 잘 나타낸 것을 고르시오.
① 편집자 – 작가   ② 기자 – 작가
③ 학생 – 교사   ④ 도서관 사서 – 학생
⑤ 서점 직원 – 손님

04
대화를 듣고, 그림에서 대화의 내용과 일치하지 <u>않는</u> 것을 고르시오.

05
대화를 듣고, 여자가 할 일로 가장 적절한 것을 고르시오.
① 집 청소하기   ② 집 가격 비교하기
③ 이삿짐 정리하기   ④ 친구에게 전화하기
⑤ 부동산 업체 방문하기

06
대화를 듣고, 남자가 태블릿 PC를 사려는 이유를 고르시오.
① 큰 화면을 원해서   ② 예전 것을 잃어버려서
③ 용량이 큰 것을 원해서   ④ 새로운 디자인을 원해서
⑤ 가벼운 것이 필요해서

07
대화를 듣고, 여자가 지불할 금액을 고르시오.
① $162   ② $216   ③ $270
④ $324   ⑤ $360

08
대화를 듣고, 화성에 관해 언급되지 <u>않은</u> 것을 고르시오. 3점
① 색깔   ② 지구로부터의 거리
③ 위성의 수   ④ 크기
⑤ 하루 길이

09
Slow Food Festival에 관한 다음 내용을 듣고, 일치하지 <u>않는</u> 것을 고르시오. 3점
① 이틀간 개최되었다.
② 농산물들이 저렴한 가격에 판매되었다.
③ 지역 음식점들의 요리법이 공유되었다.
④ 유기농 재배에 관한 강의가 있었다.
⑤ 축제 기간 동안 음악 밴드의 공연이 있었다.

10

다음 표를 보면서 대화를 듣고, 남자가 선택할 프로그램을 고르시오.

State College Exchange Programs 2019

	Program	Language	Length	Price	Housing
①	A	English	One semester	$9,000	On-campus
②	B	English	One year	$12,000	Off-campus
③	C	Chinese	One semester	$7,000	Off-campus
④	D	English	One year	$13,000	On-campus
⑤	E	English	One year	$16,000	On-campus

11

대화를 듣고, 남자의 마지막 말에 대한 여자의 응답으로 가장 적절한 것을 고르시오.

① This store isn't too crowded today.
② Look! There he is over by that display.
③ How cute! I think I've met your son before.
④ Why don't you play hide-and-seek at home?
⑤ I think you should find a new job at another store.

12

대화를 듣고, 여자의 마지막 말에 대한 남자의 응답으로 가장 적절한 것을 고르시오.

① Only if you promise to drive carefully.
② Sorry, but I'm busy on Saturday night.
③ Our car isn't big enough to take camping.
④ Okay. I can take you shopping tonight after dinner.
⑤ I'd love to, but I don't need to buy anything at the shopping center.

13

대화를 듣고, 여자의 마지막 말에 대한 남자의 응답으로 가장 적절한 것을 고르시오.

Man: _____

① Why don't you start exercising right away?
② I'm sure you'll pass the exam next month.
③ It's good to hear that you got your old job back.
④ Why do you want to work for a computer company?
⑤ Wow! Your misfortune turned into a good opportunity.

14

대화를 듣고, 남자의 마지막 말에 대한 여자의 응답으로 가장 적절한 것을 고르시오. 3점

Woman: _____

① It doesn't sound like a very realistic book.
② I think there is another copy at the library.
③ I'm glad he was able to cure all his patients.
④ The stories are probably from his imagination.
⑤ His experience must've helped him write the book.

15

다음 상황 설명을 듣고, Samantha가 Kelly에게 할 말로 가장 적절한 것을 고르시오.

Samantha: _____

① I'm on a diet.
② You're a good cook.
③ I'm sure he'll like it.
④ What are friends for?
⑤ Sandwiches are $1 each.

[16~17] 다음을 듣고, 물음에 답하시오.

16

여자가 하는 말의 주제로 가장 적절한 것은?

① places that animals hibernate
② how animals survive in winter
③ how hibernation affects animals
④ which animals hibernate in winter
⑤ ways to decrease your metabolism

17

언급된 신체적 변화가 <u>아닌</u> 것은?

① body temperature
② breathing
③ heart rate
④ hormone levels
⑤ metabolism

녹음을 다시 한 번 듣고, 빈칸에 알맞은 말을 쓰시오.

01

M: Ladies and gentlemen, thank you for shopping at Rich's Sporting Goods. It is now 8:50, and we will _____ _____ in 10 minutes. If you have any merchandise to purchase, please go to the cashier now. Lines 5 and 6 are _____ _____. If you have not finished your shopping, we will be open tomorrow at 10 a.m. We will be having _____ _____ _____ this weekend. Please pick up a circular _____ _____ _____ _____ tonight, or check your newspapers tomorrow morning. We hope to see you again soon. Thank you for shopping at Rich's.

02

W: _____ _____ _____ _____ the biology test next week?

M: Not really. I've been studying all week at the library, but I still feel like I'm not ready.

W: How about studying together?

M: I think _____ _____ _____ _____ _____.

W: No, it will be good. We can quiz each other on the various topics. And, if one of us has a question, the other might be able to answer it.

M: I guess you're right. Sometimes I need a little help.

W: And _____ _____ _____ _____ _____ with the two of us.

M: All right. Then what time should we meet?

W: How about 10 o'clock tomorrow morning?

03

W: Congratulations on your new novel. It's another best-seller.

M: Thank you. I'm just _____ _____ _____ _____ _____.

W: What do you think is the secret to your success?

M: I don't think there really is a secret. I'm just a normal guy who _____ _____ _____ experiences.

W: So, are all the storylines in your novel from your _____ _____?

M: You could say that. I'm not creating something new. I write what I know.

W: That's amazing. You must've traveled all over the world to write this book.

M: Well, I haven't been everywhere, but I've been to a lot of places. If I'm not writing, I'm _____ _____ _____.

W: That's why your book is so amazing. Anyway, thank you so much for your time.

M: No, thank you for this talk.

04

M: This is my favorite café. Let's study here.

W: Oh, what are those books behind that glass door?

M: This is a book café. You can choose a book to read.

W: Wow, that's nice. Where do you want to sit?

M: How about on that sofa?

W: Well, _____ _____ _____ _____ _____ _____ where people pick up their orders. That might disturb us.

M: Okay. Then what about the long rectangular table in the center? We won't _____ _____ _____.

W: Sounds good. And there is a cute flower-shaped lamp on the table! I think I'll turn it on.

M: Okay. There's no one _____ _____ _____ _____. I'll go and order a drink for us.

W: Yes, it's a good time to order.

05

M: Kim, have you talked to a real estate agent yet?

W: No. I thought we should talk about _____ _____ _____ _____ first.

M: Well, I want to live somewhere convenient, so I'd

like to find a place close to a subway station.

W: The only thing I care about is size. Our kids need their own rooms now.

M: But wouldn't a large apartment cost much more?

W: Maybe. But being close to a subway station is expensive, too.

M: Why don't you _____ _____ first? I heard there are some websites where you can compare different apartment prices. I don't have enough time to search right now.

W: Okay, I will. It's probably a good idea to compare prices online before visiting a real estate agent.

M: I think so, too. I'll call an agent _____ _____ _____.

W: Great.

06

M: Lisa, will you go to the electronics market with me tomorrow?

W: Sure. What do you need to buy?

M: I want to get a new tablet PC.

W: But you just bought one last year. Did you lose it already?

M: No, I still have it. But _____ _____ _____ _____ it.

W: You're not? Why? Is it too slow?

M: No, it's pretty fast. And it has a large screen, which I really like.

W: Then what's the reason? Do the batteries die too quickly?

M: They don't last very long, but I don't mind. _____ _____ _____ its weight.

W: Oh, I see. You carry it around with you all the time, don't you?

M: Yes, and it's quite heavy. So I find it very inconvenient.

W: Then we can find _____ _____ _____ for you tomorrow.

07

[Telephone rings.]

M: Z Tix. How can I help you?

W: Hello. I'd like to _____ _____ _____ for the Yay Sound concert.

M: Okay. We have three kinds of seats; side, balcony, and orchestra. They cost $80, $60, and $100 respectively.

W: Which seats are closest to the stage?

M: The orchestra seats are closest, but _____ _____, _____ _____. The side seats are still pretty close, though, and they're reasonably priced.

W: Okay. I'll take three side seats. But isn't this the slow season?

M: Yes. During peak season, you would pay $120 for the same seat. It's a bargain!

W: Oh, nice! I also read online that I could get an early bird discount. Is it too late for that?

M: It's _____ _____ _____ _____ the discount, but you have to be a member.

W: I am! Here is my membership number: QX665.

M: Let me see... *[Typing sound]* Okay. You'll get a 10% discount.

W: Good. Can I pay by phone?

M: Yes. Just give me your credit card number, please.

08

M: I'm writing a report on the red planet, Mom.

W: The red planet? You must mean Mars.

M: That's right. Did you know that its color _____ _____ _____ _____ in its soil?

W: No, I didn't. What else have you learned?

M: Well, it was named for the Roman god of war. And it's _____ _____ _____ from the sun.

W: I see. Does it have a moon like Earth's?

M: Actually, it has two. But they're _____ _____ our moon.

W: Mars itself is smaller than Earth, isn't it?

M: Yes, it is. It's about half the size. But it's also

similar in some ways.

W: Can you give me an example?

M: Sure. A day on Mars lasts just a half hour longer than a day on Earth.

W: That's interesting. Maybe you can go there someday.

09

W: Last week, the town of Seaview held its first ever Slow Food Festival. The event _____ _____ _____ Elm Park on Saturday and Sunday, and attracted more than 5,000 visitors. The highlight of the festival was the farmers' market, where locally-grown produce was _____ _____ _____ _____. There were also a number of food booths where local restaurants offered their healthiest menu items for tasting. Separate presentations were held on _____ _____ _____ _____ _____, which were educational and entertaining. And throughout both days, some of the best bands in town _____ _____ _____ that kept the crowds dancing until the sun went down.

10

M: Guess what! I was accepted into the exchange program!

W: That's great! Have you decided which program to choose yet?

M: _____ _____. This one looks good, but the classes are in Chinese.

W: Your Chinese isn't good enough. You need a program with classes in English.

M: You're right. And I think I should pick one _____ _____ _____ _____ _____.

W: Yes. One semester isn't enough time to learn about a culture.

M: But I also have to pay attention to my budget.

W: That's true. How much can you spend?

M: My parents said $15,000 is the maximum amount.

W: That's not bad. You should choose a program with

off-campus housing. It'd be more exciting.

M: No, I don't think so. _____ _____ _____ living on campus.

W: Okay. Well, then this one looks like the best fit for you.

11

M: I don't know what to do. I can't find our son anywhere!

W: Did you _____ _____ _____ _____ _____? He could be hiding.

M: Yes, of course. Oh, I think I should go to the center for lost children.

W: (Look! There he is over by that display.)

12

W: Dad, can you _____ _____ _____ _____ on Saturday night?

M: I don't know. Where do you want to go?

W: I want to drive my friends to the shopping center.

M: (Only if you promise to drive carefully.)

13

M: Tina! Long time no see!

W: It's nice to see you, Jaemin. _____ _____ _____ _____?

M: Fine. I've just been working and exercising every day, as usual.

W: I see. Are you still working at the same computer company?

M: Yes, I am. What about you? How's your work going?

W: Um... Actually, I _____ _____ some hard times. I lost my job.

M: Oh, really? What happened?

W: The financial situation of my company wasn't good, so they fired me along with some other employees.

M: That's terrible.

W: Yes, it was quite a shock. But luckily, it _____ _____ _____ for me.

M: Oh, great. But how? Did you find a new job?

W: Yeah. I took the teacher recruitment examination and passed it. I'll be working in a school next month.

M: (Wow! Your misfortune turned into a good opportunity.)

14

W: Hi, Mark. What are you reading?

M: Hi, Susan. It's a really interesting novel that I _____ _____ _____ _____ _____. Actually, I just finished it.

W: What is it about?

M: It's about a doctor whose patients have mysterious diseases.

W: Mysterious diseases? Is the doctor able to cure them _____ _____ _____?

M: Yes. He has some difficulties at first, but eventually he cures all of them.

W: It sounds really exciting. The writer must have a great imagination.

M: He is definitely a good writer, but I don't think the book is completely fictional.

W: Really? _____ _____ _____?

M: The writer was a doctor for 30 years.

W: (His experience must've helped him write the book.)

15

M: Kelly is Samantha's best friend. One day, Samantha noticed that Kelly always brings dry bread for lunch. Kelly's family has been having _____ _____ since her father _____ _____ _____. Knowing Kelly's family situation, Samantha asked her mother to _____ _____ _____. Samantha would bring the two lunches and give one to Kelly. Kelly was _____ _____

for Samantha's consideration. In this situation, what would Samantha say when Kelly thanked her?

Samantha: (What are friends for?)

16~17

W: Hello, class. Last time we talked about some animals that _____ _____ _____. But what is hibernation? Today, we are going to talk about what happens to animals when they hibernate. One thing that hibernating animals do is _____ _____ _____ _____. Sometimes, their body temperature can _____ _____ _____ 18 degrees Celsius. Hibernating animals also slow down their breathing. They might take just one breath per minute. The same is true for their heart rate. For some hibernating animals, it slows down as much as 75%. In addition, hibernating animals decrease their metabolism. This means that they do not use as much energy. As a result, they can survive for a very long time _____ _____ _____ _____ or water. Isn't hibernation amazing? Are there any other things you want to know about hibernation? Now, we're going to talk about where some animals hibernate.

1번부터 17번까지는 듣고 답하는 문제입니다.
1번부터 15번까지는 한 번만 들려주고, 16번부터 17번까지는 두 번 들려줍니다. 방송을 잘 듣고 답을 하기 바랍니다.

01
다음을 듣고, 여자가 하는 말의 주제로 가장 적절한 것을 고르시오.
① 물의 중요성
② 물 부족 문제
③ 지구 환경 보호
④ 신진대사의 원리
⑤ 수질 오염의 심각성

02
대화를 듣고, 남자의 의견으로 가장 적절한 것을 고르시오.
① 역사 교육을 강화해야 한다.
② 잘못된 내용의 역사책이 많다.
③ 소설을 통해 상상력을 길러야 한다.
④ 다양한 분야의 책을 읽는 것이 좋다.
⑤ 판타지 소설은 문학적 가치가 떨어진다.

03
대화를 듣고, 두 사람의 관계를 가장 잘 나타낸 것을 고르시오.
① 비서 - 임원
② 변호사 - 의뢰인
③ 비서 - 외부인
④ 직장 동료 - 직장 동료
⑤ 영업사원 - 고객

04
대화를 듣고, 그림에서 대화의 내용과 일치하지 않는 것을 고르시오.

05
대화를 듣고, 여자가 남자에게 부탁한 일로 가장 적절한 것을 고르시오.
① 일찍 귀가하기
② 집안일 도와주기
③ 자명종 사다 주기
④ 새벽에 일찍 깨워주기
⑤ 아침에 태워다 주기

06
대화를 듣고, 여자가 초상화를 가져갈 수 없는 이유를 고르시오.
① 초상화 가격이 비싸서
② 초상화가 완성되지 않아서
③ 물감이 채 마르지 않아서
④ 초상화가 집에 어울리지 않아서
⑤ 초상화가 마음에 들지 않아서

07
대화를 듣고, 여자가 지불한 금액을 고르시오.
① $100
② $150
③ $200
④ $250
⑤ $300

08
대화를 듣고, 스터디 모임에 관해 언급되지 않은 것을 고르시오.
① 모이는 요일
② 개설 시기
③ 모임 장소
④ 학습 과목
⑤ 인원수

09
Northern Mornings에 관한 다음 내용을 듣고, 일치하지 않는 것을 고르시오. 3점
① 주말에는 방송되지 않는다.
② 국제 정치 뉴스를 전문적으로 보도한다.
③ 진행자는 이전에 비슷한 프로그램을 진행한 적이 있다.
④ 세 명의 기자가 프로그램에 참여할 것이다.
⑤ 지난 방송들은 인터넷으로 다시 들을 수 있다.

10

다음 표를 보면서 대화를 듣고, 두 사람이 시청할 TV 프로그램을 고르시오.

	Program	Genre	Time	Language
①	A	Documentary	5:30 - 7:00 p.m.	English
②	B	Talk show	7:05 - 9:30 p.m.	Spanish
③	C	Talk show	9:35 - 10:40 p.m.	English
④	D	Reality show	10:45 - 11:55 p.m.	Spanish
⑤	E	Reality show	12:00 - 1:30 a.m.	English

11

대화를 듣고, 여자의 마지막 말에 대한 남자의 응답으로 가장 적절한 것을 고르시오.

① Yes, just this black bag please.
② No, I already have many bags.
③ No, I have never seen that bag.
④ Yes, I'd like that large brown bag.
⑤ I'm not sure. Let me check my passport.

12

대화를 듣고, 남자의 마지막 말에 대한 여자의 응답으로 가장 적절한 것을 고르시오.

① I heard that's not very healthy.
② There is a café across the street.
③ Yes, Berlin has many restaurants.
④ Yes, there is a dining car that way.
⑤ Yes, we can transfer to an express.

13

대화를 듣고, 남자의 마지막 말에 대한 여자의 응답으로 가장 적절한 것을 고르시오. 3점

Woman: _____

① I guess it really is worth making a resolution.
② Maybe you can give me some weight loss tips.
③ I'm sorry, but I did not meet my goals last year.
④ All right, but please don't disappoint me again.
⑤ You should have set a more difficult goal last year.

14

대화를 듣고, 여자의 마지막 말에 대한 남자의 응답으로 가장 적절한 것을 고르시오.

Man: _____

① Sure, I'd be happy to help.
② What time does the choir start?
③ I'm afraid I have plans this spring break.
④ Sorry, but I don't know how to play the piano.
⑤ Do you want me to find someone who can play the piano?

15

다음 상황 설명을 듣고, Lori가 반 친구들에게 할 말로 가장 적절한 것을 고르시오. 3점

Lori: _____

① I'm sure we'll all learn from our failures.
② I know everyone faced the same challenges.
③ I'm grateful to the teachers for the work they do.
④ I want to thank you for helping me along the way.
⑤ I'm so proud that you're graduating with good grades.

[16~17] 다음을 듣고, 물음에 답하시오.

16

여자가 하는 말의 목적으로 가장 적절한 것은?

① 지역 축제를 홍보하려고
② 행사 일정을 안내하려고
③ 행사가 취소됨을 공지하려고
④ 재즈 밴드의 수상을 축하하려고
⑤ 콘서트 표 예매 방법을 설명하려고

17

언급된 행사가 아닌 것은?

① 발레 공연
② 노래 경연 대회
③ 재즈 공연
④ 플루트 만들기
⑤ 자신만의 CD 녹음하기

녹음을 다시 한 번 듣고, 빈칸에 알맞은 말을 쓰시오.

01

W: It is well known that 98% of our body is _____ _____ _____. When our body doesn't have a sufficient amount of water, we can experience serious health problems. It can even _____ _____ _____. By consuming enough water, we can _____ _____ _____ _____ normally. This is more important than any vitamin or nutrient. Not only our body but also our world is mainly made of water. It is not wrong to say that we, human beings, can simply not live without _____ _____ _____.

02

W: I finished my book.

M: The one about King Sejong? What did you think?

W: One of _____ _____ _____ _____ _____ in a while. Want to borrow it?

M: No, thanks. I prefer to read fiction.

W: Oh, but there's so much you can learn from reading historical nonfiction.

M: Yeah, but I think the purpose of reading is _____ _____ _____ _____. That's why I read novels.

W: But you can imagine what life was like during King Sejong's time.

M: It's still based on real life. _____ _____ _____, you read about things no human has ever seen.

W: Like aliens and dragons?

M: Yeah. That challenges the imagination — and everyone needs more of that.

03

[Phone rings.]

W: Mr. Thompson's office. How can I help you?

M: This is Frederick Jones. Is Mr. Thompson in?

W: He's on the phone right now. Would you like to _____ _____ _____ _____ _____?

M: How long will he be on the phone?

W: I don't know how long the call will take. He's been on this call for about 10 minutes already.

M: Then I'd like to leave a message.

W: Yes. Go ahead.

M: You have my name. I'm an old friend of his who _____ _____ _____ _____ before and I'll be back in town next week. I'm hoping to see him when I return.

W: Does he _____ _____ _____?

M: I'm pretty sure he does, but I'll give it to you again.

04

M: My family and I went to a really comfortable campsite last week. You can see it in this picture.

W: Oh, _____ _____ _____ big rocks.

M: Yes, that helped to keep the edges clear.

W: Oh, that big tree _____ _____ _____ good for shade.

M: Right. That's why we put our tent under it.

W: I can also see two comfortable chairs with armrests. I guess they were good for reading books.

M: Yes, they were.

W: What are _____ _____ _____ _____ for next to the chairs?

M: We built a fire every night. It was really exciting.

W: You must have had a lot of fun there.

M: Yes. We even had a hammock between two trees. We took naps there, like my sister is doing in the picture.

05

W: Sweetheart, when are you getting up tomorrow morning?

M: I don't know. I don't even know what time I'm going to bed. Why?

W: I need to get up really early tomorrow. Do you think you can _____ _____ _____?

M: What time do you need to wake up?

W: I need to get ready and leave home by 5 a.m.

M: That's really early. Why don't you _____ _____ _____ _____?

W: I did, but I want to make sure that I'm able to get up.

M: I'm going to be up late working, so I can't promise you.

W: Then, why don't you wake me up when you go to bed?

M: Are you sure you want that?

W: Sure. It's fine as long as I get up in the morning.

M: Okay. Go to bed now, so you can _____ _____ _____.

06

M: All right. _____ _____ _____. I'm doing the final part of your portrait now.

W: I'm so excited. I can't wait to see it.

M: All right... I'm finished! Here it is. What do you think?

W: Oh my gosh! Is that really me?

M: Yes. _____ _____ _____? You don't like it?

W: No, I love it. But do I really look that beautiful?

M: Yes, you do. I think I captured your kind heart and cheerful personality.

W: I can't wait to take it home tonight!

M: Actually, you can't. _____ _____ at least 24 hours for the paint to dry.

W: Oh. That's too bad, but I understand. I'll come back in a couple of days.

07

M: Susan, did you get a gift for Mom?

W: Yeah, I just went to _____ _____ _____ from the jewelry store. Here it is.

M: A pearl necklace! How much was it?

W: It wasn't that expensive. You owe me just $150.

M: $150? That's too much, Susan! What were you thinking?

W: Well, you said your budget was $200. Since we're paying 50-50, it's not that expensive.

M: I meant $200 total _____ _____ _____ _____. Anyway, that's all right. It is a nice necklace.

W: Yes. _____ _____! Mom will love it.

M: Yeah, I think so too. I'll give you my share tomorrow.

W: All right.

08

W: Where are you going, Paul?

M: I'm going to meet my study group. We _____ _____ twice a week, every Tuesday and Thursday.

W: Oh! I didn't know you were part of a study group.

M: Yes. My friend Kim started it _____ _____ _____ _____ _____.

W: Do you study anything in particular?

M: Yes, math and science. Those are two of my most difficult classes.

W: I see. To be honest, I _____ _____ _____ in the library.

M: Really? Why do you say that?

W: When too many people study together, I get distracted easily.

M: Personally, I find it helpful. There are six of us, and we answer each other's questions.

W: Well, everyone has their own studying style.

M: Exactly. Anyway, I have to run. I'll talk to you later.

09

M: Welcome to our newest radio program, *Northern Mornings*. We'll _____ _____ _____ _____ every weekday morning from 6 a.m. to 8 a.m., bringing you the news, playing the latest hits, and, hopefully, making you laugh. My name is Charles Lee, and I'll be your host. You may _____ _____ _____ from my previous show on global politics, *Weekends with Charles*. Together with a team of three of Canada's top radio reporters, I'll be _____ _____ _____ from the Toronto area, and also giving you timely updates on international news. And if you miss the show, you can always log in to our website and _____ _____ _____ _____.

10

[Phone rings.]

W: Hello.

M: Hi, Rachel. I'm _____ _____ _____ _____. What are you doing?

W: You're coming home early! Oh, it's 7 p.m. already. I've been watching a documentary.

M: Really? What happened to you? You don't like documentaries!

W: You're right, but I had to watch it for my homework. Now I should start working on my essay based on it.

M: Oh... When do you think you can finish it?

W: Hmm... _____ _____ _____?

M: How about watching TV then?

W: Hmm, I don't like the show _____ _____ _____ _____ _____. That type of show is so boring.

M: Then, what kind of show do you want to watch?

W: I wouldn't mind watching a reality show to _____ _____ _____ _____ my essay and the documentary.

M: Great! Well, I'm fine with anything. Just make sure it's not the one in Spanish. I don't feel like reading subtitles.

11

W: Good morning, sir. Could I have your ticket and passport?

M: Yes, here you are.

W: Thank you. Your flight will be leaving from gate 7. Will you be _____ _____ _____?

M: (Yes, just this black bag please.)

12

M: Excuse me, is this the train to Berlin?

W: Yes, it is. _____ _____ _____ in just a few minutes.

M: Thank you. By the way, is there food on the train?

W: (Yes, there is a dining car that way.)

13

M: It's December 29. I can't believe a new year is almost here.

W: Neither can I. I always feel excited before the start of a new year.

M: So do I. What are your _____ _____ _____?

W: Oh, I don't make resolutions.

M: Why not?

W: I think people set unrealistic goals. They can't keep them for very long, and then they just _____ _____ _____ _____.

M: I see what you mean, but my resolution worked out well last year. I decided to lose 5 kg by August.

W: Really? Were you successful?

M: I didn't meet my goal. I only lost 3 kg, but _____ _____ _____ _____ I accomplished something.

W: Oh, I guess resolutions can be beneficial after all, even if you don't meet your goal.

M: That's what I think.

W: (I guess it really is worth making a resolution.)

14

M: What are you doing this Christmas?

W: Actually I'm planning on having a small concert with some friends.

M: I didn't know you _____ _____ _____.

W: I play the piano rather well. Robby will also come and play the classic guitar with her husband, Steve.

M: It seems there are many people involved; it sounds great.

W: It should be fun. I'm getting my old school band back together to play some Christmas carols.

M: Everyone _____ _____ _____ _____ _____. Don't you have any singers?

W: We can't think of anyone who can sing well. Do you know someone?

M: I sing in _____ _____ _____ every week.

W: Wow! That sounds great! Would you like to sing with us at the concert?

M: (Sure, I'd be happy to help.)

15

M: Lori Cummings is graduating at the top of her class. She had a difficult school life because she's _____ _____ _____ _____. There were many moments when she just wanted to give up. But whenever she faced a dilemma, her friends and classmates were there for her. They helped her believe in herself. Lori feels that their support is _____ _____ _____ _____ _____. She is happy to be finishing school, but she is also grateful to her classmates. Today is her graduation day, and she wants to _____ _____ _____ _____ her graduating class. In this situation, what would Lori say to her class?

Lori: (I want to thank you for helping me along the way.)

16~17

W: Ladies and gentlemen, welcome to the first ever International Music Festival. Thank you to all of the volunteers who worked so hard to make this event possible. Our opening ceremonies will begin at 10 a.m. with a dance by the Whirlwind Ballet Club. That _____ _____ _____ _____ a singing contest on the main stage at 11. Everyone _____ _____ _____ _____ _____ the contest, and the winner will receive a cash prize. The Neilson Jazz Trio will perform at 2 p.m. They're one of the best bands in the country, so you _____ _____ _____ _____ that. Please note that following the performance, you'll be able to purchase the band's CDs at the booth next to the stage. At 4 p.m., visit the picnic area to learn how to make your own wooden flute. All of the materials are provided _____ _____ _____ _____. And at 6 p.m., you can hear a variety of rock bands on the main stage. We hope everyone has a great time today.

1번부터 17번까지는 듣고 답하는 문제입니다.
1번부터 15번까지는 한 번만 들려주고, 16번부터 17번까지는 두 번 들려줍니다. 방송을 잘 듣고 답을 하기 바랍니다.

01

다음을 듣고, 남자가 하는 말의 목적으로 가장 적절한 것을 고르시오. [3점]

① 졸업식 순서를 알려주려고
② 졸업 앨범 파티에 초대하려고
③ 졸업 앨범 제작 참여를 권유하려고
④ 졸업 앨범 대금 납부를 촉구하려고
⑤ 졸업 앨범 운영위원회를 소개하려고

02

대화를 듣고, 여자의 의견으로 가장 적절한 것을 고르시오.
① 디자인 업계는 현재 심각한 불황이다.
② 회사를 자주 옮기는 것은 바람직하지 않다.
③ 숙련된 디자이너는 그에 맞는 처우가 필요하다.
④ 급여보다는 적성에 맞는 직장을 선택해야 한다.
⑤ 무급 인턴십 제도는 경력상 도움이 되지 않는다.

03

대화를 듣고, 두 사람의 관계를 가장 잘 나타낸 것을 고르시오.
① 모델 - 매니저
② 편집장 - 사진작가
③ 사진작가 - 모델
④ 직장 동료 - 직장 동료
⑤ 식당 종업원 - 손님

04

대화를 듣고, 그림에서 대화의 내용과 일치하지 않는 것을 고르시오.

05

대화를 듣고, 남자가 할 일로 가장 적절한 것을 고르시오.
① 집안 청소하기
② 초대장 나눠주기
③ 간식 사러 가기
④ 사람들에게 전화하기
⑤ 장 볼 목록 작성하기

06

대화를 듣고, 여자가 기분이 언짢은 이유를 고르시오.
① 귀금속 가게가 모두 문을 닫아서
② 주문한 물건이 제때 오지 않아서
③ 시계에 잘못된 메시지가 쓰여 있어서
④ 주문한 것과 다른 물건을 받게 되어서
⑤ 사촌과 연락이 되지 않아서

07

대화를 듣고, 남자가 지불할 금액을 고르시오.
① $30   ② $32   ③ $34   ④ $36   ⑤ $38

08

대화를 듣고, 캠프 활동으로 언급되지 않은 것을 고르시오.
① 낚시   ② 카약   ③ 수영
④ 승마   ⑤ 하이킹

09

Kew Gardens에 관한 다음 내용을 듣고, 일치하지 않는 것을 고르시오. [3점]
① 유네스코 세계 문화유산이다.
② 세계에서 가장 많은 현존하는 식물을 보유하고 있다.
③ 1700년대에 영국 국립식물원이 되었다.
④ 희귀 식물의 종자를 보존하고 있다.
⑤ 연간 2백만 명의 관광객이 방문한다.

10

다음 표를 보면서 대화를 듣고, 여자가 참가할 워크숍을 고르시오.

Workshops for Teaching Yoga

	Session	Period	Place	Tuition Fee
①	Teaching Experts	Apr. 1-3	St. Petersburg	$300
②	Teaching Experts	Apr. 9-11	Tampa	$400
③	Teaching Standing Poses	Apr. 15-17	Lakeland	$300
④	Teaching All Levels	Apr. 22-24	Lakeland	$350
⑤	Teaching All Levels	Apr. 28-30	St. Petersburg	$350

11

대화를 듣고, 남자의 마지막 말에 대한 여자의 응답으로 가장 적절한 것을 고르시오.

① That sounds like a lot of fun.
② I've never met your girlfriend.
③ The beach is too crowded today.
④ It must have still been cold in the sea.
⑤ I should have gone to that beach with you.

12

대화를 듣고, 여자의 마지막 말에 대한 남자의 응답으로 가장 적절한 것을 고르시오.

① Yes, that will help me.
② This medicine should help you.
③ No, I cannot do anything for you.
④ Can you bring me some hot soup?
⑤ No, I haven't caught a cold this year.

13

대화를 듣고, 남자의 마지막 말에 대한 여자의 응답으로 가장 적절한 것을 고르시오.

Woman: _____

① Can you give me a ride home?
② Where did you park your car?
③ It was a really nice ride, wasn't it?
④ You'll probably have to postpone that.
⑤ I should have checked the weather forecast.

14

대화를 듣고, 여자의 마지막 말에 대한 남자의 응답으로 가장 적절한 것을 고르시오. 3점

Man: _____

① No. I'd rather watch you play soccer than baseball.
② I'm sorry. Next time we can watch the drama you like.
③ Of course not. Playing sports is more fun than watching TV.
④ No problem. I'll buy tickets for tomorrow's game this afternoon.
⑤ Sure. Once you understand it, the game will be more interesting.

15

다음 상황 설명을 듣고, 코치가 운동선수에게 할 말로 가장 적절한 것을 고르시오.

Coach: _____

① Better luck next time.
② Easier said than done.
③ All your hard work has paid off.
④ You'll have your chance one day.
⑤ Cheer up! I don't want you to get discouraged.

[16~17] 다음을 듣고, 물음에 답하시오.

16

남자가 하는 말의 목적으로 가장 적절한 것은?

① 여름철 질병에 관해 경고하려고
② 방학 계획의 중요성을 강조하려고
③ 전염병의 증상에 관해 설명하려고
④ 여름철 질병을 피하는 방법을 알려주려고
⑤ 다양한 질병을 치료하는 방법을 가르치려고

17

언급된 증상이 <u>아닌</u> 것은?

① food poisoning ② heat stroke ③ chicken pox
④ heart attacks ⑤ skin rashes

녹음을 다시 한 번 듣고, 빈칸에 알맞은 말을 쓰시오.

01

M: I'd like to welcome all the parents this afternoon. I appreciate your taking time to be here. I'm Mr. Rainis from _____ _____ _____. With the upcoming release of this year's yearbook, I'd like to remind all parents that this is your last chance to _____ _____ _____ _____ of congratulations to your children. It's a great way to remind your loved ones how special they are. All messages will be printed in the yearbook, and some of them will be read at the yearbook party. Because of _____ _____ _____, it is necessary for all messages to be submitted no later than February 15. Thank you for your attention, and we look forward to hearing from you.

02

M: I just got off the phone with Lee from DNP Design.

W: And?

M: Good news and bad news. They _____ _____ _____ _____ their design team...

W: Good for you! But?

M: But it would be a volunteer position.

W: What? They can't think you'd work for them for free.

M: I don't know what to do. Maybe if I volunteered for a few months they'd hire me on and _____ _____ _____ _____.

W: Sam, you're an experienced designer. You're _____ _____ _____ _____ as an unpaid intern.

M: I know, but it's a good opportunity — DNP Design is a well-respected firm.

W: Which is why they don't need to rely on interns. They can afford to pay their employees.

03

M: It's almost lunchtime. Are you ready to go?

W: Not yet. I need to look at these photographs first.

M: Are these the ones you're going to use for the magazine cover?

W: Yes. I need to choose two of them that _____ _____ _____.

M: Hmm... In my opinion, these two are a perfect match. They even match with this month's theme.

W: I agree. I'll use those photos.

M: Great. Now that _____ _____ _____ the cover, would you like to get something to eat?

W: Sorry, but I have to finish something else now. I need to design the layout and get it to my editor in an hour.

M: In that case, I'll pick up a sandwich and bring it to you.

W: That would be great. Thanks a lot!

M: No problem. I'm happy to _____ _____ _____.

04

M: We need a picture of someone at this festival for our magazine cover.

W: I like the boy playing drums _____ _____ _____ _____.

M: No, we already used a photo of that band last year.

W: Then what about the student carrying a big sign with his school's name on it?

M: I think the boy wearing a superhero costume looks more interesting.

W: Oh, the guy _____ _____ _____ _____ like he's flying into the sky?

M: Yes. Oh, I also like those two dressed up like a dog and a cat and holding hands.

W: They look adorable. But, I decided to use a picture of that boy.

M: Who?

W: The one _____ _____ _____ _____ _____. I think that's the best shot.

05

W: _____ _____ _____. I don't know if I'll be ready for the party on Saturday.

M: What can I do to help?

W: Well, I was just about to move this sofa...

M: No problem. Let me help.

W: But I'm still not sure where I want everything to go.

M: How about invitations? Should I _____ _____ _____?

W: Not right now. *[Pause]* I know. We need some snacks.

M: Do you mean potato chips, pretzels, and drinks?

W: Yes. I can't get to the store anytime soon.

M: Well, I'll go right away. Just _____ _____ _____ _____.

W: That would be a great help.

M: It's no trouble at all. It's going to be a great party.

06

M: Hi Kate. How are you?

W: Hello Billy. I'm fairly annoyed, to _____ _____ _____ _____.

M: What's wrong?

W: Well, I ordered a new watch for my mom, but the jewelry shop _____ _____ _____ _____.

M: What happened?

W: I'm not sure exactly, but they said it wouldn't be ready today.

M: That's terrible. Have you decided what you're going to do?

W: Not yet. I guess I could just buy another watch, but I wanted one _____ _____ _____ _____ _____ on it.

M: My cousin works at a jewelry shop. He might be able to help you. I'll call him now if you like.

W: Oh, that would be great. Thanks a lot.

07

[Telephone rings.]

W: Special Cake Service. How can I help you?

M: Hi. I'd like to _____ _____ _____ for my wife's birthday.

W: Sure. What kind of cake do you want?

M: I saw a rose cake on your website. What is it?

W: We make cream with rose blossoms. The cake is also _____ _____ _____. It looks and smells like roses.

M: Interesting. How much is it?

W: It's $30. But you have to pay a little more if you want to add decorations.

M: Okay. I want you to write "Happy Birthday, Elizabeth!" on the cake and add a queen sculpture in the middle.

W: So that will be $4 for the lettering and $6 for the sculpture.

M: Hmm... Is there any way I can get a discount?

W: If you _____ _____ _____, you'll get a 20% discount.

M: Okay! I want to join.

W: Great. Please tell me your name and address.

08

[Phone rings.]

M: Hello. How can I help you?

W: Yes, I'm looking for a summer camp for my son and would like some information about yours.

M: Of course. Our camp is _____ _____. There are a lot of activities.

W: Such as?

M: We have a lake, so there are _____ _____ _____ _____, such as fishing.

W: How about something more active?

M: There's kayaking, too. And we also have other kinds of boats, like canoes.

W: What about horseback riding?

M: We take the children once or twice to nearby

stables. _____ _____.

W: I see.

M: Also, the hiking is beautiful here. It's a lovely area.

W: It sounds nice. I'll discuss it with my husband.

M: We'd love to have your son. Thank you for calling.

09

W: Kew Gardens is located in London, and is a UNESCO World Heritage Site. It's an important center for research and education about plants. Kew Gardens _____ _____ _____ _____ _____ of living plants. The collection was first begun in the 1700s, and Kew became a British national botanical garden in 1840. Today it's an important location for _____ _____ _____ _____ _____ _____, and researching plant science. The site also has a library that contains one of the world's largest collections of books on plants. Kew Gardens is famous for its huge Victorian greenhouses, _____ _____ _____ the first structures of their kind in Europe. With all these famous attractions, Kew Gardens receives two million visitors annually.

10

M: Yoga Center. How may I help you?

W: Yes, I'm interested in _____ _____ one of your workshops.

M: Great. When exactly would you like to come?

W: After the first week of April would be best.

M: We have _____ _____ _____ excellent workshops. Are you doing it for teacher training?

W: Yes, definitely. I'm going to start teaching soon.

M: Well, how about "Teaching Experts" on April 9 and 10?

W: Experts? No, I'm not ready for that.

M: Okay. Well, the "Teaching All Levels" workshop is in St. Petersburg at the end of the month. That will be _____ _____ _____.

W: That does sound good. But St. Petersburg is far

from me.

M: Well, we have the same workshop earlier in the month at a different location. And we also have a workshop there on teaching standing poses.

W: Hmm... Then I'll start with the cheaper one. If that workshop is good, I'll take the "Teaching All Levels" one, too!

11

M: Would you like to go with me _____ _____ _____ _____ tomorrow?

W: That depends. Where does your friend live?

M: He lives in Busan. We could also visit the beach.

W: (That sounds like a lot of fun.)

12

W: Hello Jake. Are you okay? You don't look very well.

M: Actually, I've _____ _____ _____ all week.

W: I'm sorry to hear that. Is there anything I can do?

M: (Can you bring me some hot soup?)

13

M: I'm glad work is over.

W: Yes, it's finally Friday.

M: Oh, no, look at the weather.

W: _____ _____ _____.

M: I'm going to get soaked.

W: Didn't you drive today?

M: No, I walked. It was nice this morning.

W: Didn't you _____ _____ _____ _____?

M: No, I don't think so.

W: It said it's going to rain all weekend.

M: Oh, no. I had big plans.

W: Really? For what?

M: I was going to _____ _____ _____ _____ and try out my new bicycle.

W: (You'll probably have to postpone that.)

14

M: Hey, Jasmine.

W: Hi, Troy. What are you watching?

M: It's a baseball game. I suppose you're not _____ _____ _____ _____.

W: Why do you say that? I love sports.

M: Oh, I'm sorry. You just didn't seem very excited when I mentioned baseball.

W: Well, to be honest, I really don't know all of the rules of baseball.

M: Yeah, it's _____ _____. Well, there's also a soccer game on Channel 13.

W: Really? Now, soccer is a sport I know well. I used to play on my high school team.

M: I didn't know that you were an athlete. Shall I change the channel?

W: No, you don't need to. You can _____ _____ _____ to me, can't you?

M: (Sure. Once you understand it, the game will be more interesting.)

15

W: A young athlete — a weightlifter — has trained very hard, but he's still not doing well in competition. He has tried to _____ _____ _____ _____ and spend more time training than any of his rivals. Despite the intense training, he still only _____ _____ _____ _____ fifth or sixth place. Frustrated, but not ready to quit, he vows that he will win one day. And, finally, he does. He comes _____ _____ _____! In this situation, what is his coach most likely to say to him?

Coach: (All your hard work has paid off.)

16~17

M: Hello, class. Summer vacation is a great time to have fun and relax, but you also have to be careful. There are many serious health problems that are common in summer. One of these problems is food poisoning. It makes you feel awful! People get it when they consume food or water _____ _____ _____ _____ with harmful bacteria. Another summer health problem is heat stroke. It is caused by spending too much time in direct sunlight. It makes you _____ _____ _____ _____.
Chicken pox is also common in summer. People with chicken pox have itchy red bumps on their skin. They catch this disease when they are around infected people _____ _____ _____ _____.
Many people also get skin rashes in summer. These show up when people sweat a lot. The sweat irritates people's skin and _____ _____ _____ _____ _____. How can we avoid these health problems this summer? I'm interested to hear your ideas.

1번부터 17번까지는 듣고 답하는 문제입니다.
1번부터 15번까지는 한 번만 들려주고, 16번부터 17번까지는 두 번 들려줍니다. 방송을 잘 듣고 답을 하기 바랍니다.

01

다음을 듣고, 여자가 하는 말의 목적으로 가장 적절한 것을 고르시오.
① 본인의 사임을 알리려고
② 올해의 신제품을 광고하려고
③ 직원들의 노고를 칭찬하려고
④ 새 임원의 업적을 소개하려고
⑤ 신제품 개발의 필요성을 강조하려고

02

대화를 듣고, 두 사람이 하는 말의 주제로 가장 적절한 것을 고르시오.
① 디지털카메라 촬영 기법
② 최신형 전자제품의 장단점
③ 최신형 전자제품의 출시 시기
④ 전자제품별 구매하기 좋은 시기
⑤ 전년 대비 전자제품 가격의 인상 폭

03

대화를 듣고, 두 사람의 관계를 가장 잘 나타낸 것을 고르시오.
① 기자 - 연예인　　　② 라디오 진행자 - 가수
③ 사회자 - 배우　　　④ 심사위원 - 오디션 참가자
⑤ 사회자 - 작곡가

04

대화를 듣고, 그림에서 대화의 내용과 일치하지 않는 것을 고르시오.

05

대화를 듣고, 여자가 남자에게 부탁한 일로 가장 적절한 것을 고르시오.
① to browse travel sites
② to pick up her friend
③ to lend her his car
④ to take her to the airport
⑤ to help her out with homework

06

대화를 듣고, 남자가 여자를 찾아간 이유를 고르시오.
① 돈을 빌리려고
② 자동차 수리비를 지불하려고
③ 아들의 성적 문제를 상의하려고
④ 여자의 아들을 야구 경기에 데려가려고
⑤ 여자의 아들이 저지른 잘못을 이야기하려고

07

대화를 듣고, 남자가 지불할 금액을 고르시오. **3점**
① $34　　② $36　　③ $40　　④ $44　　⑤ $60

08

대화를 듣고, Blueport에 관해 언급되지 않은 것을 고르시오.
① 주민 수　　　　　② 최고 관광지로 선정된 이유
③ 일일 방문객 수　　④ 지역 주요 산업
⑤ 주요 교통수단

09

Mountain Scouts에 관한 다음 내용을 듣고, 일치하지 않는 것을 고르시오.
① 자원봉사자들로 구성되어 있다.
② 3일간 야영을 할 예정이다.
③ 강변 청소를 할 것이다.
④ 물고기 먹이 주기 활동이 포함되어 있다.
⑤ 목요일 오후까지 신청할 수 있다.

10

다음 표를 보면서 대화를 듣고, 여자가 빌릴 차를 고르시오.

	Model	Capacity	Luggage Space	Pick-up Location	Air Conditioning
①	A	3	1 bag	Airport	Yes
②	B	4	2 bags	Airport	No
③	C	5	2 bags	Airport	Yes
④	D	5	1 bag	Downtown	No
⑤	E	6	3 bags	Downtown	Yes

11

대화를 듣고, 여자의 마지막 말에 대한 남자의 응답으로 가장 적절한 것을 고르시오.

① Go diving in the Philippines someday.
② Follow the safety rules, and you'll be fine.
③ Everyone gets nervous before they go abroad.
④ I've gone on vacation there many times before.
⑤ Don't worry about me. I'm an experienced diver.

12

대화를 듣고, 남자의 마지막 말에 대한 여자의 응답으로 가장 적절한 것을 고르시오.

① We have to look for another venue.
② You are the right person for the presentation.
③ I really wanted to see the presentation.
④ You can find the hall on the second floor.
⑤ The hall will be really crowded on the weekend.

13

대화를 듣고, 남자의 마지막 말에 대한 여자의 응답으로 가장 적절한 것을 고르시오.

Woman: _____

① Of course. I'd enjoy the company.
② No, thank you. I don't like jogging.
③ Sure. I hate going to the doctor alone.
④ No problem. See you Tuesday morning.
⑤ I'm sorry, but I have an appointment that day.

14

대화를 듣고, 여자의 마지막 말에 대한 남자의 응답으로 가장 적절한 것을 고르시오. 3점

Man: _____

① That's too cheap. How about $15?
② I don't know. Hawaii is pretty far away.
③ Thanks for the tip. I'll come back tomorrow.
④ That's great. I'll take three pineapples today.
⑤ Thanks for the pineapples. They're delicious.

15

다음 상황 설명을 듣고, Tim이 Fiona에게 할 말로 가장 적절한 것을 고르시오.

Tim: _____

① Someone ran over my bicycle.
② You shouldn't ride your bicycle in the dark.
③ I was parking my bicycle and got hit by a car.
④ I tried to avoid the cat, but I was going too fast.
⑤ I'm sorry. I couldn't see your bicycle in the dark.

[16~17] 다음을 듣고, 물음에 답하시오.

16

여자가 하는 말의 주제로 가장 적절한 것은?

① different energy sources
② benefits of sustainable energy
③ disadvantages of using fossil fuels
④ ways to conserve energy resources
⑤ an alternative source of energy in the future

17

언급된 에너지의 특성이 아닌 것은? 3점

① renewable   ② carbon-free   ③ abundant
④ cheap   ⑤ flexible

녹음을 다시 한 번 듣고, 빈칸에 알맞은 말을 쓰시오.

01

W: Over the 15 years that I have worked here, I have been honored to see our company achieve many great things. And I'm very proud to have been a leader here, particularly _____ _____ _____ _____ _____ the product we developed last year is now the best-selling of its kind. Therefore, this has not been an easy decision, but I must announce that I will no longer be serving as CEO of ACN Industrial, Inc. Ivan Sloan, our current vice president, will _____ _____ _____ _____. Considering his passion and experience, I am confident that ACN Industrial, Inc., has a bright future under Mr. Sloan's leadership. I want to thank all of you for your hard work and support _____ _____ _____.

02

W: Hey, Mark. I'm considering buying a new flat-screen TV. What do you think?

M: I wouldn't buy one now if I were you.

W: What makes you say that?

M: Well, TVs usually _____ _____ _____ _____ in November, when all of the holiday sales are starting.

W: That's true. Maybe it would be worth it to wait a bit longer.

M: In the meantime, weren't you thinking about _____ _____ _____ _____ _____?

W: Is November a good time to buy that as well?

M: Actually, stores usually have the lowest prices on cameras in January and February.

W: But that's not the holiday season.

M: No, it isn't. However, that's when _____ _____ _____ _____ _____ come out. So, if you buy an older model at that time, you can get a really good price.

03

M: Hello, everyone. This is Doctor Cool and you're listening to *KSPR*. Our special guest tonight is Michele White. Welcome to the show, Michelle.

W: Hello, Doctor. It's great to be here. I'm excited to sing some songs for your listeners.

M: That's great! But before we start, I have a few questions for you.

W: Okay.

M: Can you tell us more about the rumors concerning you and the famous actor, Joe Pitcher?

W: Oh, that? It was nothing. We just met at a party once.

M: Okay, and how about your health? Didn't you _____ _____ _____ at your last concert?

W: Yes. It hurt a lot! But it's much better now.

M: I'm glad to hear that. Finally, what new album are you working on?

W: It's a collection of pop songs. It will come out on August 15.

M: We're all _____ _____ _____ it! So, can you sing some songs from it now?

W: Sure. I'd be happy to.

M: Listeners, we'll be right back with the pop sensation's live performance _____ _____ _____. Stay tuned!

04

W: Shall we go to the modern music hall? The sign says it's _____ _____ _____.

M: Not yet. Our daughter is still listening to the curator explain what is in that display case.

W: Yes, she seems to be _____ _____ _____ _____ that people made these instruments in the 1700s.

M: Yes, she's very interested in making instruments. Wow, look at this violin! It looks very old.

W: Yes, it does. The sign on the right says that it's 500 years old.

M: It's amazing. Oh, who is that in the picture in the

round frame?

W: I guess he is the person who actually played the instruments.

M: He does look like a musician. *[Pause]* Oh, look! That man is taking a picture of the harp.

W: I'm not sure he's _____ _____ _____ _____ of the instruments.

M: Maybe we should ask someone who works here.

05

W: David, do you have some free time this evening?

M: Sure. I don't have any plans. What's up?

W: I have to write a ten-page essay for history class. _____ _____ _____, and I'm not even halfway finished.

M: Well, I was always good at history. I'd be glad to help.

W: Actually, I don't need help with the essay. All I need is some quiet _____ _____ _____ on it.

M: So what do you need me for?

W: Well, I promised Alexis that I'd _____ _____ _____ from the airport tonight. She's getting back from her trip to Mexico.

M: I didn't know Alexis was in Mexico.

W: Yes. She was visiting her family in Mexico City. Her _____ _____ _____ at 9 o'clock.

M: Oh. Well, I'd be glad to go get her. Can I borrow your car?

W: Sure. I really appreciate it.

06

[Knocking on door]

W: Oh. Hello, Andy.

M: Hi, Wendy. Sorry for coming by so early.

W: No problem. I've been up for hours. What can I do for you?

M: I wanted to talk to you about your son Sam.

W: Uh-oh. What did he do?

M: Well, I don't like to complain, but he and his friends _____ _____ _____ on my new car this morning.

W: You're kidding. How did that happen?

M: They were _____ _____ _____ to each other as they walked to school. It hit my car and cracked the windshield.

W: Are you sure it was Sam?

M: I saw him quite clearly before he ran away.

W: I'm so sorry. I'll _____ _____ _____ _____ and make sure he's punished.

07

M: Honey! Have you decided _____ _____ _____ _____ to use for the curtains?

W: Yes, I think this white one will look nice. And let's buy that checkered fabric for the tablecloths.

M: Good idea. I think we will need eight yards of fabric to make the curtains.

W: Are you sure? I think we will need _____ _____ _____ _____ for our room.

M: Okay. What about the tablecloths?

W: Hmm... Two yards should be enough, don't you think?

M: Yes, that should be fine.

W: Well, the white fabric costs $3 per yard, and the checkered fabric costs $5.

M: _____ _____ _____ _____ the sales tax.

W: You're right. How much is that?

M: 10%.

W: Okay. That's not bad for all that fabric.

M: Right. I'll go to the counter and pay.

08

M: Welcome to the town of Blueport.

W: Thank you, Mayor Johnson. Tell us a little about your town.

M: Sure. Blueport is rather small, with a population of

_____ _____ _____.

W: Right, but it was recently selected as the best tourist attraction in the country because of its beautiful location between the mountains and the sea.

M: Yes, the number of visiting tourists has increased steadily over the last 15 years.

W: So, is tourism the main industry?

M: Yes. Most of the townspeople are employed at local hotels and restaurants.

W: I see. I've also heard that Blueport is now a car-free area.

M: Yes. This helps the environment and _____ _____ _____ _____.

W: How do people get around?

M: Many people use bicycles, but we do _____ _____ _____ they can use.

W: That's very interesting. Thanks for talking with us.

09

M: Mountain Scouts are _____ _____ _____ _____ to keep the forests around our town clean and safe for both people and wildlife. For this year's parent-child project, we are going to spend three days camping next to Rocky River. While we're there, we are going to spend our days _____ _____ _____ _____ _____ along the river's edge. A clean river will be better for all of us, especially the fish and other animals that live in the water. Anyone who is interested in joining us on this trip should _____ _____ _____ _____ _____ by Thursday afternoon.

10

M: City Rental Center, how can I help you?

W: I need to rent a car for one week, starting this Friday.

M: All right. How many _____ _____ _____ _____?

W: Four. My husband and two kids will be with me.

M: Okay. And how much space will you need for luggage?

W: We'll be bringing two bags. Where can I pick the car up?

M: We have two offices: One is downtown, and the other _____ _____ _____ the airport.

W: Great. I'll pick it up at the airport. Do all of your cars have air conditioners?

M: Not all of them. But the weather is cool this time of year, so you don't need one.

W: Actually, I _____ _____ _____ _____. So I'd prefer a car with one.

M: That's not a problem. A suitable car will be waiting for you at the airport when you arrive.

W: Excellent. Thank you for your help.

11

W: I'm going to go scuba diving when I visit the Philippines next month.

M: That sounds great. Have you gone diving before?

W: No, this is _____ _____ _____, and I'm a little worried. It looks dangerous.

M: (Follow the safety rules, and you'll be fine.)

12

M: Kate, were you able to book the presentation hall?

W: Bad news. It is already _____ _____ for the weekend.

M: I see. So what do you think we should do?

W: (We have to look for another venue.)

13

M: Erica, where are you going?

W: Hi, Dan. I'm _____ _____ _____ go jogging.

M: I didn't know you were a jogger.

W: Well, I just started recently. I went to the doctor last month, and she said I need to exercise more often.

M: Yeah, I know _____ _____ _____. I spend too much time sitting at my desk. I'm starting to gain weight.

W: You could come jogging with me if you'd like.

M: How often do you go?

W: Three times a week. I go every Monday, Wednesday, and Friday at 10 in the morning.

M: Hmm. I could do that. Is it okay _____ _____ _____ _____?

W: (Of course. I'd enjoy the company.)

14

W: Hi. Can I help you find something?

M: Yes. Do you have any pineapples?

W: We do. They're right over here, next to the bananas.

M: Ah. I see them. Wow, $5 for one pineapple? Isn't that sort of expensive?

W: Not really. _____ _____ all the way from Hawaii.

M: They do look good. I need several for a party I'm having this weekend.

W: Well, if you don't need them today, you'd better drop by again tomorrow. _____ _____ _____ _____ on fruit and vegetables then.

M: Are pineapples going to be on sale?

W: Yes. _____ _____ _____ will be three for $10.

M: (Thanks for the tip. I'll come back tomorrow.)

15

M: Fiona is a high school student. Every day, she rides her bicycle to school and back. One day after school, she _____ _____ _____ _____ in the driveway. It is dark when her older brother Tim comes home from work that evening. He _____ _____ Fiona's bicycle while parking his car. Fiona comes running out of the house and sees _____ _____ _____. She asks Tim what happened. In this situation, what is Tim most likely to say to Fiona?

Tim: (I'm sorry. I couldn't see your bicycle in the dark).

16~17

W: I'm Dr. Kristine Hoover, and today I'd like to talk to you about energy. We all know about _____ _____ _____, such as fossil fuels. And you've probably heard a lot about sustainable energy that comes from the wind or the sun. But you might not know much about biomass energy. It is made by using plants or parts of plants, such as leaves. _____ _____ _____ _____, biomass is considered a form of renewable energy. Also, there is an abundant supply of plants available. We don't have to search for them like we do for fossil fuel. Because of this, biomass is also cheaper to produce. There is no need to drill into the ground or build pipelines. Finally, biomass is a flexible energy source. It can _____ _____ _____ many different forms, and it can be made from a wide variety of organic materials. It might be the clean energy source of the future.

1번부터 17번까지는 듣고 답하는 문제입니다.
1번부터 15번까지는 한 번만 들려주고, 16번부터 17번까지는 두 번 들려줍니다. 방송을 잘 듣고 답을 하기 바랍니다.

01

다음을 듣고, 남자가 하는 말의 목적으로 가장 적절한 것을 고르시오.
① 미아를 찾으려고
② 영업시간을 안내하려고
③ 아동복 세일을 알리려고
④ 분실물 습득을 알리려고
⑤ 야구 경기 일정을 알리려고

02

대화를 듣고, 여자의 의견으로 가장 적절한 것을 고르시오.
① 컴퓨터를 깨끗하게 사용해야 한다.
② 오류 없이 서류를 작성하는 것이 중요하다.
③ 전자제품 무상 서비스 기간을 늘려야 한다.
④ 키보드가 고장 나면 새로 구입하는 것이 낫다.
⑤ 전자제품 구매 시 사용설명서를 읽어봐야 한다.

03

대화를 듣고, 두 사람의 관계를 가장 잘 나타낸 것을 고르시오.
① 기자 - 편집장   ② 변호사 - 판사
③ 작가 - 독자   ④ 교사 - 학부모
⑤ 학생 - 도서관 사서

04

대화를 듣고, 그림에서 대화의 내용과 일치하지 않는 것을 고르시오.

05

대화를 듣고, 남자가 할 일로 가장 적절한 것을 고르시오.
① 부엌을 칠하는 것을 도와주기
② 여자에게 새 컴퓨터 사주기
③ 컴퓨터에 인터넷 설치하기
④ 컴퓨터 수리점에 연락하기
⑤ 사촌에게 도움 요청하기

06

대화를 듣고, 여자가 남자의 개를 돌볼 수 없는 이유를 고르시오.
① 식당을 예약해놔서
② 개를 좋아하지 않아서
③ 주말에 일 해야 해서
④ 가족들과 여행을 가야 해서
⑤ 아르바이트 면접을 보러 가야 해서

07

대화를 듣고, 여자가 지불할 금액을 고르시오.
① $5   ② $10   ③ $15   ④ $20   ⑤ $25

08

대화를 듣고, redwood에 관해 언급되지 않은 것을 고르시오. 3점
① 최대 높이   ② 수명   ③ 분포지
④ 멸종 이유   ⑤ 보호가 필요한 이유

09

Ansari X Prize에 관한 다음 내용을 듣고, 일치하지 않는 것을 고르시오. 3점
① 천만 달러의 상금이 수여되었다.
② 우승하려면 2주 안에 같은 우주선을 재발사해야 했다.
③ 출품된 우주선은 정부 자금으로 제작되었다.
④ 설립 후 8년 만에 수여되었다.
⑤ 민간 우주선 개발에 대한 관심을 불러일으켰다.

10

다음 표를 보면서 대화를 듣고, 두 사람이 선택할 장소를 고르시오.

	Venue	Renovated	Meeting Rooms	Guest Rooms	Price
①	A	2010	8	225	$4,500
②	B	2012	8	100	$2,500
③	C	2013	10	150	$2,000
④	D	2014	12	200	$3,000
⑤	E	2015	15	250	$4,000

11

대화를 듣고, 남자의 마지막 말에 대한 여자의 응답으로 가장 적절한 것을 고르시오.

① No, I just don't eat raw fish.
② It's okay. I had a big breakfast.
③ You should try it. It's not that bad.
④ No, I brought my lunch box today.
⑤ Yes, I know a lot about eating fish.

12

대화를 듣고, 여자의 마지막 말에 대한 남자의 응답으로 가장 적절한 것을 고르시오.

① Sorry, I don't take that class anymore.
② Thanks. I'll see you in class tomorrow.
③ The speaker delivered an impressive speech.
④ Definitely. Application forms are available online.
⑤ For sure. It also taught me how to organize my materials.

13

대화를 듣고, 남자의 마지막 말에 대한 여자의 응답으로 가장 적절한 것을 고르시오. 3점

Woman: _____

① Are you sure it's okay if I borrow it?
② Right. My lawn mower is better than yours.
③ I'm not sure if we can afford a new lawn mower.
④ If you needed to borrow it, you should have asked me earlier.
⑤ My car broke down yesterday. Can you bring it back to my house?

14

대화를 듣고, 여자의 마지막 말에 대한 남자의 응답으로 가장 적절한 것을 고르시오.

Man: _____

① When are these books due?
② Okay, I'll come back next week.
③ Okay, here's my driver's license.
④ When can I register for my classes?
⑤ But I've lived in this town for years.

15

다음 상황 설명을 듣고, Alex가 Monica에게 할 말로 가장 적절한 것을 고르시오.

Alex: _____

① Please make an outline before you start.
② When can you finish gathering information?
③ Make the presentation file whenever you like.
④ Take more responsibility for this team project.
⑤ If you are busy, I will make the presentation file.

[16~17] 다음을 듣고, 물음에 답하시오.

16

남자가 하는 말의 주제로 가장 적절한 것은?

① what the American Red Cross does
② what the Red Cross symbol means
③ how the American Red Cross was founded
④ who founded the International Committee of the Red Cross
⑤ the relationship between each country's Red Cross branches

17

the American Red Cross에 관해 언급되지 <u>않은</u> 것은?

① 설립 시기　　　　② 설립자
③ 재정 상태　　　　④ 첫 지사가 있었던 주(州)
⑤ 현재 미국 지사의 수

녹음을 다시 한 번 듣고, 빈칸에 알맞은 말을 쓰시오.

01

M: Attention, shoppers. We have an announcement to make about _____ _____ _____. A young boy named Orlando Miller is missing. He was at the Children's Corner 10 minutes ago. He is six years old and of _____ _____ _____ _____. He has brown hair and brown eyes. He is wearing a blue and white striped shirt and a New York Yankees baseball cap. His father is _____ _____ _____ at the Customer Service Center. If you see him, please assist him to the Customer Service Center so that he can _____ _____ with his father. Thank you.

02

W: This keyboard is _____ _____ again.

M: What's wrong?

W: Some of the letters keep getting stuck while I'm typing. I think it's too old.

M: Well, why don't you try to clean it? Maybe _____ _____ _____ into the keyboard will get the dust out and solve the problem.

W: I already tried that. It didn't seem to do any good.

M: Then, I'll take it to _____ _____.

W: I don't think shops even offer that service. Keyboards are so inexpensive these days that it's easier to just replace them.

M: So I guess you'd prefer a new one.

W: I think that's the easiest solution.

03

M: Excuse me, Kristin? Do you have a minute?

W: Yes, but please _____ _____ _____. I'm extremely busy.

M: Sure, this will just take a second. I have a question about my _____ _____ _____.

W: Your article? Ah, yes. The one on the effects of air pollution. It was quite good.

M: Thanks. I was just wondering why you changed the last sentence. Was there something wrong with it?

W: If I remember correctly, there was nothing wrong with the grammar of the sentence. I just thought it _____ _____.

M: Really? Okay, I guess maybe you're right. How about the rest of the article?

W: Like I said, it was quite good. There were _____ _____ _____ _____ _____, but I like your style. Keep up the good work.

M: Thank you. I'll let you get back to work now.

W: Thanks. I have so many things to do today.

04

W: Hi, Cory! I'm glad we could do a video chat today. Is that your new room?

M: Yes, Mom. My homestay family and I just finished decorating it.

W: Are those pictures above your bed?

M: Yeah, they're pictures of our family and my friends. They'll _____ _____ _____ while I'm studying abroad.

W: Great! Where did you put your desk?

M: It's under this big window.

W: That's nice. Is there any place to put your books?

M: There's a bookcase with four shelves next to my desk. The bottom shelf is empty. I'm going to _____ _____ _____ _____ my new books.

W: Good idea. Is that your bookbag on the couch?

M: Yep! It's right between the two cushions.

W: Most of the floor is covered by that rug. Is your room cold?

M: No, I just wanted to make my room cozier. So I bought the rug. _____ _____ _____ _____?

W: The striped pattern looks nice! Have a great time at school!

M: Thanks, Mom! I'll talk to you again soon!

05

W: Mark, do you remember when I helped you paint your kitchen?

M: I sure do. You were a big help. I still _____ _____ _____ _____ for that.

W: Yes, you do. So I was wondering if you could help me with my new computer.

M: I didn't know you had a new computer.

W: Yes, my parents bought me a new laptop for Christmas. The problem is I can't _____ _____ _____ _____ to the internet.

M: I'd love to help you, but I don't know anything about the internet.

W: I know you don't. But your cousin Danny does. He's a _____ _____.

M: Yeah, he really is. Would you like me to call him for you?

W: I'd really appreciate it.

M: No problem. I'll see if he can _____ _____ _____ _____ _____ tonight.

06

W: Hi, Tom. What are you going to do this weekend?

M: I'm _____ _____ _____ _____ to Jeju Island with my family.

W: Wow. You must be really excited.

M: Of course I am. But could you do me a favor?

W: Sure. What is it?

M: I need someone to _____ _____ _____ my dog for two days while we're away.

W: Hmm... I wish I could, but I can't.

M: Why is that?

W: Actually, I have _____ _____ _____ on the weekend.

M: I didn't know you had a part-time job. Where do you work?

W: I serve food at an Italian restaurant near my house. Anyway, how about asking someone else?

M: I guess I'll have to.

07

M: Hi! Welcome to Donut World. What can I get for you today?

W: I'm going to a parents' meeting at the elementary school, and I need to bring some donuts. How much would _____ _____ _____?

M: _____ _____ _____ what kind of donuts you want.

W: Hmm. Those chocolate ones look good. How much are they?

M: They're 50 cents each.

W: Okay. And how about the jelly-filled ones?

M: The jelly donuts are a little more expensive. They cost _____ _____.

W: Well, those look good, too. I'll take ten chocolate donuts and ten jelly donuts.

M: All right. Is there anything else I can get you?

W: No, that's all. Thanks for your help.

08

W: These are the giant redwoods. They're the tallest trees in the world.

M: How tall can they grow?

W: Well, the world record is more than 115 meters.

M: It must take a long time to get so big.

W: Yes, it does. But redwoods can live for _____ _____ _____ _____.

M: Do they only grow here in California?

W: This is where most redwoods are found, but there is also a species that grows in China.

M: Really? I didn't know that.

W: Unfortunately, all redwood species are endangered. Since 1850, _____ _____ _____ _____ have disappeared.

M: That's too bad. They're beautiful trees.

W: They're more than that. They also remove a lot of carbon from the atmosphere.

M: Yes. And that helps _____ _____ _____. We really need to protect them.

09

W: The Ansari X Prize, formally known as the X Prize, was created in the spring of 1996 by the X Prize Foundation. The foundation offered ten million dollars to the first group that could launch a successful spacecraft. The spacecraft had to be able to _____ _____ _____ within a two-week period. Additionally, it had to be manned and couldn't be built _____ _____ _____. After years of attempts, someone finally won it in 2004. The prize was created in order to demonstrate that private corporations and groups were capable of contributing to the field of space transport. _____ _____ _____ in the private space race, the Ansari X Prize is helping to create the next generation of explorers.

10

W: Our company is hosting the industry conference this year.

M: That's right. We need to choose a hotel _____ _____ _____ _____.

W: The boss said that she wants a modern place.

M: A modern place? What does she mean by that?

W: She wants us to choose a hotel _____ _____ _____ in 2012 or later.

M: Okay. Well, how many meeting rooms will we need?

W: We need a place with at least ten meeting rooms. Eight isn't enough.

M: All right. How about hotel rooms? Is 200 too few?

W: No, 200 would be fine. But _____ _____ _____ _____.

M: I think we have a couple of options. What's our budget?

W: It's pretty small. Let's go for the cheaper option.

M: All right. I'll call the hotel right now.

11

M: What are you going to _____ _____ _____?

W: I think I will get fish and chips.

M: Really? I thought you didn't like to eat fish!

W: (No, I just don't eat raw fish.)

12

W: I started taking a class on _____ _____ _____ _____ _____. It's an important skill.

M: Really? I took that class last year.

W: Oh, I didn't know that. Do you think it helped you become a better public speaker?

M: (For sure. It also taught me how to organize my materials.)

13

[Phone rings.]

M: Hello?

W: Hi, Chuck. It's your sister.

M: Hi, Vicky. What's going on?

W: Not much. Do you remember when you _____ _____ _____ _____ last week?

M: Sure. Mine hasn't been working very well, and the grass in my yard was getting pretty high.

W: I know. But your grass is short and neat now, right?

M: It sure is. Your lawnmower works great.

W: Then don't you think you should _____ _____ _____?

M: I'm sorry. I didn't know you needed it. You can _____ _____ and get it any time.

W: (My car broke down yesterday. Can you bring it back to my house?)

14

W: Hi. How _____ _____ _____ _____ today?

M: This is a great library! I just moved here a few

days ago with my family.

W: Oh. Welcome to Northville.

M: Thanks. I was wondering if I could _____
_____ a library card today.

W: Of course. I just need to see some sort of ID.

M: Well, I don't have a _____ _____ yet. But I'm a
student at the university.

W: Do you have your student ID card with you?

M: No, I won't get those until next Monday.

W: Oh. I'm sorry, but I can't _____ _____ a library
card until you have some ID.

M: (Okay, I'll come back next week.)

15

W: Alex and Monica are working together at a
marketing department. They have gotten a team
project to make a new marketing plan. They
decided to divide the work in half. Alex's job
is to gather information, and Monica's is to
_____ _____ _____. Alex already collected
all the required information from the internet.
Furthermore, he wrote down an outline for the
presentation. But Monica hasn't started making
the presentation file yet. She just says that she
is _____ _____ _____ other things. Alex
doesn't think he can continue his work without the
presentation file. He also feels that, even though
it is a team project, he has been doing most of the
work. He wants her to _____ _____ _____
_____. In this situation, what would Alex most
likely say to Monica?

Alex: (Take more responsibility for this team project.)

16~17

M: Good morning, everyone! Are you interested in
donating money to _____ _____ _____ in
the US, but don't know who to give to? I have
the solution for you. Donate your money to the
American Red Cross. You might think that this
is the same as the International Red Cross, but

it isn't. In fact, the International Committee of
the Red Cross _____ _____ _____ 1863.
The American Red Cross was started about 20
years later by an American woman named Clara
Barton. She got the idea from the international
organization. Barton began her career as a teacher.
But during the American Civil War, she _____
_____ _____ _____. After the war, she went
to Europe for a vacation. That's _____ _____
_____ _____ the International Red Cross.
When she came back to America, she convinced
the US government of the need for an American
division. It started with just a few branches in New
York State. Now, there are more than 650 locations
across the country. The American Red Cross and
the International Red Cross have become global
leaders for charity work.

1번부터 17번까지는 듣고 답하는 문제입니다.
1번부터 15번까지는 한 번만 들려주고, 16번부터 17번까지는 두 번 들려줍니다. 방송을 잘 듣고 답을 하기 바랍니다.

01

다음을 듣고, 여자가 하는 말의 목적으로 가장 적절한 것을 고르시오.
① 박람회 연장 시간을 알리려고
② 박람회 폐막 시각을 안내하려고
③ 관련 업계의 대표들을 소개하려고
④ 취업 박람회의 개최 장소를 알리려고
⑤ 박람회 부스 이용자들에게 정리를 요청하려고

02

대화를 듣고, 남자의 의견으로 가장 적절한 것을 고르시오.
① 제품 구입 전에 사전 조사가 필요하다.
② 스마트폰은 디자인보다 기능이 중요하다.
③ 브랜드 인지도와 제품의 질은 관련이 없다.
④ 제품 구입 후기만 믿고 구입해서는 안 된다.
⑤ 제품 구입 전에 인터넷에서 가격 비교를 해야 한다.

03

대화를 듣고, 두 사람의 관계를 가장 잘 나타낸 것을 고르시오.
① 요리사 - 손님   ② 요리사 - 웨이터
③ 대행사 직원 - 고객   ④ 진행 요원 - 관광객
⑤ 호텔 매니저 - 웨이터

04

대화를 듣고, 그림에서 대화의 내용과 일치하지 않는 것을 고르시오.

05

대화를 듣고, 여자가 할 일로 가장 적절한 것을 고르시오.
① 태국에서 전화하기   ② 고양이에게 먹이 주기
③ 손님 응대하기   ④ 쌓인 눈 치우기
⑤ 기념품 선물하기

06

대화를 듣고, 남자가 휴대전화 통신 업체를 바꾸려는 이유를 고르시오.
① 약정 기간이 만료되어서
② 서비스 품질이 떨어져서
③ 타사보다 기본요금이 비싸서
④ 사용 요금이 잘못 청구되어서
⑤ 무료 인터넷 데이터 제공량이 적어서

07

대화를 듣고, 남자가 지불할 금액을 고르시오.
① $12   ② $14   ③ $18   ④ $20   ⑤ $22

08

대화를 듣고, 선거에 관해 언급되지 않은 것을 고르시오.
① 투표자 수   ② 득표율   ③ 선거 공약
④ 유세 기간   ⑤ 선거 비용

09

slow-food movement에 관한 다음 내용을 듣고, 일치하지 않는 것을 고르시오. 3점
① 1980년대에 이탈리아에서 시작되었다.
② 패스트푸드 식당 개업이 시발점이 되었다.
③ 150여 개국 이상에서 10만여 명의 회원이 참여하고 있다.
④ 사교 활동으로서의 음식에 중점을 둔다.
⑤ 식품의 정확한 영양 성분 표기에 노력을 기울이고 있다.

10

다음 표를 보면서 대화를 듣고, 여자가 구입할 드레스를 고르시오.

Fashion Mart - Prom Dresses

	Dress	Length	Style	Color	Price
①	A	Short	A-Line	Green	$79
②	B	Short	A-Line	Black	$109
③	C	Medium	A-Line	Gold	$99
④	D	Medium	Ball gown	Red	$99
⑤	E	Long	Ball gown	Pink	$129

11

대화를 듣고, 여자의 마지막 말에 대한 남자의 응답으로 가장 적절한 것을 고르시오.

① I know. I think I met him before.
② You should see him during the week.
③ The hospital is right around the corner.
④ I can tell you where the best clinic in town is.
⑤ I did. I took some medicine, but it didn't help.

12

대화를 듣고, 남자의 마지막 말에 대한 여자의 응답으로 가장 적절한 것을 고르시오.

① It takes 30 minutes by train.
② I have waited since last month.
③ It's 6:30 now. I can easily get there in time.
④ It's okay. Tickets will be available tomorrow.
⑤ In about three hours. I'll need to hurry to get there.

13

대화를 듣고, 여자의 마지막 말에 대한 남자의 응답으로 가장 적절한 것을 고르시오.

Man: _____

① Your final grade will be an A.
② Give it a shot. He might accept it.
③ I admit your attendance was good.
④ No need. Your work has been excellent.
⑤ I'm sorry. You'll have to repeat the class.

14

대화를 듣고, 남자의 마지막 말에 대한 여자의 응답으로 가장 적절한 것을 고르시오.

Woman: _____

① It's my pleasure to sign autographs.
② I never thought I'd sit next to a celebrity.
③ Sorry, I'm not available for volunteer work.
④ You should have attended the concert I just gave.
⑤ Not at all. I should thank you for the opportunity to help.

15

다음 상황 설명을 듣고, Jane이 그녀의 오빠에게 할 말로 가장 적절한 것을 고르시오. 3점

Jane: _____

① Taking the bus to school is your best option.
② I'm in a hurry. Could you drive me to school?
③ We have to tell Mom and Dad what happened.
④ Why didn't you answer the messages I sent you?
⑤ Please concentrate on the road when you're driving.

[16~17] 다음을 듣고, 물음에 답하시오.

16

여자가 하는 말의 주제로 가장 적절한 것은? 3점

① the different features of moles
② self-examinations for skin cancer
③ new breakthroughs in skin cancer prevention
④ the advantages of laboratory tests for skin cancer
⑤ advice about how to report skin cancer symptoms

17

언급된 피부암의 형상이 아닌 것은? 3점

① mirror images ② uneven sides
③ irregular borders ④ different colors
⑤ changing sizes

녹음을 다시 한 번 듣고, 빈칸에 알맞은 말을 쓰시오.

01

W: Good afternoon, everyone. Today's career fair will be coming to a close in an hour. I'd like to thank you very much for _____ _____, and remind you to make sure you've visited all the booths you're interested in before _____ _____. If there are some booths you didn't have time to visit, the fair will be here until Friday, and it is open from 10 a.m. until 6 p.m. each day. With representatives from a wide variety of industries and professional fields, our career fair is the best place to find your first job after graduation, so please make the most of _____ _____ _____. Thank you very much.

02

W: Oh, look at that smartphone. It's really cheap.

M: Yes, but I don't _____ _____ _____.

W: But it looks nice, and it comes with a case.

M: Well, the quality might not be very good. You should look into that brand before buying it.

W: I guess you're right, but won't it be difficult?

M: Not really. You can _____ _____ _____ on the internet or talk to a salesperson at a phone store.

W: That's not a bad idea. I don't want it to break right away.

M: Exactly. Being informed helps you _____ _____ _____ _____.

W: Yeah, I suppose you are right.

03

M: Good morning, Mrs. Reed. I have more information for you about the 18th.

W: Oh, did you _____ _____ _____ at the Plaza Hotel?

M: Yes, I did. And what about serving the buffet

outside?

W: Outside? I hadn't thought of that.

M: Well, more and more of my clients are choosing to host outside, especially in summer.

W: Okay, sounds good. Did you _____ _____ _____ with the buffet?

M: Yes. I found a catering company that deals in organic food as you requested.

W: Well done. Most of the guests are environmental campaigners, so we really can't have it any other way.

M: Agreed. All right, I'll email you _____ _____ _____ tonight.

W: Thanks, Toby. You're the best.

04

[Cell phone rings.]

W: Hello?

M: Ms. Jenkins, I _____ _____ _____ _____ in my office. Can you look for them?

W: Sure. *[Opening the door]* Oh! It looks like someone wrote your meeting schedule on the whiteboard.

M: Right, I have several meetings next week.

W: So, where did you put the tickets?

M: They're probably on my desk between my laptop and my tablet PC.

W: No, they're not. And they're not on your desk chair, either.

M: Can you _____ _____ _____ _____ by the window?

W: I still don't see them.

M: Oh, sorry. I just found the tickets in my wallet. It's my mistake. But while you're there, can you also make sure _____ _____ _____ _____?

W: Yes, it's closed. Do you need anything else?

M: No. Sorry for troubling you. See you tomorrow!

05

[Phone rings.]

W: Hello, Julia speaking.

M: Hi, Julia! It's Mike. I'm calling from Thailand!

W: Mike! How are you?

M: I'm doing great. My vacation is going wonderfully. How about you?

W: I'm good, but I envy you. We ___ ___ ___ here.

M: I know, I read about it online.

W: So, what's up? Don't worry — I fed your cat just like you asked me to.

M: Oh, thank you! But there's something else I need to ask you.

W: What's that?

M: As you know, there's a bus stop in front of my shoe store, and I'm not there to ___ ___ ___ ___ from it.

W: Ah, I got it. You're worried people are going to ___ ___ ___ on the snow?

M: That's right. Please take care of it. If you do, I'll bring you back a wonderful present!

W: Okay. I'll do it for you.

06

W: Hi, James. How's your new phone?

M: The phone is fine, but to be honest, I am thinking of changing mobile service providers.

W: Really? Do you like another company's service plan better than the one you have now?

M: Not really. I'm just ___ ___ ___ ___ ___.

W: What's the problem?

M: Well, last week they were doing maintenance on their network, so I couldn't get any reception.

W: That's not good.

M: Right. Also, their network is very slow, so sometimes it's ___ ___ ___ the internet.

W: That must be very frustrating.

M: Yeah. I want to switch to a company that provides ___ ___.

07

M: Wow, there are ___ ___ ___ ___ here.

W: Yes. Please take a look around.

M: How much is this Impressionist art calendar?

W: That's $12.

M: It's perfect for my mom. I'll take it.

W: Good choice. And did you see our range of mugs over there?

M: Yes, they're beautiful. I like the one with Van Gogh's artwork on it.

W: Me too. Each one of those mugs is $10.

M: They're a little expensive. How about this smaller cup with the museum logo on it?

W: That one's $6.

M: Okay, ___ ___ ___.

W: Yes. Is that all?

M: Oh, I'll also take these two pens. They cost ___ ___ ___.

W: Okay.

08

W: Tonight's guest is Ron Green. He was just elected to the National Assembly. Welcome, Ron.

M: Thanks, Sarah. I'm glad to be here.

W: So, how many people ___ ___ ___ last week's election?

M: Approximately 16,000 people in the district voted.

W: I see. And of those votes, how many did you receive?

M: I received ___ ___ ___, which is more than 70% of the total.

W: Why do you think your campaign was so successful?

M: I think people liked the fact that I promised to ___ ___ and improve the highway system.

W: May I ask how much the campaign cost you?

M: Certainly. We spent about four million dollars. That's a lot of money, but it was worth it.

W: Well, I wish you the best of luck.

M: Thank you for having me on the show.

09

M: In a world of busy schedules and fast food, some people are looking for ways to slow down. The slow-food movement is aimed at restoring the joy of sharing meals with others. The movement got started in the 1980s in Italy _____ _____ _____ a fast-food restaurant opening there. Members of the movement focus on the joy of _____ _____ _____ _____ around the dinner table. Since its founding, the movement has spread to more than 150 countries, and it now has about 100,000 members. With a focus on food _____ _____ _____ _____, the movement reminds us that food can be a social pleasure as well as source of nutrients.

10

M: Are these the dresses that you're considering for the prom?

W: Yes. I can't decide _____ _____ _____ _____.

M: Well, what length do you want?

W: Anything except long. It's too hard to dance in a long dress.

M: That makes sense. And it looks like there are two different styles.

W: Yes. The simple ones are A-line dresses, and the fancy ones are ball gowns.

M: Ball gowns are beautiful, but they're too much for a prom. I like simple ones.

W: So do I. Now I need to choose a color. To be honest, I like them all.

M: Don't get a green dress. I don't think green _____ _____ _____ your hair color.

W: Actually, you're right. And I can't spend more than $100 on it.

M: Then this one right here is the best choice.

W: Yes! I'll order it now. I hope it _____ _____ _____ _____!

11

W: Hey, Joshua, are you all right? You don't look so good today.

M: I have _____ _____ _____ since last weekend.

W: Why haven't you seen a doctor?

M: (I did. I took some medicine, but it didn't help.)

12

M: Hi, Mary. You're going to the airport today, aren't you?

W: Yes, it's my first trip abroad. I've been waiting for this moment for a long time.

M: What time does _____ _____ _____ _____?

W: (In about three hours. I'll need to hurry to get there.)

13

W: Good afternoon, sir. I'd like to _____ _____ _____ in your English class.

M: Okay. Come in. You're Maria, right?

W: Yes, that's right. Unfortunately, I failed your class.

M: Yes. It looks like you did very _____ _____ _____ _____ _____.

W: I know. I understood the material, but I got nervous and panicked. I get too stressed out during tests.

M: I'm sorry to hear that. Let me look at your other grades.

W: Thank you. I _____ _____ _____ on the quizzes. I really would like to pass your class.

M: Well, you're right. Your quiz scores are much better than your final exam score.

W: I worked very hard in your class. Do you think you can change my grade?

M: I'm sorry, but I can't. I'm looking at your attendance now, and I see that you missed eight classes.

W: I know, but...

M: The school's policy is that you can be absent _____ _____ _____ five times.

W: So there's nothing you can do?

M: (I'm sorry. You'll have to repeat the class.)

14

M: Hello, I'm Dr. Lee. It looks like we'll be sitting next to each other during this flight.

W: Pleased to meet you.

M: Wait a minute, _____ _____ _____! You're a famous singer, aren't you?

W: I'm surprised you recognize me. My fans are usually much younger than you.

M: Well, I'm a cancer specialist at a children's hospital, and many of my patients are your fans.

W: Ah, now I understand.

M: Actually... Could I ask you something?

W: Certainly.

M: How about _____ _____ _____ to give a small concert at the hospital? The children would be delighted.

W: Oh, it would _____ _____ _____.

M: Thank you. I'm so grateful.

W: (Not at all. I should thank you for the opportunity to help.)

15

M: Jane is a high school student, and she lives _____ _____ _____ her school. There is no subway system in her town, and there are few buses and taxis. Her parents used to drive her to school every day, but her older brother recently _____ _____ _____ _____. Now he often drives her to school and to the mall when she needs to go shopping. He always drives slowly and carefully, but one thing is making Jane worried.

When her brother gets text messages on his phone while driving, he usually reads them right away. Sometimes he even answers them. She's afraid that it may eventually _____ _____ _____. In this situation, what would Jane most likely say to her brother?

Jane: (Please concentrate on the road when you're driving.)

16~17

W: When it comes to cancer, prevention and early detection are your best tools. Though most forms of cancer require a laboratory test to be identified, skin cancer can often be detected by sight alone. Individuals can examine themselves for _____ _____ _____ on their skin. The entire body should be inspected on a regular basis. This allows you to see if any changes have occurred. Look for moles that are uneven, where the two sides are not mirror images of each other. Irregular borders around moles could be a sign of trouble. You should also _____ _____ _____ the color of the mole. If different colors appear in the same mole, you should report it to your doctor. Also, a mole that changes in size or shape over time is another reason for concern. Any of the above symptoms should be reported to your doctor immediately. That way, _____ _____ can be carried out. Remember, by diligently following these tips, you can stop skin cancer before it gets _____ _____ _____.

1번부터 17번까지는 듣고 답하는 문제입니다.
1번부터 15번까지는 한 번만 들려주고, 16번부터 17번까지는 두 번 들려줍니다. 방송을 잘 듣고 답을 하기 바랍니다.

01

다음을 듣고, 남자가 하는 말의 목적으로 가장 적절한 것을 고르시오.
① Abba의 콘서트를 소개하려고
② 티켓이 매진되었음을 알리려고
③ 뮤지컬 앙코르 공연을 홍보하려고
④ 뮤지컬 감상평 이벤트를 공지하려고
⑤ 뮤지컬 배우들과의 팬 미팅을 안내하려고

02

대화를 듣고, 여자의 의견으로 가장 적절한 것을 고르시오.
① 역사 교육을 강화해야 한다.
② 잦은 시험은 학습 의욕을 떨어뜨린다.
③ 발표 수업은 효율적인 학습 방법이다.
④ 예습과 복습이 성적 향상에 도움이 된다.
⑤ 필기시험보다 더 나은 평가 방법을 시행해야 한다.

03

대화를 듣고, 두 사람의 관계를 가장 잘 나타낸 것을 고르시오.
① 강사 - 학생   ② 기자 - 음악가
③ 영업사원 - 고객   ④ 교사 - 학부모
⑤ 경비원 - 입주민

04

대화를 듣고, 그림에서 대화의 내용과 일치하지 않는 것을 고르시오.

05

대화를 듣고, 여자가 할 일로 가장 적절한 것을 고르시오.
① TV 보기   ② 빨래하기   ③ 먼지 닦기
④ 설거지하기   ⑤ 옷 정리하기

06

대화를 듣고, 남자가 아르바이트를 그만두려는 이유를 고르시오.
① 급여가 적어서   ② 일이 지루해서
③ 일이 늦게 끝나서   ④ 학교 성적이 떨어져서
⑤ 취미 생활을 할 시간이 없어서

07

대화를 듣고, 남자가 지불할 금액을 고르시오. 3점
① 75만 원   ② 85만 원   ③ 95만 원
④ 105만 원   ⑤ 115만 원

08

대화를 듣고, 새로 개봉한 영화에 관해 언급되지 않은 것을 고르시오.
① 개봉일   ② 관객 평점   ③ 장르
④ 등장인물   ⑤ 상영 시간

09

Mrs. Winter에 관한 다음 내용을 듣고, 일치하지 않는 것을 고르시오.
① 대학원 과정을 수료했다.
② 중학교 교사 경력이 있다.
③ 음악 교사로 부임한다.
④ 취미로 합창단에서 노래 공연을 한다.
⑤ 슬하에 다섯 명의 자녀가 있다.

10

다음 표를 보면서 대화를 듣고, 두 사람이 선택한 여행지를 고르시오.

	Destination	Price	Days	Tour Guide	Complimentary Gift
①	Bali	$1,600	5	O	Spa tickets
②	Phuket	$1,000	4	X	Bottle of wine
③	Maldives	$1,800	4	O	Bottle of wine
④	Boracay	$1,500	6	O	Spa tickets
⑤	Sydney	$2,600	7	X	Spa tickets

11

대화를 듣고, 남자의 마지막 말에 대한 여자의 응답으로 가장 적절한 것을 고르시오.

① You almost lost your wallet.
② I can buy you lunch tomorrow.
③ You should come to class on time.
④ Don't forget to visit me at lunchtime.
⑤ Don't worry. I'm sure we can find it.

12

대화를 듣고, 여자의 마지막 말에 대한 남자의 응답으로 가장 적절한 것을 고르시오.

① No, my band only plays on Fridays.
② No, I play the electric guitar these days.
③ Yes, I quit the band but still play the drums.
④ Yes, I think the drums are priced reasonably.
⑤ I'm afraid that we don't have any instruments.

13

대화를 듣고, 여자의 마지막 말에 대한 남자의 응답으로 가장 적절한 것을 고르시오.

Man: _____

① Do you need anything else?
② Would you eat there again with me?
③ I should have listened to you carefully.
④ I guess I should find another restaurant.
⑤ I guess I'd better make a reservation quickly.

14

대화를 듣고, 남자의 마지막 말에 대한 여자의 응답으로 가장 적절한 것을 고르시오. 3점

Woman: _____

① I prefer my old scarf to the new one.
② That's beyond my budget. I can't afford it.
③ I wish I knew where she bought her new coat.
④ I can't find my new scarf. Do you know where it is?
⑤ I just wish she would ask me before borrowing my stuff.

15

다음 상황 설명을 듣고, Mike가 그의 할머니에게 할 말로 가장 적절한 것을 고르시오.

Mike: _____

① I'm glad you enjoyed your trip.
② Don't worry about it. Just get well soon.
③ I want to apologize for my terrible mistake.
④ It was so sweet of you to send me this present.
⑤ I'd like to, but I can't. Maybe some other time.

[16~17] 다음을 듣고, 물음에 답하시오.

16

여자가 하는 말의 주제로 가장 적절한 것은? 3점

① the purposes and functions of flags
② the evolution of flags throughout history
③ the importance of flags for self-expression
④ the most interesting flags in modern times
⑤ the process of making flags in the Middle Ages

17

언급된 조직이 아닌 것은?

① the US army ② the Girl Scouts
③ state governments ④ the Olympics
⑤ the United Nations

녹음을 다시 한 번 듣고, 빈칸에 알맞은 말을 쓰시오.

01

M: I have some good news for all you fans who missed the last performance! *Mamma Mia* is back! All tickets to _____ _____ _____ based on the songs of Abba were sold out during its last three-month run in Seoul. The cast for *Mamma Mia*'s upcoming encore season will feature world-famous actors, and you may never get a chance to see them all on one stage again if you _____ _____ _____. The encore season will run from January 14 to January 31 at Korea Music Hall. Tickets are expected to _____ _____ fast, so get yours today!

02

M: We have another test tomorrow. I'm _____ _____ _____ _____.

W: Oh, me too. We have too many tests. There is a better way for teachers to check our progress.

M: What sort of assessment are you thinking of?

W: Research projects. _____ _____ _____ _____ _____, we could research the things we are taught in class.

M: That might be more interesting, I suppose.

W: I did a history project like that once. It gave me a deeper understanding of historical events.

M: I guess we could _____ _____ in class, too.

W: Yeah, there are certainly better ways we could be assessed at school.

03

M: Okay, today's lesson is over now.

W: Thanks, Mr. Anderson. I'm enjoying playing the piano so much that I want to _____ _____ _____ too.

M: That would be great. Do you have a piano at home?

W: No, but I will get one soon.

M: Are you going to buy an acoustic piano or a digital one?

W: Actually, I wanted to _____ _____ _____ about that.

M: Do you live in a house or an apartment?

W: An apartment, with my family.

M: Then I suggest you buy a digital one.

W: Why?

M: Digital pianos allow you to practice using headphones so that you won't _____ _____ _____ _____.

W: That's a good point.

04

W: I can't believe we're seeing Big Ben.

M: What? That clock at the top of the tower? I've seen it before.

W: But only in pictures. Most people want to see it _____ _____.

M: That must be why those tourists are taking pictures in front of it.

W: Of course. Look at that boat on the River Thames.

M: Yes, it's _____ _____ _____ _____. Do you see that bus the woman is getting on?

W: That's a double-decker bus. They're common in London.

M: A bus with two stories? It looks strange.

W: I guess. Do you see the two birds by the tower?

M: I only see one bird. It's _____ _____ _____.

W: There's another one sitting on the tower. I think they live there.

05

M: Jane, all you do is watch TV!

W: Well, I'm tired when I get home after a long day at work.

M: I am, too. _____ _____ _____ that I always do the housework?

W: Well, I suppose not. But I don't know how.

M: Have you ever used the dishwasher?

W: No, I haven't.

M: Well, have you ever used the washing machine?

W: I did once, but I _____ _____ _____ _____. You were really angry, remember?

M: Look, just take this cloth and go around all the rooms _____ _____ _____ _____ the tables and shelves.

W: Sure. I can do that.

06

W: Long time no see, Thomas!

M: Hi, how are you?

W: I'm fine. How have you been lately?

M: Pretty good, except that I've been busy working two part-time jobs during the week.

W: Wow, that must be tiring! How can you manage to study and work _____ _____ _____ _____?

M: Actually, it's quite difficult. I'm thinking of _____ _____ _____.

W: Really? But there are some advantages to having a job. For example, you can earn money and gain practical experience.

M: That's true. But work takes time away from studying. So _____ _____ _____ _____ _____ this semester.

W: In that case, quitting is probably the right decision.

M: I agree. Anyway, I have to go to work. See you later!

07

W: Good morning, sir. Welcome to PC World.

M: Thanks. I want to buy a computer for my daughter. Could you help me?

W: Of course. Which model are you interested in?

M: How much is _____ _____ with the 17-inch LCD monitor?

W: That one is ₩750,000. The same computer but with a larger 19-inch monitor is ₩50,000 more.

M: I see. I think she'll also _____ _____ _____. Do you have any?

W: We have two models. The inkjet printer is ₩100,000 and the laser printer is ₩250,000.

M: Okay. I'll _____ _____ _____ and the 19-inch monitor.

W: Would you like to buy a printer, too?

M: Yes. I'll take _____ _____ _____, please.

08

W: Do you have any plans tonight?

M: No, I don't.

W: Well, a new movie was released last Thursday. Let's go see it.

M: What's it called?

W: It's called *Pirate Attack*. It _____ _____ _____ on my favorite movie website. People who saw it gave it nine and a half stars out of 10.

M: Is it an action movie?

W: Yes. It's supposed to _____ _____ _____. And it's in 3D!

M: Sounds exciting. What time is it showing?

W: There are three shows. The first one is at 6, and then there's one at 8 and 10. How about having dinner and then going to see the 8 p.m. show?

M: Well, do you know how long it is? My parents won't let me _____ _____ _____ _____.

W: It's 90 minutes long.

M: That sounds good then.

09

W: Please welcome our new teacher, Mrs. Winter. Let me _____ _____ _____ _____ _____. She graduated from the University of California

and went on to complete her graduate work there as well. Before coming to Harper, she taught at Norwood Junior High in North Sacramento. She will be teaching English here at Harper High School. _____ _____ _____ _____,
Mrs. Winter has been singing with the amateur city choir and has been performing in concerts at the City Convention Center. She _____ _____ _____, has five children and loves grading papers!

10

M: Let's look at this brochure to see where we should _____ _____ _____ _____.

W: Oh, I've always dreamed of going to Sydney.

M: So have I, but it's too expensive. Our budget is limited to $2,000.

W: I know. Well, that's all right. How many vacation days can you get?

M: I can't _____ _____ more than five days because it's our busy season.

W: I see. Well, that's fine, but we need a tour guide.

M: Hmm... I think it would be fun to decide what to do and what to eat ourselves.

W: But this is our honeymoon. I want to be able to rest without having to worry about planning things.

M: Okay, that's fine. Then there seems to be only two places we can go. Which destination do you prefer?

W: Oh, let's look at the free gifts. Hmm... I would definitely prefer a massage rather than wine.

M: I agree. I'll _____ _____ _____.

11

M: Lindsay, have you seen my wallet?

W: No, I haven't. Did you have it before lunch?

M: I think so. I guess I _____ _____ _____ _____ with me when I was leaving the classroom.

W: (Don't worry. I'm sure we can find it.)

12

W: Hello, Jackie! Are you still _____ _____ _____ _____?

M: Yes, we practice every Friday after school. Would you like to come and listen to us play?

W: That sounds great. You still play the drums, don't you?

M: (No, I play the electric guitar these days.)

13

W: What are you reading, Martin?

M: I'm looking for a good restaurant to take my parents to this Saturday. This magazine _____ _____ _____ _____. Look, have you been here?

W: Which one?

M: Here. It says this is one of the best Italian restaurants in Seoul.

W: The Sweet Garden? Oh, that's ridiculous. I've been there, and it was one of the worst Italian restaurants _____ _____ _____.

M: What makes you say so?

W: Well, the atmosphere was good, as it says. But the dishes were just average, especially considering the high prices. But the worst thing was the service!

M: What was wrong with it?

W: The waiters stood together at the counter and just _____ _____ _____ _____ without even taking care of customers. My food was served cold. I complained about it after dinner.

M: (I guess I should find another restaurant.)

14

M: What are you doing, Sarah?

W: I've been looking for my new scarf for over half an hour. Have you seen it, Dad?

M: Isn't that it over there?

W: Where?

M: _____ _____ _____ _____ next to your

coat.

W: No, that's my old scarf. My new scarf is red with _____ _____.

M: Oh. I think I saw Kelly wearing it this morning.

W: Really? Kelly was wearing my new scarf?

M: Yes. She said it _____ _____ _____ her new coat.

W: I can't believe it! I just bought that scarf yesterday.

M: Uh-oh. Try not to be too angry with her.

W: (I just wish she would ask me before borrowing my stuff.)

15

M: Mike and his grandmother are very close. She loves him very much. Next week, Mike will graduate from university. His whole family is planning to be there, including his grandmother, who must _____ _____ _____ _____ to get there. Unfortunately, two days before his graduation, Mike's grandmother becomes sick. It isn't serious, but it is enough to _____ _____ _____ _____ to the graduation ceremony. She calls Mike to explain her situation and to _____ _____ not being able to go. In this situation, what is Mike most likely going to say to his grandmother?

Mike: (Don't worry about it. Just get well soon.)

16~17

W: For more than 4,000 years, flags have helped people and organizations express themselves, send messages, and represent ideas. In the Middle Ages, knights went to battle wearing heavy armor, so it was difficult to tell who was who. So they started carrying flags _____ _____ _____ from their enemies. Today, almost every organization has a flag. The Girl Scouts, for example, use a golden trefoil on a blue background to symbolize the sun shining over the world's children, thus encouraging young women and girls to _____

_____ _____ throughout the world. Each state government of the United States has its own flag to represent the core values of the state. On the Olympic flag, the five circles correspond to the five continents. On the United Nations' flag, olive branches carrying the earth signify world peace. Even individuals may have flags. Kings and queens often have a personal flag that _____ _____ _____ _____ _____ they are currently in.

1번부터 17번까지는 듣고 답하는 문제입니다.
1번부터 15번까지는 한 번만 들려주고, 16번부터 17번까지는 두 번 들려줍니다. 방송을 잘 듣고 답을 하기 바랍니다.

01

다음을 듣고, 여자가 하는 말의 주제로 가장 적절한 것을 고르시오.
① 가정 교육의 중요성
② 집중력을 높이는 방법
③ 개인별 학습 방법의 차이
④ 정서 발달에 효과적인 놀이
⑤ 음악이 학습에 도움을 주는 이유

02

대화를 듣고, 남자의 의견으로 가장 적절한 것을 고르시오.
① 충동구매를 자제해야 한다.
② 사무실은 항상 깔끔해야 한다.
③ 쓰레기 매립지를 확장해야 한다.
④ 일회용 제품 사용을 줄여야 한다.
⑤ 가능한 한 물건을 재활용해야 한다.

03

대화를 듣고, 두 사람의 관계를 가장 잘 나타낸 것을 고르시오.
① 영양사 - 학생
② 심판 - 운동선수
③ 코치 - 육상 선수
④ 의사 - 환자
⑤ 연주자 - 지휘자

04

대화를 듣고, 그림에서 대화의 내용과 일치하지 않는 것을 고르시오.

05

대화를 듣고, 남자가 여자에게 부탁한 일로 가장 적절한 것을 고르시오.
① 입장권 받기
② 동료 병문안 가기
③ 시험 점수 매기기
④ 파티복 골라 주기
⑤ 파티장에 태워다 주기

06

대화를 듣고, 여자가 신용카드를 사용할 수 없는 이유를 고르시오.
① 신용카드를 분실해서
② 남자가 호텔 숙박료를 지불해서
③ 여자가 지난 달 결제 대금을 연체해서
④ 남자가 친구에게 줄 선물을 많이 사서
⑤ 남자가 여행 중에 신용카드를 갖고 가서

07

대화를 듣고, 여자가 지불할 금액을 고르시오. 3점
① $54 ② $63 ③ $64 ④ $69 ⑤ $75

08

대화를 듣고, 강연에 관해 두 사람이 언급하지 않은 것을 고르시오.
① 강연 시기
② 강사
③ 강연 주제
④ 강연 장소
⑤ 강연 길이

09

Angkor Wat에 관한 다음 내용을 듣고, 일치하지 않는 것을 고르시오. 3점
① 12세기 초 힌두교 사원으로 건설되었다.
② 외벽의 길이는 3.5km가 넘는다.
③ 안쪽에 5개의 탑이 둥글게 서 있다.
④ 처음에는 지금과 다른 이름이었다.
⑤ 캄보디아 국기에 그려져 있다.

10

다음 표를 보면서 대화를 듣고, 남자가 구입할 커피 원두를 고르시오.

	Coffee	Bean Type	Style	Roast Date	Price
①	A	Arabica	Whole	6 days ago	$18
②	B	Robusta	Whole	3 day ago	$22
③	C	Robusta	Ground	6 days ago	$25
④	D	Robusta	Ground	4 days ago	$30
⑤	E	Robusta	Ground	10 days ago	$35

11

대화를 듣고, 여자의 마지막 말에 대한 남자의 응답으로 가장 적절한 것을 고르시오.

① I'd love to. How old are they?
② You should take good care of them.
③ Really? How much did you pay for them?
④ It's fantastic. My cats like playing with me.
⑤ Right. Dogs do not get along well with cats.

12

대화를 듣고, 남자의 마지막 말에 대한 여자의 응답으로 가장 적절한 것을 고르시오.

① This building is open to all students.
② Since the rules are strict, nobody violates them.
③ They can use the vending machines in the lobby.
④ Yes. They should remind the students of the rules.
⑤ Sure. They will help the students check out books.

13

대화를 듣고, 남자의 마지막 말에 대한 여자의 응답으로 가장 적절한 것을 고르시오.

Woman: _____

① That's a good idea. I'll try it.
② I'm sure my parents will love you.
③ All of my friends have nice parents.
④ Why are you nervous? You'll be fine.
⑤ I hope I make a better impression this time.

14

대화를 듣고, 여자의 마지막 말에 대한 남자의 응답으로 가장 적절한 것을 고르시오.

Man: _____

① Sorry, but I've made other plans.
② I don't think our professor can come.
③ But we can't eat snacks in the library.
④ I think we should talk to our professor.
⑤ That sounds much better than the library.

15

다음 상황 설명을 듣고, Josh가 Ryan에게 할 말로 가장 적절한 것을 고르시오. 3점

Josh: _____

① That's not my hammer!
② There's a hole in the roof!
③ Thanks for being so helpful!
④ I'm not sure how to fix this hole!
⑤ Please walk slowly and carefully!

[16~17] 다음을 듣고, 물음에 답하시오.

16

여자가 하는 말의 목적으로 가장 적절한 것은?

① 사업 성공 비법을 조언하려고
② 새로 개원한 학원을 홍보하려고
③ 더 좋은 언어 학습법을 설명하려고
④ 제2 언어 학습의 중요성을 강조하려고
⑤ 새로운 비즈니스 영어 수업을 소개하려고

17

언급된 내용이 아닌 것은?

① 강좌 난이도 　② 수업 내용 　③ 보충 수업 제도
④ 등록비 　⑤ 할인 정책

녹음을 다시 한 번 듣고, 빈칸에 알맞은 말을 쓰시오.

01

W: When I was in school, I liked to listen to music while studying. My parents thought it was _____ and wouldn't allow it. What they didn't know was that it helped me study. Many people don't realize there is no _____ _____ to study. People have different _____ _____ _____. When I see my son studying, I get tense because he keeps moving around. To me it seems like a _____ _____ _____, but that's how he learns.

02

M: Don't throw out that shoebox.

W: Why? I don't need it for my shoes anymore.

M: I know. But we can use it for other things. It would be good for _____ _____ _____ _____.

W: You really hate throwing things out, don't you?

M: Yes. I always try to _____ _____ _____ for old items.

W: Is that because you like saving money?

M: That's part of it. But I'm more worried about _____ _____ _____. Everyone should try to get the most use out of things so they throw away fewer items.

W: Then I guess you don't like plastic forks and spoons.

M: Actually, I just wash them and use them again!

03

W: The big race is coming up in less than a month.

M: I know. I'm going to be doing a lot of training.

W: Now, don't _____ _____. I only want you to jog two miles in the morning.

M: All right.

W: And in the evenings, when it's cool, we'll work on your speed.

M: That sounds like a good plan.

W: Yesterday, you ran faster than you have in a while.

M: Yes. If I can do that again, I'll win the race.

W: You certainly will. Now, exercise isn't everything. You need to _____ _____ _____ over the next few weeks, too.

M: I know, lots of proteins and lots of carbohydrates.

W: That's right. I've got a special diet written out for you. It will help you _____ _____ _____ _____ your training and get ready for the big race.

M: Thanks. Well, let's get started.

04

M: It's so nice to take a walk in the park with you.

W: Yes, it is. We should come here more often.

M: Oh, there are three students in taekwondo uniforms _____ _____ _____ _____ _____ their coach.

W: The coach has the same kind of uniform, but his has a Korean flag on the sleeve.

M: Yeah, maybe we can _____ _____ _____ _____ and watch them.

W: Do you mean the bench by the tree?

M: Yes, it is in the shade.

W: It is, but I don't think we can sit there.

M: Why not? I think it would be fun if we pick some apples from the tree, too.

W: Yes, but that woman with a dog is _____ _____ _____ _____.

M: Oh, I see. She's probably going to sit here.

05

M: Those students are really excited about the big dance tonight.

W: Yes, they're all going to look so lovely in their dresses and suits.

M: Are you going to be working there?

W: No. I was _____ _____ _____ _____ to

volunteer to help out.

M: Oh, I see.

W: But I'm finished grading now, so I plan on stopping by to _____ _____ _____.

M: Well, did you hear that Arthur is ill?

W: No. Wasn't he supposed to volunteer?

M: Yes, now we are _____ _____ _____. Do you think you could take tickets at the door?

W: Oh sure. That would be fun. I don't know what I'm going to wear though.

M: Well, you don't need to dress up. I'll just wear my normal work clothes.

W: Good. Then I'll just do the same.

06

W: I tried to use our credit card today, but the cashier said we've reached our credit limit. Do you know what happened?

M: Oh, I'm sorry. That's my fault. I had to use the credit card unexpectedly last weekend. I'll get it _____ _____ tomorrow.

W: Last weekend? When you went camping with your friends? What happened?

M: Well, remember I told you that it rained all weekend?

W: Yes. You said _____ _____ _____ _____.

M: That's right. So we had to get a couple of hotel rooms.

W: Yes, you already told me all this. Are you saying you paid for everything?

M: Well, nobody brought a lot of cash. We didn't expect to need much while camping.

W: That makes sense.

M: So I paid for everything with our credit card. The guys will _____ _____ _____ when I see them tomorrow.

W: Okay. Now I understand. Just make sure to pay our credit card bill as soon as you get the money!

07

M: Hi, can I help you?

W: Yes, I'd like two tickets for tomorrow's play. How much are they?

M: That depends. Normal tickets are $20 each, but seats in the first three rows cost $10 more.

W: I'd like seats in the front. I don't mind paying extra.

M: All right. Are you a student? Students get a 20% discount.

W: No, I'm not. But _____ _____ _____ a theater membership card.

M: Okay. That gives you a 10% discount on each ticket.

W: Great. Here's my card.

M: Thanks. *[Pause]* Hmm... You can get a discount with your card now, but _____ _____ _____ it expires next month.

W: Does it? How much does it cost to renew my membership?

M: It's $10 for six months or $15 for one year.

W: Okay. _____ _____ _____ _____ _____ and renew my membership for a year.

08

W: I went to a fascinating lecture last Tuesday, Matthew. You would have liked it.

M: Really? _____ _____ _____ _____?

W: It was a professor from a university in Germany. He studies the universe.

M: Oh, I see. So did he talk about space travel?

W: Actually, his lecture was about whether it could be possible to _____ _____ _____.

M: It sounds like a science fiction movie.

W: Exactly. His ideas were very creative.

M: I'm surprised you liked it. Science isn't your strong point.

W: Yes, but he explained things very clearly. It was nearly _____ _____ _____, but I wasn't bored.

M: I wish I could have gone.

W: It was fun. Afterwards, a group of us went to the library and discussed it.

M: Next time you go to an interesting lecture, let me know!

09

M: Welcome to Angkor Wat, the largest religious temple in the world. It was built in the early 12th century _____ _____ _____ _____.
Later, however, it was changed into a Buddhist temple. Its outer wall is more than three and a half kilometers long. Inside, there are three long passageways, _____ _____ _____ five towers arranged in the shape of a cross. Originally, it was named Preah Pisnulok, but later this was changed to Angkor Wat, which means "city of temples".
It has become the national symbol of Cambodia and is even featured on the country's official flag. Now let's go inside and _____ _____ _____ _____.

10

W: Welcome to the Bean Factory. How can I help you?

M: Hi. I'd like to buy _____ _____ _____ _____ coffee.

W: Sure. The chart behind me shows today's choices.

M: Oh! I prefer coffee with a strong flavor. Is Arabica a good choice?

W: Actually, Robusta coffee has a stronger flavor. _____ _____ _____.

M: Really? Then I'll get that. And I don't have a coffee grinder, so I don't want whole beans.

W: All right. Do you have a freshness preference?

M: Yes. I'd like something that _____ _____ _____ _____ _____ _____ _____.

W: That's not a problem. We have a couple of choices that fit your needs.

M: Yes, I can see that. I guess I'll just get the cheaper one.

W: Certainly. Here you go.

11

W: Do you want a pet?

M: Sure, I would love to have a cat or a dog.

W: That's good. My dog had puppies, so I was wondering _____ _____ _____ _____.

M: (I'd love to. How old are they?)

12

M: Lucy, look at those students! They're taking drinks into the reading room.

W: Really? The library rules say that no food or _____ _____ _____.

M: Shouldn't the librarians do something?

W: (Yes. They should remind the students of the rules.)

13

W: I'm really nervous.

M: Why?

W: I'm meeting my boyfriend's parents for the first time tonight.

M: Meeting his parents? _____ _____ _____.

W: I know. I really hope they like me.

M: Of course they'll like you.

W: What if I _____ _____ _____? You know how I am when I'm nervous.

M: I've never heard you say anything stupid. Relax.

W: I just know it's going to be awful.

M: Well, _____ _____ _____ it will be. Just imagine you're meeting one of your girlfriend's parents.

W: (That's a good idea. I'll try it.)

14

M: We're going to talk to the professor about the
_____ _____. Will you be able to come?

W: When are you going?

M: During his office hours on Wednesday.

W: When are his office hours?

M: From 2 p.m. to 4 p.m.

W: Okay, I'll come. I have a lot of questions about this
assignment.

M: Yeah, _____ _____ _____. After we talk to
him, we should go to the library and get to work.

W: Good idea. But if you think it would be more
comfortable, we could _____ _____ _____
_____.

M: Actually, that would be great. I'll ask the others.

W: Then we can drink coffee and _____ _____.

M: (That sounds much better than the library.)

15

M: Josh and his son Ryan are on the roof _____
_____ _____. Josh is a little worried that Ryan
is going to hurt himself, since he has never worked
on a roof before. After a little while, Josh asks
Ryan to get a hammer from his toolbox on the
other side of the roof. _____ _____ _____
his father, Ryan gets up and runs across the roof
to the toolbox. Josh watches him nervously. When
he finds the hammer and stands up to come back,
Josh decides _____ _____ _____ to him. In
this situation, what will Josh most likely say to
Ryan?

Josh: (Please walk slowly and carefully!)

16~17

W: In today's world, speaking a second language is
an _____ _____ _____ _____. China is
quickly becoming an international business power.
So why not use your free time to learn business
Chinese? Little's Language Institute recently
opened and is offering classes in both Mandarin
and Cantonese. If you're just getting started, there
are several beginner courses that teach language
basics. And for those who already speak Chinese,
we offer business classes and _____ _____. All
of our classes focus on the specific vocabulary and
structures used in business situations, including
job interviews, presentations and negotiations.
While some schools teach business language skills
in a traditional classroom, our classes are held in
an office environment. And if your busy schedule
forces you to _____ _____ _____ _____,
you can take up to three make-up classes at no
extra charge. What's more, if you register for two
or more classes a month, you'll _____ _____
_____ _____. If you want to learn more, stop
by any evening to sample one of our classes.

1번부터 17번까지는 듣고 답하는 문제입니다.
1번부터 15번까지는 한 번만 들려주고, 16번부터 17번까지는 두 번 들려줍니다. 방송을 잘 듣고 답을 하기 바랍니다.

01

다음을 듣고, 남자가 하는 말의 목적으로 가장 적절한 것을 고르시오.
① 반려동물 가게를 홍보하려고
② 반려동물 기르기를 장려하려고
③ 조깅 동호회 모임에 초대하려고
④ 동물 보호 협회 가입을 권유하려고
⑤ 우울증에 대한 경각심을 고취하려고

02

대화를 듣고, 여자의 의견으로 가장 적절한 것을 고르시오.
① 주차장 보수가 시급하다.
② 건물 관리인을 충원해야 한다.
③ 고객 불만 사항은 즉시 처리해야 한다.
④ 건물에 충분한 주차 공간을 확보해야 한다.
⑤ 건물 주인에게 자동차 수리비를 요구해야 한다.

03

대화를 듣고, 두 사람의 관계를 가장 잘 나타낸 것을 고르시오.
① 변호사 - 고객   ② 호텔 접수원 - 투숙객
③ 교수 - 학생   ④ 부동산 중개인 - 손님
⑤ 고용인 - 고용주

04

대화를 듣고, 그림에서 대화의 내용과 일치하지 않는 것을 고르시오.

05

대화를 듣고, 두 사람이 할 일로 가장 적절한 것을 고르시오.
① 은행 가기   ② 돈 빌리기   ③ 이력서 쓰기
④ 선물 사러 가기   ⑤ 아르바이트 가기

06

대화를 듣고, 여자가 소풍에 갈 수 없는 이유를 고르시오.
① 지붕을 수리해야 해서
② 가족 중 한 명이 다쳐서
③ 집에 소지품을 두고 와서
④ 소풍 도시락을 준비해야 해서
⑤ 수리점에 맡긴 물건을 가져와야 해서

07

대화를 듣고, 남자가 지불할 금액을 고르시오. 3점
① $13   ② $16   ③ $17   ④ $18   ⑤ $20

08

대화를 듣고, Peter and the Wolf에 관해 언급되지 않은 것을 고르시오. 3점
① 해설자의 역할   ② 악기의 역할   ③ 등장인물
④ 줄거리   ⑤ 평가

09

earthship에 관한 다음 내용을 듣고, 일치하지 않는 것을 고르시오.
3점
① 폐타이어를 쌓아 올려 벽을 세운다.
② 벽면은 진흙과 짚으로 채워져 있다.
③ 사방에 유리창이 있어 집 안에 햇빛이 잘 들어온다.
④ 햇빛 이외의 난방원은 필요하지 않다.
⑤ 저소득층 가정에 아주 유용하다.

10

다음 표를 보면서 대화를 듣고, 여자가 구입할 여행안내서를 고르시오.

Top 5 Guidebooks for Spain

	Book	Year	Map	Customer Rating	Availability
①	A	2014	X	★★	Out of stock
②	B	2015	X	★★★★★	In stock
③	C	2016	O	★★★	In stock
④	D	2016	O	★★★★	In stock
⑤	E	2017	O	★★★★	Out of stock

11

대화를 듣고, 남자의 마지막 말에 대한 여자의 응답으로 가장 적절한 것을 고르시오.

① Yes, I can lend you my bike.
② Sure. Let's go for a long ride.
③ Yes, I'd love to go shopping with you.
④ Yes, swimming in the river sounds fun.
⑤ No, motorcycles are not allowed in this park.

12

대화를 듣고, 여자의 마지막 말에 대한 남자의 응답으로 가장 적절한 것을 고르시오.

① I'll use a special program.
② Thank you for your compliment.
③ It was hard to answer that question.
④ I'm sure I'm going to pass this math exam.
⑤ It took me one week to finish this homework.

13

대화를 듣고, 남자의 마지막 말에 대한 여자의 응답으로 가장 적절한 것을 고르시오.

Woman: _____

① All right, we'll meet you there.
② Okay, we can make plans then.
③ Well, you should apologize to her.
④ I'm sorry for changing the schedule.
⑤ Sure, she'll be happy to hear the news.

14

대화를 듣고, 여자의 마지막 말에 대한 남자의 응답으로 가장 적절한 것을 고르시오.

Man: _____

① I'd like to have the chicken soup and salad.
② Oh, I didn't copy it. It's my original recipe.
③ What time does your housewarming party start?
④ If you give me the recipe, we can make it together.
⑤ Then I'll show you how to make it next time you come over.

15

다음 상황 설명을 듣고, 아버지가 Emily와 그녀의 남동생에게 할 말로 가장 적절한 것을 고르시오.

Father: _____

① What a mess you made!
② The play was terrible. What a day!
③ I cannot believe you kids ate all of that food.
④ It was a nice party. Thank you for inviting us.
⑤ How sweet! I won't say a word to your mother.

[16~17] 다음을 듣고, 물음에 답하시오.

16

남자가 하는 말의 주제로 가장 적절한 것은?

① the reason why sugar causes cavities
② various tooth-strengthening exercises
③ the best way to have your cavities filled
④ food and drinks that damage your teeth
⑤ how to keep your teeth clean effectively

17

언급된 음식이 <u>아닌</u> 것은?

① ice   ② alcohol   ③ bread
④ potato chips   ⑤ chocolate

녹음을 다시 한 번 듣고, 빈칸에 알맞은 말을 쓰시오.

01

M: People often _____ _____ even when they are with other people. And when children leave home or when parents or partners die, people often feel _____ _____. Pets will not only fill up your emptiness but also make you happy. Some people say they don't like pets because they are dirty and _____ _____ _____ _____ to take care of their needs. But owning a pet will make you more active and diligent, since you have to give them food, bathe them and take them for walks. You may even become friends with neighbors who have pets. This is why I think people should have pets.

02

W: Have you seen the state of the parking lot lately?
M: Yeah. It's full of holes.
W: I _____ _____ _____ _____ in one this morning.
M: It's very inconvenient.
W: That's not all. It makes a bad first impression on our customers. When they drive up to our building, the first thing they see is a _____ _____ _____.
M: So what are you going to do about it?
W: I'm going to let the building manager know that this is not acceptable. He needs to _____ _____ _____ _____ as soon as possible.
M: I hope he listens to you.
W: Me too. I'm going to call him right now.

03

W: How can I help you?
M: I'm looking for a one-bedroom apartment. I saw your advertisement.
W: Let me see if _____ _____ _____ _____ _____.

[Pause] No, sorry. All the one-bedrooms are taken.
M: Already?
W: Well, most contracts _____ _____ at least two months before school starts.
M: Are there any other options?
W: We still have several studios.
M: A studio would be fine. How much are they per month?
W: The _____ _____ _____ is $500, and you need to pay a $500 _____ _____ today.
M: Okay. Is it a one-year contract or a six-month contract?
W: We only offer one-year contracts.

04

W: This ski resort is awesome! From the top, we'll be able to see all the mountains around the resort.
M: Yeah, but it looks like we'll have to wait a long time to get on the ski lift.
W: Yeah. The line for the ski lift is quite long.
M: By the way, where are Ben and his son?
W: Over there. He is helping his son _____ _____ _____ _____.
M: Oh, isn't his son cute? That snowboard next to Ben looks new. I like _____ _____ _____ on it.
W: Ben said his wife, Carrie, gave it to him for his birthday. Oh, here she comes.
M: Oh, is she the woman _____ _____ _____?
W: Yes. Let's go say hi to her.

05

W: Ethan, do you know that Parents' Day _____ _____ _____?
M: Yeah, I know. What are you going to get Mom and Dad?
W: Well, actually, I don't have any money for gifts.
M: You're so bad at _____ _____!

W: I know. Maybe we can give them a gift from both of us. I can _____ _____ _____ later.

M: Sorry, but no. I don't want to keep giving you money! Why don't you get a part-time job?

W: Okay, fine. But I'm not good at finding jobs. I don't know where to start.

M: I'll help you out. Let's start with _____ _____ _____.

W: Okay.

06

[Cell phone rings.]

W: Hello, Blake.

M: Hello, Christine. Are you ready to _____ _____ _____ _____?

W: Sorry, but I don't think I can go.

M: Why? What's going on?

W: A part of my house's roof collapsed, so I need to _____ _____ _____.

M: That's really shocking. I hope no one was hurt.

W: Fortunately, everyone is okay. But we have to fix it as soon as possible.

M: That's for sure. Actually, I made food for us last night. I'm sorry we can't eat it together.

W: You did? _____ _____ _____. You should share it with some of your friends.

M: I will. So, have you contacted a repairman?

W: Yes, I have. He's on his way.

07

M: Hi, I need some notebooks for my class.

W: Okay. We have lots of notebooks here.

M: How much is that red one?

W: It costs $3 and has 100 pages.

M: I need _____ _____ _____. Do you have any 200-page notebooks?

W: Certainly. The orange notebooks have 200 pages and cost $5. There are two types: with lines and

without lines.

M: I see. Do you have a 300-page notebook with lines?

W: Yep. _____ _____ _____ have 300 pages and cost $7.

M: Okay. I'll take _____ _____ _____ with lines and one 300-page notebook with lines.

W: Okay, so three notebooks. Here you go.

M: Thank you.

08

M: How was the concert?

W: It was great. I really enjoyed *Peter and the Wolf.*

M: *Peter and the Wolf*? It _____ _____ _____ _____ _____.

W: It is. A narrator tells a story about a boy and a bird who catch a wolf. While the narrator talks, an orchestra plays music to _____ _____.

M: What is the music like? It must be exciting enough to entertain children.

W: Yes, it is. The different instruments of the orchestra all sound like the characters in the story.

M: Really? What instrument sounds scary like a wolf?

W: That's the French horn. The flute sounds like a bird, the oboe sounds like a duck, and the clarinet sounds like a cat.

M: What about Peter?

W: Oh, the strings play a happy tune for Peter. He's my favorite. And the bassoon plays Peter's grandfather. He sounds very old and grumpy.

M: It sounds like a lot of fun.

W: Yes, _____ _____ a children's classic.

09

W: Though they might sound like they're from outer space, earthships are a reality here on earth. These environmentally-friendly structures are made from used rubber tires. About 2,000 tires are needed for a small home. The tires _____ _____ _____ to create walls, and they're packed with dirt and straw to keep the heat in. Glass windows cover the front of the home to _____ _____ _____. The walls of the homes trap heat so well that no heat source other than sunlight is required. Earthships are equipped with solar panels so that electricity is also unnecessary. These homes cost much less than traditional homes, so they're _____ _____ _____ _____ as well as environmentalists.

10

W: I need a travel guide for my trip to Spain. What do you think of this one?

M: It looks good, but check out _____ _____ _____ _____ _____.

W: Oh, it's too old! I want one published in 2015 or later.

M: Yes. Otherwise the information will be out of date.

W: I should also get a map of Spain so _____ _____ _____ _____.

M: I agree. Maps are very useful. Some of these guidebooks have a map included.

W: That's great. And what are these stars next to the name of each book?

M: That's the website's customer rating. Five stars is the highest rating.

W: Really? Well, I want a book with a rating of more than three stars.

M: Then this one looks like the best choice.

W: Yes, but _____ _____ _____ _____. I'm leaving soon, so I want something that can be delivered right away.

M: Oh. In that case, order this one instead.

11

M: It's so nice to be by the riverside. The sound of the water makes me feel relaxed.

W: Me too. I love taking walks around places like this.

M: Look! There's a rental shop over there. Do you want to _____ _____?

W: (Sure. Let's go for a long ride.)

12

W: Max, are you using your computer to do your math homework?

M: Yes, I _____ _____ _____ the answers and make graphs.

W: Wow, that sounds complicated. How are you going to do it?

M: (I'll use a special program.)

13

[Cell phone rings.]

W: Hello? David! It's good to hear from you.

M: Hi, Mom. Sorry, it's been too long since my last call. _____ _____ _____ _____ for my exams.

W: That's okay. How do you like your new roommate?

M: He's nice. I like him. By the way, is it this weekend when you're going to take a trip with Aunt Gina?

W: Well, we were going to drive to San Diego. But Gina doesn't _____ _____, so we won't go.

M: Oh, that's too bad. Then can I come home to see you? The midterm exams are finally over.

W: Really? It would be wonderful to see you again.

M: Why don't we do something together? We can take a drive or go to the park.

W: That sounds great. But can we talk about it later? I have to run now. I _____ _____ _____ _____ in 20 minutes.

M: Okay. Then I'll give you a call tomorrow.

W: (Okay, we can make plans then.)

14

W: I really enjoyed myself at your _____ _____, Jake.

M: I'm glad you did. How did you like the food?

W: The food was delicious. What was in that salad _____ _____?

M: Oh, the salad. That was made with Cajun fried chicken.

W: _____ _____ on it was great. I've never tasted anything like it before.

M: I think everyone liked it. It was eaten pretty quickly.

W: Could you teach me how to make it?

M: I'll _____ _____ _____ _____ if you'd like.

W: I don't follow recipes well. I always get something wrong.

M: (Then I'll show you how to make it next time you come over.)

15

W: Emily and her brother decided to do something special for their parents' 20th _____ _____. In the afternoon, they sent their parents to the theater to see their favorite play. While their parents were gone, Emily and her brother carefully _____ _____ _____ _____ to serve in time for their parents' return. They bought their father's favorite wine, too. But before they had put the _____ _____ on the meal, their father unexpectedly came back home to get his wallet and saw what they had made. In this situation, what would Emily's father most likely say to Emily and her brother?

Father: (How sweet! I won't say a word to your mother.)

16~17

M: My name is Dr. Kim, and I'm a dentist. There are many things we all need to do _____ _____ _____ _____ _____ _____. These include watching what we eat. Everyone knows that sugar is bad for our teeth, but there are other surprising foods and drinks that can create problems. Even ice can cause _____ _____ _____ _____ _____ _____. Ice is simply frozen water, but biting it can damage your teeth. Alcoholic drinks are another problem. Alcohol makes your mouth dry, and cavities form more easily in dry mouths. And did you know that bread causes cavities? The starch in bread is turned into sugar _____ _____ _____, and sugar causes cavities. The same is true of potato chips. They get stuck between your teeth, and their starch gradually becomes sugar. Avoiding these foods and drinks as much as possible is a good idea. And, of course, _____ _____ _____ after every meal!

1번부터 17번까지는 듣고 답하는 문제입니다.
1번부터 15번까지는 한 번만 들려주고, 16번부터 17번까지는 두 번 들려줍니다. 방송을 잘 듣고 답을 하기 바랍니다.

01

다음을 듣고, 여자가 하는 말의 목적으로 가장 적절한 것을 고르시오.
① 공연 프로그램을 소개하려고
② 공연 일정에 대해 안내하려고
③ 비상시 대피 요령을 알려주려고
④ 공연장 내 편의시설 이용을 권장하려고
⑤ 공연장에서 지켜야 할 규칙을 설명하려고

02

대화를 듣고, 두 사람이 하는 말의 주제로 가장 적절한 것을 고르시오. 3점
① 풍력 에너지의 장단점
② 다양한 친환경 에너지
③ 에너지원 고갈의 심각성
④ 환경 오염을 줄이는 방법
⑤ 야생 동물 보호를 위한 노력

03

대화를 듣고, 두 사람의 관계를 가장 잘 나타낸 것을 고르시오.
① 교사 – 학생 ② 교사 – 교장 ③ 학생 – 교장
④ 교사 – 학부모 ⑤ 학생 – 학부모

04

대화를 듣고, 그림에서 대화의 내용과 일치하지 <u>않는</u> 것을 고르시오.

05

대화를 듣고, 남자가 여자에게 부탁한 일로 가장 적절한 것을 고르시오.
① 이사 도와주기 ② 이사할 집 알아보기
③ 영화 보러 함께 가기 ④ 동생 돌보기
⑤ 집들이 준비 같이 하기

06

대화를 듣고, 여자가 경찰관을 찾아간 이유를 고르시오.
① 꽃을 배달해 주기 위해서
② 꽃가게 위치를 알려주기 위해서
③ 손님이 가게에서 난동을 부려서
④ 이웃 주민들 간에 싸움이 벌어져서
⑤ 십 대들이 가게 앞에서 소란을 피워서

07

대화를 듣고, 남자가 지불할 금액을 고르시오. 3점
① $780 ② $810 ③ $1,080
④ $1,200 ⑤ $1,700

08

대화를 듣고, Art World에 관해 언급되지 <u>않은</u> 것을 고르시오.
① 노래 길이 ② 노래 내용 ③ 원곡 가수
④ 음원 판매량 ⑤ 인기 있는 이유

09

internship fair에 관한 다음 내용을 듣고, 일치하지 <u>않는</u> 것을 고르시오. 3점
① 오전 10시부터 오후 5시까지 진행된다.
② 기업 대표들과 대화할 기회가 있다.
③ 온라인 사전 등록을 해야 참여할 수 있다.
④ 행사 이전에 이력서 견본을 제출해도 된다.
⑤ 기업 대표들이 제출된 이력서를 검토할 것이다.

10

다음 표를 보면서 대화를 듣고, 남자가 신청할 봉사활동 프로그램을
고르시오.

International Volunteer Programs

	Program	Region	Type of Service	Length	Cost
①	A	Africa	Child Care	4 weeks	$700
②	B	Africa	Education	7 weeks	$1,000
③	C	Europe	Arts	3 months	$2,200
④	D	Europe	Nursing	1 year	$5,000
⑤	E	Asia	Medical Aid	6 months	$3,000

11

대화를 듣고, 여자의 마지막 말에 대한 남자의 응답으로 가장 적절한
것을 고르시오.

① I want to play another game.
② Not everyone is good at teaching.
③ Chess pieces are either black or white.
④ You can use my chessboard and pieces.
⑤ Don't worry. The game's rules are simple.

12

대화를 듣고, 남자의 마지막 말에 대한 여자의 응답으로 가장 적절한
것을 고르시오.

① I'll go and talk to them.
② They are mopping the floor now.
③ I'm looking for a new place to live.
④ You'd better walk quietly in your room.
⑤ Getting good sleep makes you feel better.

13

대화를 듣고, 남자의 마지막 말에 대한 여자의 응답으로 가장 적절한
것을 고르시오.

Woman: _____

① How much did that app cost?
② I can't believe how hard it's raining!
③ Maybe we should go to the beach instead.
④ Well, weather forecasts are not always accurate.
⑤ May I have the paper when you're done with it?

14

대화를 듣고, 여자의 마지막 말에 대한 남자의 응답으로 가장 적절한
것을 고르시오.

Man: _____

① I don't need to study for this test.
② I don't know where the library is.
③ Derrick called to cancel our plans.
④ I can't concentrate with all the noise there.
⑤ I'm going to meet some friends and study with
them.

15

다음 상황 설명을 듣고, Jason이 승무원에게 할 말로 가장 적절한 것
을 고르시오.

Jason: _____

① That's great. Thanks so much.
② No, thank you. This seat is just fine.
③ Thanks anyway. I appreciate your effort.
④ I'm sorry, but there aren't any empty seats.
⑤ That's okay. My baby will stop crying soon.

[16~17] 다음을 듣고, 물음에 답하시오.

16

남자가 하는 말의 주제로 가장 적절한 것은?

① the necessity of a policy change
② the importance of family and friends
③ the efforts against canceling a concert
④ the reasons for changing a festival venue
⑤ the conflicts surrounding an entrance fee

17

언급된 정보가 아닌 것은?

① 개최 시기　　② 개최 장소　　③ 입장료
④ 행사 내용　　⑤ 경품 내용

녹음을 다시 한 번 듣고, 빈칸에 알맞은 말을 쓰시오.

01

W: Can I have your attention, please? I've been asked to _____ _____ _____ _____ to you before we go inside. First of all, you must remain quiet during the performance. When the orchestra is playing, there is to be no talking. Between songs, you may speak, but only if you whisper. You _____ _____ _____ _____ at any time. Also, all cell phones should be turned off now. Any student who doesn't _____ _____ _____ will be asked to leave the auditorium.

02

M: They're putting up some wind turbines in the fields near my grandmother's house.

W: I'm seeing more and more turbines these days.

M: A lot of people are _____ _____ _____ wind energy because it doesn't create any pollution.

W: Right. And since the source of energy is the wind, it never runs out.

M: Of course, there are some _____ _____ _____ _____.

W: I can't think of any.

M: For starters, the amount of wind in an area is not very constant. That makes it difficult to estimate how much energy can be produced.

W: It's still better than nothing. And it doesn't harm anyone.

M: Well, maybe not people, but what about wildlife? Some studies show that birds and bats can be _____ _____ _____ _____.

W: I guess I didn't think of that.

03

M: Excuse me, Ms. Thompson. Can I speak with you for a moment?

W: Of course. Come into my office. And please, call me Barbara.

M: Thank you, Barbara.

W: So how was your first week of classes here at Stratford High School?

M: It _____ _____ _____. The students are great, and the other teachers have been very helpful.

W: That's good to hear. I've worked very hard to _____ _____ _____ _____ _____ _____ for everyone.

M: You've done a great job. I'm very happy I took this job.

W: Good. But is there something wrong? You look concerned.

M: Well, it's just my schedule. I'm worried that teaching six classes a day is going to _____ _____ _____ _____ _____.

W: I understand that it's difficult. But I think it'll get easier as time goes by.

M: I hope so. I really want to do my best for all my students.

04

M: We made it! Isn't the view amazing?

W: Yes. That was a tough climb, but we're finally at the top. The palace down there looks so beautiful with its traditional architecture and roofs.

M: And look at all the buildings behind it!

W: Those skyscrapers _____ _____ from up here.

M: Hmm... The main road is almost empty. That's surprising.

W: It has been closed because of a marathon.

M: I see. Oh, can you see that building on the left? The tall apartment building? It's very famous _____ _____ _____ _____ _____.

W: It is pretty neat. But you know what my favorite part of the view is?

M: What's that?

W: The tower _____ _____ _____ _____ the mountain behind the buildings.

M: Yes, I like that too.

05

M: Lily, can you do me a big favor?

W: I don't know. What's up?

M: I'm moving to a new apartment this Friday and I need somebody to help me.

W: Do you have a lot of _____ _____ _____?

M: Not so much. And my new apartment is only a few blocks from my old one.

W: What floor is your new apartment on?

M: Well... It's on the fourth floor. And _____ _____ _____.

W: No elevator? You're kidding!

M: I wish I were. That's why I really need your help.

W: Okay. I'll help you, but you have to do me _____ _____ _____ _____.

M: Sure, anything.

W: Good. Can you _____ _____ _____ _____ tomorrow night while I'm at the movies?

06

W: Excuse me, officer. Can I speak to you for a moment?

M: Sure. What can I do for you?

W: My name is Carol Harris, and I own the flower shop on the corner.

M: First Avenue Flowers? I know the place. I bought my wife roses there on Valentine's Day.

W: Ah. I thought you looked familiar. Anyway, I have _____ _____ _____ _____ _____.

M: What's wrong?

W: There's a large group of teenagers on the sidewalk in front of my shop.

M: Have they been causing trouble?

W: Not really. But they've been playing music and talking loudly all afternoon. I think they're _____ _____ _____ _____ _____.

M: I see. I'll go talk to them. If they want to _____ _____, they should go to the park.

W: Thank you very much, officer.

07

[Telephone rings.]

W: Capitol Travel. What can I do for you?

M: I'd like to reserve a round-trip plane ticket from Incheon to Madrid.

W: When are you leaving?

M: I'm leaving on January 15. I'll stay there for ten days.

W: I see. An economy ticket for a direct flight is $1,200. It's $500 more for business class.

M: I definitely _____ _____ _____ _____, but that's still expensive.

W: Are you a student? We're offering a 10% discount on direct flights to students in January.

M: Great! I'm a university student. But it's still _____ _____ _____ _____.

W: Hmm... If your schedule permits, how about changing the flight to include a stopover in Istanbul? It'll cost $300 less than the discounted price, and the layover is only three hours.

M: _____ _____ _____ _____ a better deal. I'd like to reserve that ticket.

W: Okay. Would you like to purchase travel insurance as well? It's only $50.

M: No, thanks. I'm already insured.

08

W: I just got the Guitar Girls' new album.

M: Is it good?

W: Yes. The best song is "Art World". It's nearly _____ _____ _____.

M: "Art World"? Do you mean the song about how beautiful the earth is?

W: Yes, that's it!

M: That's not an original song. It was recorded by

George Kim about 20 years ago.

W: I didn't know that. Did he write it?

M: He wrote the music, but I don't know _____
_____ _____ _____. Anyway, his version
wasn't very popular.

W: Really? Well, it's going to be a big hit for the
Guitar Girls. Everyone is talking about it.

M: I wonder why.

W: It's not surprising. It's a song about the
environment, which is something people are really
interested in these days.

M: I guess _____ _____ _____.

09

M: Winter vacation is quickly approaching. So our
university is having an internship fair to help
you _____ _____ _____. If you want to get
some work experience during the winter break, be
sure not to miss this event. The fair will be held
on January 25, from 10 a.m. to 5 p.m. in Wilson
Auditorium. Representatives from companies in
many different fields will be present at the event.
The fair will give you a chance to talk with them to
get _____ _____ _____. There is no entrance
fee, but you should register online by January 18 to
participate. If students submit a sample résumé to
the career services office _____ _____ _____
_____, it will be reviewed by a career counselor
and returned with helpful suggestions.

10

W: What are you looking at, Jacob? Is that from the
volunteer service?

M: Yes. It's a list of their _____ _____ _____. I'm
trying to choose one.

W: I see. Is there a certain region that you want to go
to?

M: Not really. But I don't want to go to Asia. It's too
far away.

W: What kind of volunteer service do you want to get
involved in? You're interested in education, aren't

you?

M: Yes, but anything is okay, except arts. I don't have
_____ _____ _____.

W: All right. Some of these programs last a long time.
Six months is too long, isn't it?

M: No. _____ _____ _____ _____ _____ is
fine.

W: Well, then you should choose between these two.
This one is cheaper.

M: Yes, but I don't think four weeks is enough time.
I want something longer than that.

W: Then this is your best choice. It looks exciting,
doesn't it?

M: Yes! I think this will be an amazing experience.

11

W: Can you teach me _____ _____ _____
_____?

M: Of course. Chess is a very exciting game. I'm sure
you'll become good at it.

W: But I don't even know the basic rules!

M: (Don't worry. The game's rules are simple.)

12

M: Who's staying on the second floor of our
dormitory?

W: A group of new students moved in. Why?

M: They _____ _____ _____ _____ _____
every night. I can't get to sleep!

W: (I'll go and talk to them.)

13

M: Uh-oh. It looks like we're going to have to _____
_____ _____ this afternoon.

W: Why? It's such a perfect day.

M: Yes, but it's going to rain later.

W: _____ _____ _____. Look out the window.
There isn't a cloud in the sky.

M: Hmm. I suppose there isn't. That's strange.

W: I thought you wanted to go on a picnic. You said you thought it would be romantic.

M: I do think it would be romantic. I just don't want to _____ _____ _____.

W: I don't get it. What makes you think it is going to rain?

M: It says so right here on this weather app.

W: (Well, weather forecasts are not always accurate.)

14

W: Are you going to study for the math test tonight?

M: Of course. That test is half of our _____ _____. I need to do well.

W: Tell me about it. If I don't get a good grade in this class, I _____ _____ _____.

M: I know you'll do great.

W: I hope so. I'm meeting Derrick tonight to study together. Do you want to join us?

M: Sounds good. Where are you meeting?

W: At the campus coffee shop.

M: Oh, no thanks, then. I think I'll just _____ _____ _____ _____.

W: What made you change your mind?

M: (I can't concentrate with all the noise there.)

15

W: _____ _____ _____ _____ from a long business trip to Moscow, Jason is sitting in front of a couple with a baby. As soon as the airplane takes off, the baby begins to cry and doesn't stop. Jason asks the flight attendant if he can _____ _____ _____ _____. The flight attendant tells him there are no empty seats, but she can talk to the other passengers and see if someone will _____ _____ _____ him. After 10 minutes, she comes back and says she's sorry but no one wants to sit near the crying baby. In this situation, what is Jason most likely to say to the flight attendant?

Jason: (Thanks anyway. I appreciate your effort.)

16~17

M: May I have your attention, please? As you already know, our annual school festival will be held next Saturday at the town park. This is a popular event that is attended by hundreds of students and their families each year. We've _____ _____ _____ this year, however, about our decision to charge visitors $2 to enter the festival. In the past there was no fee, and people have been asking me why this has changed. The simple answer is that the festival _____ _____ _____ _____ _____. It used to be held on the school soccer field, but there simply wasn't enough room. Now we need to _____ _____ _____ to use the park. What's more, there will be three concerts, lots of food and several fun games and rides at the festival this year. None of this would be possible without the money we will collect from people _____ _____ _____. Thank you for understanding, and I hope you all enjoy the festival!

1번부터 17번까지는 듣고 답하는 문제입니다.
1번부터 15번까지는 한 번만 들려주고, 16번부터 17번까지는 두 번 들려줍니다. 방송을 잘 듣고 답을 하기 바랍니다.

01

다음을 듣고, 남자가 하는 말의 목적으로 가장 적절한 것을 고르시오.
① 책을 추천하려고
② 진로에 대해 조언하려고
③ 일자리 정보를 제공하려고
④ 독서의 중요성을 강조하려고
⑤ 효과적인 구직 요령을 소개하려고

02

대화를 듣고, 여자의 의견으로 가장 적절한 것을 고르시오.
① 옥상에 채소를 심자.
② 새 아파트로 이사하자.
③ 채소를 좀 더 자주 먹자.
④ 다양한 취미 활동을 즐기자.
⑤ 전원생활을 할 수 있는 곳을 찾아보자.

03

대화를 듣고, 두 사람의 관계를 가장 잘 나타낸 것을 고르시오.
① 집주인 - 세입자   ② 공사장 인부 - 건축가
③ 수리공 - 집주인   ④ 부동산 중개인 - 고객
⑤ 판매원 - 관리자

04

대화를 듣고, 그림에서 대화의 내용과 일치하지 않는 것을 고르시오.

05

대화를 듣고, 여자가 남자에게 부탁한 일로 가장 적절한 것을 고르시오.
① 전화 대신 받아주기
② 여자의 남편 간호하기
③ 여자를 병원에 데려다주기
④ 사장의 연락처 알려주기
⑤ 사장에게 문자메시지 보내기

06

대화를 듣고, 남자가 여자를 방문한 이유를 고르시오.
① 면담 시간을 변경하려고
② 시험 날짜를 확인하려고
③ 수업 내용에 대해 질문하려고
④ 수업에 늦은 이유를 설명하려고
⑤ 과제 마감 시간 연장을 부탁하려고

07

대화를 듣고, 여자가 지불할 금액을 고르시오.
① $23   ② $26   ③ $28   ④ $33   ⑤ $38

08

대화를 듣고, Mr. Peterson에 관해 언급되지 않은 것을 고르시오.
① 존경하는 인물   ② 교사가 된 이유
③ 출신 학교   ④ 취미
⑤ 교사로서의 경력

09

daylight saving time에 관한 다음 내용을 듣고, 일치하지 않는 것을 고르시오. 3점
① 1900년대 캐나다에서 처음으로 시행되었다.
② 1차 세계대전 중에 인기를 얻었다.
③ 모든 전문가가 에너지 절약 효과가 있다는 데 동의한다.
④ 자동차 사고는 일광 절약 시간 중에 감소한다.
⑤ 심장마비 발병률은 제도 시행 첫 주에 증가한다.

10

다음 표를 보면서 대화를 듣고, 남자가 구매할 텐트를 고르시오.

	Tent	Style	Size	Set Up	Weight
①	A	Tunnel	2-person	Pop-up	2.8 kg
②	B	Tunnel	3-person	Pop-up	3.6 kg
③	C	Dome	3-person	Regular	6 kg
④	D	Dome	4-person	Pop-up	4.8 kg
⑤	E	Pyramid	4-person	Regular	6.5 kg

11

대화를 듣고, 남자의 마지막 말에 대한 여자의 응답으로 가장 적절한 것을 고르시오.

① I think I know where it is.
② How long is the ferry ride?
③ Let's just walk. It's a nice day.
④ I think we're on the wrong bus.
⑤ Well, there is nothing to see there.

12

대화를 듣고, 여자의 마지막 말에 대한 남자의 응답으로 가장 적절한 것을 고르시오.

① Sure. I'll show you a cheaper one.
② Yes. I got a 10% discount on my purchase.
③ Sorry. The price has already been reduced.
④ Wow! I didn't know you had an expensive painting.
⑤ You can look around for free if you are an art student.

13

대화를 듣고, 여자의 마지막 말에 대한 남자의 응답으로 가장 적절한 것을 고르시오.

Man: _____

① No, I don't think age is important.
② Right. Some people are just very lucky.
③ Yes. Never stop trying and you'll succeed.
④ It is. You need to learn the secrets of business.
⑤ I don't know. I've received a lot of good advice.

14

대화를 듣고, 남자의 마지막 말에 대한 여자의 응답으로 가장 적절한 것을 고르시오.

Woman: _____

① Do you mind if I try them out?
② I promise I won't touch anything!
③ Shh! The filming is about to begin.
④ I'll get you some autographs from the cast.
⑤ How much did you pay for the equipment?

15

다음 상황 설명을 듣고, Steven이 학생에게 할 말로 가장 적절한 것을 고르시오. 3점

Steven: _____

① Why didn't you take any notes during class?
② Do you know how I can register for this class?
③ When does this year's summer vacation begin?
④ Do you have a translation program for Chinese?
⑤ Please tell me what we have to do for the next class.

[16~17] 다음을 듣고, 물음에 답하시오.

16

여자가 하는 말의 주제로 가장 적절한 것은? 3점

① the paved roads of ancient Rome
② the water system of ancient Rome
③ the features of an ancient Roman city
④ the architectural style of ancient Rome
⑤ an efficient way to find drinking water

17

구조물에 관해 언급된 내용이 아닌 것은?

① 제작 시기　　　② 길이　　　③ 제작 목적
④ 구성 재료　　　⑤ 작동 원리

녹음을 다시 한 번 듣고, 빈칸에 알맞은 말을 쓰시오.

01

M: There was a period in my life, when I had no idea what I was going to do. If you feel the same way, this is the book for you. It was this book that helped me start _____ _____ what was really important. Keep in mind that it is not a crystal ball. But it does provide a lot of useful information about who you are. I strongly recommend it to anyone who _____ _____ _____ _____ a few answers about themselves.

02

W: I love our new apartment. I've never lived somewhere with a rooftop patio before.

M: Yeah, isn't it great? We'll be able to sit out here and _____ _____ _____ _____ of the city.

W: Actually, I was thinking it could be more useful than that.

M: Oh really? What was your idea?

W: Well, _____ _____ _____ some carrots and lettuce up here? And maybe some peas, too.

M: Wouldn't that be a lot of work?

W: Not really. I could do it on the weekends. It would be a fun hobby, and we'd also have _____ _____ _____ _____.

M: Okay. I like that idea.

W: Great! How about going shopping for seeds this afternoon?

03

M: So, what do you think?

W: It's beautiful. Was the kitchen recently renovated?

M: Yes, the owner _____ _____ _____ less than a year ago.

W: I see. Can you tell me what kind of heating it has?

M: It uses natural gas, so your bills will _____ _____.

W: That's good to know. So what are they asking for it?

M: The asking price is currently $225,000.

W: I don't know... That seems a bit expensive for this neighborhood.

M: Well, the price is likely to come down _____ _____.

W: Yes... Let me think about it, and I'll call you in the afternoon.

M: That's fine. Here's my business card. Please feel free to _____ _____ _____ _____ at any time.

04

M: Hello, Olivia. Is there a problem?

W: Yes, Mr. Simmons, there is. My dressing room isn't very clean.

M: I'm sorry to hear that. I told Mrs. Jackson to hang your dress _____ _____ _____.

W: She did, but there are empty boxes on the floor.

M: Oh, really? Maybe the person who used the room before forgot to put them away.

W: And _____ _____ _____ boots on the floor.

M: I'm sorry about that. Is there anything else?

W: As a matter of fact, there is. There's _____ _____ _____ _____ _____ on the round table as well.

M: I see. I'll get someone to clean the room right away.

W: Thank you. And could you get rid of the photographs on the mirror? I can't see my face well because of them.

M: Okay. I'll make sure the cleaner takes them down.

05

[Phone rings.]

M: AP Shopping, this is Phil. How can I help you?

W: Hi, Phil. It's Whitney.

M: Oh, hi. I noticed you weren't at your desk, so I've been _____ _____ _____.

W: I appreciate it. Anyway, the boss changed his mobile phone number, didn't he?

M: Yes, last week. Don't you have it?

W: No, and I guess he's away on business this week. I need to tell him _____ _____ _____ _____.

M: Is something wrong?

W: Well, there's nothing wrong with me. But my husband hurt his knee.

M: I'm sorry to hear that. Is he in serious pain?

W: It's pretty bad, so I need to take him to the hospital. It would be great if you could _____ _____ _____ _____ to me.

M: Okay. I'll do it right away. And I hope your husband is okay.

W: Thanks, Phil.

06

M: Hello, Prof. Williams. May I speak to you for a moment?

W: Well, Simon, my office hours are usually in the afternoon.

M: I know, but it's an emergency.

W: I see. What's the problem?

M: I won't be able to submit my essay _____ _____ _____ today.

W: I don't understand. You have had more than a week to complete it.

M: I finished it, Prof. Williams, but _____ _____ _____.

W: Well, you should have backed up your work.

M: I did, but I left my USB drive at home. If I can have an extension of two hours, I could _____ _____ _____ _____ later today.

W: All right, Simon, but don't let this happen again.

M: I won't. Thank you.

07

M: How was your dinner?

W: It was great, but I don't think _____ _____ _____ _____.

M: Really? Let me take a look at it.

W: We had two steak dinners and two glasses of wine. That's just $28.

M: Yes, I see them on your bill.

W: But look, there's also a _____ _____ _____ _____ on our bill.

M: Yes, a piece of chocolate pie for $5. So that's $33 in total.

W: But we didn't have any chocolate pie.

M: Oh, you didn't? I'm sorry. I'll _____ _____ _____ _____ your bill.

W: Thank you.

08

W: Guess what! I just met the new science teacher. He's really nice.

M: Oh, really? What's his name?

W: It's Mr. Peterson. He has a big poster of Einstein in his classroom.

M: Why? Is that because he's his hero?

W: Exactly. He said he became a teacher so he could discover the next Einstein.

M: Maybe it's you.

W: I doubt it. Anyway, he _____ _____ Harvard, so he must be smart.

M: Wow, Harvard. That's a good school. He must spend all his time studying.

W: Actually, he seems really active. He said he _____ _____ _____ _____ _____.

M: He sounds interesting. I hope I get to meet him.

W: The principal is going to _____ _____ _____ _____ tomorrow morning.

M: Great. I can meet him then.

09

W: Daylight saving time is the practice of moving the local time ahead by one hour during the summer months. The system was first put into use in the Canadian city of Port Arthur in 1908. It _____ _____ in many countries during World War I. It was considered a way to save energy, though some experts do not believe it is effective for doing so. Interestingly, the number of vehicle accidents is reduced _____ _____ _____ _____. This is most likely because it's safer to drive when it is light outside. The rate of heart attacks increases during the first week of the time change. _____ _____ _____ _____ the loss of one hour of sleep when the clocks are moved forward.

10

W: Good afternoon, sir! Are you looking for a tent?

M: Yes, I am. I need one for a camping trip next week.

W: I see. Well, we have several styles to choose from. Our dome tents are very popular.

M: I like them. I also like the tunnel tents.

W: But not the pyramid tent?

M: No, it's _____ _____ _____. Anyway, I need something big enough for me and my two friends.

W: All right. If you're beginners, I recommend getting a pop-up tent.

M: Actually, this will be the first camping trip for all of us. Are pop-up tents easier to set up?

W: Yes, they are. And will you be _____ _____ _____ _____ or driving?

M: Hiking. It's a five-kilometer hike through the forest.

W: Then I recommend you choose one of our lighter tents.

M: Great idea. The lighter, the better!

W: In that case, this is the tent I suggest you buy. _____ _____ _____ _____!

11

M: I'm really excited you came to visit me in New York!

W: Yes, I want to see Times Square. How can we get there?

M: We could _____ _____ _____. Or we could walk if you like.

W: (Let's just walk. It's a nice day.)

12

W: Excuse me. How much is this painting?

M: That one costs $500. The artist is quite famous.

W: That's too expensive. Can you _____ _____ _____ _____?

M: (Sorry. The price has already been reduced.)

13

W: Before we begin, thank you for giving me a bit of your time.

M: No, thank you _____ _____ _____. I've always enjoyed your show.

W: Thank you. So, can you begin by telling us the secret of your business success?

M: Well, I don't really have any secrets. I don't do anything different from anyone else.

W: Yet you've accomplished more than most people and earned millions of dollars.

M: I think that's just because I always _____ _____ _____ and never give up.

W: I see. So you don't think you were just lucky?

M: _____ _____. Persistence and hard work are the key, not luck.

W: So is that the advice you'd give to young people today?

M: (Yes. Never stop trying and you'll succeed.)

14

M: Hi, Brenda. Come in.

W: Hi, Paul. Thanks for inviting me. Wow, I can't believe you work here.

M: Yes, it's a great job. I've always _____ _____ _____ _____.

W: Is this where they film dramas?

M: Some of them. I can show you where they film *Golden Princess*, if you'd like.

W: That would be great! That's _____ _____ _____ _____ _____.

M: Follow me, then. It's right down this hall.

W: Do you think we'll be able to see any famous actors or actresses?

M: No. They're not filming right now. Nobody will be there.

W: That's okay. It will still be interesting.

M: Just be careful. _____ _____ _____ _____ is very expensive.

W: (I promise I won't touch anything!)

15

M: Steven grew up in Chicago, but he's always been interested in other cultures. He _____ _____ beginner and intermediate Chinese courses in school. He really enjoys learning Chinese, so he decides to study in China as an exchange student. But when he flies to China and goes to his first class, he struggles to understand what the teacher is saying. It's tough for him to _____ _____ _____ the other students, because the lecture is in Chinese. When the class finishes, he understands that there's an assignment, but he's not exactly sure what it is. So he decides to ask _____ _____ _____ _____ _____ about the homework. In this situation, what would Steven most likely say to the student?

Steven: (Please tell me what we have to do for the next class.)

16~17

W: Good morning, everyone, and welcome to our tour. If you look behind me, you will _____ _____ _____ _____ an ancient aqueduct built by the Romans. The first aqueducts appeared around 300 BC, but the ancient Romans built many _____ _____ _____. The city of Rome alone had about 800 km of aqueducts. Why did the Romans need these structures? The city developed as more and more people moved to Rome. Thus, more clean water had to be carried all around the city. So the Romans built the aqueducts, which used gravity to _____ _____ _____ _____. This system also allowed the Romans to keep their cities clean and bring water to places that were far from rivers or lakes. The Romans could thus build cities anywhere and still _____ _____ _____ _____ _____. Although most Roman aqueducts were actually underground, some were very high, such as the one you see today.

1번부터 17번까지는 듣고 답하는 문제입니다.
1번부터 15번까지는 한 번만 들려주고, 16번부터 17번까지는 두 번 들려줍니다. 방송을 잘 듣고 답을 하기 바랍니다.

01

다음을 듣고, 여자가 하는 말의 주제로 가장 적절한 것을 고르시오.
① 당뇨병의 주요 원인
② 미국의 세법 제정 절차
③ 미국 청소년들의 비만 정도
④ 청량음료가 건강에 끼치는 영향
⑤ 청량음료 소비 감소를 위한 법 추진

02

대화를 듣고, 남자의 의견으로 가장 적절한 것을 고르시오.
① 음식은 신선도가 가장 중요하다.
② 유기농 식품은 지나치게 비싸다.
③ 지나친 당분 섭취는 건강에 해롭다.
④ 잘못된 식습관이 암의 주요 원인이다.
⑤ 식품 첨가물은 건강상 문제를 일으킬 수 있다.

03

대화를 듣고, 두 사람의 관계를 가장 잘 나타낸 것을 고르시오.
① 모델 – 디자이너　　② 기자 – 디자이너
③ 직원 – 사장　　　　④ 학생 – 교수
⑤ 편집장 – 기자

04

대화를 듣고, 그림에서 대화의 내용과 일치하지 않는 것을 고르시오.

05

대화를 듣고, 여자가 남자에게 부탁한 일로 가장 적절한 것을 고르시오.
① 인터넷에 구인 글 올리기
② 이력서 쓰는 것을 도와주기
③ 웹사이트 추천하기
④ 샌드위치 주문하기
⑤ 남자의 삼촌에게 전화하기

06

대화를 듣고, 남자가 좋은 성적을 받으려는 이유를 고르시오.
① 장학금을 받으려고
② 부모님께 칭찬받으려고
③ 목표한 대학에 진학하려고
④ 원하는 선물을 받으려고
⑤ 친구와의 경쟁에서 이기려고

07

대화를 듣고, 여자가 지불할 금액을 고르시오.
① $35　　　　② $47　　　　③ $70
④ $95　　　　⑤ $105

08

대화를 듣고, 분실된 지갑에 관해 언급되지 않은 것을 고르시오.
① 색상　　　② 분실 장소　　　③ 크기
④ 재질　　　⑤ 내용물

09

Through the Eyes of Children에 관한 다음 내용을 듣고, 일치하지 않는 것을 고르시오. 3점
① 일주일간 열리는 사진 전시회이다.
② 빈곤한 아이들이 작품의 소재이다.
③ 작품의 주요 배경은 남미이다.
④ 작품을 통해 신체의 강인함을 표현하려고 한다.
⑤ 작가가 가장 좋아하는 작품은 'Fire'이다.

10

다음 표를 보면서 대화를 듣고, 여자가 구입할 노트북을 고르시오.

	Model	Screen Size	Storage Capacity	Weight	Color
①	A	15 inches	500 GB	1.1 kg	Black
②	B	15 inches	500 GB	2.4 kg	Black
③	C	14 inches	256 GB	970 g	White
④	D	14 inches	128 GB	1.6 kg	White
⑤	E	13 inches	256 GB	1.3 kg	White

11

대화를 듣고, 여자의 마지막 말에 대한 남자의 응답으로 가장 적절한 것을 고르시오.

① I'll travel around Europe next year.
② My uncle lives in Germany, so I visited him.
③ Yes, the scenery was as beautiful as expected.
④ I will contact you by email when you move there.
⑤ Yes, but it will take some time to get ready to move.

12

대화를 듣고, 남자의 마지막 말에 대한 여자의 응답으로 가장 적절한 것을 고르시오.

① You can buy a computer on the internet.
② Most computers have viruses these days.
③ You can download one from the internet.
④ My computer is more up-to-date than yours.
⑤ This site does not allow you to copy its material.

13

대화를 듣고, 남자의 마지막 말에 대한 여자의 응답으로 가장 적절한 것을 고르시오.

Woman: _____

① Please give my regards to your family.
② I'll scold my children for their behavior.
③ Thank you. I don't want my sleep disturbed.
④ What a lovely party! Thank you for inviting me.
⑤ It's unreasonable to complain at this time of night.

14

대화를 듣고, 여자의 마지막 말에 대한 남자의 응답으로 가장 적절한 것을 고르시오. 3점

Man: _____

① I'm upset that you cheated.
② I'm not going to reward my students with gifts.
③ You should study for yourself, not just for a reward!
④ I'm very impressed by your new attitude towards learning.
⑤ That's because your mother told me to give you extra help.

15

다음 상황 설명을 듣고, Martin의 어머니가 Martin에게 할 말로 가장 적절한 것을 고르시오.

Martin's mother: _____

① It's not as important as you think.
② Please clean up the cookie crumbs.
③ Don't worry. You'll do better next time.
④ Why didn't you donate all your allowance?
⑤ You'll have other opportunities to help out.

[16~17] 다음을 듣고, 물음에 답하시오.

16

여자가 하는 말의 주제로 가장 적절한 것은? 3점

① air travel with pets
② the best products for pet owners
③ a newly launched airline service
④ packing techniques for long trips
⑤ the importance of pet vaccination

17

언급된 준비물이 아닌 것은?

① proof from a vet   ② a pet carrier
③ drugs   ④ a towel
⑤ vaccinations

녹음을 다시 한 번 듣고, 빈칸에 알맞은 말을 쓰시오.

01

W: Many countries around the world _____ _____ _____ _____ items such as beer and cigarettes. They do this to reduce consumption of these items. Now, some lawmakers in the US are considering a new type of tax: a soft drink tax. Doctors have known the _____ _____ _____ soft drinks for decades. Soft drinks contain huge amounts of sugar and caffeine. These can _____ _____ and even diabetes. By placing an extra tax on soft drinks, lawmakers hope to encourage citizens to drink less cola and become healthier as a result.

02

W: Let's get some snacks for our study group later.
M: Sure, but we should be careful about the snacks we buy. Many of them _____ _____ _____.
W: Additives? You mean like salt and sugar?
M: No, chemicals that _____ _____ _____ _____ _____ and artificial colors.
W: What's so bad about them?
M: They can be really harmful if you eat them over a long period of time.
W: Oh, I remember hearing some food dyes can _____ _____.
M: Well, that's not 100% certain, but they can be dangerous.
W: I know, but unfortunately there are very few foods without artificial additives these days.

03

M: Excuse me. May I ask you some questions about today's show?
W: Sure. Which magazine are you with?
M: I'm not with a magazine. I _____ _____ the local newspaper.
W: Oh, I see. Please go ahead.

M: Thank you. First question: What's the theme for your collection?
W: The theme is "strength". That's why I used strong fabrics like leather.
M: I see. And are you excited to have this year's top supermodel _____ _____?
W: You mean Francesca Romano? Yes, I'm delighted she's modeling my collection.
M: And finally, there's a rumor that this is your final collection. Is that true?
W: Not at all. I'm _____ _____ _____, so I will design clothes forever.
M: Thank you for your time.

04

M: Working as a security guard is boring. Everything is the same every day. Even the employee at the information desk is always _____ _____ _____ jacket and hat.
W: Yes, but the people are always different. Do you see that woman pulling her luggage?
M: Yes, I see her.
W: Look at her dog. It's wearing glasses. You don't see that every day.
M: Well, not every day.
W: Also, that man at the information desk is wearing a very unique T-shirt. It has a _____ _____ _____.
M: I'm starting to see your point.
W: And there are two people _____ _____ _____ _____ in line behind him.
M: Maybe you're right. This job is more interesting than I thought.

05

M: Hi, Paula. What are you doing?
W: Hey, Daniel. I'm just looking for a part-time job.
M: You're looking online?

W: Of course. There are plenty of websites where
_____ _____ _____ .

M: I see. So what kinds of jobs are available?

W: Well, most of the jobs are waiting tables.

M: Have you _____ _____ any of them?

W: No. There's a problem. Most of the employers want someone over the age of 19.

M: Oh, that's a shame. Hey, wait a minute! My uncle is opening a sandwich shop at the mall next week. Perhaps he'll _____ _____ _____ .

W: That'd be great! Could you call him up and ask him if he's looking for a worker?

M: No problem. I'd be happy to help.

06

M: What a nice laptop! Is it yours, Kelly?

W: Yes, I got it from my dad for my birthday.

M: _____ _____ _____ _____ ! I wish I had one, too.

W: Isn't your birthday coming up next month?

M: Yes. You have a good memory.

W: Why don't you _____ _____ _____ _____ get you one for your birthday? They might do it.

M: Probably not. They said they would only buy a gift if I got good grades this semester.

W: That's not a big deal. Just study hard and do your homework.

M: I will. And I'll ask for a laptop if I _____ _____ _____ .

W: Good luck.

07

M: How can I help you?

W: I want to make video calls on my computer, but I don't have the necessary equipment.

M: Well, first you _____ _____ _____ .

W: You mean a set of headphones with a small microphone?

M: Right.

W: Okay. How much are those?

M: Our standard model is $20, and our best model is $55. Which do you want?

W: _____ _____ _____ , please. What else do I need?

M: You have a built-in webcam in your laptop, don't you?

W: No, my laptop is really outdated. So, I guess I need a webcam. Do you sell them, too?

M: Yes. We have everything from this simple $15 one to this $40 model.

W: Well, since it's just for chatting, I'll take the basic one.

M: Okay. Anything else? Our USB memory sticks are _____ _____ for just $10.

W: No, that's everything.

08

[Phone rings.]

W: Lost and Found Center. Can I help you?

M: Did anyone bring in a blue wallet today?

W: Not yet, but it's still early. Did you leave it on _____ _____ _____ _____ ?

M: Yes, on the number 43 bus this morning.

W: All right. Is it a large wallet?

M: No, _____ _____ .

W: Okay. Was there a lot of money in it?

M: Not really. Just $7. But it was a gift, so I really want to find it.

W: I understand. Is there anything inside that will _____ _____ _____ _____ ?

M: Yes. There's a picture of my mother. And my library card too.

W: I see. Please give me your name and phone number. We'll give you a call if someone finds it.

M: Okay. My name is Donald Jones, and my number is 812-111-1520.

09

M: Good afternoon and welcome to the opening of my exhibition. My exhibition is entitled *Through the Eyes of Children*, and it will be featured here for the ___ ___ ___. The theme of this collection is ___ ___. I've traveled throughout South America taking pictures of disadvantaged children in the hopes of sharing their stories with you. Some of you have asked me what I ___ ___ ___ when taking a photograph. The answer to that is "courage". I see courage in every child I photograph. For that reason, my favorite picture in the collection is the one entitled "Fire". As you can see, the eyes of that young Bolivian girl are burning with ___ ___ ___ ___.

10

M: Welcome to Computer City. What can I help you with today?

W: I heard you're ___ ___ ___ ___ ___. Can you help me choose one?

M: Of course. We have laptops with three different screen sizes.

W: I don't like the smallest size. I often ___ ___ ___ ___ ___.

M: Okay. How about storage space? Is 256 GB enough?

W: Yes, but no less than that. I need enough space for all my programs.

M: All right. And will you be ___ ___ ___ ___?

W: Yes. I plan on taking it with me on business trips, so it should weigh less than 2 kg.

M: That's not a problem. Take a look at these two models. They're both very popular.

W: They both look good, but I like the white one better. It will match my office.

M: That's a great choice. You're getting a terrific bargain today.

W: I hope so!

11

W: Hi, Tom. Where did you go during the holidays?

M: I went on a family trip to Europe. My family ___ ___ ___ there.

W: Really? You mean you will live in Europe soon?

M: (Yes, but it will take some time to get ready to move.)

12

M: My computer is ___ ___ ___. It takes a long time to open files.

W: Try running an anti-virus program. My computer had the same problem, but that really helped.

M: That's a good idea! But how should I do it?

W: (You can download one from the internet.)

13

[Knocking on door]

M: Hello?

W: Hello, I'm Melissa. I live downstairs.

M: Nice to meet you, Melissa.

W: Well, I'm here ___ ___.

M: Oh, really? What about?

W: The noise coming through my ceiling from your apartment is too much.

M: I'm sorry. My family has invited some guests over for a small party. It'll be over before long.

W: The worst part is that I can hear children jumping around and ___ ___, right above my bedroom.

M: I'm very sorry about that. I'll ___ ___ ___.

W: (Thank you. I don't want my sleep disturbed.)

14

M: Ah, Miyoung, there you are. Can you come into my office please?

W: Yes, Mr. O'Reilly.

M: Please take a seat. Now, I want to talk to you about something.

W: What is it, sir?

M: Well, actually, I'm very pleased with you. Your English marks have _____ _____ this semester, haven't they?

W: Yes, they have. Thank you.

M: Did you _____ _____ _____ _____ or something?

W: Not exactly. My mom said she'd buy me a new skateboard if I got an A in English.

M: What? So, you've only been _____ _____ _____ _____ in English?

W: Well... Yes. I really want it.

M: (You should study for yourself, not just for a reward!)

15

M: Martin is at home on a Monday afternoon, talking to his mother about his day at school. He explains that in the morning there was _____ _____ _____. The teachers and students collected donations for UNICEF to help children in Africa. Having spent most of his allowance during the weekend and not having much money in his pocket that day, Martin donated only a dollar. Now he _____ _____ _____ _____ _____. He tells his mother that he feels guilty enjoying delicious snacks when there are children in Africa _____ _____ _____. In this situation, what would Martin's mother most likely say to Martin?

Martin's mother: (You'll have other opportunities to help out.)

16~17

W: While pets may be great companions, an airline trip with them can be difficult. Before you travel, carefully check your airline's regulations regarding pets, as policies vary. Pets less than two months old are usually not allowed. And you may need proof from your veterinarian that your pet is _____ _____ _____ travel. All animals are required to be transported in a pet carrier, which is usually stored under the plane with the luggage. The carriers must be labeled clearly with your contact information and the pet's photo. They should be lined with a towel on the bottom. The carrier should be at least _____ _____ _____ _____ _____ the animal to allow for movement during the trip. Additionally, it's important to _____ _____ _____ _____ so that they can be accessed if an emergency arises. For those of you who are traveling internationally, be sure that your pet receives the _____ _____ required by the destination country. Planning ahead can ensure the safety and comfort of your pet.

1번부터 17번까지는 듣고 답하는 문제입니다.
1번부터 15번까지는 한 번만 들려주고, 16번부터 17번까지는 두 번 들려줍니다. 방송을 잘 듣고 답을 하기 바랍니다.

01

다음을 듣고, 남자가 하는 말의 목적으로 가장 적절한 것을 고르시오. 3점

① 기상 예보를 알리려고
② 안전 수칙을 설명하려고
③ 수업이 지연됨을 공지하려고
④ 폭설로 인한 휴교를 알리려고
⑤ 학생들과 교직원들의 귀가를 요청하려고

02

대화를 듣고, 여자의 의견으로 가장 적절한 것을 고르시오.
① 경험자에게서 듣는 조언은 값지다.
② 오랜 친구들과 계속 연락해야 한다.
③ 회사에서 동료 간의 관계는 중요하다.
④ 외향적인 사람들이 친구들 사이에서 인기가 많다.
⑤ 친구의 고충을 잘 들어주는 것이 도움이 될 수 있다.

03

대화를 듣고, 두 사람의 관계를 가장 잘 나타낸 것을 고르시오.
① 고객 – 수의사
② 학생 – 교사
③ 기자 – 동물 사육사
④ 관광객 – 여행 가이드
⑤ 관광객 – 동물 조련사

04

대화를 듣고, 그림에서 대화의 내용과 일치하지 않는 것을 고르시오.

05

대화를 듣고, 두 사람이 할 일로 가장 적절한 것을 고르시오.
① 책 읽기
② 운동하기
③ 일기 쓰기
④ 담배 끊기
⑤ 일찍 일어나기

06

대화를 듣고, 남자가 전화를 건 이유를 고르시오.
① 여자의 낚싯대를 빌리려고
② 함께 낚시하러 가고 싶어서
③ 생선 요리하는 법을 물어보려고
④ 낚시 여행 계획에 대한 조언을 얻으려고
⑤ 다음 주에 여자와 만날 약속을 잡으려고

07

대화를 듣고, 여자가 지불할 금액을 고르시오.
① $2.00
② $3.50
③ $4.00
④ $5.00
⑤ $10.00

08

대화를 듣고, 우주선에 관해 언급되지 않은 것을 고르시오.
① 자금의 출처
② 제조사
③ 이름
④ 목적지
⑤ 시사점

09

Free Rice website에 관한 다음 내용을 듣고, 일치하지 않는 것을 고르시오. 3점
① 게임으로 세계 기아문제를 도울 수 있다.
② 게임은 다양한 형태의 문제로 제시된다.
③ 정답을 맞힐 때마다 20알의 쌀이 기부된다.
④ 사이트 내 광고 비용은 쌀알로 지급된다.
⑤ 사이트 출시일에 1,000알 미만의 쌀이 기부되었다.

10

다음 표를 보면서 대화를 듣고, 남자가 들을 수영 강좌를 고르시오.

	Class	Level	Day	Time	Class Size
①	A	Basic	Tue./Thurs.	6 a.m.	15
②	B	Basic	Sat.	7 p.m.	15
③	C	Intermediate	Mon./Wed./Fri.	10 a.m.	20
④	D	Intermediate	Tue./Thurs.	8 p.m.	20
⑤	E	Advanced	Sat.	11 a.m.	20

11

대화를 듣고, 남자의 마지막 말에 대한 여자의 응답으로 가장 적절한 것을 고르시오.

① You can order one online.
② I've never been into rock music.
③ There is a newspaper stand over there.
④ Thanks for getting these concert tickets.
⑤ The copy machine in this office is out of order.

12

대화를 듣고, 여자의 마지막 말에 대한 남자의 응답으로 가장 적절한 것을 고르시오.

① So did you enjoy your ice cream?
② You'd better hurry. They are closing soon.
③ Not at all. It was good to hear you found it.
④ If it weren't for you, I couldn't have found it.
⑤ You're welcome. I can lend it to you anytime.

13

대화를 듣고, 여자의 마지막 말에 대한 남자의 응답으로 가장 적절한 것을 고르시오. 3점

Man: Yes. _____

① I can't wait to tell her I won the lottery.
② She'd rather use the money to help others.
③ She'll buy a new sports car with the money.
④ I hope she doesn't get too upset about my plan.
⑤ I just wish she'd stop spending so much money.

14

대화를 듣고, 남자의 마지막 말에 대한 여자의 응답으로 가장 적절한 것을 고르시오.

Woman: That's a shame. _____

① You should get yourself a good guitar.
② You should take good care of the violin.
③ You shouldn't let your talent go to waste.
④ But I can't teach you how to play the guitar.
⑤ Then why don't you let your kids teach themselves?

15

다음 상황 설명을 듣고, Caroline이 그녀의 어머니에게 할 말로 가장 적절한 것을 고르시오.

Caroline: _____

① Thanks, Mom! You're the best.
② Remember, this concert is a must.
③ I hate that band. Give the passes to Joey.
④ But I've done well in school. Why can't I go?
⑤ How did you know I wanted front row tickets?

[16~17] 다음을 듣고, 물음에 답하시오.

16

여자가 하는 말의 주제로 가장 적절한 것은?

① the evolution of swimsuits
② the variety of natural fabrics
③ inventions inspired by nature
④ products invented by accident
⑤ animals used in product testing

17

언급된 동물이 아닌 것은?

① sharks   ② birds   ③ lizards
④ squirrels   ⑤ mice

녹음을 다시 한 번 듣고, 빈칸에 알맞은 말을 쓰시오.

01

M: Can I have your attention, please? This is an important announcement for all students and staff. According to the forecast, temperatures are expected to ＿＿＿ ＿＿＿ ＿＿＿ tonight, shortly after sunset. The heavy snowfall we are now experiencing is expected to stop soon, but the snow already on the roads is likely to ＿＿＿ ＿＿＿ ＿＿＿ ＿＿＿ within a couple of hours. Because of the potential for dangerous road conditions, we are asking everybody to leave the school by 6:00. At 6:30, the school building will be locked up. Our students' safety is paramount, so all teachers are asked to assist in guiding students home. I wish everyone ＿＿＿ ＿＿＿ ＿＿＿ ＿＿＿ this evening. Thank you for listening.

02

W: Matthew, why the long face?

M: I just met my best friend, Tom. He said his boss doesn't like him.

W: Oh, did you ＿＿＿ ＿＿＿ ＿＿＿ ＿＿＿?

M: Well, I'm still in college, so I didn't know what to say. I just listened.

W: That's good. He probably didn't want any advice anyway.

M: Really? But he complained so much.

W: He probably just wanted to ＿＿＿ ＿＿＿ ＿＿＿ ＿＿＿ to you because you're his best friend. He must feel comfortable talking to you.

M: Do you think he ＿＿＿ ＿＿＿ and less stressed now?

W: Yes. Sometimes just listening can be a big help for people.

M: Oh I see. That's probably why he thanked me.

03

W: Thanks for your time today.

M: Oh, it's my pleasure.

W: Here's my first question. Do you ＿＿＿ ＿＿＿ ＿＿＿?

M: Sure do. I love all the animals, especially the chimpanzees and monkeys that I work with.

W: How do you take care of them?

M: First, I make sure they are comfortable around people. Some of them are brought here from the wild. So I need to ＿＿＿ ＿＿＿ ＿＿＿ they don't get stressed out by their new and unfamiliar environment.

W: What about the ones that were born here?

M: They are much easier to deal with. They are used to having people around them.

W: Did you have to ＿＿＿ ＿＿＿ ＿＿＿ for this job?

M: Yes. To work here, a college degree in animal science, zoology, or a related field is required.

W: All right. Thank you for your answers. I believe they will be of great help to those who want to be like you.

04

W: Hi, Honey! How was your trip?

M: Very productive. *[Sniffing]* What's that amazing smell?

W: Oh, I was cooking something. Let's go to the kitchen.

M: Sure. *[Pause]* Hmm... Why is the refrigerator door open?

W: Oh, I ＿＿＿ ＿＿＿ ＿＿＿ ＿＿＿ when I came to welcome you.

M: Oh okay. The sink is full of dishes. You must be cooking a lot of food.

W: No, actually I'm just making pork chops on our gas range.

M: Oh! I love pork chops.

W: I know. Do you notice anything different about the kitchen?

M: ＿＿＿ ＿＿＿ ＿＿＿ are new. They look nice.

W: Yes, I'm glad you like them. Now, why don't you

take a shower while I finish making dinner?

M: Sounds good. Let me grab a banana from the table first.

W: Okay, don't take too long!

M: _____ _____ _____ _____. Then I'll tell you more about my trip.

05

W: What's your New Year's resolution?

M: Well, I haven't decided yet. I don't know if it will work, because I didn't _____ _____ _____ _____.

W: What was it?

M: Two things. To quit smoking and exercise regularly.

W: How long were you able to stick to each resolution?

M: Maybe till the middle of January. I feel so ashamed, but it was too hard to _____ _____ _____. How about you?

W: Me? My resolutions were to keep a journal and to read classics, at least one book per month.

M: And?

W: Actually, I stuck to them. Sometimes I was _____ _____ _____ _____, but I kept going. So my two resolutions finally became habits.

M: Wow, that's inspiring. I know what my resolutions will be for this year. I will try to quit smoking again, and I'll exercise at least three times a week.

W: Sounds like a plan! I'll join you at the gym.

06

[Phone rings.]

W: Hello?

M: Hi, Julie. It's Nick.

W: Hey, what's up?

M: You told me you like to go fishing, right?

W: Yeah, it's one of my hobbies.

M: Actually, I'm going fishing with my friends next week. I'm wondering if you could _____ _____ _____ _____.

W: Sure, you can come by my place and pick it up tomorrow.

M: Thank you so much. I'm a bit worried because it's my first time.

W: Fishing is not that difficult. All you have to do is just put the pole into the water and wait. You'll _____ _____ _____ _____ soon.

M: Oh, I see. I'm so excited to go.

W: Don't forget to _____ _____ _____ _____ with some of the fish you caught. It's delicious!

07

W: Hi. I notice that you have _____ _____ _____?

M: Oh yes, our eggs are the best because they come from free-range chickens!

W: Really? Do you sell the chickens too?

M: Yes, the chickens are $5 _____.

W: And how much are the eggs?

M: The eggs are $2 for a dozen or $3.5 for two dozen.

W: Well, how many omelets can I _____ _____ _____ _____ of eggs?

M: You could make six omelets with two eggs in each omelet.

W: Okay then, I'll _____ _____ _____ _____. Do you have change for a $5 bill?

M: Sure, here you are. Enjoy the omelets.

08

W: Did you hear about the most recent spacecraft to visit the International Space Station?

M: No. What's so special about it?

W: It's the first time a private company _____ _____ _____ _____.

M: That's really interesting. I suppose before space flights were always funded by governments.

W: That's right. This private ship is called SpaceX Dragon, and it landed at the space station safely as planned.

M: How would they _____ _____ from doing that?

W: I'm not sure. I think they wanted to prove they could do it first.

M: I suppose it's a good first step.

W: Whatever happens, we are now in the age of private space travel.

M: That's really going to change things. It'll be exciting to see _____ _____ _____ space travel.

09

W: If nobody has told you about the Free Rice website, then it's time you paid it a visit. This site is amazing: For just a few minutes of your time, you can play a game to improve your vocabulary and simultaneously _____ _____ _____ _____ _____ around the world with the help of the United Nations (UN). The game is simple — choose the correct definition of a word from four choices. For every correct answer, 20 grains of rice are donated to _____ _____ _____ _____. The ads on the site pay for the rice. Then the UN World Food Program distributes the rice to those in need. This site launched on October 7. That day, fewer than _____ _____ _____ _____ were sent. But by October 24, more than 30 million grains were being donated each day.

10

W: Why are you looking at the fitness center's schedule, Kevin?

M: I want to take a swimming class. I don't know which one to choose.

W: Oh. I thought you already knew _____ _____ _____.

M: Actually, I do. I've taken lessons before.

W: Then you should probably take an advanced class.

M: No. It was a long time ago. Basic or intermediate _____ _____ _____.

W: I see. How about this Saturday class?

M: Once a week isn't enough. I want to learn quickly. I need a class that meets _____ _____ _____ _____ _____.

W: Well, you have some options. What time is best for you?

M: I work from 9 to 5. So I can only take classes that start before 8 a.m. or after 7 p.m.

W: Then you can choose one of these two.

M: I'll take the one with the smaller class size. Thanks for your help, Maria!

11

M: Who is this? It sounds familiar.

W: This is my favorite band, The Rock Stars. This is their newest album.

M: I really like it. Where can I _____ _____ _____ _____ _____?

W: (You can order one online.)

12

W: Excuse me. Can you tell me where the ice cream store is?

M: Yes, it's on the other side of the shopping center. _____ _____ _____ and turn left.

W: Thanks. I think I can find it now.

M: (You'd better hurry. They are closing soon.)

13

W: I heard _____ _____ _____ _____. Congratulations!

M: Thanks.

W: What are you going to do with the money? Buy a car?

M: No, I already have a car. I want to buy something that I never thought I would have — a sailboat.

W: What a wonderful idea! Do you know _____ _____ _____ _____?

M: No, but I'll learn.

W: Did you tell your wife about your plan? She must be happy.

M: Yes, I did. But she actually wants to give _____ _____ _____ _____ _____.

W: Oh, really? That's very generous of your wife.

M: Yes. (She'd rather use the money to help others.)

14

W: How long have you been _____ _____ _____?

M: About a year.

W: Wow, you're so good! Where did you take lessons?

M: I didn't take any lessons. I just _____ _____.

W: Wow, I could never do that.

M: Well, I've been playing music since I was a kid, so I learn new instruments easily.

W: What other instruments do you play?

M: I learned to play the piano when I was small. Then when _____ _____ _____ _____ I played the violin.

W: Oh, I love the violin. Do you still play?

M: No. I'd like to, but _____ _____ _____ I have right now is a guitar.

W: That's a shame. (You shouldn't let your talent go to waste.)

15

W: Caroline is very upset. Her favorite band is coming to her city, but she couldn't get tickets. Caroline's mother _____ _____ _____ her. Caroline has done really well in school this year, and her mother wants to do _____ _____ _____ _____. She knows someone in the music industry and was able to _____ _____ _____ for Caroline. Now Caroline can meet her favorite band. In this situation, what is Caroline most likely to say to her mother?

Caroline: (Thanks, Mom! You're the best.)

16~17

W: Good morning, class. As you know, we can learn many valuable things from nature. Today, we'll look at inventions that drew inspiration from animals. First, did you know that designers have developed a new material for swimsuits? _____ _____ _____ shark skin. Sharks are one of the fastest creatures in the ocean. This material will help people move more quickly through the water. Next, think about _____ _____ _____.
The first people to build airplanes probably got their ideas from birds. Then, there are the small, strong pads that _____ _____ _____ _____ to another surface. Where did inventors get the idea for them? Their idea came from the feet of geckoes. These small lizards have special skin on their feet that helps them to climb on walls. Finally, the wingsuit is another great example. People _____ _____ _____ _____ _____ this idea on their own. They based the design on flying squirrels. Aren't these inventions impressive? Have any other inventions come from animals? Think for a few minutes, and then you'll discuss your ideas as a class.

1번부터 17번까지는 듣고 답하는 문제입니다.
1번부터 15번까지는 한 번만 들려주고, 16번부터 17번까지는 두 번 들려줍니다. 방송을 잘 듣고 답을 하기 바랍니다.

01
다음을 듣고, 여자가 하는 말의 주제로 가장 적절한 것을 고르시오.
① 실내 조경의 이점
② 수도관 동파 예방법
③ 난방비를 줄이는 방법
④ 겨울철 에너지 절약 방법
⑤ 동상에 걸렸을 때의 응급처치

02
대화를 듣고, 남자의 의견으로 가장 적절한 것을 고르시오.
① 파티 계획을 미리 세워야 한다.
② 졸업 파티는 성대하게 해야 한다.
③ 정기적인 치아 검진을 받아야 한다.
④ 탄산음료를 가끔 마시는 것은 괜찮다.
⑤ 어렸을 때 올바른 식습관을 형성해야 한다.

03
대화를 듣고, 두 사람의 관계를 가장 잘 나타낸 것을 고르시오.
① 디자이너 - 모델   ② 손님 - 손님
③ 가게 주인 - 점원   ④ 점원 - 고객
⑤ 세탁소 주인 - 고객

04
대화를 듣고, 그림에서 대화의 내용과 일치하지 않는 것을 고르시오.

05
대화를 듣고, 남자가 할 일로 가장 적절한 것을 고르시오.
① 저녁 식사 만들기
② 식당에 전화해서 주문하기
③ 약속 장소에 나가기
④ 식당에 들르기
⑤ 여자를 데리러 가기

06
대화를 듣고, 남자가 전화를 건 이유를 고르시오.
① 데리러 와달라고 요청하려고
② 엄마의 발목 상태를 확인하려고
③ 가까운 병원 위치를 알려주려고
④ 농구 경기 관람을 허락받으려고
⑤ 준비물을 가져와 달라고 부탁하려고

07
대화를 듣고, 여자가 지불할 금액을 고르시오. 3점
① $60   ② $80   ③ $90   ④ $95   ⑤ $105

08
대화를 듣고, Big Ben에 관해 언급되지 않은 것을 고르시오.
① 소재지   ② 설계자   ③ 높이
④ 완공 연도   ⑤ 시계의 개수

09
Circle Cruise day tour에 관한 다음 내용을 듣고, 일치하는 것을 고르시오. 3점
① 승객에게 다과가 제공된다.
② 부두에서 점심이 제공된다.
③ 해돋이 관광을 포함한다.
④ 어린이들은 탑승이 허용되지 않는다.
⑤ 구명조끼는 신청하면 제공된다.

10

다음 표를 보면서 대화를 듣고, 남자가 신청할 패키지를 고르시오.

Leyland Sports Center Summer Specials

	Package	Class	Cost per Month	Frequency of Classes	Pool Access
①	A	golf	$90	1 per week	O
②	B	golf	$115	2 per week	X
③	C	tennis	$75	2 per week	O
④	D	tennis	$90	3 per week	X
⑤	E	yoga	$75	3 per week	O

11

대화를 듣고, 여자의 마지막 말에 대한 남자의 응답으로 가장 적절한 것을 고르시오.

① They are too strict sometimes.
② The science teacher is really kind.
③ The new school is down the street.
④ It makes me angry when that happens.
⑤ My teachers usually give me good grades.

12

대화를 듣고, 남자의 마지막 말에 대한 여자의 응답으로 가장 적절한 것을 고르시오.

① We can take my car there.
② I already signed up for it.
③ Sure, but I might need help studying.
④ There are many students in my class.
⑤ Okay, I will take the advanced chemistry course.

13

대화를 듣고, 여자의 마지막 말에 대한 남자의 응답으로 가장 적절한 것을 고르시오.

Man: _____

① Please don't be angry with me.
② Just bring some food that you like.
③ I like doing dangerous activities, too!
④ You don't need to worry. I'll help you!
⑤ In that case, you should cook for yourself.

14

대화를 듣고, 남자의 마지막 말에 대한 여자의 응답으로 가장 적절한 것을 고르시오.

Woman: _____

① Really? I can't wait to see your tricks.
② What did you buy at the shopping center?
③ This is my skateboard. I didn't steal yours.
④ Are you sure you locked up your bicycle?
⑤ That's too bad. You must have been sad!

15

다음 상황 설명을 듣고, Jane이 그녀의 남편에게 할 말로 가장 적절한 것을 고르시오.

Jane: Honey, _____

① let's order pizza.
② it's not your fault.
③ this is very thoughtful.
④ thanks for reminding me.
⑤ congratulations on your promotion.

[16~17] 다음을 듣고, 물음에 답하시오.

16

남자가 하는 말의 목적으로 가장 적절한 것은? 3점

① 개인위생의 필요성을 강조하려고
② 새로운 바이러스 백신을 홍보하려고
③ 전염병 발생 경로와 예방법을 설명하려고
④ 신종 전염병의 원인과 위험성을 경고하려고
⑤ 유람선 여행 시 유의사항에 대해 공지하려고

17

언급된 증상이 <u>아닌</u> 것은?

① a stomachache   ② a rash   ③ vomiting
④ a headache   ⑤ a fever

24 DICTATION 🎧

녹음을 다시 한 번 듣고, 빈칸에 알맞은 말을 쓰시오.

01

W: Good evening, everyone. This is an announcement from the Daerim Apartment security office. Tonight we are expecting _____ _____ _____. So to protect the pipes from freezing, we ask that all of you leave _____ _____ _____ _____ from the faucet all night. If you do not do this, the pipes may _____ _____ _____, requiring major repairs. This would cause you to have no running water or heat for hours. Therefore it's strongly recommended that you _____ _____ _____ _____. Thank you, and have a good evening.

02

M: Will you _____ _____ _____ the grocery list for me, Jenny?

W: Sure, what is it?

M: A bottle of soda.

W: Soda? But you always tell me that stuff is terrible for you. That's why we never buy it.

M: Well, yes. It's full of sugar, and it's bad for your teeth. But I wanted to _____ _____ _____ for the party on Saturday.

W: Well, okay.

M: Don't worry. Drinking soda is only a problem if you _____ _____ _____ _____.

W: So it's okay on special occasions?

M: Right. And your sister's high school graduation is a very special occasion.

03

M: The line is very long, isn't it?

W: Yes, I've been standing here for 30 minutes!

M: Why are you waiting here?

W: Well, I bought this sweater yesterday, but look! There's _____ _____ _____ _____!

M: Oh, yeah, I see it. Are you going to exchange it or _____ _____ _____?

W: Well, I really like the color and style, so I think I'll just exchange it.

M: I'm usually happy with the quality of the clothes I get here.

W: I usually am, too. But this time I'm _____ _____. I hope they're not going to start selling low-quality things here.

M: I don't think they will. I know the owner, and she's _____ _____ about the items she brings in.

W: I hope so.

04

M: Let's look for a pet in this pet shop. That white cat is cute.

W: Well, I don't think I'd like owning a cat.

M: What about a dog? Look at that little dog _____ _____ _____.

W: Isn't it adorable? But I don't really want a dog, either.

M: Well, what do you want?

W: I'm not sure. Look at these birds. I could get one of those small birds _____ _____ _____ _____. They really like to sing.

M: Yes, but that would get annoying. A fish would be quieter.

W: But those fish bowls on the shelf look expensive.

M: Oh, what about the rabbit _____ _____ _____ _____ _____?

W: It's cute. Now I know what I want.

05

[Phone rings.]

W: Hello?

M: Hi, honey. So you're finally home. How was your day?

W: Exhausting. It feels like _____ _____ _____

_____ since the moment I woke up.

M: Sorry to hear that. I just wanted to let you know I'll be home around 7 tonight.

W: All right. But to be honest, I'm _____ _____ _____ _____ tonight.

M: That's fine. Why don't we just eat out?

W: Oh! I'd love to have some dumplings from that little place near your office.

M: Sounds great. Why don't you meet me here at 6:30?

W: I don't think I have the energy. What if I call them and order some to go?

M: Sure, no problem. I can _____ _____ _____ _____ _____ home.

W: Perfect. See you soon, dear!

06

[Mobile phone rings.]

W: Hello?

M: Hi, Mom. Are you at home?

W: Yes, I just got in. Is everything okay?

M: No. I just _____ _____ _____ playing basketball.

W: Oh dear. Is it serious?

M: I can't walk by myself. I need you to come and pick me up.

W: Okay. Where are you now?

M: I'm outside the community center, next to the basketball court.

W: I'll be there in 10 minutes. Sit down and _____ _____ _____ _____ your bad ankle.

M: Thanks Mom. Could you bring some ice and a towel too? I want the swelling of my ankle to go down.

W: Of course. Just relax, and I'll be there _____ _____ _____ _____.

07

W: My mother's birthday is tomorrow, and I still haven't bought anything for her.

M: Why don't you get her flowers? Wait a second. [Pause] _____ _____ _____ _____ _____.

W: Wow, there are lots of flowers! I like this flower basket.

M: What basket size do you want? The small ones are $40, the medium ones are $50, and the large ones are $60.

W: I want to get the largest one. And _____ _____ _____ _____ _____ in addition to the flowers.

M: Look at this. You can add chocolates or flower soap to your basket. Oh, you can write a message for her, too.

W: There are too many options. I like the flower soap, but they're $5 each. That's too expensive.

M: But _____ _____ _____ _____ _____ _____. Chocolate is $30, and a card is $10.

W: Hmm... You're right. I'll just add four bars of soap, and I'll make a card myself.

M: Great choice! Oh, when do you want to have it delivered?

W: Tomorrow around 1 p.m.

M: Delivery for that time is $15.

W: Okay.

08

M: Hi, Jihyun. How was your trip to England?

W: It was great. We went to London and got to see Big Ben.

M: Big Ben, a clock tower in the Palace of Westminster?

W: Yes. It was designed by Augustus Pugin after _____ _____ _____ _____ by a fire in 1834.

M: I think _____ _____ _____ _____ _____. It's very tall, right?

W: Yes. It's nearly 100 meters tall, and each clock face is _____ _____ _____ _____.

M: There's more than one clock?

W: Yes. There's one on each of the tower's four sides.

M: It sounds interesting. Did you take any pictures of it?

W: Yes, I took many pictures! I'll show them to you tomorrow.

09

M: Thank you for signing up for the Circle Cruise day tour. The boat leaves tomorrow at 10 a.m. promptly, so please don't be late. Lunch will _____ _____ and refreshments will be provided, but you're welcome to bring _____ _____ _____ on board. The cruise will last approximately three hours, starting and ending right here at the Eastside docks. As we sail around the city, you'll have _____ _____ all morning, so bring a camera. Also, the forecast is for sunshine, so be sure to wear your sunscreen. Parents bringing children on board are urged to keep an eye on their kids, and everyone is required to _____ _____ _____ _____ at all times. Again, thanks for signing up, and we'll see you in the morning.

10

W: Hey! Is that a flyer for the sports center I go to? I thought you didn't want to join.

M: I didn't, but now they're offering some _____ _____ _____.

W: Let me see. *[Pause]* Package E is what I have. That's a great deal.

M: I know, but yoga isn't exactly what I had in mind. I'd prefer to do _____ _____.

W: Oh, well then, I'd recommend tennis or golf, but those packages are more expensive.

M: Right, but that's okay. As long as it's under $100 a month, it's fine.

W: Well, there are _____ _____ _____ _____. How about this one?

M: I want to have classes more than once a week.

Once a week is not enough.

W: Yeah, I understand. I like to go to my yoga class as often as possible.

M: I also want to be able to use the pool. I really like swimming, so I'd use it a lot.

W: You have _____ _____ _____ _____! But they do have one class that fulfills all of them.

M: Yes, this one seems fine. I think I'll sign up today!

11

W: _____ _____ _____ _____ your new school, Michael?

M: It's okay, but I don't really like the teachers much.

W: Why not?

M: (They are too strict sometimes.)

12

M: Have you _____ _____ _____ your classes yet?

W: No, I'm having trouble choosing. What do you recommend?

M: I'm taking an advanced math course. Do you want to take it with me?

W: (Sure, but I might need help studying.)

13

M: Michelle, why don't we _____ _____ this weekend?

W: Sure, I don't have anything special planned. What do you have in mind?

M: Well, a group of us are _____ _____ _____. Would you like to come with us?

W: That sounds nice. Where are you going?

M: Have you ever been to Mount Seorak? It's beautiful this time of year.

W: No, but I'd love to see it! What are you going to do there?

M: Well, we thought we'd _____ _____ _____

_____ and sleep in one of the cabins there.

W: What will you eat?

M: I'm bringing a gas burner, and everyone is bringing some food.

W: It sounds fun, but I don't really like mountain climbing. I get scared when _____ _____ _____ _____ _____.

M: (You don't need to worry. I'll help you!)

14

W: Hey, Ted, look at my new skateboard!

M: Wow! It's really cool! Where did you get it?

W: My dad bought it for me for my birthday.

M: It looks really expensive. Can you _____ _____ _____?

W: Not yet, I'm still practicing. I want to learn to jump while riding it.

M: I used to have a skateboard like that one, but _____ _____ _____.

W: Really? Where was it stolen?

M: From the shopping center. I _____ _____ _____, and when I came out, it was gone.

W: Did you _____ _____ _____?

M: No, that was my mistake. I'd just left it outside near the bicycles.

W: (That's too bad. You must have been sad!)

15

W: Jane has had a long day at work. She was very busy and her boss _____ _____ _____ her. She is exhausted. The last thing she wants to do is go home and make dinner. 'Maybe we could just order a pizza,' she thinks to herself. After climbing the stairs to their apartment, she opens the door to find that the _____ _____ _____ and her husband isn't there. "Where could he be?" she wonders as she gets out her cell phone. But as she enters the dining room she finds a candlelit dinner, _____ _____ _____. Her husband greets her _____ _____ _____ and says, "I'm sorry you had a bad day." In this situation, what would Jane most likely say to her husband?

Jane: Honey, (this is very thoughtful.)

16~17

M: You've packed your bags and you're ready for a dream vacation _____ _____ _____ _____ _____. But a nightmare could be hiding on the ship. It's called the norovirus, and it's commonly found on cruise ships and other places _____ _____ _____ _____ in a small space. It is spread through contact with an infected person. It is also spread through food and water that has _____ _____ by the virus. The symptoms of norovirus infections are usually not severe. They include an upset stomach and vomiting. Sometimes those suffering from a norovirus infection get a headache and a high fever. The symptoms usually last for only a few days, but a person can be contagious for up to 14 days. Fortunately, prevention of norovirus infections is easy. Washing your hands regularly and thoroughly will significantly _____ _____ _____ _____ _____ the illness. Also, frequently touched surfaces, such as handrails and doorknobs, should be cleaned regularly.

1번부터 17번까지는 듣고 답하는 문제입니다.
1번부터 15번까지는 한 번만 들려주고, 16번부터 17번까지는 두 번 들려줍니다. 방송을 잘 듣고 답을 하기 바랍니다.

01

다음을 듣고, 남자가 하는 말의 목적으로 가장 적절한 것을 고르시오.
① 할인 행사를 알리려고
② 수면의 중요성을 강조하려고
③ 올바른 세안법을 공유하려고
④ 새로운 화장품을 홍보하려고
⑤ 이벤트 참여 방법을 안내하려고

02

대화를 듣고, 여자의 의견으로 가장 적절한 것을 고르시오.
① 건강을 위해서는 채식이 필수적이다.
② 채식주의자는 섭취하는 음식량이 적다.
③ 패스트푸드 음식의 섭취를 줄여야 한다.
④ 채식주의자가 되는 것은 큰 노력이 필요하다.
⑤ 채식주의자 식단은 건강에 좋지 않을 수 있다.

03

대화를 듣고, 두 사람의 관계를 가장 잘 나타낸 것을 고르시오.
① 비서 - 임원
② 호텔 직원 - 투숙객
③ 경비원 - 주민
④ 전화 상담원 - 고객
⑤ 여행사 직원 - 여행객

04

대화를 듣고, 그림에서 대화의 내용과 일치하지 않는 것을 고르시오.

05

대화를 듣고, 남자가 할 일로 가장 적절한 것을 고르시오.
① 과일 배달해주기
② 과일값 깎아주기
③ 싱싱한 과일 골라주기
④ 과일 따로 챙겨 두기
⑤ 블루베리 파이 만들어주기

06

대화를 듣고, 여자가 기분이 좋은 이유를 고르시오.
① 시험이 끝나서
② 기말고사 성적이 좋아서
③ 최신형 자동차를 구입해서
④ 부모님께서 해외여행을 허락하셔서
⑤ 방학 동안 자동차 여행을 할 수 있어서

07

대화를 듣고, 여자가 지불할 금액을 고르시오. 3점
① $28.00
② $28.50
③ $29.00
④ $29.50
⑤ $30.50

08

대화를 듣고, 그림에 관해 언급되지 않은 것을 고르시오.
① 제목
② 가격
③ 사용된 물감
④ 완성 연도
⑤ 크기

09

wireless system에 관한 다음 내용을 듣고, 일치하지 않는 것을 고르시오. 3점
① 설치 공사는 금요일 저녁에 시작한다.
② 일요일 오전까지 공사가 완료될 예정이다.
③ 공사 중에는 인트라넷 원격 접속이 불가능하다.
④ 월요일에 접속 방법에 관한 설명이 있을 것이다.
⑤ 건물 내 모든 층에서 이용이 가능할 것이다.

10

다음 표를 보면서 대화를 듣고, 남자가 구입할 러닝머신을 고르시오.

	Model	Max. Speed	Courses	Belt Length	Folding
①	A	10 mph	12	56 inches	Yes
②	B	11 mph	12	58 inches	Yes
③	C	11 mph	8	55 inches	Yes
④	D	12 mph	14	55 inches	No
⑤	E	12 mph	12	58 inches	No

11

대화를 듣고, 남자의 마지막 말에 대한 여자의 응답으로 가장 적절한 것을 고르시오.

① Could you email me?
② Thank you for lending it to me.
③ I didn't know you sent me a letter.
④ I think you need to check your mailbox more often.
⑤ Thank you so much. There must have been a mistake.

12

대화를 듣고, 여자의 마지막 말에 대한 남자의 응답으로 가장 적절한 것을 고르시오.

① No, thanks. I've already tasted it.
② Then will we go take a ship from there?
③ Yes, I want some chocolate for Valentine's Day.
④ No, I've already shipped my parcel at the post office.
⑤ Yes, please. I'm curious about how you package your chocolate.

13

대화를 듣고, 여자의 마지막 말에 대한 남자의 응답으로 가장 적절한 것을 고르시오.

Man: _____
① I've never studied there, either.
② I agree. Their security is quite good.
③ Oh, you must have been embarrassed.
④ My father goes to the library sometimes.
⑤ I can't believe they treated your father like that.

14

대화를 듣고, 남자의 마지막 말에 대한 여자의 응답으로 가장 적절한 것을 고르시오.

Woman: _____
① You already have high grades.
② Yes. Maybe you can go back next year.
③ You should think about the volleyball team instead.
④ An injury shouldn't stop you from running.
⑤ I think your parents may not approve.

15

다음 상황 설명을 듣고, Lisa가 남자에게 할 말로 가장 적절한 것을 고르시오.

Lisa: _____
① Now you've paid off everything.
② I've never been so nervous in my life.
③ Please take your hands off my painting.
④ Give it to him if you want. I don't mind.
⑤ Thanks for your purchase and your generosity.

[16~17] 다음을 듣고, 물음에 답하시오.

16

남자가 하는 말의 주제로 가장 적절한 것은?
① choosing a new home wisely
② how to move furniture easily
③ keeping your home organized
④ useful tips for home decorating
⑤ advice for packing household items

17

언급된 물건이 아닌 것은?
① electronics   ② kitchenware   ③ small furniture
④ clothes   ⑤ books

녹음을 다시 한 번 듣고, 빈칸에 알맞은 말을 쓰시오.

01

M: Try new Sosoft Lotion today! After two weeks, we guarantee you'll _____ _____ _____ _____ in the tone and color of your skin. Freckles, acne and _____ _____ _____ and your skin will be softer and smoother. Sosoft Lotion is available for all skin types: dry, normal and oily. Choose the lotion that suits your skin, and feel confident that this product is working for you. _____ _____ every night, and we guarantee you'll be satisfied. Don't miss out on this _____ _____ on Sosoft Lotion, here today.

02

M: So what do you want to have for lunch?
W: There's a new hamburger restaurant down the street. How about going there?
M: I don't eat hamburgers. I've decided to _____ _____ _____.
W: Are you sure about that? You'd be giving up a lot of good food.
M: Yes, but I heard that a vegetarian diet _____ _____.
W: Hmm. I don't ever want to be a vegetarian.
M: Why not? Don't you want to be healthier?
W: It takes longer for vegetarians to heal, and they can easily become _____ _____ _____.
M: Oh, I didn't know that. Maybe I ought to think again about being a vegetarian.

03

[Phone rings.]
M: Good afternoon, how can I help you?
W: Hi. I'm calling from Room 204, and I have a question.
M: All right. Let me know what your question is about, and I'll _____ _____ _____ to the appropriate person.
W: I'm trying to make an international call from my room, but I can't.
M: Oh, in that case, I'm the person you need to talk to.
W: I've been pressing "9" and then the number, but _____ _____ _____.
M: For an international call, you need to press "0" first, and then "9".
W: Oh, I see. And do I need to _____ _____ _____ _____?
M: No. You can dial directly, and the charges will be added to your bill _____ _____ _____ _____.
W: Great. Thanks for your help.

04

M: I'm tired of this quiz show.
W: Then let's watch the nature show. It starts at 8.
M: What time is it now? Oh, it'll be _____ _____ _____ _____.
W: Great! [Pause] Oh, I can't get this remote control to work.
M: Maybe it needs a new battery.
W: I put a new battery in the remote control yesterday.
M: Did you get it from the striped box next to the plant?
W: Yes. Why?
M: All of _____ _____ _____. I'm going to recycle them later. The new ones are in the middle drawer under the TV.
W: Oh, really? I'll get a new one, then.
M: Okay. And can you _____ _____ _____ on the right? It looks like I left it open.

05

W: Mr. Wilson, are there any fresh blueberries left?
M: Sorry, Ms. Carter. We've already sold them all

today.

W: But you'll be getting more in tomorrow?

M: Yes. They get picked in the mornings and _____ _____ in the afternoons.

W: How many boxes do you usually get?

M: Oh, about 30 boxes. And we always _____ _____ by the end of the day.

W: That's what I'm worried about! I really want some of your blueberries, but I can't get here early.

M: Why don't I _____ _____ _____ for you, then?

W: Would you do that? I'll drop by at about 8 to pick them up.

M: That's fine. How many would you like?

W: I'll take two boxes. I'm going to make a blueberry pie.

M: I'll _____ _____ _____ so I don't forget.

06

M: Joan. You look like you're in a good mood. Did you do well on your final exams?

W: I haven't got my grades back yet.

M: Oh. Well, I guess you're happy to have three months of summer vacation _____ _____ _____.

W: Actually, I'm already a bit bored after a few days of no school.

M: That's too bad. Do you have any special plans to keep you entertained this summer?

W: Yes, I'm going to _____ _____ _____.

M: Road trips? How?

W: Because I _____ _____ _____ _____ yesterday. My mom said I can borrow her car whenever I want.

M: That's awesome! That's why you look excited. You can go wherever you want to.

07

[Phone rings.]

M: Home Chicken, _____ _____ _____ _____ _____?

W: Yes, I'd like two orders of fried chicken and one order of spicy chicken.

M: Would you like our combo deal _____ _____ _____ _____ _____?

W: What kind of combos do you have?

M: You can get French fries and cola with an order of fried chicken for $9.50. If you order the chicken only, it's $9.

W: How much is the spicy chicken?

M: The regular spicy chicken is $10. And the spicy chicken combo is also just 50 cents more.

W: I see. I think I'll take two fried chicken combos, and one regular spicy chicken.

M: And will you be _____ _____ _____ or would you like delivery?

W: _____ _____ _____ _____, please.

M: Your order should be there in 30 minutes. Thank you for calling Home Chicken.

08

W: This is a lovely painting. _____ _____ _____ _____?

M: Yes, it is. We're asking $500 for it.

W: I see. Was it painted by a famous artist?

M: Yes. Her name is Annabella Longo. She's _____ _____ _____ these days.

W: I've heard of her. She's famous for her watercolor paintings, isn't she?

M: That's right, although this one is oil on canvas.

W: I see. Who is the man in the painting?

M: That's her brother who died in 2005. It was just a year after she _____ _____ _____.

W: That's too bad. But I think the painting would look great on my living room wall.

M: I'm sure it would. It's 50 centimeters wide and 75

centimeters tall.

W: That's the perfect size. I'll take it.

M: Excellent. We accept both cash and credit cards.

09

W: May I have everyone's attention, please? Starting on Friday at 6:30 p.m., our internet connection will be disabled while _____ _____ _____ _____ is installed throughout the building. This is a major project and isn't expected to be completed until sometime Sunday morning. This will, of course, affect anyone who plans on working over the weekend. You also will be _____ _____ _____ to our intranet site remotely, so please plan accordingly. We will be holding a meeting first thing on Monday to show you how to log in to the new system. Starting on Monday, the system will _____ _____ _____ in the office except for the basement levels. If you have any questions, please contact Dave in the IT department on his cell phone.

10

M: Excuse me. Do you have _____ _____ _____ _____?

W: Yes, we do. These five models are all on sale at 20% off.

M: Oh, great! But I don't know which one to choose.

W: I can help you. What features are you looking for?

M: Well, I want one with a maximum speed higher than ten miles per hour.

W: _____ _____ _____ _____. Do you care about how many course settings it has?

M: Yes, I do. I need at least 10 different courses to choose from.

W: Okay. And you should also consider the length of the belt.

M: I'm pretty tall, so I think it should be longer than 55 inches.

W: All right. Are there any other features you require?

M: Yes. I want to store it under my bed, so _____ _____ _____ _____ _____ _____.

W: In that case, I have the perfect treadmill for you.

11

M: Hello, I'm looking for Olivia Murray.

W: Yes, that's me. Oh, you must be the new neighbor who _____ _____ _____.

M: Yes. I dropped by to give you this letter. It was in my mailbox.

W: (Thank you so much. There must have been a mistake.)

12

W: On the left is the machine that makes our chocolate.

M: Wow, the air _____ _____ _____ in here!

W: On the right is our shipping department. Do you want to look around?

M: (Yes, please. I'm curious about how you package your chocolate.)

13

M: Megan, you don't look so good.

W: Yeah, I had a terrible night last night.

M: What happened?

W: Somebody _____ _____ _____ from my backpack!

M: Really? Where were you?

W: I was _____ _____ _____ _____, studying after school.

M: Did you lose much?

W: No, there wasn't much money in the wallet. And there weren't any credit cards.

M: That's good.

W: But my father went to the library and yelled at the librarians for _____ _____ _____ _____.

M: (Oh, you must have been embarrassed.)

14

M: I'm thinking about _____ _____ _____
_____, Jessica.

W: Why? You're one of the top runners on the team.

M: My grades in English and science are dropping.
My parents think I'm doing too much.

W: You also _____ _____ _____, don't you?

M: Yes, I do that three times a week.

W: Do you enjoy that more than being on the track
team?

M: It's difficult to say. I like them both.

W: Hmm... How are you going to decide?

M: I think I'll talk to my track coach first and see if I
can _____ _____ _____ _____.

W: (Yes. Maybe you can go back next year.)

15

W: Lisa is a well-known artist and _____ _____
in Rocky Bay. She gives her time and money to
environmental causes. Through her art, she tries
to _____ _____ of ways people can help the
environment, such as participating in recycling
programs. In her paintings, she likes to show the
beauty of nature, and how human activity can
_____ _____ _____. During one of her art
exhibits, a new resident of Rocky Bay was so
impressed with Lisa's art that he not only bought
two paintings, but also _____ _____ _____
_____ _____ to the environmental group. In
this situation, what would Lisa most likely say to
the man?

Lisa: (Thanks for your purchase and your generosity.)

16~17

M: In today's workshop I want to talk about something
practical. We've all moved to a new house or
apartment at _____ _____ _____ _____
_____. It can be stressful, especially when
we're getting ready to move. Packing up all of our
belongings is _____ _____ _____, but doing
it wisely can make moving much easier. When
packing electronics, for example, be sure to place
each device in the same box as its manual. _____
_____ _____ is also important. Kitchenware
should be wrapped up in newspaper, and small
pieces of furniture should be covered in bed
sheets. This will _____ _____ _____. Also,
heavy items such as books should be placed in a
few small boxes rather than one large one. This
will make them much easier to move. Remember,
a little bit of careful planning can turn a stressful
move into a smooth and stress-free experience.

1번부터 17번까지는 듣고 답하는 문제입니다.
1번부터 15번까지는 한 번만 들려주고, 16번부터 17번까지는 두 번 들려줍니다. 방송을 잘 듣고 답을 하기 바랍니다.

01

다음을 듣고, 여자가 하는 말의 목적으로 가장 적절한 것을 고르시오.
① 분실물을 찾아주려고
② 행사 진행을 알리려고
③ 영업시간을 공지하려고
④ 방문 고객을 환영하려고
⑤ 신상품 출시를 예고하려고

02

대화를 듣고, 두 사람이 하는 말의 주제로 가장 적절한 것을 고르시오.
① 과로의 위험성
② 규칙적인 운동의 필요성
③ 컴퓨터 보안 프로그램 설치의 필요성
④ 컴퓨터 타자 치는 속도를 높이는 방법
⑤ 컴퓨터 사용 시 손목 부상을 예방하는 법

03

대화를 듣고, 두 사람의 관계를 가장 잘 나타낸 것을 고르시오.
① 호텔 직원 - 투숙객   ② 고객 - 안내원
③ 관광객 - 관광객   ④ 상인 - 손님
⑤ 고객 - 판매원

04

대화를 듣고, 그림에서 대화의 내용과 일치하지 <u>않는</u> 것을 고르시오.

05

대화를 듣고, 남자가 여자에게 부탁한 일로 가장 적절한 것을 고르시오.
① to have dinner with his family
② to lend her wedding gown to his sister
③ to design a wedding gown for his sister
④ to help his sister choose a wedding gown
⑤ to work for a fashion magazine with him

06

대화를 듣고, 여자가 팀을 바꾸려는 이유를 고르시오.
① 팀원들과 사이가 좋지 않아서
② 프로젝트 기간이 너무 길어서
③ 팀원들과 시간 맞추기가 어려워서
④ 함께 하고 싶은 다른 친구가 있어서
⑤ 자신이 원하는 프로젝트 주제가 아니어서

07

대화를 듣고, 남자가 지불할 금액을 고르시오.
① $13   ② $14   ③ $15   ④ $17   ⑤ $18

08

대화를 듣고, Georgia Jones에 관해 두 사람이 언급하지 <u>않은</u> 것을 고르시오. 3점
① 출생지   ② 사망 연도   ③ 설계한 건물 수
④ 건축 양식   ⑤ 대표작 준공 시기

09

volunteer program에 관한 다음 내용을 듣고, 일치하지 <u>않는</u> 것을 고르시오. 3점
① 미술 전공자는 지원 시 우대받는다.
② 80여 명의 자원봉사자들이 있다.
③ 대부분의 자원봉사자들은 미술 전시관에서 일한다.
④ 방문객을 대상으로 한 설문조사를 돕는다.
⑤ 기념품 가게에서 일하기도 한다.

10

다음 표를 보면서 대화를 듣고, 여자가 수강할 강좌를 고르시오.

Monthly Schedule for Japanese Class

	Class	Day	Time	Level	Price
①	A	Mon. / Wed.	20:00-22:00	From Beginner to Intermediate	$140
②	B	Tue. / Thur.	09:00-11:00	From Beginner to Intermediate	$110
③	C	Fri.	20:00-22:00	From Beginner to Advanced	$120
④	D	Sat.	09:00-11:00	From Beginner to Advanced	$110
⑤	E	Sun.	20:00-22:00	From Intermediate to Advanced	$110

11

대화를 듣고, 여자의 마지막 말에 대한 남자의 응답으로 가장 적절한 것을 고르시오.

① I can't find a repair shop nearby.
② I got the TV fixed just a month ago.
③ You should watch this program as well.
④ I can't remember. Let me get new batteries.
⑤ The grocery store downstairs sells batteries.

12

대화를 듣고, 남자의 마지막 말에 대한 여자의 응답으로 가장 적절한 것을 고르시오.

① You should go there with your friends.
② My classmates decided to go to the park.
③ I want to be a zookeeper when I grow up.
④ I was disappointed there weren't many animals.
⑤ The giant pandas were the most interesting to me.

13

대화를 듣고, 남자의 마지막 말에 대한 여자의 응답으로 가장 적절한 것을 고르시오.

Woman: _____

① Oh, really? Let's go right now.
② I know. Even babies look in the mirror.
③ You should read that science magazine.
④ Oh, you can't judge people by their IQs.
⑤ It's okay. I can see the dolphins on TV tonight.

14

대화를 듣고, 여자의 마지막 말에 대한 남자의 응답으로 가장 적절한 것을 고르시오.

Man: _____

① I won't. I promise.
② I'll turn the music up for you.
③ My parents will buy me a new one.
④ I like your cell phone, too. It's cool.
⑤ Don't you know you should not make calls while driving?

15

다음 상황 설명을 듣고, 상사가 남자에게 할 말로 가장 적절한 것을 고르시오.

Boss: _____

① There's no solution.
② I'll let you work on night duty.
③ Be at work by 6 a.m. tomorrow.
④ You don't need to work overtime today.
⑤ Make sure you are on time from now on.

[16~17] 다음을 듣고, 물음에 답하시오.

16

남자가 하는 말의 주제로 가장 적절한 것은?

① how to cure insomnia
② why having a routine is important
③ how different energy drinks work
④ what foods can improve sleep quality
⑤ what methods are best to fight sleepiness

17

언급된 음식이 **아닌** 것은?

① coffee ② water ③ energy drinks
④ fruit ⑤ nuts

녹음을 다시 한 번 듣고, 빈칸에 알맞은 말을 쓰시오.

01

W: I'd like to make a short announcement for customers shopping at our department store. We've prepared a small event _____ _____ _____ _____ our appreciation for our customers. At 5 o'clock, which is just one hour from now, we will raffle off to give various gifts for ten of our customers. Anyone who _____ _____ _____ _____ today is eligible to win. Just cut out the entry form from the bottom of your receipt, and submit it to the box _____ _____ _____ _____ _____. We encourage everyone to take part in this event. We'll always do our best to make your shopping experience as enjoyable as possible. Thank you very much.

02

W: It seems like you're spending a lot of time on the computer these days, James.

M: Yes, that's true. I'm working _____ _____ _____ _____ to get all of my work reports done on time.

W: Well, if you're using the keyboard that much, you should be careful.

M: What makes you say that? I can't _____ _____ from typing, can I?

W: Actually, you can. When you repeat the same motion over and over, it can damage your wrists.

M: I didn't know that.

W: It's important that you _____ _____ _____. Even if they're just for a few minutes, they can help your body recover.

M: Okay. What else can I do?

W: _____ _____ _____ _____. Don't let them rest on the desk while typing.

M: Thanks for the tips.

03

M: Hi, do you know if _____ _____ _____?

W: Probably not. There's a bag on it. You didn't get a bed number from the reception?

M: Yes. This is supposed to be my bed. But as you can see, there's a backpack on it.

W: Oh, maybe someone just put it on the bed for a second. If you have the number, it's yours.

M: Thanks. By the way, I'm Martin from the US.

W: Lovely to meet you. I'm Sophie, from Croatia.

M: _____ _____ _____ _____ _____?

W: For about three weeks. Are you new to London?

M: Yes, it's my first day. Are there _____ _____ _____ _____ _____?

W: I'd strongly recommend Camden Market on a weekend. You can experience the lively atmosphere there.

04

W: Come over here! There seem to be lots of grass-eating animals here.

M: Yeah. Oh! The sheep is just _____ _____ _____ _____ _____.

W: Yes, it is so cute. Look at that little girl. She is holding her mom's hand, and she can't take her eyes off of the sheep.

M: She's probably just curious because the sheep is wearing a shirt. It looks very funny.

W: Hey, that zookeeper is holding a baby rabbit in his hands.

M: Those little boys are petting the rabbit. They seem to be amazed by _____ _____ _____ _____.

W: Oh look! There is a zebra over there. They're my favorite.

M: Let's get closer and look at it.

W: But we should wait until that guy finishes taking a picture of his son _____ _____ _____ _____ _____.

05

M: Hello, Jennifer. Hey, are you free tomorrow?

W: Yeah, sure. Why?

M: Actually, I'm going shopping for wedding gowns tomorrow, and I need your help.

W: Wedding gowns? _____ _____ _____? Why didn't you tell me!

M: Calm down. My sister is getting married next month. She asked me to help her choose a dress.

W: What about her fiancé? _____ _____ _____ _____ _____ for the bride's brother to help her choose it?

M: He's busy these days, so I said I would go with her. Anyway, I need your advice. Since you work for a fashion magazine, I thought you would _____ _____ _____ _____ gowns.

W: I'd be happy to help you. This will be fun.

M: Thanks. Is 3 o'clock tomorrow afternoon okay with you? I can pick you up at home.

W: That's fine. See you then.

06

W: Hello? I'm sorry to bother you, professor.

M: It's okay, Amy. Come on in and sit down. What's the problem?

W: I want to know if I can _____ _____ _____ _____.

M: Why? Is there a problem with your team members?

W: No, they're okay. But I have a different class schedule from them. So it's _____ _____ _____ _____ _____.

M: I understand. Let's see... Okay. How about joining Craig and Madison's team? They need one more person because Peter _____ _____ _____.

W: That's great. Craig and I have the same major, and Madison takes a psychology class with me. Thank you, professor.

M: You're welcome.

07

W: Good afternoon. Can I help you with something?

M: Yes. Let's see. I'll take *The Fallen City* and *Lost Paradise*.

W: Both movies you've chosen are _____ _____. They're $5 each.

M: All right. Oh, has the *X-Men* DVD been returned?

W: Let me check. *[Typing sound]* Yes, it's in. It's not a new release, so it's $3.

M: Great. _____ _____ _____ _____ _____.

W: Will that be all for you today? You have two new release DVDs and one non-new release.

M: Actually, I also want _____ _____ _____ _____ of this fantasy novel. Here they are.

W: Okay, the books are $1 each.

M: All right. That'll be all for me today.

08

W: What an unusual building! I've never seen anything like it.

M: Yes. It _____ _____ _____ Georgia Jones, the famous architect.

W: Oh, I've heard of her. Is she from around here?

M: Well, she was born in Paris, but she lived here most of her life.

W: So she's _____ _____ _____?

M: No, she died in 2007. But while she was alive, she designed nearly 100 buildings.

W: Her style is really unique.

M: She studied in Asia while she was a student, so her work _____ _____ _____ _____ _____.

W: I see. So is this building her most famous work?

M: No. It's quite interesting, but not as well-known as some of her other buildings.

W: Can we go inside? I want to see the interior.

M: Sure. Let's go.

09

M: The Royal Museum is one of the most impressive museums in the world. It contains a breathtaking collection of art and antiques from various cultures, making it one of Europe's top tourist attractions. In order to _____ _____ _____ _____ well, we operate an ongoing volunteer program for people who are interested in helping run the museum. Our volunteers, who generally number around 80, are extremely valuable to us. Most of them work in the art galleries, where they give information to visitors _____ _____ _____ _____ their museum experience. Some help in the museum's administrative departments by assisting with _____ _____ _____. Others serve our visitors in the gift shop and the cafeteria. We really couldn't achieve so much without them.

10

W: Oh, look. Here's the new schedule for Japanese classes.

M: Oh, that's right! You want to _____ _____ _____, don't you? Well, there seems to be plenty of beginner classes available.

W: Hmm... Fridays are impossible for me. I go to my reading club every Friday.

M: Then how about _____ _____ _____?

W: I work part-time from 9 to 5 on Saturdays.

M: Well, they have a class on Sundays too.

W: Yeah, but it's for people from intermediate to advanced.

M: Oh, right. How about weekday evenings?

W: Hmm... I _____ _____ _____ to be under $120. I already have a lot of monthly bills to pay.

M: Then can you make it on a weekday morning? There's a cheap morning one.

W: Oh yes, you're right. That one works.

M: Perfect! Let's get you signed up.

11

W: What's wrong with the TV, Jake?

M: I'm not sure, but it's not working. It won't even turn on!

W: Maybe something's wrong with the remote control. When did you last _____ _____ _____?

M: (I can't remember. Let me get new batteries.)

12

M: Hello, Claire. Where did you go for _____ _____ _____ _____?

W: We went to Whitewater Park. The zoo there was very big.

M: That's nice. I went there, too, last year. What did you like there the most?

W: (The giant pandas were the most interesting to me.)

13

W: Wow, take a look at this article!

M: What's it about?

W: It's about how dolphins can _____ _____ in the mirror.

M: Really? I thought only humans and chimpanzees could do that.

W: It's hard to believe, but apparently dolphins can recognize themselves, too.

M: Right. I read in a science magazine that dolphins have a high IQ.

W: What magazine was it? I'd like to read more about it.

M: It also has a special article _____ _____ _____ this month.

W: That sounds interesting.

M: I can _____ _____ _____ _____ with you and show it to you.

W: (Oh, really? Let's go right now.)

14

M: Ms. Peterson, please give me my cell phone back.

W: You _____ _____ _____, Paul.

M: I'm so sorry. I thought I had set my cell phone to silent, but I hadn't. It was just an _____ _____.

W: The sound of your phone really disturbed the class. Why did you have it so loud?

M: I'm sorry. I can't hear it very well at home, that's why.

W: Anyway, as a punishment, I'm going to tell your parents what happened.

M: Please don't... Couldn't you forgive me just this one time, Ms. Peterson? If my parents find out, they'll take the phone away from me. I worked a whole month to pay for it, and I've only had it for three days.

W: Well, all right. _____ _____ _____ to class again, or I'll call your parents next time.

M: (I won't. I promise.)

15

W: A young man was excited to get his first job. There was only one problem. Although he hated getting up early, he was supposed to _____ _____ _____ _____, not the night shift. No matter what he did to wake up, he always fell asleep again. But he _____ _____ _____ _____ and was able to make it to work on time every day during the first week. But after that, he started arriving late. After the third time, the boss said he _____ _____ _____ if he arrived late again. The young man told his boss the problem. The boss thought about it and told him that he _____ _____ _____. In this situation, what would the boss most likely say to the young man?

Boss: (I'll let you work on night duty.)

16~17

M: Do your parents drink a cup of coffee first thing each morning? Or perhaps you have friends who buy energy drinks to stay awake while studying. Both coffee and energy drinks help people _____ _____ _____ because they contain something called caffeine. It's effective, but it can be unhealthy if you _____ _____ _____ _____ _____. What's worse, it can interrupt your sleep cycle and stop you from getting rest when you really need it. However, there are also many natural methods you can use to _____ _____ _____ _____ _____. The simplest way is to get up and move around. Doing so will bring more oxygen to your brain and keep you awake. Eating a healthy snack, such as fruit or nuts, can also help. These foods contain the protein that a sleepy body needs to get energy. Finally, bright lights can be an easy way to _____ _____ _____. If you're in a dark place, your body might think it's time for bed.

1번부터 17번까지는 듣고 답하는 문제입니다.
1번부터 15번까지는 한 번만 들려주고, 16번부터 17번까지는 두 번 들려줍니다. 방송을 잘 듣고 답을 하기 바랍니다.

01

다음을 듣고, 남자가 하는 말의 목적으로 가장 적절한 것을 고르시오.
① 인터넷 쇼핑몰을 홍보하려고
② 컴퓨터 프로그래머를 모집하려고
③ 새 홈페이지 이용 방법을 설명하려고
④ 홈페이지 접속 오류에 대해 사과하려고
⑤ 바이러스 백신 프로그램 설치를 권장하려고

02

대화를 듣고, 여자의 의견으로 가장 적절한 것을 고르시오.
① 출근 시간을 연장해야 한다.
② 대중교통 이용을 장려해야 한다.
③ 지하철 환승 구간의 개조가 필요하다.
④ 통근 시간대에 지하철 운행을 더 자주 해야 한다.
⑤ 지하철이 혼잡할수록 질서를 지키는 것이 중요하다.

03

대화를 듣고, 두 사람의 관계를 가장 잘 나타낸 것을 고르시오.
① 매니저 - 배우   ② 기자 - 배우   ③ 팬 - 연예인
④ 서점 주인 - 기자   ⑤ 배우 - 감독

04

대화를 듣고, 그림에서 대화의 내용과 일치하지 않는 것을 고르시오.

05

대화를 듣고, 남자가 할 일로 가장 적절한 것을 고르시오.
① 문서 번역하기
② 직장 추천하기
③ 이력서 검토하기
④ 여행 정보 제공하기
⑤ 프랑스어 선생님 소개하기

06

대화를 듣고, 여자가 전화를 건 이유를 고르시오.
① 돈을 빌려달라고 부탁하려고
② 친구 병문안을 같이 가자고 하려고
③ 프랑스 여행에 대한 조언을 들으려고
④ 프랑스어 수업을 같이 듣자고 제안하려고
⑤ 수학여행에 자리가 났다는 것을 알려주려고

07

대화를 듣고, 남자가 지불할 금액을 고르시오. 3점
① $15   ② $16   ③ $17   ④ $18   ⑤ $19

08

대화를 듣고, 여자가 옮길 객실에 관해 두 사람이 언급하지 않은 것을 고르시오.
① 이용 가능 시기   ② 크기   ③ 층수
④ 전망   ⑤ 비용

09

Red Creek Camp에 관한 다음 내용을 듣고, 일치하지 않는 것을 고르시오.
① 십 대들이 체력을 강화할 수 있도록 도와준다.
② 전문가들로 구성된 강사진을 두고 있다.
③ 8월 첫 3주간 운영된다.
④ 필요하면 위험한 활동을 진행할 수도 있다.
⑤ 건강과 관련하여 의사의 확인서가 필요하다.

singer, isn't it?

W: Yes. This would be a great first step.

M: I'm sure they'll choose you. You're a great singer.

W: Thanks, but I think I sang poorly during the audition.

M: I doubt it. _____ _____ _____ _____ _____?

W: The band members didn't look impressed. They won't pick me.

M: You don't know that. When will they announce the audition results?

W: Not until this Friday. _____ _____ _____ _____ _____ they don't choose me?

M: (Forget about it until Friday. There's nothing you can do now.)

15

M: Joseph is a recent college graduate and is currently searching for his first job. One day, _____ _____ _____, he decides to search for his own name on the internet. He is surprised to find many pictures of himself. When he clicks on them, he discovers that they all come from his friend Meg's blog. _____ _____ _____ _____, and she used his name, so anyone can find them. Joseph has heard that nowadays interviewers look for information about job candidates on the internet. He doesn't want interviewers to _____ _____ _____ _____ of him based on the pictures that have been posted online. In this situation, what would Joseph most likely say to Meg?

Joseph: (Could you delete the pictures of me from your blog?)

16~17

W: Good morning, class. Today, I'd like to talk about Shakespeare's plays of the 16th and 17th centuries. I am guessing that you think watching Shakespeare plays was _____ _____ _____ _____ _____. However, in the 17th century, even people from the lower classes would attend performances. Shakespeare's plays were often performed at the Globe Theater in London, where people could watch them _____ _____ _____ _____ _____ seats on the balcony. Instead, they could pay just a penny to get in and stand on the ground. That's why those people were nicknamed "groundlings". The groundlings were known _____ _____ _____ _____ the plays. They would shout at the characters they didn't like and sometimes even throw things at them. I hope you will become as excited about Shakespeare's works _____ _____ _____ _____.

1번부터 17번까지는 듣고 답하는 문제입니다.
1번부터 15번까지는 한 번만 들려주고, 16번부터 17번까지는 두 번 들려줍니다. 방송을 잘 듣고 답을 하기 바랍니다.

01

다음을 듣고, 여자가 하는 말의 목적으로 가장 적절한 것을 고르시오.
① 신작을 발표하려고
② 작가를 소개하려고
③ 간담회를 시작하려고
④ 판촉 행사를 진행하려고
⑤ 새로 나온 책을 홍보하려고

02

대화를 듣고, 남자의 의견으로 가장 적절한 것을 고르시오.
① 부모님의 조언을 새겨들어야 한다.
② 어렸을 때부터 용돈 관리를 해야 한다.
③ 아르바이트는 학생들에게 도움이 된다.
④ 일정에 맞춰 일을 끝내는 것이 중요하다.
⑤ 학생은 공부 외에도 여러 취미 활동을 해야 한다.

03

대화를 듣고, 두 사람의 관계를 가장 잘 나타낸 것을 고르시오.
① 직원 - 상사   ② 집주인 - 세입자
③ 수리공 - 고객   ④ 전화 상담원 - 고객
⑤ 영업사원 - 고객

04

대화를 듣고, 그림에서 대화의 내용과 일치하지 않는 것을 고르시오.

05

대화를 듣고, 여자가 남자에게 부탁한 일로 가장 적절한 것을 고르시오.
① 여자를 병원에 데려다주기
② 자동차 수리하기
③ 재택근무 신청하기
④ 근무 일정 조정하기
⑤ 여자와 함께 출퇴근하기

06

대화를 듣고, 남자가 집에서 늦게 나온 이유를 고르시오.
① 아침에 늦잠을 자서
② 택시가 늦게 와서
③ 열차 시각을 착각해서
④ 여행 준비가 늦어져서
⑤ 친구와의 통화가 길어져서

07

대화를 듣고, 여자가 지불할 금액을 고르시오.
① $80   ② $99   ③ $110   ④ $117   ⑤ $130

08

대화를 듣고, Extra에 관해 언급되지 않은 것을 고르시오.
① 출시 시기   ② 최대 속력   ③ 판매 가격
④ 할부 판매 여부   ⑤ 시승 가능 여부

09

North Valley English Camp에 관한 다음 내용을 듣고, 일치하는 것을 고르시오.
① 강사진은 대부분 원어민이다.
② 7월 내내 운영된다.
③ 야외 활동 후 수업이 진행된다.
④ 주당 가격은 $700이다.
⑤ 온라인으로 신청할 수 있다.

10

다음 표를 보면서 대화를 듣고, 두 사람이 예약할 숙소를 고르시오.

Brisbane's Best Hostels

	Hostel	Price per Day	Pool	Barbecue	Breakfast
①	A	$15	X	O	Not included
②	B	$20	O	X	Not included
③	C	$30	O	O	Not included
④	D	$40	O	O	Included
⑤	E	$60	X	X	Included

11

대화를 듣고, 여자의 마지막 말에 대한 남자의 응답으로 가장 적절한 것을 고르시오.

① All right. Let's go to a gift shop.
② Sure, I can pick you up tomorrow.
③ Sorry, I already bought a gift for you.
④ Linda told me to come in the evening.
⑤ I will be ready for the party by tomorrow.

12

대화를 듣고, 남자의 마지막 말에 대한 여자의 응답으로 가장 적절한 것을 고르시오.

① No. I like reading paper books.
② No. This magazine is not mine.
③ Yes. We should focus on its advantages more.
④ Yes. It says e-books might damage your eyesight.
⑤ Yes. It says e-books will replace paper books in the near future.

13

대화를 듣고, 남자의 마지막 말에 대한 여자의 응답으로 가장 적절한 것을 고르시오. 3점

Woman: _____

① That's okay, I can give you change.
② All right. Sign on the screen please.
③ Thanks, I really appreciate your help.
④ Maybe you shouldn't drink so much coffee.
⑤ Yes, and bring back this box of dark coffee.

14

대화를 듣고, 여자의 마지막 말에 대한 남자의 응답으로 가장 적절한 것을 고르시오. 3점

Man: _____

① Well, be sure to notify me next time.
② I guess I can just go to another floor.
③ Thank you for being so understanding.
④ I'm just glad that you're finally finished.
⑤ Hopefully the email server will be fixed soon.

15

다음 상황 설명을 듣고, Mr. Wallace가 세입자에게 할 말로 가장 적절한 것을 고르시오. 3점

Mr. Wallace: _____

① I'm no longer the owner of that building.
② Don't worry, I'll come over to fix it right away.
③ I'll take care of everything after my vacation is over.
④ You'll have to call the building manager from now on.
⑤ Thank you, it's nice to have my hard work appreciated.

[16~17] 다음을 듣고, 물음에 답하시오.

16

남자가 하는 말의 주제로 가장 적절한 것은?

① 경력 관리의 중요성　　② 커버레터 작성 방법
③ 면접에 성공하는 비결　　④ 뛰어난 인재 고용하기
⑤ 입사 지원 시 유의사항

17

언급된 항목이 <u>아닌</u> 것은? 3점

① 지원 업무　　② 희망 연봉　　③ 관련 업무 능력
④ 본인의 특성　　⑤ 수상 경력

녹음을 다시 한 번 듣고, 빈칸에 알맞은 말을 쓰시오.

01

W: Can I have your attention, please? I'd like to thank you all for coming here to our store on such a rainy Tuesday morning. Of course, the reason you're here is the woman sitting at the table behind me. Charlotte Chapman has certainly _____ _____ _____ _____ herself in the literary field, and she's done so at an amazingly early age. There are even some who credit her with reviving our nation's slumping publishing industry all on her own. She'll _____ _____ _____ of her latest work of fiction for the next hour, so if you can all _____ _____ _____ _____, we can get started right away.

02

W: My mom wants me to quit my part-time job. She thinks it's _____ _____ _____ _____ _____.

M: Well, maybe, but it also teaches you to be more responsible for yourself.

W: Right. I have to finish my work before the deadline.

M: Plus, you earn money. Jobs gives you a chance to learn _____ _____ _____ your money.

W: Yeah. I have already saved quite a bit.

M: You can also tell your mom that you are learning useful skills for when you're an adult.

W: That's true. And my grades are still pretty good.

M: Right. See? There are _____ _____ _____ you can make to convince her.

W: You're right. Thanks for the tips.

03

[Phone rings.]

M: Thank you for calling RCI. How can I help you?

W: I have a few questions about the international call

program _____ _____ _____ _____.

M: Certainly. Can I first ask if you currently have an account with our company?

W: Not now, no. I did in the past, but I switched over to TeleSmart when I moved.

M: Well, actually, we _____ _____ _____ _____ to return customers.

W: Do you? I didn't realize that.

M: Yes, you'll get 5% off on the initial installation charges.

W: That's good to know. Now can I ask a few questions about the international plan?

M: Let me transfer you to _____ _____ _____ _____. Do you mind holding for a few seconds?

W: Sure, no problem.

04

[Telephone rings.]

W: Hello?

M: Hi, Jennifer. This is your brother. What are you doing?

W: I'm setting the table for Mom's birthday dinner.

M: Oh, that's great. Which table are we going to eat at?

W: The round table in the living room. I put out four chairs and _____ _____ _____ for each person.

M: Did you make the table look nice?

W: Yes, there is a napkin folded into the shape of a bird on each plate.

M: That sounds nice. Did you put a basket of flowers _____ _____ _____ _____ the table like I suggested?

W: Yes. That was a great idea. But I didn't put any candles on the table.

M: Why not? Mom loves candles.

W: Yes, but I think they _____ _____ _____ _____ _____.

05

M: Hi, Gwen. Welcome back to work. Sorry to hear about your accident the other day.

W: Thanks. I'm happy to _____ _____ _____.

M: Yeah, you were quite lucky. Let me know if there's anything I can do for you.

W: Actually, my car's going to be in the shop for two weeks.

M: All right. Do you need me to drive you to and from work?

W: Well, I can use my husband's car on Tuesdays and Thursdays. Those are my husband's days to work from home.

M: Wow, I envy him.

W: So, why don't I drive us on those days, and _____ _____ _____ on the others?

M: Sure, that would be great.

W: Yes. It would make _____ _____ _____ _____.

06

W: Hi. Where would you like to go?

M: Please take me to Seoul Station as quickly as possible.

W: Okay. _____ _____ _____ _____ _____, please.

M: How long will it take to get there?

W: It will take about 20 minutes if we don't _____ _____ _____ _____.

M: It's going to be tight. If I don't get there within 30 minutes, I'll miss my train.

W: I'll do my best, but I can't guarantee that you'll _____ _____ _____ _____.

M: That's okay. It's my fault. I'm going on a trip with my friend. Unfortunately, I misread the train schedule and left home too late.

W: Let's just hope the traffic is not heavy.

07

W: So, have you figured out what's wrong with my TV?

M: Yes, ma'am. Unfortunately, the screen is damaged. I'll need to replace it.

W: It _____ _____ _____ when we moved. How much will it cost?

M: Well, let's see... For a 50-inch television...

W: Sorry, I think it's only 47 inches.

M: Ah, right. Then a new screen will cost $80. And _____ _____ _____ _____.

W: How much is that?

M: It's $30. And a $20 installation charge will be added.

W: Hmm... It's more expensive than I expected. Do I get a discount if I _____ _____ _____?

M: Yes. You'll get 10% off the total.

W: That's good. Will it take long?

M: The screen will take three days to arrive.

W: All right. I guess I can wait.

08

W: Hi. I'm looking for a new car. I need something inexpensive.

M: Well, we have a special budget model. It's the cheapest one we have.

W: What's it called?

M: The model is called the Extra. It was just _____ _____ _____.

W: I see. How much does it cost?

M: The normal price is $12,000. But we also have an installment plan.

W: Can you tell me about that?

M: If you pay $1,000 now, you'll pay just $200 a month for the next five years.

W: That _____ _____. Can I take it for a test drive?

M: Yes. I think you'll find that it drives quite well.

W: I hope so. Where is it?

M: It's the red car over there. I'll just need to _____ _____ _____ _____ first.

09

M: Do you want to make the most of your summer vacation? Then, why don't you come to North Valley English Camp? Located on the shore of a beautiful mountain lake, our camp features a staff made up entirely of _____ _____. Programs start in July and they _____ _____ _____ _____. Campers will attend fun and informative classes in the evening, and take part in _____ _____ _____ during the day, including sports, crafts and boating, all of which will be conducted entirely in English. Best of all, North Valley English Camp is affordable: only $700 for a two-week program. Registration forms and additional information can be found on our website at www. nvecamp.net.

10

M: We need to book a room at a hostel in Brisbane soon.
W: You're right. We _____ _____ _____ _____ _____ _____. What's our budget?
M: We can afford up to $50 per night.
W: Okay. Here's a list of popular hostels. We can afford most of them.
M: Oh, good. But it's the middle of summer, so we shouldn't stay at a place without a pool.
W: I agree. How about a barbecue? Is that important?
M: Yes! Barbecue parties are a great way _____ _____ _____ _____.
W: Okay. How about breakfast? Should we get a place that includes breakfast?
M: Yeah. I like to start my day _____ _____ _____, _____ _____.
W: Okay. Then let's stay at this place.
M: All right. I'll call and reserve a room right now.

11

W: Hey, James! You know Linda's birthday party is tomorrow, right?
M: Yes. It's going to be really exciting. Did you _____ _____ _____ for her?
W: Not yet. Would you go out and look for a good one with me?
M: (All right. Let's go to a gift shop.)

12

M: Hi, Chloe. What are you reading?
W: It's a magazine article about e-books. It says e-books are very handy and have many other advantages.
M: Does it _____ _____ _____ of using e-books?
W: (Yes. It says e-books might damage your eyesight.)

13

W: Good afternoon. Is this everything?
M: Yes, just these three boxes of instant coffee, please.
W: All right then. Do you _____ _____ _____ _____?
M: Yes, here it is.
W: Thank you. That comes to $24.
M: Shouldn't it be $16? I thought they were "_____ _____, _____ _____ _____."
W: Ah, I see. This one box is dark coffee, which is more expensive. Only the mild coffee is on sale.
M: Oh, I thought all three were mild coffee. I must have _____ _____ _____ _____ by mistake.
W: That's okay. I will cancel it and scan one more box of mild coffee instead.
M: Thank you. Should I get another box of mild coffee from the shelf then?
W: (Yes, and bring back this box of dark coffee.)

14

M: Jisun, can I ask you a quick question?

W: Sure, go ahead.

M: I was wondering if you were almost finished with _____ _____ _____. I've been waiting for almost 10 minutes.

W: Oh, no. Not even close. I still have nearly 100 copies of this project to make. It should take me about another half an hour.

M: Half an hour? You should let people know if you're going to be using the copy machine for such a long time.

W: Actually, I _____ _____ _____ to everyone suggesting they use machines on the fourth or fifth floor.

M: Did you? I'm sorry. I haven't checked my e-mail this morning.

W: That's okay. I know this is _____ _____ _____, but it was unavoidable.

M: (I guess I can just go to another floor.)

15

W: Mr. Wallace is an elderly man. In his youth, he purchased several apartment buildings. As the neighborhood improved over the years, rents rose higher and higher, and now Mr. Wallace is extremely wealthy. For several decades he has worked hard, personally taking care of all the problems of _____ _____ _____ _____ _____. Now, however, he has decided to move to Spain and live a life of comfort and relaxation. He has _____ _____ _____ _____ to take care of any problems in his buildings in the future. Several days before he is scheduled to leave, one of Mr. Wallace's tenants calls to tell him that _____ _____ _____ _____. In this situation, what would Mr. Wallace most likely say to the tenant?

Mr. Wallace: (You'll have to call the building manager from now on.)

16~17

M: When applying for a job, your résumé is one way for _____ _____ to learn about your work history. However, many people forget that a cover letter is also an important part of the process. A cover letter is an introduction of yourself, and it's usually the first information an employer will have about you. Your cover letter should explain which job you would like to apply for. You should then include comments about why you are _____ _____ _____ _____. For example, you could explain the skills and personality traits you have that would be useful in the company's work environment. If you have any special achievements, _____ _____ _____ _____, they should also be mentioned. While it's fine to send the same résumé to multiple companies, you should never do this with cover letters. Instead, personalize the letter to highlight the specific traits of each company you are interested in. Employers like to see that you have taken the time to research their company _____ _____ _____ _____ _____.

1번부터 17번까지는 듣고 답하는 문제입니다.
1번부터 15번까지는 한 번만 들려주고, 16번부터 17번까지는 두 번 들려줍니다. 방송을 잘 듣고 답을 하기 바랍니다.

01

다음을 듣고, 남자가 하는 말의 목적으로 가장 적절한 것을 고르시오.
① 서평 작성을 요청하려고
② 온라인 서점을 홍보하려고
③ 이벤트 지원자를 모집하려고
④ 도서 반납 절차를 안내하려고
⑤ 책을 반송할 주소를 알려주려고

02

대화를 듣고, 여자의 의견으로 가장 적절한 것을 고르시오.
① 인터넷 광고에 현혹되지 말아야 한다.
② 불법 다운로드 행위를 제재해야 한다.
③ 자극적인 인터넷 기사를 규제해야 한다.
④ 인터넷 정보를 무조건 신뢰해서는 안 된다.
⑤ 외계인의 존재 여부는 뉴스로서 가치가 없다.

03

대화를 듣고, 두 사람의 관계를 가장 잘 나타낸 것을 고르시오.
① 판사 - 변호사   ② 경찰관 - 용의자
③ 변호사 - 용의자   ④ 경찰관 - 피해자
⑤ 변호사 - 경찰관

04

대화를 듣고, 그림에서 대화의 내용과 일치하지 않는 것을 고르시오.

05

대화를 듣고, 두 사람이 할 일로 가장 적절한 것을 고르시오.
① 이삿짐 싸기   ② 중고 매장 방문하기
③ 인터넷 설치하기   ④ 전자 상가 방문하기
⑤ 온라인으로 물건 구입하기

06

대화를 듣고, 남자가 수업에 불평하는 이유를 고르시오.
① 스페인어 시험에 낙제해서
② 강사의 춤 실력이 좋지 않아서
③ 강사가 학생을 배려하지 않아서
④ 생각했던 것만큼 실력이 늘지 않아서
⑤ 수업이 너무 많아 여가 시간이 없어서

07

대화를 듣고, 여자가 지불할 금액을 고르시오.
① $1,350   ② $1,400   ③ $1,450
④ $1,500   ⑤ $2,000

08

대화를 듣고, 목걸이에 관해 언급되지 않은 것을 고르시오. 3점
① 발견 장소   ② 재질   ③ 발견 시기
④ 금전적 가치   ⑤ 용도

09

drone에 관한 다음 내용을 듣고, 일치하지 않는 것을 고르시오.
① 원격으로 조종할 수 있다.
② 현재 약 3만 대가 사용 중이다.
③ 전쟁 지역에서 자주 볼 수 있다.
④ 정보를 수집하는 데 사용되기도 한다.
⑤ 민간용으로 쓰기 위한 점검이 진행 중이다.

10

다음 표를 보면서 대화를 듣고, 남자가 구입할 차를 고르시오.

	Model	Convertible	Air Conditioner	Leather Seats	CD Player
①	A	X	X	X	X
②	B	X	O	X	O
③	C	X	O	O	O
④	D	X	O	O	X
⑤	E	O	X	O	O

11

대화를 듣고, 남자의 마지막 말에 대한 여자의 응답으로 가장 적절한 것을 고르시오.

① I'm on my way to get my phone fixed.
② I'm using it now, and I'm pretty satisfied.
③ I'm fine. I can buy that phone anytime I want.
④ It's not very expensive, and it has a wide screen.
⑤ I searched the internet. You should look for it, too.

12

대화를 듣고, 여자의 마지막 말에 대한 남자의 응답으로 가장 적절한 것을 고르시오.

① I will decorate your garden.
② I can come to the party. See you there!
③ I want you to help. This table is quite heavy.
④ My uncle's family will be there. We haven't met up in a while.
⑤ My father will buy some food. He is also going to cook.

13

대화를 듣고, 여자의 마지막 말에 대한 남자의 응답으로 가장 적절한 것을 고르시오.

Man: _____

① It happens all the time. We look alike.
② I can introduce you to him sometime.
③ That's okay. I didn't remember you either.
④ No problem. I look very different these days.
⑤ I'm sorry, but I don't know what you're talking about.

14

대화를 듣고, 남자의 마지막 말에 대한 여자의 응답으로 가장 적절한 것을 고르시오. 3점

Woman: _____

① Cheer up. You'll do better next time.
② What time do you have to catch the bus?
③ Then I guess you should hurry up and study.
④ Don't worry. I can fix your broken computer.
⑤ Then you'll have to go to the office and explain.

15

다음 상황 설명을 듣고, George가 소녀에게 할 말로 가장 적절한 것을 고르시오. 3점

George: _____

① Wow, you must really love cats.
② I think you have the wrong address.
③ I'm sorry. I couldn't give up one of my cats.
④ Thank you! I didn't even know that she was gone!
⑤ Would you like to babysit my cats when I'm not home?

[16~17] 다음을 듣고, 물음에 답하시오.

16

남자가 하는 말의 목적으로 가장 적절한 것은?

① 자원봉사자를 모집하려고
② 수강 신청 방법을 설명하려고
③ 새로운 그림 강좌를 소개하려고
④ 일정이 변경되었음을 공지하려고
⑤ 주민센터에 기부할 것을 요청하려고

17

언급된 정보가 아닌 것은?

① the contents   ② the duration
③ the time   ④ the location
⑤ the tuition fee

녹음을 다시 한 번 듣고, 빈칸에 알맞은 말을 쓰시오.

01

[Answering machine beeps.]

M: Hello, this message is for Mr. Ron Waterman. This is the online bookshop Book World. We haven't been able to reach you, so I'm leaving this message. Thank you for _____ _____ _____ for the book reviewer last week. I'm calling to let you know you have been selected to write a review for the book *The Tipping Point*. Please write a review on our website by June 24. You may freely _____ _____ _____ about the book. Please note that the review has to be at least 200 words in length. If you have any questions, _____ _____ to call us at 080-262-5544. Thank you.

02

M: I saw an amazing news story last night. It said scientists found _____ _____ _____ _____ _____ on Earth.

W: Do you actually believe that?

M: Well, of course _____ _____ _____ at first. But when I looked at other articles on the site, I found that there have been a lot of real UFO sightings.

W: I'm not so sure. I think those sites are dangerous.

M: Really? I think they're fun!

W: They're entertaining to you, but some people really believe fake news like that. It can be harmful to our society.

M: I can see _____ _____ _____. And I guess those sites just want to make money from advertising.

W: Right. We shouldn't be so quick to trust every news website we find.

03

W: Good morning. Are you Daniel Lohan?

M: Yes. Can I help you?

W: Do you recognize the person in this photograph?

M: No, I've never seen her.

W: She claims you _____ _____ _____.

M: I never did any such thing.

W: Were you downtown yesterday at 2 p.m.?

M: No, I was here at home, watching TV.

W: Can anybody _____ _____?

M: Well, my wife was out... But I was here!

W: She's _____ _____ you. I'm afraid we have to at least take you in for more questions. Come with me.

M: I want my lawyer to meet me.

04

M: What are you looking for in the biology section?

W: I'm looking for a book about cellular biology. Actually, the book I want is _____ _____ _____ _____. Can you help me get it down?

M: Sure. Here you go.

W: Thank you. You know, it's funny. The top shelf is _____, but that shelf second from the top is empty.

M: Right. I think the books should be spread out.

W: I agree. So, what are you looking for?

M: I'm looking for information about famous biologists, but I'm not sure where to look.

W: Maybe the librarian could help. Isn't she _____ _____ _____ _____?

M: No, she's talking with the man with the big backpack. I think he has some questions too.

W: Well, you'd better ask her when she's done helping him.

05

W: I'm so _____ _____.

M: What's the matter?

W: I've been searching the internet looking for

electronic items, but it's so difficult.

M: Why are you looking for them now?

W: I move at the end of this week, and I realized all the electronic items I use _____ _____ _____ _____.

M: Why don't you go to the electronics market in Yongsan? The prices there are usually cheap.

W: I've already been there, but the prices were higher than I expected.

M: Then, why not buy _____ _____?

W: That's a great idea, but from where?

M: I know a great store. I can _____ _____ _____ tomorrow.

W: Oh, thanks.

06

W: How are your salsa dance lessons going?

M: Actually, I'm thinking of _____ _____ _____.

W: What? You said you loved it, and you even began to learn Spanish!

M: Yes, I still like learning salsa, but did I tell you about my new salsa teacher, Mariana?

W: Yes, I heard your class got a new teacher. What's she like?

M: You know, our last teacher, Sophie, was very _____ _____ _____. Unfortunately, Mariana's not.

W: I heard she's a talented dancer, and she's _____ _____ _____.

M: Yes, she dances very well. But as a teacher, she doesn't care much about the students.

W: What do you mean?

M: She _____ _____ _____ _____ in front of the other students. It's so embarrassing. She just doesn't have patience, and that's exactly what a teacher needs.

W: Oh, that's not good. How about finding another class?

M: I think I'll have to.

07

W: Excuse me, I'm looking for a washing machine.

M: Well, we have _____ _____ _____ _____ models to choose from. Do you have anything specific in mind?

W: I'm looking for something with an allergy care cycle.

M: How about this one? It's a popular large-capacity, drum-type model with an allergy care cycle.

W: Hmm... It's $2,000. That's more than I wanted to pay.

M: Well, if you _____ _____ _____ our store's credit card, you can get a 25% discount.

W: That's not bad. Can I also apply the 10% discount from the summer sale?

M: No, if you choose the credit card discount, the summer sale discount is not applied.

W: I see. I'll buy it anyway, though. How can I sign up for the credit card?

M: Just _____ _____ _____ _____.

08

M: Excuse me. Can I ask you some questions about that necklace?

W: Certainly. What would you like to know?

M: Well, first of all, where does it come from? Is it Egyptian?

W: Yes, it is. An explorer found it in an Egyptian pyramid and _____ _____ _____ _____.

M: I see. It looks like it's made out of gold.

W: It is. It is believed to be _____ _____ _____ _____.

M: That's amazing. Did it belong to an Egyptian king?

W: Actually, it belonged to Cleopatra, the famous queen.

M: No wonder it's worth so much money. Does it have any special meaning?

W: Well, it was only worn _____ _____ _____

_____.

M: That's very interesting. Thank you for answering my questions.

W: My pleasure. Enjoy the rest of the exhibit.

09

W: The U.S. military is increasing its use of drones, unmanned aircraft that can be flown remotely. Experts predict that ten years from now, as many as 30,000 drones _____ _____ _____ _____. Though often seen in war zones, not all drones attack the opposition's target. Many _____ _____ _____ _____ _____ through video cameras attached to the drone. In fact, so much information is produced that the U.S. Air Force employs about 70,000 people to _____ _____ _____. Drones are being tested for civilian use as well. This means that police officials, firefighters and emergency medical responders could be using the technology in their own operations. Love them or hate them, it seems that drones are here to stay.

10

W: What's that?

M: It's a brochure from a car company. I just _____ _____ _____ at work, so I'm going to buy a new car.

W: That's great. What kind of car?

M: Well, I've _____ _____ the new Falconi.

W: Wow, Falconis are great cars. So what's the problem?

M: There are _____ _____ _____, and I can't decide which one I want.

W: Do you want a convertible?

M: Well, no. But it has to have _____.

W: Okay. What about leather seats?

M: Yes, I want them. And I want a CD player, but not a DVD player.

W: Well, I guess you should get that one then.

M: You're right. Thanks for your help.

11

M: Why don't you have a smartphone, Lucy?

W: Actually, I am going to buy one very soon. I already _____ _____ _____ _____.

M: Really? What is it like?

W: (It's not very expensive, and it has a wide screen.)

12

W: Hi, John. What are you doing?

M: Oh, I'm just _____ _____ _____. My family is hosting a dinner party in our garden tonight.

W: That's nice. Who's coming to the party?

M: (My uncle's family will be there. We haven't met up in a while.)

13

W: Excuse me. Haven't we met before?

M: I don't think so.

W: You _____ _____ _____. I'm sure we've met. Your name is Tom, isn't it?

M: No, my name is David.

W: Oh. You _____ _____ _____ a man named Tom I met in Greenville.

M: I have a brother who lives in Greenville. But his name is Tim.

W: That's it! Tim! He worked for the electric company.

M: Yes. And did he have a big black dog?

W: Yes, a big black, hairy dog.

M: That's my brother.

W: _____ _____ _____ _____. I'm sorry about that.

M: (It happens all the time. We look alike.)

14

M: Carrie, I've done something terrible!

W: What's wrong?

M: I _____ _____ _____ from school and I left it on the bus!

W: Oh, no! You have to try to find it as soon as possible.

M: I know. I put the laptop under my seat because I was reading a book. Then, I suddenly realized it was my stop, so I _____ _____ _____ _____ quickly!

W: And forgot the laptop.

M: What am I going to do?

W: Look. Here is my cell phone. Call the bus company. Tell them what happened, and ask them to check _____ _____ _____ _____.

M: What if they don't find it?

W: (Then you'll have to go to the office and explain.)

15

W: George has two pet cats. He does his best to _____ _____ _____ them despite his busy schedule. Before he leaves for work and after he comes home, he _____ _____ and plays with them. But he knows the cats need more care and attention. One day, one of his cats is missing when George comes back from work. It has been a very busy week, and George is tired. He is not sure what to do. Just then, there is a knock at the door. When George opens it, he sees a little girl. She says she lives next door and has found his cat. She explains that she played with it _____ _____ _____ _____. George thanks the girl, but he can see she likes the cat very much. In this situation, what is George likely to say to the girl?

George: (Would you like to babysit my cats when I'm not home?)

16~17

M: Attention, everyone. I'm pleased to see that so many people signed up for Painting Basics here at the Cyrus Community Center. This course will help you learn painting techniques for beginners. From choosing the right colors and brushes to creating realistic shadows, you'll learn everything you need to know to get started with this fun and relaxing hobby. The class will _____ _____ _____ _____. We'll meet every Tuesday and Thursday for two hours from 6:30 to 8:30 p.m. Every class will be held right here in Room 203. After each class, you'll be free to _____ _____ and work on your paintings. However, please note that the community center closes at 10 p.m., so you must _____ _____ _____ by then. The course fee that you paid covers the cost of all the supplies you'll need for the class. So, the only thing you need to bring is your imagination. Before we get started this evening, I'd like each of you to tell the class your name and _____ _____ _____ _____.

1번부터 17번까지는 듣고 답하는 문제입니다.
1번부터 15번까지는 한 번만 들려주고, 16번부터 17번까지는 두 번 들려줍니다. 방송을 잘 듣고 답을 하기 바랍니다.

01
다음을 듣고, 여자가 하는 말의 주제로 가장 적절한 것을 고르시오.
① 행사 순서 안내
② 운동회 팀 편성 기준
③ 대회 시상 내역과 심사 기준
④ 부상 방지를 위한 안전 수칙
⑤ 경기 관람 시 지켜야 할 사항

02
대화를 듣고, 남자의 의견으로 가장 적절한 것을 고르시오.
① SUV는 여러모로 유용하다.
② 현재 SUV 가격은 너무 비싸다.
③ 대형차보다는 소형차가 실용적이다.
④ 연비가 높은 자동차를 구입해야 한다.
⑤ 환경을 위해 자동차 사용을 줄여야 한다.

03
대화를 듣고, 두 사람의 관계를 가장 잘 나타낸 것을 고르시오.
① 은행원 – 채무자   ② 배달원 – 고객
③ 직원 – 가게 주인   ④ 전화 교환원 – 고객
⑤ 여행 가이드 – 관광객

04
대화를 듣고, 그림에서 대화의 내용과 일치하지 않는 것을 고르시오.

05
대화를 듣고, 남자가 할 일로 가장 적절한 것을 고르시오.
① 쓰레기양 줄이기
② 재활용 통 나눠주기
③ 재활용 쓰레기 분류하기
④ 쓰레기를 지정일에 버리기
⑤ 가난한 사람들을 위해 기부하기

06
대화를 듣고, 여자가 내일 시험을 볼 수 없는 이유를 고르시오.
① 시험이 취소되어서
② 수업을 받지 못해서
③ 부모님이 허락하지 않아서
④ 본인의 졸업식에 가야 해서
⑤ 가족 행사에 참여해야 해서

07
대화를 듣고, 여자가 지불할 금액을 고르시오. 3점
① $125   ② $130   ③ $175
④ $190   ⑤ $200

08
대화를 듣고, club meeting에 관해 언급되지 않은 것을 고르시오.
① 진행자   ② 안건   ③ 참가 인원
④ 모임 장소   ⑤ 모임 시각

09
red thread disease에 관한 다음 내용을 듣고, 일치하지 않는 것을 고르시오. 3점
① 잔디에 발생하는 질병이다.
② 표면이 젖고 뿌리가 건조하면 발생한다.
③ 물을 너무 자주 주면 상태가 악화된다.
④ 손상된 잎사귀에는 가늘고 빨간 줄이 생긴다.
⑤ 현재는 치료법이 없다.

10

다음 표를 보면서 대화를 듣고, 여자가 구매할 반지를 고르시오.

Fashionable Designer Rings

	Ring	Color	Karats	Engraving	Shipping
①	A	Rose gold	10K	X	in 2 days
②	B	Rose gold	14K	O	in 1 week
③	C	Silver	N/A	X	in 1 week
④	D	Yellow gold	14K	O	in 3 days
⑤	E	Yellow gold	14K	X	in 2 weeks

* N/A: Not Available

11

대화를 듣고, 여자의 마지막 말에 대한 남자의 응답으로 가장 적절한 것을 고르시오.

① Okay, give me the file then.
② Your file cabinet is over here.
③ I think I already sent it to you.
④ All right. I'll do it as soon as I can.
⑤ Don't worry. It will be fixed this afternoon.

12

대화를 듣고, 남자의 마지막 말에 대한 여자의 응답으로 가장 적절한 것을 고르시오.

① Okay, I'll take you there next weekend.
② Well, they do not take any orders online.
③ Oh, I didn't know you imported that item.
④ Sorry, but I can't share the recipe for them.
⑤ I would love to take a trip to France someday!

13

대화를 듣고, 여자의 마지막 말에 대한 남자의 응답으로 가장 적절한 것을 고르시오. 3점

Man: _____

① Thanks! I owe you one.
② I know better than that.
③ I don't think we can afford it.
④ You never yield a single point.
⑤ I didn't want to go there in the first place.

14

대화를 듣고, 남자의 마지막 말에 대한 여자의 응답으로 가장 적절한 것을 고르시오.

Woman: _____

① Can I have a sandwich, too?
② Wait, I only ordered two hot dogs.
③ Sorry, but I think you forgot my fries.
④ Wow! Maybe I will try that one next time.
⑤ I don't think I would like cheese on a hot dog.

15

다음 상황 설명을 듣고, Rachel의 어머니가 Rachel에게 할 말로 가장 적절한 것을 고르시오. 3점

Rachel's mother: _____

① I have the sweetest kids in the world.
② This is going to be a fantastic vacation.
③ Christmas dinner was terrible this year.
④ How can you say you can't give me a hand?
⑤ I'm so happy to hear you can help with the cooking like last year.

[16~17] 다음을 듣고, 물음에 답하시오.

16

여자가 하는 말의 주제로 가장 적절한 것은?

① 최신 스마트폰의 기능 소개
② 새로운 컴퓨터 백신 프로그램
③ 전력 효율성이 좋은 제품 고르는 법
④ 비상시 컴퓨터를 보호하는 전자 기기
⑤ 휴대폰 배터리를 효율적으로 사용하기

17

언급된 내용이 <u>아닌</u> 것은?

① 충전 시간   ② 제품의 특징   ③ 사용법
④ 할인 가격   ⑤ 구매 방법

녹음을 다시 한 번 듣고, 빈칸에 알맞은 말을 쓰시오.

01

W: Attention, everyone. As you know, we've organized several events for sports day, but before we begin, I want to _____ _____ _____. After we've divided you into teams, the first thing we'll do to _____ _____ _____ is go through a physical warm-up together. And any sport can be dangerous if you play rough. Please don't kick, push or hit other players. We don't want any injuries on a day that's supposed to be _____ _____ _____. Now, I will call out the names of the students on team one.

02

M: See that white SUV over there? That's my dream car.

W: You've got to be kidding.

M: Why do you say that? It's such a nice car. I'll probably never be able to afford one, though.

W: But they're so wasteful, Eric. _____ _____ _____ _____, SUVs require much more gasoline than the average vehicle. They're so pointless.

M: I disagree. SUVs can do things that the average car can't.

W: Like go off-road or _____ _____?

M: Yeah, exactly.

W: When have you ever seen an SUV used for that purpose? People use their SUVs for driving around town.

M: Well I wouldn't. I want mine for _____ _____ _____ _____ and driving across rough countryside.

03

[Knocking on door]

M: Hello, I'm delivering the food you ordered. Please make sure everything is here.

W: Thank you, it looks good. How much do _____ _____ _____?

M: It comes to ₩30,000.

W: Here you are. I just want to tell you that I was confused when I was _____ _____ _____ tonight.

M: Oh, why's that?

W: The person taking the order spoke only French. For a minute, I thought I had the wrong number.

M: Ah yes. She answers the phone if there's no one else available.

W: I understand. It's just that I _____ _____ _____ _____ _____.

M: I'll mention it to the owners.

W: No, it's okay now that I know what's going on.

M: Yes, but other customers might get confused too.

04

W: These movie posters are cool!

M: They are! These are all popular movies from the 1990s.

W: Have you seen any of them?

M: I saw that one with the ant wearing a suit. How about you?

W: I'm familiar with this one about a pitcher. He is _____ _____ to throw the ball.

M: I saw that movie, too. Oh, I heard this movie got an award. Two guys are staring at the Great Wall of China.

W: Do you know the story from the movie?

M: No, but I want to see it. Which one would you like to see?

W: I might like this one with the girl _____ _____ _____ with her hands in front of her face.

M: I want to see the one about those three aliens _____ _____ _____ _____.

05

W: Thanks for bringing your bags down early on Tuesday morning.

M: Of course. It's _____ _____ _____, and I wanted to get them out early.

W: The only thing is you've put all the bottles, cans, plastic, and cardboard in the same bag.

M: Is there a problem with that?

W: Well, we ask _____ _____ _____ _____ into different bags.

M: I see. I just moved here, so I didn't know that. I'm sorry.

W: That's okay. But by separating them, they can go to the proper plant, meaning, the paper can go to the paper center and the plastic bottles can go to the plastics plant.

M: I'll remember that next time.

W: Great. Did you get a recycling bin?

M: No. What kind of recycling bin?

W: It's to hold your recyclables. You can _____ _____ _____ _____ _____. I can get you one.

M: Sounds good. Thanks.

06

W: Excuse me, Mr. Wilson.

M: Yes, Leah? What can I do for you today?

W: I know we have a math test tomorrow, but...

M: Yes, you've been having problems in math these days. Do you think you may _____ _____ _____?

W: I'll have to think about that. But for now, I've come to ask if I can take the test _____ _____ _____ _____.

M: That's not a very common request. What's going on?

W: My brother is graduating from university tomorrow, and my parents really want me to attend the ceremony.

M: Hmm, I guess that is a _____ _____ _____. Okay then, I expect you here at lunchtime on Wednesday to take the test.

W: Thank you, Mr. Wilson.

M: Okay, but study hard. It's not going to be easy.

W: I will. I'll get my brother to help me!

07

M: Do you like the plates?

W: Yes, I'm considering buying some of them, but I haven't decided which ones yet.

M: We're selling them _____ _____ _____ _____, since we're closing down.

W: Oh, wow. How much do these plates cost?

M: They were $300 for a set. But now, a set is just $100. There are _____ _____ _____ _____ _____.

W: That's good. Can I choose any color I want?

M: Yes, except for blue. We only have three blue plates left, so we can't offer a set.

W: Oh, but I really like those. How much are they individually?

M: Today, they are $30 each.

W: Hmm... I'd like to get a set of white plates for $100. And maybe I'll add a blue plate.

M: If you add all three blue plates, I'll give you a $5 discount on each.

W: Great! I'll take them all, then. Thank you so much!

08

W: Will you be attending tomorrow's club meeting, Simon?

M: I'm not sure. I have a violin lesson at 6 p.m.

W: That's not a problem. We should be finished by 5:30 _____ _____ _____.

M: All right. Will you be leading the meeting this week?

W: Yes, it's my turn. We have a few things to talk

about.

M: Yes. We need to _____ _____ _____ _____ for this year's festival.

W: And we also need to think of a way to _____ _____ _____ _____.

M: Will all 12 current members be there?

W: Everyone except Sarah and Todd. They'll be away on a field trip.

M: I see. So we'll be having it in Room 211 as usual?

W: That's right. Make sure to come.

M: Don't worry. See you tomorrow!

09

M: If you have a yard at your house, you'd better _____ _____ _____ red thread disease. This disease can now be found throughout the whole country. It attacks grass, and it occurs when the surface of a lawn is wet and the grass's roots are dry. _____ _____ _____ will also cause the disease to get worse. The first symptoms of red thread disease are white-colored patches. These patches can be from two inches to three feet wide. Over time, they may join to form large areas of damaged grass. If you _____ _____ _____ a blade of infected grass, you will see small red lines running across it. If you think your lawn has red thread disease, the best treatment is to use fertilizer that has nitrogen. It will help your lawn recover and prevent the disease in the future.

10

W: The rings on this website are so pretty. I'm going to order one.

M: Yes, they are nice. Will you order a gold one or a silver one?

W: I love silver, but it _____ _____ _____ _____. So I'll get a gold one.

M: Do you prefer rose gold or yellow gold?

W: I like them both. But the gold has to be 14 karats.

M: Why is that?

W: I think it looks brighter and shinier than ten-karat gold.

M: Really? _____ _____ _____ _____.

W: It's true. And I also want to have my name engraved on the inside of the ring.

M: They offer that service for a few of these styles.

W: Yes. So it looks like I need to decide between these two.

M: You want to _____ _____ _____, don't you? This one ships sooner.

W: You're right! I'll order that one.

11

W: Jake, are you done with that file I gave you?

M: No, I'm really sorry, Brianna. I haven't had time to do it yet.

W: Please _____ _____ _____ _____ this afternoon.

M: (All right. I'll do it as soon as I can.)

12

M: These pickles are so delicious. Where do you buy them?

W: I get them at a specialty shop. _____ _____ _____ France.

M: Wow, I'd like to go there with you some time.

W: (Okay, I'll take you there next weekend.)

13

W: Where should we go on vacation this year?

M: I'd like to _____ _____ _____, like Vietnam or Guam.

W: There's a lot to do in Vietnam. We would be busy seeing all the sites.

M: I don't really feel like being a tourist. I just want to relax.

W: Then Guam would be a better choice.

M: What do you think of that?

W: To be honest, I was thinking of a place that's _____ _____ _____ _____.

M: Yes, I know how uncomfortable you can get in the heat.

W: But Guam is a small island, so there's probably _____ _____ from the ocean.

M: It's okay. We can go somewhere cooler.

W: No, let's go to Guam. I know how much you've been wanting to relax on a beach.

M: (Thanks! I owe you one.)

14

W: Hi, I'd like two jumbo chili cheese dogs, please.

M: Sure, it'll just take _____ _____ _____.

W: How much is it?

M: It's $5 total.

W: Here you are. Could you put mustard and ketchup on those _____ _____?

M: Sure. Anything else?

W: Could I also have some onions, please?

M: Sure. *[Pause]* Here you are — two chili cheese dogs.

W: Thanks. Oh, and I just noticed that there is something called a "quesadoga" on your menu. _____ _____ _____ _____?

M: Oh, that is one of our special types of hot dogs. It is a hot dog wrapped in a tortilla with spicy Mexican cheese and hot sauce.

W: (Wow! Maybe I will try that one next time.)

15

M: Every Christmas, Rachel's mother makes Christmas dinner, _____ _____ _____, and cleans up after the meal. Sometimes, Rachel, her brother, and their father clean up, but they don't usually help with the cooking. This year, Rachel's mother is very sick and can only stand up for a few minutes at a time. It will _____ _____ _____ for her to take care of everything. Rachel has

decided that she will _____ _____ _____ this year, and she's going to ask her brother to help, too. In this situation, what will her mother most likely say to Rachel when she finds out?

Rachel's mother: (I have the sweetest kids in the world.)

16~17

W: A sudden loss of electricity to your computer is harmful to the system, and you could also lose the important files you were working on. Generator-X is the solution for you! Generator-X is a battery that supplies an additional 15 minutes of power to your computer. This gives you time to save your files and _____ _____ _____ _____ properly. And, Generator-X is easy to use. Simply plug it in to any standard outlet, and plug your computer into the device. _____ _____ _____ _____, Generator-X will beep to indicate that the battery is in use. That alerts the user to the problem. When the power is restored, the battery _____ _____ _____. This amazing device usually sells for over $100, but we're offering it at the special price of just $74.99. You can visit our website at www.generatorxsales.com to place an order. It is also available at department stores around the country. Get Generator-X today and _____ _____ _____ _____ _____ again!

1번부터 17번까지는 듣고 답하는 문제입니다.
1번부터 15번까지는 한 번만 들려주고, 16번부터 17번까지는 두 번 들려줍니다. 방송을 잘 듣고 답을 하기 바랍니다.

01

다음을 듣고, 남자가 하는 말의 목적으로 가장 적절한 것을 고르시오.

3점

① 홍수 피해 상황을 알리려고
② 수질 오염의 심각성을 알리려고
③ 물 부족 상황에 대해 경고하려고
④ 홍수 발생 시 안전 수칙을 설명하려고
⑤ 수해 복구 작업에 대한 도움을 요청하려고

02

대화를 듣고, 여자의 의견으로 가장 적절한 것을 고르시오.
① 학습에 유용한 게임이 필요하다.
② 게임 광고는 전면 금지되어야 한다.
③ 정부는 민원을 신속하게 처리해야 한다.
④ 게임 중독에 대한 해결 방안이 필요하다.
⑤ 청소년용 게임에 대한 판매 규정이 있어야 한다.

03

대화를 듣고, 두 사람의 관계를 가장 잘 나타낸 것을 고르시오.
① 사회자 – 신부
② 신랑 – 신부
③ 신랑 – 웨딩플래너
④ 사회자 – 웨딩홀 직원
⑤ 웨딩플래너 – 웨딩홀 직원

04

대화를 듣고, 그림에서 대화의 내용과 일치하지 않는 것을 고르시오.

05

대화를 듣고, 두 사람이 할 일로 가장 적절한 것을 고르시오.
① 파티 준비하기
② 쇼핑몰에 가기
③ 바비큐 파티에 가기
④ 주말 계획 세우기
⑤ 부모님과 식사하기

06

대화를 듣고, 남자가 새벽에 일찍 일어난 이유를 고르시오.
① 이웃이 찾아와서
② 공원에 가기로 해서
③ 자명종 시계가 울려서
④ 이웃이 소란을 피워서
⑤ 근무일이라고 착각해서

07

대화를 듣고, 여자가 지불할 금액을 고르시오.
① $11.6
② $12.0
③ $12.2
④ $16.2
⑤ $17.2

08

대화를 듣고, triathlon에 관해 언급되지 않은 것을 고르시오.
① 참가 인원
② 경기 시각
③ 참가비
④ 경기 장소
⑤ 경기 내용

09

New City Zoo에 관한 다음 내용을 듣고, 일치하지 않는 것을 고르시오.
① 나라에서 가장 큰 동물원이다.
② 500종 이상의 동물들을 보유하고 있다.
③ 동물원 내에서 버스 투어를 할 수 있다.
④ 안내소는 식당 옆에 있다.
⑤ 플래시를 터뜨리며 사진을 찍어서는 안 된다.

10

다음 표를 보면서 대화를 듣고, 남자가 선택할 수업을 고르시오.

	Class	Level	Class Type	Classes per Week
①	Beginner English	Basic	Video Chat	Three
②	One-on-One Free Talking	Intermediate	Phone	Three
③	Business English	Intermediate	Phone	Two
④	Group Free Talking	Advanced	Phone	Two
⑤	TV Drama English	Advanced	Video Chat	Two

11

대화를 듣고, 남자의 마지막 말에 대한 여자의 응답으로 가장 적절한 것을 고르시오.

① Yes, I was sorry to hear that.
② I'm afraid I'm awfully busy today.
③ Sure, I have plenty of time right now.
④ It's just a few minutes before 4 o'clock.
⑤ I don't think that entrance is open today.

12

대화를 듣고, 여자의 마지막 말에 대한 남자의 응답으로 가장 적절한 것을 고르시오.

① I might get too homesick.
② My parents have studied abroad.
③ Yes, let's meet tonight to study together.
④ Sorry, but I prefer to study at the library alone.
⑤ Actually, I didn't have a chance to think about it.

13

대화를 듣고, 여자의 마지막 말에 대한 남자의 응답으로 가장 적절한 것을 고르시오. 3점

Man: _____

① Thanks for letting me go on vacation.
② Yeah, I'm glad we finally got organized.
③ I'm sorry, but you need to work overtime.
④ Me too. This is such a relaxing vacation.
⑤ Then you don't need to take time off.

14

대화를 듣고, 남자의 마지막 말에 대한 여자의 응답으로 가장 적절한 것을 고르시오.

Woman: _____

① I hope I didn't spoil your day.
② I'm sorry about everything, too.
③ I already know you're pregnant.
④ I can't believe you're doing this to me now.
⑤ No problem. There's nothing more important than family.

15

다음 상황 설명을 듣고, Andrea가 이웃에게 할 말로 가장 적절한 것을 고르시오. 3점

Andrea: _____

① I'm so glad we're good friends now.
② I think you're right. I'll never do it again.
③ You should pick up trash on the mountain.
④ You're not supposed to leave your garbage in the hallway.
⑤ Please stop by anytime. I'll treat you to a lovely dinner.

[16~17] 다음을 읽고, 물음에 답하시오.

16

여자가 하는 말의 주제로 가장 적절한 것은?

① the first bacteria
② how viruses spread
③ the benefits of some viruses
④ the effects of bacteria on humans
⑤ the differences between viruses and bacteria

17

박테리아에 관해 언급된 내용이 <u>아닌</u> 것은?

① 발견한 사람   ② 최초 발견 시기   ③ 종류
④ 번식 과정   ⑤ 유해성 여부

녹음을 다시 한 번 듣고, 빈칸에 알맞은 말을 쓰시오.

01

M: 60 residents in three Calgary neighborhoods _____ _____ _____ their homes Wednesday evening as the Glenmore _____ _____ _____ for the first time in history. The surge of water down the Elbow River threatened many homes and apartments. Residents who stayed were _____ _____ _____ and pools of water in their basements. Damage to property could cost millions. Weather forecasters warn that the worst is _____ _____ _____. Thunderstorms and strong winds are expected throughout the next few days.

02

W: Look at this video game they're advertising.
M: Wow... That's a bloody image on the cover.
W: Can you believe that this kind of game _____ _____ _____ _____ _____?
M: What? There must be some kind of age restriction.
W: That seems sensible, but there aren't _____ _____ _____ the sale of video games in our country.
M: I didn't realize that.
W: After seeing the cover of this game, I'm _____ _____ _____ to our local politician.
M: You should. If I had kids, I wouldn't want content like this to be marketed to them.
W: I agree. It's much too violent for youngsters. We should do something about it.

03

M: How much would it cost for you to _____ _____ _____?
W: The price varies based on the services you need.
M: What is included in the basic package?
W: My basic package is to help pick out a dress, find a good catering service, and _____ _____ _____ for the wedding hall.
M: Well, honestly, we can't afford to take time off from work to focus on our wedding.
W: In that case, in addition to the basic package, I will include picking up your flowers, doing the decorations, and handling guest services on that day.
M: Oh, that's great. You'll take full charge of our wedding, so we don't have to _____ _____ _____ _____.
W: That's right.
M: Okay, then please give us the full package and make it a perfect day.
W: Sure. I'll do my best _____ _____ _____.

04

W: What a nice picture!
M: Yes, that's my family in our yard.
W: Your father looks very serious. Why does he _____ _____ _____ _____?
M: Oh, he always does that when he talks.
W: I see. Then the woman _____ _____ _____ _____ sitting next to your father is your mother?
M: No, that's my oldest sister. As she gets older, she's starting to look just like my mother. My mother is standing on the path to the house. She's the one in the flower-print apron.
W: Oh, I like her smile. And that row of flowers _____ _____ _____ is very beautiful!
M: I agree. And do you see the large plant in the round pot next to the door?
W: Yes, I see it.
M: My grandparents gave that to us as a special gift.

05

M: I guess you're going to spend this weekend _____ _____ _____ _____.

W: Why do you say that?

M: It's your father's birthday, isn't it?

W: So? We're supposed to spend this weekend together.

M: I thought you would at least go to your dad's birthday dinner.

W: No, I'm just going to stop by and _____ _____ _____ _____.

M: And then we'll go to Jack's barbeque party?

W: I don't see why not. But first, help me buy my dad's present.

M: Sure. Let's go to the mall. There's _____ _____ _____ _____ _____.

W: Really? Great! Let's go now.

06

W: Hello, Bruce. Are you ready to go to the park?

M: No, I'm not.

W: What's wrong? You look exhausted. Were you up late last night?

M: No, I went to bed at 11 o'clock. But I _____ _____ at 5 a.m.

W: 5 a.m. on a Saturday morning? I thought you were going to sleep late.

M: That was my plan.

W: Did you forget to _____ _____ your alarm clock?

M: No, my alarm clock didn't wake me up. It was my neighbors. They were _____ _____ all morning.

W: That's terrible. Why didn't you go over and ask them to be quiet?

M: I was too tired. Instead, I just lay in bed all morning and stared at the ceiling.

W: You should go home and _____ _____ _____ _____. We can go to the park tomorrow.

07

M: Hello. May I _____ _____ _____?

W: Hi. I'd like to order a family meal.

M: Is that for a family of four or six?

W: _____ _____ _____?

M: The family meal for four is $10. The family meal for six costs $15.

W: Well, I have three kids and a husband.

M: Then I suggest that you get the family meal for four and order two extra biscuits for 60 cents each. Children usually prefer our biscuits.

W: How about drinks? How many sodas _____ _____ the family meal for four?

M: There are four, so you'll need to order one more for _____ _____ _____.

W: Okay, I'll do that. Thank you.

08

[Phone rings.]

M: Hello, this is Chris speaking.

W: Hi, Chris. This is Jane from the Triathlon Club.

M: Yes, hi.

W: I'm calling to _____ _____ about tomorrow's triathlon.

M: I was waiting for your call. This is going to be my first time _____ _____ _____ _____.
I'm just curious, but how many people are participating?

W: There will be about 100 people. Well, the start time is 6 a.m., so you should arrive at World Cup Stadium no later than 5:40.

M: Okay. Does the race begin right away?

W: There's a bit of time _____ _____ _____, but yes, it'll start at 6 a.m. on the dot.

M: And the swimming part comes first, right?

W: Yes, that's correct. And it's followed by biking and then running.

M: All right. Thanks for the information.

W: You're welcome.

09

W: Hello everyone, and welcome to the New City Zoo. This is the biggest zoo in the state and the second biggest in the nation. It features thousands of animals from _____ _____ _____ _____. There are three popular methods of getting around the zoo: You can follow our pedestrian pathways, _____ _____ _____ _____ _____, or ride on one of the gondolas located high overhead. More information on all of these options is available at the help desk located right over there, next to the cafeteria. On behalf of the zoo, I'd like to request that you follow some basic guidelines. First, don't feed the animals unless you receive permission in advance. Second, don't _____ _____ _____ _____ _____. And third, please always throw trash away in a trash can.

10

M: Did you see this advertisement? A new school is offering English lessons by phone or live video chat.

W: That sounds really convenient. Are you going to _____ _____ _____ _____?

M: Yes, but I'm not sure which one is best for me.

W: Well, basic level would be too easy. Either intermediate or advanced would suit you.

M: I agree. And I've already taken a business English class.

W: Have you? Well, all the other classes look good.

M: Yes, they do. But I really don't like using live video chat. My _____ _____ _____ _____ enough.

W: Okay. So how many classes do you want to take per week? I think three is best.

M: Actually, I think that's too many. I _____ _____ _____ _____ _____ these days.

W: All right. Then this class looks like the best choice.

M: Yes, it does. I can't wait to get started!

11

M: Excuse me. Is this the _____ _____ _____ _____?

W: No, the entrance is over there.

M: Oh, thank you. And do you have the time?

W: (It's just a few minutes before 4 o'clock.)

12

W: Do you think it would be interesting _____ _____ _____?

M: Yes, I've thought about it a lot. But I don't know if it's for me.

W: What makes you say that?

M: (I might get too homesick.)

13

M: Have you asked your boss about taking a vacation yet?

W: No. We _____ _____ _____ when to take our vacation yet, have we?

M: I guess not. I've just been _____ _____ _____ _____ on it.

W: So have I. This is starting to stress me out. Why don't we just decide now?

M: Sure. How about May?

W: No, that's too soon. It has to be after June 15. I'm going to be busy until the middle of June.

M: Okay. Then how about the first week of July?

W: That's fine with me.

M: Great. Let's both _____ _____ _____ _____ that week off. Then we can book a flight and a hotel.

W: All right. I feel better now.

M: (Yeah, I'm glad we finally got organized.)

14

M: Hello. Is this Susan?

W: Yes, this is Susan speaking.

M: Susan, this is Charles Kim. I'm really sorry but I _____ _____ _____ to our meeting today.

W: Why? Is something wrong?

M: Well, my wife is _____ _____ at the hospital.

W: Really? I didn't even know your wife was pregnant.

M: Yes, she is. So can we _____ _____ _____?

W: Of course. You should be with your wife.

M: Thanks. And I'm really sorry about today.

W: (No problem. There's nothing more important than family.)

15

M: It's been nine months since Andrea _____ _____ her new apartment. Her next-door neighbor _____ _____ _____ in the hallway every night. He is a young college student and seems to expect a cleaning person to pick it up. Andrea is shocked by his behavior, but she tries to be patient. Day by day, however, Andrea grows angrier. She hoped that the young man would _____ _____ _____, but he hasn't. Finally, she decides to confront him. In this situation, what will she say to her neighbor?

Andrea: (You're not supposed to leave your garbage in the hallway.)

16~17

W: Hello, class. This week, you'll be writing a report on Anton van Leeuwenhoek, who _____ _____ in 1676. So for your report, you'll need to know what bacteria are. I suspect many of you think they are the same as viruses. However, this is not the case. Bacteria are single-celled organisms that _____ _____ _____ in two, whereas viruses don't have any cells at all. Since viruses are much smaller and simpler than bacteria, van Leeuwenhoek could only see bacteria and not viruses through his microscope. Bacteria may sound bad, but some bacteria are actually good for us. They reproduce by splitting in two, meaning that they don't necessarily cause any harm since they are able to reproduce on their own. On the other hand, viruses _____ _____ _____ _____ them to create more of their own DNA, which is why most viruses are harmful. Make sure you understand what bacteria and viruses are before you write your report. _____ _____ Anton van Leeuwenhoek, I'm sure you'll learn a lot about bacteria, as well as biology in general.

1번부터 17번까지는 듣고 답하는 문제입니다.
1번부터 15번까지는 한 번만 들려주고, 16번부터 17번까지는 두 번 들려줍니다. 방송을 잘 듣고 답을 하기 바랍니다.

01

다음을 듣고, 여자가 하는 말의 목적으로 가장 적절한 것을 고르시오. `3점`

① 신간을 홍보하려고
② 자신의 인생을 회고하려고
③ 책을 읽은 소감을 공유하려고
④ 19세기 미국 사회를 설명하려고
⑤ 후배들에게 인생에 대해 조언하려고

02

대화를 듣고, 두 사람이 하는 말의 주제로 가장 적절한 것을 고르시오.
① 고양이 몸의 유연성
② 고양이 수염의 다양한 역할
③ 고양이가 털을 관리하는 방법
④ 기분에 따른 고양이의 행동 변화
⑤ 고양이가 밤에 사물을 식별하는 법

03

대화를 듣고, 두 사람의 관계를 가장 잘 나타낸 것을 고르시오.
① 웨이터 - 손님   ② 전화 상담원 - 고객
③ 호텔 직원 - 투숙객   ④ 행사 관계자 - 손님
⑤ 백화점 직원 - 고객

04

대화를 듣고, 그림에서 대화의 내용과 일치하지 않는 것을 고르시오.

05

대화를 듣고, 남자가 여자에게 부탁한 일로 가장 적절한 것을 고르시오.
① to go to the movies with him
② to copy some reports
③ to reserve a movie ticket
④ to bind some copies
⑤ to treat him to lunch

06

대화를 듣고, 여자가 이집트 여행에 실망한 이유를 고르시오.
① 숙소가 불편해서
② 편의 시설이 부족해서
③ 음식이 입에 맞지 않아서
④ 낙타를 타는 비용이 비싸서
⑤ 관광지 주변 환경이 마음에 들지 않아서

07

대화를 듣고, 남자가 지불할 금액을 고르시오.
① $28.00   ② $31.00   ③ $31.50
④ $38.50   ⑤ $41.00

08

대화를 듣고, the Spanish Civil War에 관해 언급되지 않은 것을 고르시오.
① 전쟁 연도   ② 전쟁의 원인   ③ 사망자 수
④ 사용된 무기   ⑤ 역사적 의미

09

Beauty and the Beast에 관한 다음 내용을 듣고, 일치하는 것을 고르시오.
① 6개월 동안 공연한다.
② 공연 시간은 주중과 주말 저녁 7시이다.
③ 학생 관람료는 성인 금액의 절반이다.
④ 공연 시작 전에 기념품을 구매할 수 있다.
⑤ 오리지널 팀의 멤버들로 구성되어 있다.

10

다음 표를 보면서 대화를 듣고, 여자가 선택할 수업을 고르시오.

	Class	Length	Frequency per Week	Arrangement Types	Flower Cost
①	A	1 day	X	Bouquets	$50
②	B	2 weeks	One	Bouquets	$45
③	C	2 weeks	One	Bouquets	$30
④	D	4 weeks	Two	Wreaths	$40
⑤	E	8 weeks	Three	Wreaths	$25

11

대화를 듣고, 여자의 마지막 말에 대한 남자의 응답으로 가장 적절한 것을 고르시오.

① Let's call him Lucky.
② My friend has a nice dog.
③ I don't know his name yet.
④ I'll give him a call tomorrow.
⑤ I want a new bike for my birthday.

12

대화를 듣고, 남자의 마지막 말에 대한 여자의 응답으로 가장 적절한 것을 고르시오.

① I'd love to get a new TV soon.
② But I am pretty good at soccer.
③ Well, there is still some time left.
④ Soccer seems rather boring to me.
⑤ The big baseball game starts in a minute.

13

대화를 듣고, 남자의 마지막 말에 대한 여자의 응답으로 가장 적절한 것을 고르시오.

Woman: _____

① Don't worry about it. I'm okay.
② I'd better turn off the computer.
③ Good idea. I'll go find the number.
④ My computer is infected with a virus.
⑤ Because I'm having computer problems.

14

대화를 듣고, 여자의 마지막 말에 대한 남자의 응답으로 가장 적절한 것을 고르시오. 3점

Man: _____

① I like Thai food, too.
② Let's call and get some information.
③ You should have called them yesterday.
④ It's difficult to communicate with foreigners.
⑤ How about taking an English language course?

15

다음 상황 설명을 듣고, Michelle이 그녀의 남동생에게 할 말로 가장 적절한 것을 고르시오. 3점

Michelle: _____

① Thanks for lending me money.
② You can pay me back tomorrow.
③ I'm sorry. I made a terrible mistake.
④ You should have been more careful.
⑤ How many times have I told you not to lie?

[16~17] 다음을 듣고, 물음에 답하시오.

16

남자가 하는 말의 주제로 가장 적절한 것은?

① a house built for cows
② a futuristic eco-friendly house
③ the history of North American immigrants
④ a type of house built by North American settlers
⑤ people who lived underground in the early 1990s

17

언급된 soddies의 특성이 아닌 것은? 3점

① warm   ② cool   ③ damp
④ sturdy   ⑤ uncomfortable

녹음을 다시 한 번 듣고, 빈칸에 알맞은 말을 쓰시오.

01

W: This novel truly _____ _____ _____ into the lives of people in 19th-century America. The thing I really liked about this book was the way that the author _____ _____ _____. Even though the leading character's life was so different from my own, I could really identify with him. The themes of the book are universal, making it an enjoyable read for anyone, _____ _____ _____, culture or background. The structure of the plot was a bit frustrating at times, as it went back and forth between different stories. However, when all of the stories were tied together in the end, it made it all worthwhile.

02

W: Your cat's whiskers are so adorable, Todd.
M: You mean the hairs near its mouth? They're very useful, too.
W: Are whiskers actually used for something?
M: Cats use them to _____ _____ _____ _____.
W: Oh, so they get information about their surroundings through their whiskers?
M: You got it. And, can you see how the whiskers are about _____ _____ _____ the cat's body?
W: Yes. What's the reason for that?
M: It helps the cat easily figure out whether or not it can _____ _____ _____ _____ _____. That's useful when it's trying to get away in a hurry.
W: What else are they used for?
M: They can show the cat's mood. When it's in a good mood, its whiskers are relaxed. But, look. It's pulling its whiskers back now. I think you made it mad!

03

[Phone rings.]
M: Great Home Supplies. How can I help you?
W: Well, I ordered two bowls from your website a few days ago and they arrived yesterday. Unfortunately, one of them _____ _____ when I unpacked it.
M: Oh, I'm really sorry about that. Perhaps there was _____ _____ _____ _____. I'll handle it immediately. What is your username?
W: It's "smash1".
M: [Typing sound] Yes, I can see you ordered two Portmerion bowls.
W: That's right. So what should I do?
M: Our delivery person will come by tomorrow with a new one. Please give the broken one to him.
W: Okay. Oh, do I have to _____ _____ _____ _____ again?
M: No, you don't. And you can check the status of the delivery on the "My Page" section of our website.
W: All right. Thanks.

04

W: Hi, Mike. Are you having fun?
M: Yes, I really love this place.
W: Where is Sarah?
M: She is over there resting on a towel _____ _____ _____ _____. She looks very cute, doesn't she?
W: Yes, she does. How about Chris and Dan?
M: They are playing with a ball in the water over there.
W: Oh. I also see some people _____ _____.
M: Yes. Those boats are fun. You should try one. The view of the trees across the valley is breathtaking.
W: Yeah. But why is there a rope floating in the water?
M: It's to keep the swimmers in shallow water. The balls that make it float are big enough to see _____ _____ _____.

W: That's a good idea for safety.

05

M: Oh, no! What am I going to do?

W: What's wrong?

M: I promised my girlfriend that I would take her to the movies tonight, but I _____ _____ _____ yet!

W: You should call her now and tell her you won't be able to go.

M: She'll be angry if I _____ _____ because of my work! This would be the third time. Can you help me?

W: Well, I have a little time before I meet my mom. What can I do?

M: Could you _____ _____ of these reports?

W: Sure, that's easy. Anything else?

M: No. If you do that, I'll _____ _____ _____ tomorrow morning before the boss comes in.

W: Okay, I can do it. Look at the time! If you don't leave now, you'll be late!

M: Thanks for your help! I'll buy you lunch next week!

06

M: How was your trip, Katherine?

W: It was fabulous! I visited the Egyptian Pyramids and _____ _____ _____ _____ on camelback.

M: Wow! That sounds great!

W: It was, but I was a little disappointed by all the _____ _____ there.

M: What do you mean?

W: Well, did you know that there's a shopping mall _____ _____ _____ _____ _____?

M: No, really? I think historical sites should be preserved for their own value.

W: Yes! And there were fast-food restaurants _____ _____ _____ _____.

M: Oh no! How awful!

07

M: Are you selling all of your CDs?

W: Yes, I don't listen to them very often.

M: I want some new music. Do you have any classical guitar music?

W: Yes, _____ _____ _____.

M: Oh, here's one. How much is it?

W: You can buy that one for $6 or you can buy all of the classical music CDs _____ _____ _____.

M: How many classical CDs are there all together?

W: I have seven.

M: Okay. I'll _____ _____ _____ _____. I just love classical music! Here's my credit card.

W: If you pay in cash, I'll give you a fifty-cent discount each.

M: I'm afraid I don't have any cash. I have to pay by credit card.

W: Okay.

08

M: I can't believe this actually happened.

W: What are you talking about?

M: It's an article about _____ _____ _____ in 1937 during the Spanish Civil War. Some German bombers attacked a Spanish town for three hours.

W: Three hours! That must have been terrible.

M: It was. Almost 1,700 people died. It was quite a significant event in the history of war.

W: Really? Why is that?

M: It showed how powerful an air force could be. The bombers easily destroyed the town. Afterwards, many countries tried to _____ _____ _____ _____ so that they could defend their towns in the future.

W: That is very important.

M: The most shocking thing was the target. That was the first time a country had attacked a city without soldiers.

W: I see. People must have been really upset.

M: Yes, the public _____ _____ _____, and people are still angry about the attack today.

09

M: Starting next Friday and running nightly _____ _____ _____ _____ _____, the classic musical *Beauty and the Beast* is coming to the City Arts Center. Performances _____ _____ _____ each night at 7 p.m., with additional afternoon shows starting at 1 p.m. on the weekends. Tickets are $50 for adults and $30 for students, and can be reserved online at the Arts Center website. CDs, shirts and other related items will be on sale in the lobby _____ _____ _____. The shows will include members of the original cast and a variety of international celebrities, along with some of _____ _____ _____ _____. For the latest information on cast members, keep an eye on our site.

10

W: Hi. I'd like to sign up for one of your flower arranging classes.

M: Sure. Our summer schedule is right over here. This year we're offering a special one-day class.

W: That's not _____ _____ _____ _____.
   I want a class that lasts at least two weeks.

M: That's fine. We also have longer classes. They meet once, twice or three times per week.

W: Three days a week? That's too often. Once or twice is best.

M: Okay. Now, is there a certain type of flower arrangement that _____ _____ _____

_____?

W: I'm interested in everything. Oh, wait. I already took a wreath class.

M: So you don't want to do that again?

W: Correct. And what are these prices? I thought your classes were free.

M: They are, but you need to pay for the flowers. Some classes _____ _____ _____ _____.

W: Oh, I want to use expensive flowers. I think they'll be more attractive.

M: Yes, they usually are. I think this class will be perfect for you.

11

W: Are you ready for _____ _____ _____?

M: Wow! A puppy! I've always wanted a dog.

W: What do you want to name him?

M: (Let's call him Lucky.)

12

M: What are you watching on TV?

W: It's a big soccer game. I hope _____ _____ _____.

M: Do you think they can? They seem to be losing.

W: (Well, there is still some time left.)

13

M: What's wrong, Susan?

W: My computer _____ _____ and I have so much work to do!

M: Maybe I can fix it for you. What's wrong with it?

W: I have no idea. Every time I open this program, the computer _____ _____.

M: Did you look at how much memory you're using?

W: How do I do that?

M: Well, right-click here.

W: Okay, there's a circle, and half of it is covered in blue.

M: Oh, then there's no problem with the memory.

W: Is there any way to find out what the problem is with my computer?

M: I'm afraid I can't help you. Why don't you call the computer company's _____ _____?

W: (Good idea. I'll go find the number.)

14

W: Did you like _____ _____ about Asia?

M: Yes, I really liked it.

W: Have you ever thought about traveling around Asia?

M: Yes. I really want to go to Thailand.

W: It might be difficult. You don't speak Thai.

M: Thai people _____ _____ _____! I'll speak English there.

W: That's a great idea! Let's go together!

M: Okay. We should also take some Thai classes in advance so that we'll be able to _____ _____ with them.

W: I heard the language institute around the corner is offering a new Thai class.

M: (Let's call and get some information.)

15

W: Michelle is at lunch with some of her coworkers. When _____ _____ _____ _____, she finds that she has no money in her purse. She is very embarrassed and has to borrow some money. She is sure she had money and tries to figure out what happened. She suddenly remembers that her teenage brother asked to borrow money the day before, but she said no. She angrily calls him and _____ _____ _____ _____ the money from her purse. Though he denies it, she doesn't believe him and yells at him. When she returns to the office, however, she remembers that she _____ _____ _____ _____ a present for her colleague. She picks up her phone and calls her brother to apologize. In this situation, what would Michelle most likely say to her brother?

Michelle: (I'm sorry. I made a terrible mistake.)

16~17

M: I'd like to show you a famous photo and tell you a bit about the history it shows. This is a picture of a family in front of their soddie. When the settlers arrived in the Great Plains of North America, they didn't have _____ _____ _____ _____ build houses. So they used grass and soil instead. These houses _____ _____ _____ "sod houses" or "soddies". They first appeared in the early 1900s. It's estimated that more than one million sod buildings were built in North America at that time. Soddies were warm in the winter and cool in the summer, but there were some disadvantages. Soddies were often very damp. And if you look at the picture, you'll notice there is a cow on the roof. Indeed, since soddies _____ _____ _____ _____, hungry cows sometimes climb onto them to eat the grass. Cows were not the only animals drawn to the soddies. Rattlesnakes often made their homes in the walls. Life in a soddie in the early 1900s could _____ _____ _____.

33 영어듣기 모의고사

1번부터 17번까지는 듣고 답하는 문제입니다.
1번부터 15번까지는 한 번만 들려주고, 16번부터 17번까지는 두 번 들려줍니다. 방송을 잘 듣고 답을 하기 바랍니다.

01

다음을 듣고, 남자가 하는 말의 목적으로 가장 적절한 것을 고르시오.
① 싸운 학생들을 야단치려고
② 학교 소방 훈련을 공지하려고
③ 지저분한 교실에 대해 불평하려고
④ 최근 자연재해에 관해 토론하려고
⑤ 질병을 예방하기 위한 조언을 주려고

02

대화를 듣고, 여자의 의견으로 가장 적절한 것을 고르시오.
① 꿈꾸던 학교를 포기하면 안 된다.
② 남자는 다른 대학으로 편입해야 한다.
③ 여러 대학에 입학 원서를 제출해야 한다.
④ 자신의 적성에 맞는 전공 선택이 중요하다.
⑤ 대학에서 다양한 분야의 학문을 접해야 한다.

03

대화를 듣고, 두 사람의 관계를 가장 잘 나타낸 것을 고르시오.
① 연인 - 연인
② 사회자 - 가수
③ 경찰관 - 운전자
④ 사회자 - 출연자
⑤ 라디오 진행자 - 청취자

04

대화를 듣고, 그림에서 대화의 내용과 일치하지 않는 것을 고르시오.

05

대화를 듣고, 두 사람이 할 일로 가장 적절한 것을 고르시오.
① 집으로 돌아가기
② 왔던 길 되돌아가기
③ 운전 교대로 하기
④ 경찰에게 도움 청하기
⑤ 행인에게 길 물어보기

06

대화를 듣고, 남자가 약속 시간에 늦은 이유를 고르시오.
① 회의가 길어져서
② 교통 체증이 심해서
③ 자동차 사고를 당해서
④ 사고 목격자 진술을 해야 해서
⑤ 휴대전화를 가지러 집에 들러야 해서

07

대화를 듣고, 여자가 지불할 금액을 고르시오.
① $30   ② $80   ③ $90   ④ $120   ⑤ $210

08

대화를 듣고, 한국 최초의 우주 비행사에 관해 언급되지 <u>않은</u> 것을 고르시오.
① 출생연도
② 우주에 간 연도
③ 우주 방문 횟수
④ 우주에 머문 기간
⑤ 우주에서 실시한 실험 내용

09

Rainbow Park's bus tour에 관한 다음 내용을 듣고, 일치하지 <u>않</u>는 것을 고르시오.
① Hugo Smith가 공원의 이름을 지었다.
② 입장권 구입 시 지도를 제공한다.
③ 소책자에 쿠폰이 포함되어 있다.
④ 투어 시간은 총 한 시간 반이다.
⑤ 두 시간 동안의 자유 시간이 주어진다.

10

다음 표를 보면서 대화를 듣고, 남자가 구입할 게임을 고르시오.

	Model	Category	Price	Recommended Ages	Joystick
①	A	Racing	$56	All ages	O
②	B	Soccer	$35	All ages	O
③	C	Racing	$40	13 and over	X
④	D	Racing	$45	All ages	X
⑤	E	Soccer	$38	19 and over	X

11

대화를 듣고, 남자의 마지막 말에 대한 여자의 응답으로 가장 적절한 것을 고르시오.

① Can you fill in these blanks?
② How many of each will there be?
③ When were the test results posted?
④ I am not satisfied with my test score.
⑤ Congratulations! You got an A on your essay.

12

대화를 듣고, 여자의 마지막 말에 대한 남자의 응답으로 가장 적절한 것을 고르시오.

① We work to help seniors and children.
② The meeting will be at 7 tomorrow night.
③ I am in a different class than my club members.
④ I'm in the same class as you. Didn't you see me?
⑤ Actually I am thinking about joining a photo club.

13

대화를 듣고, 여자의 마지막 말에 대한 남자의 응답으로 가장 적절한 것을 고르시오. 3점

Man: _____

① I like to listen to music when I'm studying.
② Yes, I got it from the music shop down the street.
③ No thanks. I already have all of the Beatles albums.
④ Probably, but I'd have to ask my father's permission.
⑤ Sure, but there's a special late charge for new releases.

14

대화를 듣고, 남자의 마지막 말에 대한 여자의 응답으로 가장 적절한 것을 고르시오.

Woman: _____

① Silly me, I must have forgotten.
② Go away and stop bothering me.
③ That's okay. We all make mistakes.
④ My husband must have gotten hungry.
⑤ I'm sorry. He is always getting into some sort of trouble.

15

다음 상황 설명을 듣고, 교장 선생님이 John에게 할 말로 가장 적절한 것을 고르시오. 3점

Principal: _____

① Why isn't your name written inside it?
② I'm glad you accept that it's all your fault.
③ It's bad to steal another student's pencil case.
④ He was wrong, but you can't solve your problems by fighting.
⑤ Can you explain why you brought your friend into my office?

[16~17] 다음을 듣고, 물음에 답하시오.

16

여자가 하는 말의 주제로 가장 적절한 것은?

① tourist attractions in Zanzibar
② the geographic features of Zanzibar
③ the history of Zanzibar's spice business
④ the causes of the economic crisis in Zanzibar
⑤ Zanzibar's relationship with neighboring countries

17

언급된 향신료가 아닌 것은? 3점

① cloves ② saffron ③ ginger
④ cinnamon ⑤ pepper

녹음을 다시 한 번 듣고, 빈칸에 알맞은 말을 쓰시오.

01

M: I'd like to take a moment before the end of class to address an important topic. As you all know, there is a particularly dangerous form of the flu going around this winter. One of the best ways to avoid catching it or passing it on to others is by _____ _____ _____ _____ _____ several times a day. It is also a good idea to wear a mask over your mouth and nose. Everyone is _____ _____ _____ _____ the flu, but in a place like a crowded school, where we come in close contact with so many other people throughout the day, this risk is magnified. So let's all do our best to _____ _____ _____ _____.

02

M: I just found out I _____ _____ to the Smith Institute of Art.

W: Congratulations! That's a really prestigious art school.

M: It was my top college choice. Although, now that I've been accepted, I'm having second thoughts.

W: What do you mean?

M: It's just... At Smith I won't have the option to pursue other areas of study.

W: True. But if you want to study art — and I know you do — there's no better place.

M: I know.

W: If you _____ _____ _____ your dream school, you'll regret your decision later.

M: You might be right.

W: Besides, if you go to Smith and you don't like it, you can always _____ _____ _____ _____.

03

M: Okay, we're back. At this point in the show, I'd like to take a call. Are you there?

W: Yes, hi, Ryan.

M: Hi! Thanks for calling.

W: I _____ _____ _____ _____ every morning while I drive to work. I think it's great.

M: Thank you. But I hope you're not calling while you drive! That's dangerous!

W: No, no. I'm in the parking lot now. But I'd like to _____ _____ _____.

M: Sure. Would you like to send it to someone special?

W: Yes. I want to hear "You're the Best" by Jane Jones. And the song _____ _____ _____.

M: For me? Why?

W: Because you put a smile on my face every morning.

M: What a nice thing to say!

04

W: Do you want a donut? They look good.

M: Yes, they do. I think I'll have one of these chocolate donuts with sprinkles on them.

W: Let's also get _____ _____ _____ like the one on the man's tray.

M: Those look good. I guess we need to get a tray.

W: Yes. Where are they?

M: They're on the right end of the counter.

W: All right. Oh! Look at the sign! I think we should get one more donut.

M: You're right. We can get four donuts _____ _____ _____ _____ _____.

W: Great. Then what about getting one of the ones that the woman is _____ _____?

M: The woman wearing the sleeveless top? Okay.

05

M: I don't recognize this area.

W: You _____ _____ _____ _____ back there.

M: Where?

W: Over two kilometers back. You should have let me drive.

M: Why didn't you say something? Now we'll be late for the party.

W: I thought you knew a different way.

M: Well, I guess _____ _____.

W: I don't see anybody to ask directions. Just _____ _____ and go the other way.

M: But we've made five or six turns since then. I don't remember the _____ _____ _____.

W: Don't worry. I remember the way.

M: Great! Thanks.

06

W: There you are, Michael. _____ _____ _____ _____ _____?

M: I'm really sorry, Natalie. How long have you been waiting?

W: For about 30 minutes.

M: It's been a really difficult day. And I couldn't call you because I _____ _____ _____ _____ _____.

W: Did you go home to get it?

M: No. Then I would have been even later.

W: Did something happen on your way here?

M: Yes. I saw a bad car accident.

W: That's terrible.

M: Yes. And I had to wait for the police to come so that I could _____ _____ _____ _____. I was about to call you when I realized I'd left my phone at home.

W: I see. Don't worry about it.

07

M: Hello. Welcome to Sayana Hotel. Do you have a reservation?

W: No, I don't. I _____ _____ what your rates are.

M: What type of room are you looking for?

W: Well, it's just me. So a single room is all I need.

M: Okay. A basic single room is $20 a night. The deluxe rooms are $50 a night.

W: What's the difference between a basic room and a deluxe one?

M: Deluxe rooms include a Jacuzzi and a small kitchen area.

W: I don't think I need all that. I'll just _____ _____ the basic room.

M: All right. And how many nights are you staying?

W: I'm staying four nights. Oh, _____ _____ _____?

M: It's not. You can add the breakfast buffet for $10 a day.

W: Okay. I'll add the breakfast buffet, but for tomorrow only. I have plans for the other days.

08

M: I have a question, Mina. Who was the first Korean to _____ _____ _____?

W: Her name is Yi Soyeon. She's a scientist who was born in 1978.

M: Oh! It was a woman? I didn't know that.

W: Yes. She traveled to the International Space Station in 2008.

M: That's interesting. Does Korea have its own spacecraft?

W: Not yet. She was a passenger on a Russian spacecraft.

M: I see. How long was she in space?

W: She spent ten days on the space station _____ _____ _____ _____ _____.

M: What kind of experiments did she do?

W: Among other things, she studied the _____ _____ _____ _____ plants and insects.

M: Wow, she's amazing.

09

W: Welcome to Rainbow Park's bus tour. Rainbow Park is a historical park that _____ _____ _____ Hugo Smith. When Hugo Smith first came here, he saw many rainbows and therefore named it Rainbow Park. If you look at the map that was handed out when you purchased your tickets, you will find _____ _____ _____ _____ to visit in this area. And the booklet available at the information desk has some useful coupons. This tour will take about one and a half hours. We'll be driving around the park for about 30 minutes. After that, you will have _____ _____ _____ _____ _____ to take pictures and to visit places where you can use the coupons. I hope you enjoy the tour.

10

W: Hey, David. What are you doing here?

M: Hi, Jessica. Can you _____ _____ _____? I want to buy a game for my nephew, but I don't know anything about computer games.

W: Computer games? I'm _____ _____ _____ _____ in that area. Let's see... *[Pause]* How about a soccer game? Most children love soccer.

M: Sounds cool, but actually he's crazy about cars.

W: Then he'd like this racing game. It even comes with a joystick.

M: Let me see. Wow, it's more than $50. That's expensive... I wanted something for less than $50.

W: Okay. What about this one? It's $40.

M: But look at its rating. My nephew is _____ _____ _____ _____.

W: I didn't know your nephew was that young. I'll pick something else... This one! I'm sure he'll like it. But it doesn't include a joystick.

M: That's okay. Maybe I'll buy him a joystick later.

11

M: Our test will be on Friday. It will cover the entire book.

W: I have a question. Will the test _____ _____ _____?

M: Yes, there will be fill-in-the-blank questions and several essay questions.

W: (How many of each will there be?)

12

W: Where are you going, Brandon?

M: _____ _____ _____ _____ to a meeting of the volunteer club.

W: Oh, I didn't know you joined that club. What do you guys do?

M: (We work to help seniors and children.)

13

M: Hi, Maria. Thanks for coming over to my place to study.

W: Thanks for having me.

M: Come on, we can study in the living room. My mom made some snacks.

W: All right. *[Pause]* Oh, wow! Look at all these old albums.

M: Yeah, _____ _____ _____. He collects classic albums from the 70s.

W: I'm really into old rock music. How about you?

M: I am, too. I guess _____ _____ _____ _____ _____.

W: I can't believe it. Look at this. This is the best Beatles album ever.

M: I think so, too. Shall we listen to it while we study?

W: That might be distracting. But do you think I could _____ _____ _____?

M: (Probably, but I'd have to ask my father's permission.)

14

[Knocking on door]

W: Yes? Can I help you?

M: Good evening, ma'am. I have a pizza delivery for apartment 13B.

W: There _____ _____ _____ _____. I didn't order a pizza.

M: Hmm. Is this apartment 13B?

W: Yes, it is. But I certainly didn't order a pizza.

M: That's strange. Let me _____ _____ _____ and see who called in the order.

W: Okay.

M: _____ _____ _____ _____ by Eric Peterson. Is your husband's name Eric?

W: No. But I do have a 12-year-old son named Eric.

M: Oh boy. I think I know who ordered the pizza.

W: (I'm sorry. He is always getting into some sort of trouble.)

15

M: John is an elementary school student. He is very polite and works hard, but he has a _____ _____. One day, John got into an argument with another student over a pencil case. The other boy said the pencil case _____ _____ him. John insisted it was his. They yelled for a while, but then John _____ _____ _____ _____. The principal brought John into his office and asked him what happened. John explained and showed the principal _____ _____ _____ _____ the pencil case. In this situation, what would the principal most likely say to John?

Principal: (He was wrong, but you can't solve your problems by fighting.)

16~17

W: Today, we're going to learn about Zanzibar. Zanzibar is _____ _____ _____ _____ in the Indian Ocean. It is part of Tanzania, a country on the east coast of Africa. If you've heard of Zanzibar, you probably know that it has beautiful beaches and gorgeous coral reefs, making it a great _____ _____ for scuba divers. But that is not the only thing that makes this place so special. In addition, Zanzibar has an interesting history. In the 19th century, Zanzibar started producing cloves and quickly became _____ _____ _____ _____ in the world. They also started growing ginger, cinnamon, and pepper. For this reason, Zanzibar used to be called the Spice Islands. Since spices were highly valuable in the 19th century, Zanzibar became quite wealthy. However, once other countries started developing their own spice industries, its share of the market declined. Yet it remains a wonderful place to visit. So if you are _____ _____ _____ _____ there, remember its past as the center of the spice industry while you enjoy its beaches.

1번부터 17번까지는 듣고 답하는 문제입니다.
1번부터 15번까지는 한 번만 들려주고, 16번부터 17번까지는 두 번 들려줍니다. 방송을 잘 듣고 답을 하기 바랍니다.

01

다음을 듣고, 여자가 하는 말의 목적으로 가장 적절한 것을 고르시오.
① 직사광선의 위험성을 경고하려고
② 이상 기후의 발생 원인을 설명하려고
③ 뇌우 발생 시 대처 방법을 안내하려고
④ 급격한 날씨 변화의 원인을 알려주려고
⑤ 여름에 할 수 있는 야외 활동을 제안하려고

02

대화를 듣고, 남자의 의견으로 가장 적절한 것을 고르시오.
① 공원에 주차 시설이 더 필요하다.
② 정기적으로 공원 관리를 해야 한다.
③ 정부는 세금을 낭비해서는 안 된다.
④ 공원 존립을 위한 대책을 강구해야 한다.
⑤ 동네 공원의 주차 요금을 인하해야 한다.

03

대화를 듣고, 두 사람의 관계를 가장 잘 나타낸 것을 고르시오.
① 구직자 – 면접관   ② 임원 – 비서
③ 코치 – 운동선수   ④ 의뢰인 – 변호사
⑤ 고객 – 영업사원

04

대화를 듣고, 그림에서 대화의 내용과 일치하지 않는 것을 고르시오.

05

대화를 듣고, 남자가 여자에게 부탁한 일로 가장 적절한 것을 고르시오.
① 커피 만들어 주기   ② 전화 요금 내기
③ 남자가 쓴 수필 읽기   ④ 도서관에 책 반납하기
⑤ 세탁소에서 세탁물 찾기

06

대화를 듣고, 여자가 현재 집에 없는 이유를 고르시오.
① 친구와 커피숍에서 공부하고 있어서
② 친구와 저녁 식사 약속이 있어서
③ 엄마와 외출 중이어서
④ 친구와 여행을 가서
⑤ 음식을 사러 나가서

07

대화를 듣고, 남자가 지불할 금액을 고르시오.
① $29   ② $49   ③ $50   ④ $54   ⑤ $74

08

대화를 듣고, rattlesnake에 관해 두 사람이 언급하지 않은 것을 고르시오. 3점
① 천적의 종류   ② 소리를 내는 이유
③ 방울 소리의 세기   ④ 개체 수
⑤ 크기

09

Old Mine History Tour에 관한 다음 내용을 듣고, 일치하지 않는 것을 고르시오. 3점
① 관광객들은 전동차를 타고 지하로 들어간다.
② 광부가 직접 안내를 한다.
③ 점심 식사가 포함되어 있다.
④ 금광 박물관을 방문한다.
⑤ 관광 중에 기념품 구매가 가능하다.

10

다음 표를 보면서 대화를 듣고, 여자가 선택할 기숙사를 고르시오.

Johnson College Residence Halls

	Residence	Fee	Bathroom	Meal Plan	Length of Contract
①	Towns Hall	$890	Private	Optional	6-month
②	Grey Hall	$800	Shared	Optional	6-month
③	Fisk Hall	$950	Private	Included	6-month
④	Smith Hall	$980	Private	Included	12-month
⑤	East Hall	$1,050	Private	Included	12-month

11

대화를 듣고, 여자의 마지막 말에 대한 남자의 응답으로 가장 적절한 것을 고르시오.

① Anything would be a big help.
② Yes. I'll call a computer repair shop.
③ No. I am not very good with computers.
④ I don't have enough money to replace it.
⑤ Yes. I need to email a report this evening.

12

대화를 듣고, 남자의 마지막 말에 대한 여자의 응답으로 가장 적절한 것을 고르시오.

① Do you want it delivered to your office?
② I see here that you ordered the blue one.
③ I'm sorry, but your camera isn't ready yet.
④ I'm sorry, but we don't offer refunds at this store.
⑤ We can send a technician to your home right now.

13

대화를 듣고, 남자의 마지막 말에 대한 여자의 응답으로 가장 적절한 것을 고르시오. 3점

Woman: _____

① Take it if you want to. I don't mind.
② We watch too much TV. Let's play cards.
③ I'm going to buy one at the electronics market.
④ Come on over tonight. We'll watch a DVD and you'll see.
⑤ Is there any popcorn? I can't watch a movie without popcorn.

14

대화를 듣고, 여자의 마지막 말에 대한 남자의 응답으로 가장 적절한 것을 고르시오. 3점

Man: _____

① Well, I'm not available this week.
② Okay. I've got lots of money now.
③ Look at the time. We'll be late for work.
④ You don't need to be sorry about anything.
⑤ Okay, but I insist we stop by my place to get my wallet first.

15

다음 상황 설명을 듣고, Tyler가 Abby에게 할 말로 가장 적절한 것을 고르시오.

Tyler: _____

① Do we have biology on Monday?
② I'll see you Sunday morning at the library.
③ I'm sorry, but I'm really busy this weekend.
④ I'm glad to have you as a partner on this project.
⑤ If you want credit, you should participate in this project.

[16~17] 다음을 듣고, 물음에 답하시오.

16

남자가 하는 말의 주제로 가장 적절한 것은?

① natural digestive aids
② creating the perfect garden
③ the health benefits of plants
④ the dangers of home remedies
⑤ the effects of herbs on digestion

17

언급된 식물이 <u>아닌</u> 것은?

① basil   ② chamomile   ③ ginger
④ garlic   ⑤ mint

녹음을 다시 한 번 듣고, 빈칸에 알맞은 말을 쓰시오.

01

W: On nice summer days, it's fun to spend time outdoors. But summer weather can change quickly, so you must be careful. _____ _____ _____ an hour, bright sunshine can suddenly become a dangerous thunderstorm. If this happens while you're outside, hurry into the nearest building. If you can't, _____ _____ a low area and wait for the storm to pass. But make sure you're not standing near any utility poles. Open spaces should also be avoided, as the tallest thing in an area is the most likely to be hit by lightning. Keep these tips in mind and you will be sure to _____ _____ _____ _____.

02

M: It's sad this park is closing.
W: Yeah. But our state _____ _____ _____ _____ _____ _____ anymore.
M: I know, but this place has good memories for me. I used to come here with my family and have a lot of fun. It just seems like there _____ _____ _____ _____ _____.
W: Do you have any ideas?
M: Well, I have one. But I don't know how realistic it is.
W: Let's hear it.
M: The state could earn money by _____ _____ _____ at some of the parks.
W: That's not a bad idea. It would probably raise enough money to keep the park open.
M: We should talk to a park official about it.

03

M: Thank you for coming in.
W: I just wanted to meet you one more time before _____ _____ _____. Before we start, do you have any questions for me?
M: Actually, I do. If you were to _____ _____, when would I start?
W: We will make our decision by the end of this week, so you would be starting on Monday.
M: That's great. And who would I be working with?
W: You'd be working with a team of _____ _____.
M: I see. What sort of salary does the company offer?
W: You would have a five-year contract. The salary starts at $40,000 a year, and you would be eligible for periodic raises.
M: Thank you very much. Those are my questions.
W: All right. Now I have some for you.

04

M: Are you ready to start?
W: Yes, but I'm a little nervous. This is the first time I've tried to make a carrot cake.
M: Don't worry. It's simple.
W: Okay. What do I need to do first?
M: First, you need to make sure _____ _____ _____ are on the table.
W: It looks like they're all here.
M: Make sure the eggs are in the basket and the carrots have been _____ _____ _____.
W: Everything is perfect. But where's the oven?
M: It's right behind the table.
W: Oh, I see it now. But what's in the box with the rabbits on it?
M: It's empty. We'll _____ _____ _____ _____ _____ after we finish. You can take it home with you!

05

M: Where are you going, Anne?
W: I have to _____ _____ _____ for Mom. She asked me to do several things.
M: Oh really? What kind of errands?

W: First I have to _____ _____ _____ _____ .

M: Did you get the money and the bill?

W: Yes, Dad. And Mom told me to take her clothes to the dry cleaner's.

M: Are you going to the dry cleaner's by the library?

W: Yes. Actually, I was going to go to the library after that. I have to _____ _____ _____ and write a report for homework.

M: I've got _____ _____ _____ . Would you mind returning them for me?

W: Not at all. Where are they?

M: They're on the coffee table. I'll bring them to you. Wait here for a moment.

W: Okay.

06

[Cell phone rings.]

W: Hi, Dad. What's up?

M: Hi, darling. Where are you?

W: I'm at a coffee shop _____ _____ _____ _____ .

M: Oh. I thought you would be home for dinner by now.

W: Well, I had something to eat at school.

M: I see. Do you know where your Mom is?

W: I called her earlier to let her know that I wouldn't be home for dinner. She probably just _____ _____ for a few minutes.

M: Oh, here she is now. It looks like she bought some take-out food.

W: Good. I'll see you _____ _____ _____ _____ .

M: All right honey. I'll see you later.

07

W: This website has the book _____ _____ _____ _____ . There are several editions.

M: I need the fourth edition. That's the latest, right?

How much is it?

W: Yes, that's right. The paperback version is $25. Or you could get the e-book version! That's only $15.

M: I want the paperback. _____ _____ _____ _____ ?

W: It's $4 for regular delivery.

M: How long would it take to get here?

W: Well, for $4, it would get here in three to five days. But you can pay $10 to get it by tomorrow.

M: No, that's fine. I'm not in a hurry.

W: Okay. Then just regular delivery. Anything else?

M: Yes. I want to _____ _____ _____ for my friend too. His birthday is next week.

W: Oh, well, shipping is free when you spend more than $45.

M: That's great! Then go ahead and order two paperbacks.

08

W: John, did you ever see a rattlesnake in America?

M: Yes, in the desert.

W: Do they really have a rattle on the end of their tail?

M: Yes, they do. They shake it _____ _____ _____ , such as crows, coyotes and humans.

W: Have you ever heard one make that sound?

M: No. But _____ _____ _____ makes a different rattling noise.

W: How so?

M: The bigger the rattlesnake, the louder the sound. And a higher body temperature also makes the sound louder.

W: Oh, that's interesting. Do they _____ _____ _____ to kill it?

M: Yes. Their bite is so poisonous that they cause more than 80% of snakebite fatalities in North America.

W: Wow! Are they very big?

M: Yes, they can grow to be more than two meters long.

09

M: Good morning, everybody. Thank you for booking your place on today's Old Mine History Tour. The first part of the tour is riding _____ _____ _____ through the disused tunnels of the old gold mine. This is an exciting journey that will take you more than 500 meters underground! After that, a local historian will _____ _____ _____ _____ around a special part of the mine which has been converted for visitors. Then we'll come back above ground, and lunch will be provided in the beautiful museum gardens. After lunch, you will be given _____ _____ _____ _____ the gold mine museum and to visit the gift shop. The shop sells a variety of souvenirs including handmade jewelry made from local gold.

10

W: Are you going to live on campus next year, William?

M: No, I found an off-campus apartment with my brother. How about you?

W: I want to live on campus, but I can't decide on a residence hall.

M: _____ _____ _____ _____. Are you sure you can afford to live on campus?

W: Yes. Any place less than $1,000 per month is fine.

M: Oh, that's good. Then I recommend Grey Hall. I lived there last year, and I liked it a lot.

W: The rooms there are nice, but you _____ _____ _____ _____ _____.

M: That's true. So you prefer a private bathroom?

W: Yes, I do. And I want a place that includes a meal plan.

M: All right. Do you want a six-month contract or a 12-month contract?

W: I want a longer one. _____ _____ _____ _____ all the time.

M: I know how you feel. Anyway, it looks like this residence hall is your best choice.

11

W: Hey, Jack. You don't look too happy. What's wrong?

M: My computer stopped working. I think it's a virus.

W: Oh, dear. Is there something you _____ _____ _____ right away?

M: (Yes. I need to email a report this evening.)

12

M: Hello. I'm here to pick up my digital camera.

W: All right. Can you tell me your name, please?

M: My name is Thomas Butcher. I _____ _____ _____ three days ago.

W: (I'm sorry, but your camera isn't ready yet.)

13

M: What's in the box? Did you go shopping?

W: Yes, I bought a new TV today.

M: A new TV? What was wrong with the old one?

W: Nothing, I just wanted to _____ _____.

M: Where did you get it?

W: At the electronics market on the corner.

M: Oh, I know that place. Did you _____ _____ _____ _____?

W: Yes, it was only $1,000. I couldn't help but buy it at such a good price.

M: You said "only $1,000"? I think that's too expensive.

W: Well, it has a big screen.

M: I don't really _____ _____ _____ of big screen TVs.

W: (Come on over tonight. We'll watch a DVD and you'll see.)

14

W: Ah, here's the check. Let's go pay.

M: Oh no. I seem to _____ _____ _____ _____.

W: That's okay. I can pay the whole bill.

M: This is so embarrassing.

W: Don't be embarrassed. It's ＿＿＿＿ ＿＿＿＿ ＿＿＿＿.

M: I promise I'll pay you back.

W: Don't be silly. You bought dinner last time.

M: Yes, but I invited you then.

W: Then let me invite you to dinner next week, and you can pay me back then.

M: Okay, that would be all right.

W: Let me just ＿＿＿＿ ＿＿＿＿ ＿＿＿＿, and then we can go to the theater.

M: (Okay, but I insist we stop by my place to get my wallet first.)

15

W: Tyler and Abby are partners in ＿＿＿＿ ＿＿＿＿ ＿＿＿＿ for biology class. They need to get together to work on the project, but Abby always seems to be busy. The ＿＿＿＿ ＿＿＿＿ ＿＿＿＿ ＿＿＿＿ on Monday morning, and Tyler is worried that he will ＿＿＿＿ ＿＿＿＿ ＿＿＿＿ the whole project by himself. He doesn't mind doing the work, but he doesn't think Abby should ＿＿＿＿ ＿＿＿＿ ＿＿＿＿ ＿＿＿＿. Abby has just told him that she can't meet him until Sunday night. In this situation, what would Tyler most likely say to Abby?

Tyler: (If you want credit, you should participate in this project.)

16~17

M: Good afternoon. In today's gardening lecture, we're going to discuss medicine. Not the kind of medicine ＿＿＿＿ ＿＿＿＿ ＿＿＿＿ ＿＿＿＿ ＿＿＿＿, but the kind you can grow in your garden. Personally, I always plant lots of basil in my herb garden. A certain kind of basil, known as holy basil, is great for curing colds, coughs and fevers. And if you ＿＿＿＿ ＿＿＿＿ ＿＿＿＿ ＿＿＿＿, make sure to plant some chamomile. It helps your body digest the food you eat. Ginger is ＿＿＿＿ ＿＿＿＿ ＿＿＿＿. And it also strengthens your immune system, which allows your body to fight off diseases. And finally, if you're worried about your blood pressure, make some space in your garden for garlic. Besides being delicious, it helps keep your heart healthy and your blood pressure low. All of these plants are inexpensive and easy to grow, so they ＿＿＿＿ ＿＿＿＿ ＿＿＿＿ ＿＿＿＿ ＿＿＿＿ any garden.

1번부터 17번까지는 듣고 답하는 문제입니다.
1번부터 15번까지는 한 번만 들려주고, 16번부터 17번까지는 두 번 들려줍니다. 방송을 잘 듣고 답을 하기 바랍니다.

01
다음을 듣고, 남자가 하는 말의 주제로 가장 적절한 것을 고르시오.
① 천연자원 고갈의 원인
② 플로리다 해안의 산호초 규모
③ 에버글레이즈 습지대의 정수 능력
④ 미국 동식물 개체 수 감소의 원인
⑤ 에버글레이즈 습지대를 보호해야 하는 이유

02
대화를 듣고, 여자의 의견으로 가장 적절한 것을 고르시오.
① 4D 영화는 볼만한 가치가 있다.
② 영화관에서 에티켓을 지켜야 한다.
③ 인터넷의 영화평은 신뢰도가 떨어진다.
④ 4D 영화의 종류가 좀 더 다양해져야 한다.
⑤ 특수 효과가 많은 영화는 현실성이 떨어진다.

03
대화를 듣고, 두 사람의 관계를 가장 잘 나타낸 것을 고르시오.
① 웨이터 – 요리사   ② 주부 – 판매원
③ 판매원 – 관리자   ④ 평론가 – 요리사
⑤ 식당 종업원 – 손님

04
대화를 듣고, 그림에서 대화의 내용과 일치하지 않는 것을 고르시오.

05
대화를 듣고, 남자가 여자에게 부탁한 일로 가장 적절한 것을 고르시오.
① 아이 돌보기   ② 출장 대신 가기
③ 강아지에게 먹이 주기   ④ 물고기 돌보기
⑤ 부모님께 강아지 맡기기

06
대화를 듣고, 여자가 전화를 건 이유를 고르시오.
① 인터넷 접속이 불가능해서
② 인터넷 요금제를 변경하려고
③ 인터넷 서비스 신청을 하려고
④ 와이파이 공유기를 교환하려고
⑤ 수리 기사가 약속 시각에 도착하지 않아서

07
대화를 듣고, 남자가 지불할 금액을 고르시오. 3점
① $8   ② $12   ③ $14   ④ $18   ⑤ $20

08
대화를 듣고, Netherlands에 관해 두 사람이 언급하지 않은 것을 고르시오.
① 수도   ② 주요 산업   ③ 인접 국가
④ 인구   ⑤ 언어

09
학생회에 관한 다음 내용을 듣고, 일치하지 않는 것을 고르시오. 3점
① 매주 수요일에 열릴 예정이다.
② 두 명의 임원이 선출될 것이다.
③ 특별 위원회가 회의 안건을 정한다.
④ 특별 위원회는 방과 후 활동에 속한다.
⑤ 특별 위원회에 참가하려면 부모님의 동의가 필요하다.

10

다음 표를 보면서 대화를 듣고, 두 사람이 선택할 호텔을 고르시오.

	Hotel	Distance (Minute)	Internet Cost	Fitness Facility	Additional Information
①	A	10	Free	No	Spa & Pool
②	B	5	$10	Yes	Shopping Mall
③	C	15	$12	No	Spa & Pool
④	D	3	Free	Yes	Shopping Mall
⑤	E	1	Free	Yes	Spa & Pool

11

대화를 듣고, 남자의 마지막 말에 대한 여자의 응답으로 가장 적절한 것을 고르시오.

① Okay. I'll be right there.
② I really don't like video games.
③ What should we cook for dinner?
④ I knew you would like this game.
⑤ Sorry, I'm not in the mood for lunch.

12

대화를 듣고, 여자의 마지막 말에 대한 남자의 응답으로 가장 적절한 것을 고르시오.

① It will take about twenty minutes.
② I think you should transfer to line 3.
③ Right! You should get on the subway.
④ You should hurry up. The subway is too crowded.
⑤ Get off at City Hall Station. It's just three stops from here.

13

대화를 듣고, 여자의 마지막 말에 대한 남자의 응답으로 가장 적절한 것을 고르시오. 3점

Man: _____

① Actually, I don't like tennis that much.
② I'm glad we take Mr. Kim's class together.
③ Thanks. I'll never forget that you helped me out.
④ Okay, but bring it right back when you're finished with it.
⑤ The deadline was yesterday, so you have to pay a late fee.

14

대화를 듣고, 남자의 마지막 말에 대한 여자의 응답으로 가장 적절한 것을 고르시오.

Woman: _____

① Please tell me the truth.
② How can I make it up to you?
③ I promise to come back very soon.
④ No, you were supposed to pick me up.
⑤ I can't believe that you are calling me a liar.

15

다음 상황 설명을 듣고, Kate가 Janet에게 할 말로 가장 적절한 것을 고르시오. 3점

Kate: _____

① I'm moving next week.
② Why don't you work for me?
③ That would be fine. No problem.
④ How dare they ask me to do that?
⑤ You should have planned a party for your girl.

[16~17] 다음을 듣고, 물음에 답하시오.

16

여자가 하는 말의 목적으로 가장 적절한 것은?

① 주민들에게 폭풍에 대비할 것을 경고하려고
② 최근 기상 변화의 원인에 대해 알려주려고
③ 알레르기를 피하는 것에 관해 조언하려고
④ 대기 오염에 대처하는 방법을 설명하려고
⑤ 잘못된 일기예보 내용을 정정하려고

17

언급된 대상이 아닌 것은?

① city residents   ② children
③ the elderly   ④ pregnant women
⑤ people with allergies

녹음을 다시 한 번 듣고, 빈칸에 알맞은 말을 쓰시오.

01

M: One of America's most important natural resources is the Florida Everglades. Unfortunately, most of this area has been destroyed _____ _____ _____ _____. This delicate ecosystem must be protected. It is home to _____ _____ _____ and plant species, including black bears, indigo snakes and turtles. It is also an important part of Florida's environment. The clean water that leaves these wetlands _____ _____ _____ _____ _____ _____ and coral formations off the Florida coast.

02

M: Did you do anything special this weekend, Clara?

W: My friend and I went to a 4D movie near my house.

M: Oh, it was at one of those theaters _____ _____ _____?

W: That's right. There were flashing lights, wind, water spray, and even smells. And the seats moved around to match _____ _____ _____ on the screen.

M: It sounds like an interesting experience.

W: It was. You should definitely check it out sometime.

M: Well, I've never tried it because _____ _____ _____ _____ the price of a regular ticket.

W: Trust me — you won't be disappointed. In fact, I want to go to another 4D movie soon!

M: Maybe we could go together.

03

M: Are you ready to order or do you need more time with the menu?

W: I'm still _____ _____ _____.

M: Would you like me to explain them?

W: Yes. That'd be wonderful.

M: The fish of the day is fresh salmon. It's _____ _____ _____ _____.

W: I had fish for lunch. How is your steak?

M: That's a great choice as well. It's rib-eye steak, and it's very juicy. It _____ _____ _____ _____.

W: I think I'll have the steak.

M: How would you like that cooked?

W: Medium-rare, please.

04

M: Look at my picture from the camping trip.

W: Oh, there are only two people. Is it you and your friend?

M: Yes, it is. We _____ _____ _____.

W: Yeah. It looks like there are several fish in the basket.

M: That's right.

W: I guess you couldn't swim in the lake.

M: Yes, there was a sign that said we _____ _____ _____ _____ there. But we took a tent and slept in it.

W: Yes, that's a nice dome tent.

M: I just bought it last month.

W: I see. But why do you enjoy fishing there so much?

M: I just enjoy being in nature. It is very peaceful, and I like to _____ _____ _____ around the lake.

05

[Knocking on door]

W: Yes? Who is it?

M: It's Chris from next door.

W: Oh, hi Chris. What can I do for you?

M: I know we don't know each other very well, but I was wondering if I could _____ _____.

W: Sure.

M: I need to go away on an _____ _____ _____ tomorrow morning. I'll be gone for a week and I was wondering if you could take care of Sparky for me.

W: I'm sorry, Chris, but I don't think I can help. Dogs make me very nervous.

M: Oh, no.

W: _____ _____ _____ _____ dogs since I was a child.

M: I mean, no, Sparky isn't my dog. My dog's name is Jojo, and she'll be staying with my parents.

W: Then who is Sparky?

M: Sparky is my goldfish.

W: Oh! That's no problem, then.

06

[Phone rings.]

M: Good afternoon. This is Ace Powercom.

W: Hello. Is this customer service?

M: Yes, how may I help you?

W: I'm _____ _____ _____ _____ my internet connection.

M: I see. What type of service do you have?

W: I have wireless internet service, but I'm not receiving a WiFi signal. It hasn't been working for the past two hours.

M: Okay. First, please turn off your WiFi router and then _____ _____ _____ _____ after 5 minutes. That usually fixes the problem.

W: All right. But what if that doesn't work?

M: Then please call me again, and I will _____ _____ _____ _____ to visit your home.

W: I'll do that. Thanks for your help.

07

W: Hello. How can I help you?

M: I'd like some cherries and blueberries, please. How much are they?

W: Cherries are $2 a pound, and blueberries are $4 a pound.

M: Then I'll take _____ _____, please.

W: Okay. Do you need anything else?

M: I don't think so. I was just looking for cherries and blueberries.

W: Are you sure? I _____ _____ the peaches. They are really good this year.

M: Hmm... How much are they?

W: They're just $8 for a bag. Each bag has five peaches. Would you like to try a piece?

M: Yes please. *[Pause]* Hmm... That is really good. But I already bought cherries and blueberries. This is starting to cost a lot.

W: If you get a bag of peaches, I'll give you a 10% _____ _____ _____ _____.

M: Well, that's a good deal. Okay. I'll take a bag of peaches too.

08

M: Is that an airline ticket? Are you going somewhere?

W: Yes. I'm going to go to the Netherlands.

M: Are you going there for business?

W: No. My brother is studying industrial engineering in Amsterdam, so I will visit him there.

M: Amsterdam? _____ _____ _____ _____?

W: Yes. It's also the largest city in the country. I can't wait to go.

M: Hmm... I'm trying to imagine a map of Europe. It's near Germany, isn't it?

W: Yes. It _____ _____ _____ _____ and Belgium to the south.

M: That's right. If I remember correctly, it's a pretty small country.

W: Yes. It only _____ _____ _____ of about 17 million.

M: So, do they speak German?

W: No, they don't. They speak Dutch. I'm going to study it before I go.

09

W: I'd like to start by thanking everyone who signed up to _____ _____ _____ the student council. We will be having weekly meetings throughout the semester. These will _____ _____ _____ _____, except in cases when it conflicts with our holiday schedule. Our first order of business will be to elect a president and a vice-president. Once we have selected these officers, they will be _____ _____ _____ setting meeting agendas. We'll also have several different specialized committees, and everyone will be required to volunteer for one of these. And of course, as this is an after-school activity, you'll need to bring in _____ _____ _____ signed by your parents. So, are there any questions?

10

M: I just registered us for the upcoming business conference. Now we need to choose a hotel.

W: Okay. Did the conference provide a list of recommended places?

M: They did. It even _____ _____ _____ _____ from the conference center.

W: Great! Let's pick a place that we can walk to in 10 minutes or less.

M: Sure. We should also choose a place that has a free internet connection.

W: I agree. We'll need to do some work after the conference.

M: I also want to work out in the evenings. It's _____ _____ _____ _____.

W: I'm sure we can find a hotel that has a fitness center.

M: Yes, there are a few on the list. Do you have any requirements? This one has a shopping center.

W: No, I don't need to go shopping. But _____ _____ _____ _____ in a spa.

M: All right. Then I think there's one hotel that meets all of our needs.

W: Great. Let's ask the boss to book us two rooms there.

11

M: Sally, it's time for dinner!

W: Just _____ _____ _____ _____, Dad! I'm winning at this video game.

M: No, your food is getting cold. Come and eat now.

W: (Okay. I'll be right there.)

12

W: Excuse me, which subway line should I take to go to NY Department Store?

M: Take line 2.

W: Thank you. And can you tell me _____ _____ _____ _____ _____?

M: (Get off at City Hall Station. It's just three stops from here.)

13

W: Hey, Patrick!

M: Oh, hi, Alice. What's wrong?

W: I forgot to bring my tennis racket. I need it for my physical education class. Do you have one?

M: Well, I do, but it's in my locker.

W: Can I borrow it? You know, Mr. Kim is very _____ _____ _____ _____ for class.

M: Last time you borrowed it, you didn't _____ _____ _____. And I had to come to your classroom to get it, remember?

W: Oh, did you? I'm sorry. This time, I'll return it as soon as the class ends. I won't forget. You'll be doing me _____ _____ _____.

M: (Okay, but bring it right back when you're finished with it.)

14

M: Sally, _____ _____ _____ now!

W: What's the matter?

M: What's the matter? You were supposed to pick me up at 2 p.m.

W: I was? I don't remember _____ _____ _____.

M: Well, you did.

W: I really don't remember saying that. Why would I lie to you?

M: I'm not saying that _____ _____ _____. It's just that you promised to come.

W: When did I make that promise?

M: This morning when you _____ _____ _____.

W: Now I remember! I'm so sorry.

M: That's okay. At least you remember now.

W: (How can I make it up to you?)

15

M: Kate is a trial lawyer. She stands in court every day and defends her clients. Kate _____ _____ _____ in her work and pushes herself extra hard to be a better lawyer. Janet, Kate's next door neighbor, has a little girl who admires Kate and her job. She is only a third-grader but hopes to _____ _____ _____ _____ a successful trial lawyer like Kate. Knowing that Kate adores children and even does volunteer work at the teen center, Janet approaches Kate. Janet asks Kate if her little girl can _____ _____ _____ Kate's work one day and see the court. In this situation, what would Kate most likely say to Janet?

Kate: (That would be fine. No problem.)

16~17

W: May I have your attention, please? The weather forecast for this weekend calls for our unusually high temperatures to continue. Most areas are currently safe from heat warnings. However, high temperatures, _____ _____ _____ _____ _____, will create dangerously poor air quality conditions this weekend. Unless it is unavoidable, city residents should not _____ _____ _____ for the next two days. This is especially true for young children and the elderly. People who suffer from allergies or are sensitive to air pollution should also _____ _____ _____ _____ _____ the amount of time they spend outside. What's more, to prevent conditions from getting worse, everyone is asked to drive as little as possible and to _____ _____ _____ yard waste. On Monday afternoon, a cold front will pass through the area, bringing heavy rain. This should result in a significant improvement in the conditions. You can expect a high temperature of 30 degrees Celsius this afternoon. Please _____ _____ to our station throughout the weekend for further updates. Thank you.

지은이

NE능률 영어교육연구소

NE능률 영어교육연구소는 혁신적이며 효율적인 영어 교재를 개발하고
영어 학습의 질을 한 단계 높이고자 노력하는 NE능률의 연구조직입니다.

수능만만 〈영어듣기 35회〉

펴 낸 이	주민홍
펴 낸 곳	서울특별시 마포구 월드컵북로 396(상암동) 누리꿈스퀘어 비즈니스타워 10층
	㈜NE능률 (우편번호 03925)
펴 낸 날	2022년 1월 5일 개정판 제1쇄 발행
	2022년 9월 15일 제6쇄
전 화	02 2014 7114
팩 스	02 3142 0356
홈페이지	www.neungyule.com
등록번호	제1-68호
I S B N	979-11-253-3742-3 53740
정 가	17,000원

NE 능률

고객센터

교재 내용 문의 : contact.nebooks.co.kr (별도의 가입 절차 없이 작성 가능)
제품 구매, 교환, 불량, 반품 문의 : 02-2014-7114
☎ 전화문의는 본사 업무시간 중에만 가능합니다.

만만한
수능영어

NE 능률

수능만만

영어듣기
35회

수능듣기 MINI BOOK

PART 1 수능듣기 유형 탐구
PART 2 수능듣기 유형별 필수 어휘

만만한
수능영어

수능
만만

영어듣기
35회

수 능
만 만

PART 1

수능듣기 유형 탐구　　03

01 짧은 대화 응답

02 목적

03 주제

04 의견 · 주장

05 관계

06 그림 일치

07 할 일

08 부탁

09 이유

10 금액 계산

11 언급

12 내용 일치

13 도표

14 긴 대화 응답

15 상황에 적절한 말

16 세트 문항

기출문제 해석　　20

짧은 대화 응답

유형 분석	짧은 대화를 듣고, 남자나 여자의 마지막 말에 대한 상대방의 응답을 고르는 문제 유형이다.

해결 전략	1 대화의 마지막 말이 의문문인지 평서문인지 유의하여 듣는다. 2 마지막 말이 의문문인 경우, 의문사에 유의하여 묻는 내용을 정확히 파악한 후 의 　문사에 적합한 응답을 고른다. 3 마지막 말이 평서문인 경우, 대화의 전체적인 내용과 맥락을 통해 가장 적절한 응 　답을 고른다.

기출 문제
2017 수능

대화를 듣고, 남자의 마지막 말에 대한 여자의 응답으로 가장 적절한 것을 고르시오.

① I agree. But I don't have the time for it.
② You're right. Then I'll never tell anyone.
③ Trust me. You'll realize you did the right thing.
④ I understand. But let us know if it happens again.
⑤ That's true. We've been practicing for a long time.

Script

M: Mom, I'm home. Sorry I'm so late.
W: Peter! Where were you? Your dad and I were worried about you.
M: I was practicing for the musical, and I didn't realize how late it was.
W: (I understand. But let us know if it happens again.)

해법 적용 여자가 늦게 귀가한 아들에게 걱정했다고 말하자, 아들이 늦은 이유를 설명하고 있다. 따라서 다음부터
는 늦을 경우 미리 알려줄 것을 부탁하는 ④가 여자의 응답으로 가장 적절하다. 정답 ④

02 목적

유형 분석	담화를 듣고, 화자가 하는 말의 목적을 파악하는 문제 유형이다.

해결 전략	1 담화를 듣기 전에, 선택지를 미리 읽고 담화 내용을 예측한다.
	2 담화에서 반복되는 단어나 어구를 주의해서 듣는다.
	3 담화의 처음이나 끝에 화자가 말하고자 하는 바를 밝히는 경우가 많으므로 특히 유의하여 듣는다.

기출 문제
2017 수능

다음을 듣고, 여자가 하는 말의 목적으로 가장 적절한 것을 고르시오.

① 컴퓨터 사용 시 올바른 자세에 대해 조언하려고
② 컴퓨터 사용 중 휴식의 필요성을 강조하려고
③ 컴퓨터 사용 관련 절전 요령을 설명하려고
④ 회사 내 컴퓨터 보안 강화 방침을 안내하려고
⑤ 직장 내 컴퓨터 개인 용무 사용 자제를 당부하려고

Script

W: Hello, listeners! This is *One Minute Health*. Nowadays more and more office workers have been reporting troubles caused by long hours spent in front of the monitor. The biggest problems are damage to the eyes and stress on the neck and back. Improper posture is the main cause. Here are some useful tips for keeping proper posture while you use your computer. First, make sure to sit 50 to 70 centimeters away from the monitor. Sitting too close to your monitor can hurt your eyes. Second, to lessen the stress on your neck, you need to sit directly in front of your monitor, not to the left or to the right. Lastly, try to keep your knees at a right angle to reduce the pressure on your back. Remember that proper posture is the first step to healthy computer use. This has been Brenda Smith at *One Minute Health*.

해법 적용 여자는 컴퓨터 사용 시 올바른 자세를 유지하는 것에 대한 조언을 해주겠다고 말했다. 정답 ①

| 유형
분석 | 담화나 대화를 듣고, 화자가 하는 말의 주제를 파악하는 문제 유형이다. |

| 해결
전략 | 1 담화나 대화를 듣기 전에, 선택지를 미리 읽고 무엇에 관한 내용일지 예상해 본
다.
2 전반적인 내용을 통해 화자가 말하고자 하는 바를 추론해야 한다.
3 정답을 고를 때는 너무 포괄적이거나 특정 내용과 관련된 지엽적인 선택지를 고
르지 않도록 유의한다. |

기출 문제
2015 수능

대화를 듣고, 두 사람이 하는 말의 주제로 가장 적절한 것을 고르시오.

① 수면 문제를 해결하는 방법
② 휴대전화와 업무 효율성의 관계
③ 미디어 기기가 학습에 미치는 영향
④ 규칙적인 운동의 중요성
⑤ 충분한 낮잠의 필요성

Script

W: Brian, you look really tired.
M: Yeah, I've been having difficulty sleeping recently.
W: Why? Is something troubling you?
M: Not really, but I don't know how to get over this problem.
W: Well, keeping a regular sleep pattern can help.
M: I usually go to bed and wake up at the same time every day, but it doesn't work.
W: Hmm, do you do anything particular before bed?
M: Well, I use my phone to play games or text with friends for a little while.
W: Oh, that's not good. It's important not to use media devices for at least an hour before sleep.
M: Really? I thought it would help relieve my stress.
W: It can overexcite your senses and cause trouble sleeping.
M: I see.
W: One other tip for sound sleep is to drink some warm milk before bed.
M: Okay, I'll try those things and see if they work. Thanks.

해법 적용 두 사람은 수면 문제를 해결하는 방법에 관해 말하고 있다. 정답 ①

유형 분석	대화를 듣고, 화자가 어떤 의견이나 주장을 가지고 있는지 파악하는 문제 유형이다.

해결 전략	1 여자와 남자 중 누구의 의견이나 주장을 묻는지 확인하고, 그 사람의 말에 집중한다. 2 화자가 반복적으로 사용하는 특정 단어나 어구에 주목한다. 3 대화의 후반부에 화자의 생각을 다시 한 번 표현하는 경우가 많으므로 이에 유의하여 듣는다.

기출 문제
2016 수능

대화를 듣고, 남자의 의견으로 가장 적절한 것을 고르시오.

① 교사는 수업 시 학생들의 개인차를 고려할 필요가 있다.
② 원만한 교우 관계는 학습 동기를 강화시킨다.
③ 게임을 이용한 수업은 학습에 도움이 된다.
④ 효과적인 수업을 위한 게임 개발이 중요하다.
⑤ 조용한 학습 분위기가 수업 진행에 필수적이다.

Script

M: Ms. Robinson, what was your opinion of Mr. Brown's open class today?
W: It looked interesting.
M: Yes, it did. I was impressed by how active it was.
W: But honestly, the class seemed a bit noisy to me.
M: Yeah, I know. But that's because today's class was based on games.
W: Don't you think games can be a waste of class time?
M: Well, actually I think that games can help students.
W: I know students love games, but how do they help?
M: Games make students want to participate more actively in class.
W: Do you really think so?
M: Yeah. Today I saw all of the students laughing and enjoying themselves during the class. I think the games motivated everyone to learn.
W: That makes sense. I even saw shy students asking questions.
M: Right. If students enjoy themselves in class, it'll certainly help their learning.

해법 적용 두 사람은 수업 분위기에 관해 이야기하고 있는데, 남자는 게임에 기반한 수업이 학생들의 학습에 도움이 된다고 말하고 있다. 정답 ③

유형 분석　대화를 듣고, 두 사람의 관계를 유추하는 문제 유형이다.

해결 전략
1 대화 초반에 드러나는 정황을 파악한다.
2 특정 직업의 사람이 하는 일과 관련된 표현을 익혀 두고, 이에 유의하여 듣는다.
3 대화에 언급된 내용과 관련된 유사한 직업이 선택지에 오답으로 제시되는 경우가 많으므로, 혼동하지 않도록 주의한다.

기출 문제
2017 수능
대화를 듣고, 두 사람의 관계를 가장 잘 나타낸 것을 고르시오.
① 사회자 – 마술사　　② 조련사 – 관람객
③ 무대감독 – 가수　　④ 운전기사 – 정비사
⑤ 은행원 – 고객

Script
W: Welcome back! Next, we're very excited to have today's special guest. Will you please welcome Jack Wilson?
M: Hi, Laura. Thanks for having me. I watch your TV show every morning.
W: I'm flattered, Jack. So, I saw your magic performance at the theater a few days ago. It was amazing!
M: Yeah. More than 500 people came to see it. I had a wonderful time.
W: Now, we all know making things disappear is your specialty.
M: That's right.
W: Can you tell us how you make huge things like cars disappear?
M: Well, I can't reveal my secrets, but it's not as easy as pulling a rabbit out of a hat.
W: Come on! Can't you share just one little trick with us?
M: Alright, then. Just one. I'll show you a magic trick with coins.
W: Great! When we come back from the break, we'll learn a coin trick from Jack Wilson. Stay tuned!

해법 적용　TV 쇼를 진행하는 사회자가 마술사를 소개하고, 마술사의 주특기 마술에 관해 이야기하고 있다.
정답 ①

유형
분석   대화를 듣고, 그림에서 대화의 내용과 일치하지 않는 부분을 고르는 문제 유형이다.

**해결
전략**

1 대화를 듣기 전에, 그림 속의 인물이나 사물의 주된 특징과 위치를 파악해둔다.
2 모양이나 위치를 나타내는 표현을 잘 익혀 두고, 그림 속의 인물이나 사물을 묘사
하는 해당 표현을 집중해서 듣는다.

기출 문제
2016 수능

대화를 듣고, 그림에서 대화의 내용과 일치하지 <u>않는</u> 것을 고르시오.

Script

M: What are you looking at, honey?
W: My friend Lisa sent me a picture of the park we're visiting this weekend.
M: Oh, did she? Let me see.
W: She said her family had a really good time there.
M: Oh! There are two trees near the fence. That's very nice.
W: Yeah, it is. And I like the bench between the trees.
M: Me, too. We can rest in the shade there.
W: And look at the pond with the fish. It looks peaceful.
M: It does. I like the bear statue in the middle of the picture.
W: Yeah, it's cute. I think the kids will like it, too. We can take photos in front of
it.
M: They'll also have a lot of fun with the swing at the right side of the picture.
W: Definitely. They'll really enjoy it.
M: The park will be perfect for our family.

해법 적용   대화에서는 사진의 오른쪽에 그네가 있다고 했는데, 그림에는 미끄럼틀이 있다. 정답 ⑤

07 할 일

유형 분석	대화를 듣고, 남자나 여자 혹은 두 사람이 앞으로 할 일을 고르는 문제 유형이다.

해결 전략	1 대화 초반부에 드러나는 현재 상황을 파악하고, 앞으로 일어날 일을 예측하며 듣는다. 2 할 일을 당사자가 직접 언급하지 않고 상대방이 의향을 묻거나 제안하는 경우도 있으므로, 대화의 흐름을 놓치지 않도록 한다. 3 하려는 일의 계획이 대화 중간에 바뀌기도 하므로 혼동하지 않도록 유의한다.

기출 문제
2017 수능

대화를 듣고, 여자가 할 일로 가장 적절한 것을 고르시오.

① 여분의 메모리 카드 찾기
② 생일 케이크 만들기
③ 카메라 가방 구매하기
④ 공연 연습 도와주기
⑤ 아이의 무대의상 가져오기

Script

W: Honey, look at this picture of Tony from his first birthday party.
M: Can you believe how fast our son is growing up?
W: Yeah, the time is just flying by. It's hard to believe that his first school performance is this afternoon.
M: I know. Have you prepared everything that we need to record his play?
W: Well, I've packed the video camera in that bag over there.
M: Great. How about an extra memory card?
W: Oh, right. We'll probably need another one to record the whole play.
M: Better safe than sorry.
W: That's right. I'll go find one. Did you pick up Tony's costume yet?
M: I'll go pick it up from the dry cleaner while you look for the memory card.
W: Okay, sounds good.

해법 적용 두 사람은 아이의 학예회에 갈 예정이며, 남자가 세탁소에서 아이의 의상을 찾아올 동안 여자는 비디오 카메라에 쓸 여분의 메모리 카드를 찾을 것이다. 정답 ①

유형 분석	대화를 듣고, 화자가 상대방에게 부탁한 일을 고르는 문제 유형이다.

해결 전략	1 지시문이 영어로 제시된 경우, 대화를 듣기 전에 선택지를 미리 읽고 부탁하고자 하는 내용이 무엇인지 추론해 본다. 2 화자가 부탁하는 내용과 이에 대한 상대방의 반응에 주목한다.

기출 문제
[2015 수능]

대화를 듣고, 남자가 여자에게 부탁한 일로 가장 적절한 것을 고르시오.

① to remove election posters
② to take the adviser position
③ to cancel the student meeting
④ to speak out against school violence
⑤ to register as a presidential candidate

Script

M: Eva, I've decided to run for student council president!
W: Great, Matthew!
M: Now I need to start working on some campaign promises.
W: You're right. You'll have to think about what the students want. By the way, did you hear Tony Johnson is running, too?
M: Yeah, I know. I don't think it'll be easy to win without a good campaign team.
W: That's true. You'll need people to help you with posters, speeches, and everything else.
M: Exactly. So I need an adviser like you for my campaign. Can you give me a hand?
W: Me? I'd love to help, but do you really think I'm qualified for that position?
M: Of course. You'll be a great help with the election.
W: Then, okay. I'll do my best.
M: Thanks.

해법 적용   남자는 여자에게 자신의 선거 활동을 도와줄 것을 부탁하였고, 여자는 이를 수락하고 있다. 정답 ②

유형 분석	대화를 듣고, 특정 행동이나 감정의 이유를 추론하는 문제 유형이다.

해결 전략	1 대화를 듣기 전에, 지시문과 선택지를 미리 읽고 묻는 내용을 정확하게 파악하고 대화의 내용을 미리 예상해 본다. 2 이유나 문제 상황에 대해 묻고 답하는 내용에서 이유가 직접적으로 언급되는 경 우가 많으므로, 대화의 흐름을 놓치지 않도록 한다.

기출 문제
2015 수능

대화를 듣고, 남자가 야구 경기를 보러 갈 수 없는 이유를 고르시오.

① 독감에 걸려서
② 비행기를 놓쳐서
③ 동생을 돌봐야 해서
④ 회의에 참석해야 해서
⑤ 입장권을 구할 수 없어서

Script

[Telephone rings.]

W: Hello?

M: Hey, Sarah. It's Alex. I have a question for you.

W: Sure. What's up?

M: Well, I've got two tickets to the National Baseball Championship next Friday, but unfortunately I can't go. Do you want them?

W: Yes, of course, but I thought the tickets were sold out.

M: They are. But my company buys tickets to every game in advance, and I was able to get some.

W: That's cool, but why can't you go?

M: I have to fly out of town for an important meeting with a client on that day, so I can't make it. Maybe you can take your brother with you.

W: Yeah, he'd love to go. But I feel sorry that you can't go to the game because of work.

M: Don't worry about it. Enjoy the game.

해법 적용 남자는 야구 경기가 있는 날 중요한 회의에 참석해야 해서 야구 경기를 보러 갈 수 없다고 말했다.
정답 ④

유형 분석	대화를 듣고, 남자나 여자가 지불할 금액을 고르는 문제 유형이다.

해결 전략	1 대화 중에 언급된 수치들을 이용해 금액을 계산해야 하므로 대화의 흐름을 놓치지 않도록 한다.
	2 대화를 들으면서 제시되는 숫자들을 메모해 두고, 개수, 할인율 등의 변수에 유의하여 계산한다.
	3 할인율이 언급되는 경우가 많으므로 할인에 관련된 표현에 유의하며 듣는다.

기출 문제

2017 수능

대화를 듣고, 남자가 지불할 금액을 고르시오.

① $36 ② $40 ③ $45

④ $47 ⑤ $50

Script

M: Hi. I want to get some souvenirs for my family and friends. Can you show me where the souvenir magnets are?

W: Sure. They're right here. We've got a lot to choose from. What do you think of this flag magnet?

M: That one's good, but the airplane one is better. My sons love airplanes. How much is it?

W: It's ten dollars.

M: Alright, I'll take two of them. And I'd like to pick up something for my friends, too.

W: How about these animal key rings? They're five dollars each.

M: They're nice. I'll take six of them.

W: Okay. Anything else?

M: That's it. Can I use this coupon I picked up from the hotel?

W: Yeah. That gives you 10% off the total price.

M: Wonderful. Here's my credit card.

해법 적용 남자는 아들에게 줄 비행기 모양의 자석($10)을 두 개, 친구들에게 줄 동물 모양의 열쇠 고리($5)를 여섯 개 샀고 총 금액의 10퍼센트가 할인되는 쿠폰을 사용한다고 했으므로, 남자가 지불할 금액은 45달러이다. 정답 ③

유형 분석	대화를 듣고, 화자가 언급하지 않은 것을 고르는 문제 유형이다.

해결 전략	1 대화를 듣기 전에, 선택지를 미리 읽고 해당 내용을 집중해서 듣는다. 2 대화를 들으면서 언급된 내용은 선택지에서 지워나간다.

기출 문제

2015 수능

대화를 듣고, 미용실에 관해 두 사람이 언급하지 <u>않은</u> 것을 고르시오.

① 가게 이름   ② 위치   ③ 남자 이발 비용

④ 영업시간   ⑤ 미용사 이름

Script

M: Hey, Jennifer. There's something different about you today.

W: Yeah. I got my hair cut yesterday. How do I look?

M: That style really suits you. What's the name of the hair salon? I need a haircut, too.

W: It's called "Beautiful Hair, Wonderful Day."

M: Hmm, I think I've seen it before. Can you tell me where it is?

W: Sure. It's located on Main Street, near the Central Shopping Mall.

M: Oh yeah, now I remember where it is. Do you know how much a man's haircut costs there?

W: I think it's about 15 dollars, but you can check the price on the salon's website to make sure.

M: I'll check online later.

W: I can recommend a stylist if you want. She does a great job with men's hair.

M: Okay. What's her name?

W: Alice Moore. She's really good.

M: Great. Thanks.

해법 적용 미용실의 영업시간에 대해서는 언급하지 않았다. 정답 ④

유형 분석	담화를 듣고, 내용과 일치하거나 일치하지 않는 내용을 고르는 문제 유형이다.

해결 전략	1 담화를 듣기 전에, 선택지를 읽고 내용을 예측한다. 2 대개 선택지에 제시된 순서대로 담화가 전개되므로, 내용과 선택지를 차례대로 대조하며 오답을 지워나간다. 3 시각, 날짜 등의 숫자 정보를 놓치지 않도록 유의한다.

기출 문제
2017 수능

Creative Minds Science Club에 관한 다음 내용을 듣고, 일치하지 <u>않는</u> 것을 고르시오.

① 1학년과 2학년 학생이 가입할 수 있다.
② 매주 화요일 방과 후에 모인다.
③ 작년에 다수의 발명 대회에서 수상하였다.
④ 올해의 지도 교사는 물리 선생님이다.
⑤ 더 많은 정보를 학교 게시판에서 찾을 수 있다.

Script

M: Good morning, everyone. I'm Matt Adams, president of Creative Minds Science Club. I'd like to invite you to join our club. Creative Minds is open to first- and second-year students. We meet in the science lab every Tuesday after school. We have a variety of interesting activities, like doing fun experiments and making inventions as a team. In fact, last year our club won prizes at a number of invention contests. We're very proud of our achievements. This year's advising teacher is Ms. Williams, who is a chemistry teacher at our school. This is a great opportunity to learn more about science and to put your creative mind into action. If you're interested, you can find more information on our school's bulletin board. Come and join us!

해법 적용 올해의 지도 교사는 화학 선생님(chemistry teacher)이라고 했다. 정답 ④

13 도표

**유형
분석**　도표를 보면서 대화를 듣고, 대화에 제시된 정보를 토대로 질문에 맞는 답을 찾는
문제 유형이다.

**해결
전략**
1 대화를 들으면서, 대화 내용에 해당하지 않는 선택지는 지우면서 정답의 범위를
　좁혀 나간다.
2 시각, 날짜와 같은 숫자 정보를 놓치지 않도록 한다.
3 두 화자가 제안과 거절을 반복하는 대화 흐름을 따라 최종적으로 무엇을 선택하
　는지 대화 마지막까지 집중해서 듣는다.

기출 문제
2017 수능

다음 표를 보면서 대화를 듣고, 여자가 구매할 램프를 고르시오.

Floor Lamps for Sale

	Model	Height (cm)	LED Bulbs	Price ($)	Color
①	A	120	X	30	Black
②	B	140	O	40	Black
③	C	150	O	45	White
④	D	160	O	55	Black
⑤	E	170	X	55	White

Script

M: Welcome to Jay's Lighting Store. How may I help you?

W: I'm here to look for a floor lamp for my living room.

M: Here, take a look at this catalog. We have five models you can choose from. Are you looking for anything specific?

W: Yes. It shouldn't be too short. I'd like to get one that's taller than one hundred and thirty centimeters.

M: Then how about these four models? Would you like LED bulbs?

W: Yes, I would. They last longer than standard bulbs.

M: And they save energy. I definitely recommend LEDs. What's your price range?

W: Well, I don't want to spend more than fifty dollars.

M: Then you have two options left. Which color do you like better?

W: Hmm..., I'll go with the white one.

M: Good choice. Would you like to pay in cash or by credit card?

W: I'll pay in cash.

해법 적용　높이가 130센티미터가 넘어야 하고, LED 전구여야 하며, 50달러 이하의 가격인 모델 중에서 흰색인
램프는 C이다. 정답 ③

유형 분석	긴 대화를 듣고, 남자나 여자의 마지막 말에 대한 상대방의 응답을 고르는 문제이다.

해결 전략	1 대화의 전반적인 상황과 흐름을 파악하며, 대화의 마지막 부분에 단서가 제시되므로, 끝까지 집중해서 듣는다.
	2 응답자가 여자인지 남자인지 확인하고 그 사람의 입장에 맞는 선택지를 고른다.
	3 대화 중에 나온 표현을 활용하여 제시된 오답 선택지를 고르지 않도록 유의한다.

기출 문제
2017 수능

대화를 듣고, 여자의 마지막 말에 대한 남자의 응답으로 가장 적절한 것을 고르시오. 3점

Man: _____

① That's a good idea. I'll get rid of it right away.
② I hope you're right. I'll check with them.
③ Here's the wallet. Take it to the station.
④ It's too late. The tickets are all sold out.
⑤ I think it's closed. Turn in the book tomorrow.

Script

M: Alice, why didn't you come to the music festival yesterday?
W: I was busy doing my homework. I wish I could've gone. How was it?
M: It was great! My favorite band signed my ticket.
W: Wow! Can I see it?
M: Sure. It's in my wallet. *[Pause]* Wait! My wallet! It's gone!
W: Really? Look in your coat pockets. Maybe it's there.
M: It's not here. Oh, no! What should I do?
W: When was the last time you saw your wallet?
M: Umm... I remember having it at the subway station. Did I leave it there?
W: You should contact the station's lost and found right away.
M: But what if someone already took it?
W: Don't worry. Someone with a good heart might have turned it in.
M: (I hope you're right. I'll check with them.)

해법 적용

남자는 좋아하는 가수에게 사인을 받은 티켓을 여자에게 보여주려다가 지갑을 잃어버렸다는 것을 알게 된다. 여자는 누군가가 분실물 보관소에 지갑을 가져다 두었을 거라고 남자를 위로한다. 따라서 여자의 말이 맞길 바라며 분실물 보관소를 확인해 보겠다고 하는 ②가 남자의 응답으로 가장 적절하다.
정답 ②

상황에 적절한 말

유형 분석	담화를 듣고, 담화의 상황에서 해당 인물이 할 적절한 말을 고르는 문제 유형이다.

해결 전략	1 담화를 들으면서 등장 인물이 처한 상황을 파악한다. 2 담화 후반부에 해당 인물이 상대방에게 하고자 하는 말을 직접 언급하는 경우가 많으므로 이 부분에 유의하여 듣는다.

기출 문제
2017 수능

다음 상황 설명을 듣고, Brian이 Sarah에게 할 말로 가장 적절한 것을 고르시오. 3점

Brian: _____

① You can take a singing class at the local community center.
② Try to keep a close relationship with your classmates.
③ Would you give me some advice as a mentor?
④ You need to get help with your recommendation letter.
⑤ How about volunteering as a mentor at the community center?

Script

W: Sarah and Brian are university classmates. Sarah wants to be a teacher and is interested in helping others. She plans on finding volunteer work to contribute to the community while getting teaching experience. However, she's not sure what kind of volunteer work she can do. Brian is a volunteer mentor at a local community center. He feels that his students are learning a lot, and that he's benefitting from the experience as well. Sarah tells Brian about her plan and asks him to recommend some volunteer work for her. Since Brian finds his volunteer work rewarding, he wants to suggest to Sarah that she be a mentor at the community center. In this situation, what would Brian most likely say to Sarah?

Brian: (How about volunteering as a mentor at the community center?)

해법 적용 여자가 남자에게 봉사 활동을 추천해달라고 부탁하자, 남자는 자신이 하고 있는 지역 주민센터 자원봉사 활동이 보람차다고 느끼기 때문에 여자에게도 동일한 봉사 활동을 추천해주고 싶어 한다. 따라서 지역 주민센터의 자원봉사 활동을 제안하는 ⑤가 남자의 응답으로 가장 적절하다. 정답 ⑤

유형 분석	담화를 듣고, 동일한 하나의 담화와 관련된 두 가지 문제를 해결하는 유형이다.

해결 전략	1 16번은 주로 담화의 주제나 목적을 묻는 유형이고, 17번은 주로 담화의 세부 내용을 묻는 유형이 제시된다. 2 지문 내용은 연속하여 두 번 들려주므로 처음에는 전반적인 내용을 파악하고, 다시 들을 때는 세부 사항을 확인하는 데에 주목한다. 3 지문 내용이 길기 때문에 문제 유형에 따라 세부 내용을 메모하며 듣는다.

기출 문제
2017 수능

[16~17] 다음을 듣고, 물음에 답하시오.

16. 남자가 하는 말의 주제로 가장 적절한 것은?

① several ways flowers attract animals
② popular professions related to animals
③ endangered animals living on tropical islands
④ major factors that pose a threat to animals
⑤ various animals that feed from flowers

17. 언급된 동물이 <u>아닌</u> 것은?

① hummingbirds　② bats　③ lizards　④ parrots　⑤ squirrels

Script

M: Hello, class. Last time we learned about insects, their life cycles and what they eat. As you know, many insects get food from flowers, but they aren't the only creatures that do. Today, we'll learn about a variety of animals that use flowers as a food source. First are hummingbirds. These birds use their long narrow beaks to get the flower's sweet liquid called nectar. Mysteriously, they only feed from upside down flowers. We still don't know why. Next are bats. Although most bats eat insects, some get their food from flowers. These bats have a strong sense of smell and sight compared to insect-eating bats. There are also lizards that drink nectar. These lizards are found on tropical islands that have few natural enemies. Finally, there is a type of squirrel that feeds from flowers. Most nectar-drinking animals help flowers grow in numbers, but these squirrels often harm the plant. When drinking nectar, they bite through the flower, which causes damage. Interesting, huh? What other animals use flowers in their diet? Take a minute to think, and then we'll talk about it.

- -

해법 적용

16. 남자는 꽃을 먹이의 원천으로 하는 동물들에 대해 배울 것이라고 했으므로, 남자가 하는 말의 주제로 가장 적절한 것은 ⑤ '꽃을 먹고 사는 다양한 동물'이다. 정답 ⑤

17. 앵무새(parrots)는 언급되지 않았다. 정답 ④

기출문제 해석 ☁

남: 엄마, 저 왔어요. 너무 늦어서 죄송해요.
여: 피터! 너 어디 있었니? 아빠 엄마가 걱정했잖아.
남: 뮤지컬 연습을 하고 있었는데요, 얼마나 늦었는지 깨닫지 못했어요.
여: 알겠어. 하지만 또 그런 일이 있으면 알려주렴.

① 동의해. 하지만 그럴 시간이 없어.
② 네 말이 맞아. 그러면 아무에게도 말하지 않을게.
③ 나를 믿어. 네가 옳은 일을 했다는 것을 깨닫게 될 거야.
⑤ 맞아. 우린 오랫동안 연습하고 있잖아.

realize 깨닫다

여: 안녕하세요, 청취자 여러분! 일분 건강입니다. 요즘 점점 더 많은 사무실 근로자들이 모니터 앞에서 보내는 긴 시간에 의해 초래된 문제를 보고하고 있습니다. 가장 큰 문제는 눈의 손상과 목과 허리에 가는 스트레스입니다. 잘못된 자세가 주요 원인입니다. 당신이 컴퓨터를 사용하는 동안 바른 자세를 유지하기 위한 몇 가지 조언이 여기 있습니다. 첫째로, 모니터로부터 50에서 70센티미터 떨어져 앉도록 하세요. 모니터에 너무 가까이 앉는 것은 당신의 눈을 상하게 할 수 있습니다. 둘째로, 목에 가해지는 압박을 줄이기 위해, 당신은 모니터의 왼쪽이나 오른쪽이 아닌 정면에 앉아야 합니다. 마지막으로, 허리에 가는 압박을 감소시키기 위해 무릎을 직각으로 유지하려고 노력하세요. 바른 자세가 건강한 컴퓨터 사용을 위한 첫걸음이라는 것을 기억하세요. 일분 건강의 브렌다 스미스였습니다.

improper 부당한, 잘못된 posture 자세 lessen 줄다, 줄이다 pressure 압박, 압력

여: 브라이언, 너 정말 피곤해 보여.
남: 응, 최근에 잠을 자는데 어려움을 겪고 있어.
여: 왜? 뭔가 널 괴롭히니?
남: 그렇진 않아, 하지만 나는 이 문제를 극복하는 방법을 모르겠어.
여: 음, 규칙적인 수면 패턴을 유지하는 것이 도움이 될 수 있어.
남: 나는 보통 매일 같은 시간에 자러 가고 일어나는데, 효과가 없어.
여: 음, 자기 전에 특정한 무언가를 하니?
남: 글쎄, 나는 잠시 동안 게임을 하거나 친구와 문자를 하는데 내 전화기를 이용해.
여: 오, 그건 좋지 않아. 잠자기 적어도 한 시간 전에는 미디어 기기를 사용하지 않는 것이 중요해.
남: 진짜? 나는 이것이 내 스트레스를 해소하는데 도움이 될 거라고 생각했어.
여: 이것은 너의 감각을 과도하게 자극하고 수면 장애를 유발할 수 있어.
남: 그렇구나.
여: 건강한 수면을 위한 또 하나의 조언은 자기 전에 따뜻한 우유를 마시는 거야.
남: 응, 나는 그것들을 시도하고 효과가 있나 볼게. 고마워.

particular 특정한 device 장치, 기기 relieve 없애[덜어]주다, 완화하다

남: 로빈슨 선생님, 오늘 브라운 선생님의 공개 수업에 대한 선생님의 의견은 무엇이죠?
여: 재미있어 보였어요.
남: 네, 그랬죠. 저는 수업이 아주 활동적이어서 인상적이었어요.
여: 하지만 솔직히, 저한테는 수업이 약간 소란스러워 보였어요.
남: 네, 알아요. 하지만 그것은 오늘 수업이 게임을 기반으로 했기 때문이에요.

여: 게임이 수업 시간의 낭비가 될 수 있다고 생각
　　하지 않으세요?
남: 음, 사실 저는 게임이 학생들을 도울 수 있다고
　　생각해요.
여: 저는 학생들이 게임을 매우 좋아한다는 것을
　　알지만, 그것들이 어떻게 도움이 될까요?
남: 게임은 학생들이 더욱 적극적으로 수업에 참여
　　하고 싶게 만들어요.
여: 정말 그렇게 생각하세요?
남: 네. 오늘 저는 모든 학생들이 수업 시간에 웃고
　　스스로 즐기는 것을 봤어요. 저는 게임이 모두
　　를 배우는 데 동기를 부여했다고 생각해요.
여: 일리 있네요. 저는 심지어 수줍어하는 학생들
　　이 질문을 하는 것을 봤어요.
남: 맞아요. 만약 학생들이 수업에서 스스로 즐길
　　수 있다면, 그것은 그들의 학습에 확실히 도움
　　이 될 거예요.

be impressed by …에 깊은 인상을 받다 active 활동적
인, 적극적인 participate 참가하다 motivate 동기를 부
여하다 make sense 타당하다

여: 다시 오신 것을 환영합니다! 다음으로, 저희는
　　오늘 특별한 손님을 모시게 되어 무척 즐겁습
　　니다. 잭 윌슨을 환영해 주시겠습니까?
남: 안녕하세요, 로라. 초대해주셔서 감사합니다.
　　저는 당신의 쇼를 매일 아침에 시청해요.
여: 영광이네요, 잭. 저는 며칠 전에 극장에서 당신
　　의 마술 쇼를 봤어요. 정말 놀라웠어요!
남: 네. 500명 이상의 사람들이 그걸 보러 왔어요.
　　저는 멋진 시간을 보냈습니다.
여: 자, 우리 모두는 사물을 사라지게 만드는 것이
　　당신의 특기인 것을 알아요.
남: 맞아요.
여: 자동차와 같은 거대한 것을 어떻게 사라지게
　　만드는지 우리에게 알려줄 수 있나요?
남: 음, 제 비밀을 밝힐 수는 없지만, 모자에서 토
　　끼를 꺼내는 것처럼 쉬운 건 아니에요.
여: 제발요! 그냥 한가지 작은 비법만 저희와 공유
　　해 줄 수 없을까요?

남: 좋아요, 그럼. 딱 하나만요. 제가 동전으로 하
　　는 마술 비법을 보여드릴게요.
여: 좋아요! 잠시 휴식 후에 돌아왔을 때, 우리는
　　잭 윌슨 씨로부터 동전 마술을 배워봅시다. 채
　　널 고정해 주세요!

be flattered (어깨가) 으쓱해지다 disappear 사라지다
reveal (비밀 등을) 드러내다, 밝히다

남: 여보, 뭘 보고 있어요?
여: 내 친구 리사가 이번 주말에 우리가 방문할 공
　　원 사진을 나에게 보냈어요.
남: 아, 그랬어요? 어디 봐요.
여: 그녀는 자신의 가족이 그곳에서 정말 좋은 시
　　간을 보냈다고 했어요.
남: 아! 울타리 근처에 나무 두 그루가 있네요. 정
　　말 근사해요.
여: 네. 그리고 난 나무 사이에 있는 벤치가 마음에
　　들어요.
남: 나도 그래요. 우리는 그곳 그늘에서 쉴 수 있겠
　　네요.
여: 그리고 물고기가 있는 연못을 봐요. 평화로워
　　보여요.
남: 그렇군요. 난 사진 한 가운데에 있는 곰 동상이
　　좋아요.
여: 네, 귀엽네요. 아이들도 그걸 좋아할 거예요.
　　우리는 그 앞에서 사진을 찍을 수 있어요.
남: 애들은 또 사진 오른쪽에 있는 그네를 타면서
　　재미있게 놀 거예요.
여: 그렇고말고요. 애들은 정말 재미있게 놀겠죠.
남: 그 공원은 우리 가족에게 완벽할 거예요.

fence 울타리, 담 shade 그늘 pond 연못 statue 동상
swing 그네 Definitely. 분명히, 그렇고말고.

여: 여보, 첫 번째 생일 파티의 토니의 사진 좀 봐
　　요.
남: 우리 아들이 얼마나 빨리 자랐는지 믿겨져요?

기출문제 해석

여: 네, 시간이 정말 빠르네요. 그 애의 첫 학예회가 오늘 오후라는 게 믿기 힘들어요.

남: 그러게요. 그의 연극을 녹화하기 위한 모든 걸 준비했어요?

여: 음, 저기 가방에 비디오 카메라를 챙겨놨어요.

남: 좋아요. 여분의 메모리 카드는요?

여: 아, 맞아요. 연극 전체를 찍으려면 우리는 아마도 하나 더 필요하겠네요.

남: 나중에 후회하는 것보단 미리 준비하는 게 낫죠.

여: 맞아요. 내가 하나 찾아볼게요. 토니의 의상은 가져 왔어요?

남: 당신이 메모리 카드를 찾는 동안 내가 세탁소에 가서 찾아올게요.

여: 그래요.

prepare 준비하다 extra 추가의 costume 의상

유형 08

남: 에바, 나 학생회장에 입후보하기로 결정했어.

여: 잘했어, 매튜!

남: 이제 난 선거공약을 준비하는 일을 시작해야 해.

여: 맞아. 넌 학생들이 무엇을 원하는지에 대해 생각해 봐야 할 거야. 그런데, 너 토니 존슨도 출마할 거라는 이야기는 들었니?

남: 응, 알아. 좋은 선거 유세 팀 없이는 이기기 쉽지 않을 것 같아.

여: 그건 사실이야. 넌 포스터와 연설, 그리고 다른 모든 것들에 있어서 너를 도와 줄 사람들이 필요하게 될 거야.

남: 맞아. 그래서 나는 선거유세를 위해 너와 같은 자문위원이 필요해. 나를 도와 줄 수 있니?

여: 내가? 도와주고는 싶지만, 넌 정말로 내가 그 자리에 자격이 있다고 생각하니?

남: 물론. 너라면 선거에 큰 도움이 될 거야.

여: 그럼, 좋아. 최선을 다해 볼게.

남: 고마워.

① 선거 포스터 제거하기
② 자문직 수락하기
③ 학생회 모임 취소하기
④ 학교 폭력에 반대하는 목소리 높이기
⑤ 회장 후보자로 등록하기

run for …에 입후보하다 student council president 학생회장 campaign promise 선거공약 adviser 자문위원 be qualified for …에 자격이 있다 election 선거 do one's best 전력[최선]을 다하다

유형 09

[전화벨이 울린다.]

여: 여보세요?

남: 안녕, 새라. 나 알렉스야. 나 질문이 있어.

여: 응. 무슨 일이야?

남: 저기, 나 다음주 금요일에 전국 야구 결승전 표가 두 장 있는데, 안타깝게도 나는 갈 수 없어. 너 그거 원하니?

여: 응, 당연하지, 근데 나는 표가 매진인 줄 알았어.

남: 맞아. 그런데 우리 회사에서 매 경기마다 미리 모든 표를 구입해서 나도 좀 얻을 수 있었어.

여: 멋지다, 근데 넌 왜 못 가?

남: 그날 고객과의 중요한 회의가 있어서 나는 다른 도시로 가야 해, 그래서 나는 못 가. 아마도 너는 네 남동생을 데려갈 수 있겠다.

여: 응, 그 애는 가고 싶어 할 거야. 하지만 나는 네가 일 때문에 경기에 못 간다니 유감이야.

남: 걱정 마. 경기 재미있게 봐.

unfortunately 유감스럽게도 sold out 표가 매진된 client 의뢰인, 고객 make it 성공하다; *(모임 등에) 가다[참석하다]

유형 10

남: 안녕하세요. 가족들과 친구들을 위한 기념품을 좀 사고 싶은데요. 기념품 자석을 보여주실 수 있을까요?

여: 물론이죠. 바로 여기 있습니다. 다양한 것들이 있어요. 이 깃발 자석은 어떠세요?

남: 좋긴 한데, 비행기 자석이 더 좋네요. 제 아이

들이 비행기를 좋아하거든요. 얼마인가요?
여: 10달러입니다.
남: 알았어요, 그럼 비행기 자석 2개를 살게요. 그
리고 제 친구들을 위한 기념품도 사고 싶은데
요.
여: 이 동물 모양 열쇠 고리는 어떠세요? 각각 5달
러예요.
남: 좋네요. 동물 모양 열쇠 고리 6개를 살게요.
여: 알겠습니다. 더 필요한 것 있으세요?
남: 아뇨, 그거면 됐어요. 제가 호텔에서 가져온 이
쿠폰을 사용할 수 있을까요?
여: 네. 쿠폰을 쓰시면 총액에서 10퍼센트 할인됩
니다.
남: 좋네요. 여기 제 신용카드가 있어요.

souvenir 기념품 key ring 열쇠 고리 credit card 신
용카드

유형 11

남: 오, 제니퍼. 오늘 좀 달라 보이는데?
여: 응. 어제 머리를 잘랐어. 어때?
남: 그 스타일이 너에게 정말 잘 울려. 미용실 이름
이 뭐야? 나도 머리를 잘라야 해.
여: 미용실 이름은 Beautiful Hair, Wonderful
Day야.
남: 음, 그곳을 이전에 본 적이 있는 것 같아. 그곳
이 어디 있는지 말해줄래?
여: 물론이지. 그 곳은 센트럴 쇼핑몰 근처 Main
거리에 위치해 있어.
남: 오 그렇구나. 이제야 그곳이 어디 있는지 기억
이 난다. 거기 남자 머리 자르는데 가격이 얼마
인지 아니?
여: 내 생각엔 대략 15달러인 것 같은데, 확실하게
확인하려면 그 미용실 웹사이트에서 가격을 확
인할 수 있어.
남: 나중에 내가 인터넷으로 확인해볼게.
여: 네가 원한다면 헤어 디자이너를 추천해줄 수
있어. 그녀는 남자 머리를 아주 잘 해.
남: 그래. 그녀 이름이 뭐니?
여: 앨리스 무어야. 그녀는 정말 뛰어나.

남: 좋아. 고마워.

suit 어울리다 hair salon 미용실 recommend 추천하
다 stylist 미용사

유형 12

남: 여러분, 좋은 아침입니다. 저는 Creative
Minds Science Club의 회장, 맷 애덤스입니
다. 여러분을 저희 동아리에 가입하도록 초대
하고 싶습니다. Creative Minds는 1학년과 2
학년 학생들에게 열려 있습니다. 우리는 매주
화요일 방과 후에 과학실에서 만납니다. 우리
는 팀끼리 재미 있는 실험을 하거나 발명을 하
는 등의 다양한 흥미로운 활동들을 합니다. 사
실, 작년에 우리 동아리는 많은 발명 대회에서
수상을 했습니다. 우리는 우리의 성취에 대해
보람을 느끼고 있습니다. 올해 지도 선생님은
우리 학교의 화학 선생님인 윌리엄스 선생님입
니다. 저희 동아리는 과학에 대해 더 배울 수
있고 여러분의 창의적인 생각을 행동에 옮길
수 있는 좋은 기회입니다. 만약 여러분이 관심
이 있다면, 우리 학교의 게시판에서 더 많은 정
보를 확인할 수 있습니다. 와서 우리와 함께하
세요!

invite 초대하다 experiment 실험 invention 발명(품)
achievement 업적, 성취한 것 chemistry 화학 put
... into action …을 실행에 옮기다 bulletin board 게
시판

유형 13

남: 제이의 조명 가게에 오신 것을 환영합니다. 무
엇을 도와드릴까요?
여: 제 거실에 놓을 스탠드를 보러 왔는데요.
남: 여기 이 제품 안내서를 한 번 보세요. 저희는
손님이 고를 수 있는 다섯 가지 상품을 갖고 있
습니다. 특별히 찾는 상품이 있으신가요?
여: 네. 너무 짧아서는 안되고요. 130센티미터보다
높은 전등을 사고 싶어요.
남: 그렇다면 여기 네 가지 모델은 어떠십니까?

기출문제 해석

LED 전구를 선호하십니까?
여: 네, LED 전구를 선호합니다. LED 전구는 일반 전구보다 더 오래 지속되니까요.
남: 그리고 에너지를 더 절약하죠. 저는 LED 전구를 강력히 추천 드립니다. 손님이 생각하는 램프의 가격대가 어떻게 되나요?
여: 음, 저는 50달러보다는 더 많이 쓰고 싶지 않아요.
남: 그렇다면 두 가지 선택권이 남았습니다. 어떤 색을 더 좋아하십니까?
여: 음, 흰색으로 할게요!
남: 좋은 선택이십니다. 현금으로 지불하시겠습니까, 신용카드로 지불하시겠습니까?
여: 현금으로 지불할게요.

floor lamp (바닥에 세우는) 전기 스탠드 catalog (물품 · 책 등의) 목록 specific 구체적인 go with (계획 · 제의 등을) 받아들이다

유형 14

남: 앨리스, 어제 왜 음악 축제에 안 왔니?
여: 숙제를 하느라 바빴어. 갔으면 좋았을 텐데. 어땠어?
남: 대단했어! 내가 제일 좋아하는 밴드가 내 티켓에 사인을 해줬어.
여: 와! 내가 봐도 돼?
남: 물론이야. 내 지갑에 있어. [잠시 후] 잠시만! 내 지갑! 없어졌어!
여: 정말? 네 코트 주머니를 봐봐. 아마 거기 있을 거야.
남: 여기 없어. 오, 안돼! 어쩌면 좋지?
여: 네 지갑을 마지막으로 본 게 언제니?
남: 음… 지하철역에서 지갑을 갖고 있던 게 기억나. 내가 거기에 두고 왔나?
여: 지금 당장 지하철역의 분실물 보관소에 연락해 봐.
남: 하지만 누군가 이미 가져갔으면 어쩌지?
여: 걱정마. 좋은 마음씨를 갖고 있는 누군가가 지갑을 돌려 줄 거야.
남: 네 말이 맞길 바라. 내가 확인해 볼게.

① 그거 좋은 생각이네. 그걸 당장 없애야겠어.
③ 여기 지갑이 있어. 이걸 역에 가져다 줘.
④ 너무 늦었어. 표가 모두 팔렸어.
⑤ 문을 닫은 것 같아. 내일 책을 반납해.

lost and found 분실물 보관소

유형 15

여: 새라와 브라이언은 대학교 동기이다. 새라는 교사가 되고 싶어하고 다른 사람들을 돕는데 관심이 있다. 그녀는 가르치는 경험을 얻는 동시에 지역 사회에 공헌할 수 있는 자원봉사 활동을 찾기로 계획한다. 하지만 그녀는 그녀가 어떤 종류의 자원봉사 활동을 할 수 있을지 확실하지가 않다. 브라이언은 지역 주민센터에서 자원봉사자 멘토이다. 그는 그의 학생들이 많이 배우고, 이를 통해 그 또한 성장하고 있다고 느낀다. 새라는 브라이언에게 그녀의 계획에 대해 들려주고 그녀에게 적당한 자원봉사 활동을 추천해줄 것을 부탁한다. 브라이언은 그의 봉사 활동이 보람차다고 생각했기 때문에, 그는 새라에게 지역 주민센터에서 멘토가 되라고 제안하고 싶어 한다. 이 상황에서, 브라이언이 새라에게 할 말로 가장 적절한 것은 무엇인가?
브라이언: 지역 주민센터에서 멘토로 자원봉사하는 건 어때?

① 넌 지역 주민센터에서 노래 수업을 들을 수 있어.
② 네 반 친구들과 가까운 관계를 유지하려고 노력해 봐.
③ 멘토로서 내게 조언 좀 해 주겠니?
④ 너는 추천서와 관련해서 도움을 받아야 해.

contribute 기여하다 mentor 멘토 benefit from …에서 득을 보다 recommend 추천하다

유형 16

남: 안녕하세요, 학생 여러분. 지난 시간에 우리는 곤충과 곤충의 생애 주기, 그리고 그들이 무엇을 먹는지에 대해서 배웠습니다. 알다시피, 많

은 곤충들이 꽃에서 먹이를 얻지만, 곤충이 그렇게 하는 유일한 생물은 아닙니다. 오늘 우리는 먹이 원천으로 꽃을 사용하는 다양한 동물들에 대해 배울 것입니다. 첫 번째는 벌새입니다. 이 새는 길고 좁은 부리를 사용해서 꿀이라 불리는 꽃의 달콤한 액체를 얻습니다. 신비롭게도, 그것들은 거꾸로 달린 꽃에서만 먹이를 얻습니다. 우리는 아직 이유를 알지 못합니다. 다음은 박쥐입니다. 대부분의 박쥐가 곤충을 먹긴 하지만, 일부는 꽃에서 먹이를 얻습니다. 이 박쥐들은 곤충을 먹는 박쥐에 비해 강력한 후각과 시각을 지녔습니다. 꿀을 마시는 도마뱀도 있습니다. 이런 도마뱀은 천적이 거의 없는 열대 섬에서 발견됩니다. 마지막으로, 꽃을 먹고 사는 다람쥐 일종이 있습니다. 꿀을 먹는 대부분의 동물은 꽃이 개체 면에서 늘어나도록 돕지만, 이 다람쥐는 종종 식물에 해를 끼칩니다. 꿀을 마실 때, 다람쥐가 꽃을 물어뜯는데, 이는 피해를 유발합니다. 흥미롭죠? 어떤 다른 동물들이 먹이로 꽃을 이용할까요? 잠시 생각해 보세요, 그러고 나서 그에 대해 이야기해 볼게요.

① 꽃이 동물을 유인하는 몇 가지 방법
② 동물에 관련된 인기 직업
③ 열대 섬에서 사는 멸종 위기 동물
④ 동물에게 위협을 가하는 주요인
⑤ 꽃을 먹고 사는 다양한 동물

life cycles 생애 주기 creatures 생물
hummingbirds 벌새 beak 부리 nectar 과일즙, 꿀
natural enemy 천적 harm 해를 끼치다

PART 2

수능듣기 유형별 필수 어휘 26

01 목적

02 주제 · 의견

03 관계

04 그림 일치

05 할 일 · 부탁

06 금액 계산

07 언급 · 내용 일치 · 도표

08 기타 (시간 · 날짜)

01 목적

□ announce	알리다	□ book	예약하다
□ cancel	취소하다	□ ask	물어보다
□ change	바꾸다	□ celebrate	축하하다
□ warn	경고하다	□ criticize	비판하다
□ apologize	사과하다	□ entertain	즐겁게 해 주다
□ present	제시하다	□ performance	공연
□ facility	시설, 설비	□ donate	기부하다

□ Please make sure that ... 　반드시 …하세요.
□ Can[May] I have your attention, please? 　주목해 주시겠습니까?
□ For further information, contact ... 　더 많은 정보를 원하시면, …에 연락하십시오.
□ I'm here to tell you about ... 　여러분에게 …에 대해 말씀드리려고 이 자리에 나왔습니다.
□ I'd like to say a few words about ... 　…에 대해 몇 가지 말씀드리고자 합니다.
□ I want to notify you that ... 　여러분에게 …을 알려 드리고자 합니다.
□ Here are some tips for ... 　…와 관련된 몇 가지 조언이 있습니다.
□ Don't forget that ... 　…라는 걸 잊지 마세요.

02 주제 · 의견

1 건강

□ nonsmoking area	금연 구역	□ secondhand smoke	간접흡연
□ heart attack	심장마비	□ diet	식이요법
□ vegetarian	채식주의자	□ fit	건강한
□ regular exercise	규칙적인 운동	□ medical checkup	건강검진
□ gain[lose] weight	살이 찌다[살을 빼다]		

2 환경

□ acid rain	산성비	□ air pollution	대기오염
□ contamination	오염	□ pollutant	오염물질
□ greenhouse effect	온실 효과	□ atmosphere	대기
□ endangered	멸종 위기의	□ recycle	재활용하다
□ conserve	보존하다	□ public transportation	대중교통

□ qualification	자격	□ post[position]	지위, 직책
□ confidence	자신(감)	□ competent	유능한
□ require	요구하다	□ accountant	회계사
□ job opening	공석	□ recruit	채용하다, 모집하다
□ resume	이력서	□ application	지원, 신청서
□ part-time job	아르바이트	□ hire	고용하다
□ work experience	경력	□ bachelor's degree	학사 학위
□ employment	고용	□ requirement	요건
□ qualified	자격을 갖춘	□ deadline	마감일
□ working hours	근무 시간	□ substitute	대역, 대리인
□ commute	통근하다	□ notify	통지하다
□ option	선택 (사항)	□ leadership	통솔력
□ career	경력	□ branch	지점
□ interviewer	면접관	□ interviewee	면접 대상자
□ apply for	…에 지원하다	□ personal record	이력

03 관계

1 병원·약국

□ dentist	치과 의사	□ physician	내과 의사
□ vet(veterinarian)	수의사	□ pharmacist	약사
□ life guard	구조원	□ internship	인턴직, 인턴사원 근무
□ drugstore[pharmacy]	약국	□ patient	환자
□ dentist's office	치과	□ surgery[operation]	수술
□ prescription	처방전	□ dose	(약의) 1회 복용량
□ examine	진찰하다	□ first aid	응급 처치
□ sore throat	인후염	□ take medicine[a pill]	약을 복용하다

2 여행·숙박

□ manager	지배인, 매니저	□ front desk	안내소, 접수대
□ tourist	관광객	□ check-in[out]	체크인[체크아웃]
□ three nights	3박 (4일)	□ receptionist	안내원

□ tour guide	여행 가이드	□ single[double] room	1인실[2인실]
□ luggage	짐, 수하물	□ accommodation	숙박 시설
□ suitcase	여행 가방	□ stay	머무르다, 체류하다

3 식당·가게

□ customer	손님	□ bill[check]	계산서
□ treat	대접하다, 한턱내다	□ discount	할인
□ coupon	쿠폰	□ for free	무료로
□ complimentary	무료의	□ reserve[book]	예약하다
□ dessert	후식	□ recommend	추천하다
□ appetizer	전채 요리	□ takeout	가지고 가는 음식
□ clerk	점원	□ ingredient	요소, 성분, 재료
□ order	주문; 주문하다	□ serve	(음식을) 제공하다
□ try on	입어 보다	□ go well	잘 어울리다
□ receipt	영수증	□ refund	환불
□ on sale	할인 판매 중인	□ exchange	교환하다

4 공항

□ flight attendant	승무원	□ passenger	승객
□ passport	여권	□ boarding pass	탑승권
□ ticketing	티켓 구입	□ flight	항공편
□ gate	게이트, 탑승구	□ customs	세관
□ land	착륙하다	□ take off	이륙하다
□ check in	(짐을) 부치다	□ board	탑승하다
□ departure	출발	□ arrival	도착
□ one way	편도	□ round trip	왕복 여행
□ fare	요금	□ stopover	단기 체류, 경유지
□ claim	(소지품을) 찾다, 요구하다	□ claim baggage	짐을 찾다

5 도서관

□ librarian	사서	□ overdue	연체된
□ a due date	반납일	□ return	반납하다
□ check out	(책 등을) 대출하다	□ a late fee	연체료

6 우체국·배송

mail carrier	집배원	parcel[package]	꾸러미, 소포
registered mail	등기 우편	express mail	빠른 우편
extra charge	추가 요금	fragile	깨지기 쉬운

7 은행

cash	현금; 현금으로 바꾸다	check	수표
loan	대출	interest	이자
interest rate	금리, 이율	withdraw	인출하다
make a deposit	예금하다	open an account	계좌를 개설하다
loan	대출; 대출받다	fill out a form	양식을 작성하다

8 방송·출판

reporter[journalist]	기자	director	감독
actor	배우	entertainer	연예인
host	사회자, 주최자	audience	관객
publisher	출판업자	writer	작가

04 그림 일치

1 위치

next to	… 옆에
next to the door	문 옆에
in front of	… 앞에
in front of the first row	첫 번째 줄 앞에
around	… 주위에
around the table	탁자 주위에
beside	… 옆에
beside the telephone	전화기 옆에
between	… 사이에
between two pictures	두 사진 사이에
in the middle of	… 가운데에
in the middle of the room	방 한가운데에

□ **across**	…의 건너편에
across the bridge	다리 건너편에
□ **on the opposite of**	…의 반대편에
on the opposite side of the river	강 반대편에
□ **on top of**	…의 꼭대기에
on top of the shelf	찬장 꼭대기에
□ **behind**	…의 뒤에
behind the tree	나무 뒤에
□ **above**	…의 위에
above my desk	내 책상 위에
□ **upstairs**	위층; 위층에
Did you check upstairs?	위층을 확인해 봤니?

2 모양

□ **tiny**	작은, 조그마한
tiny holes	조그마한 구멍
□ **huge**	거대한
huge waves	거대한 파도
□ **pattern**	패턴, 무늬
patterns of butterfly wings	나비 날개의 무늬들
□ **wide**	넓은
a wide river	넓은 강
□ **narrow**	(폭이) 좁은
a narrow road	폭이 좁은 도로
□ **round**	둥근
a round mirror	둥근 거울
□ **square**	정사각형의
a large square room	큰 정사각형의 방
□ **rectangular**	직사각형의
a rectangular kite	직사각형의 연
□ **triangular**	삼각형의
a triangular plate	삼각형의 접시
□ **oval**	타원형의
an oval face	타원형의 얼굴

□ **dotted**	점으로 된, 점으로 뒤덮인	
along the dotted line	점선을 따라서	
□ **spot**	점, 반점	
a dog with black spots	검은 점박이 개	
□ **curved**	구부러진, 곡선의	
a curved wall	구부러진 벽	
□ **−shaped**	…모양[형태]의	
diamond-shaped	다이아몬드 모양의	
heart-shaped	하트 모양의	
□ **plain**	민무늬의, 무늬가 없는	
a plain T-shirt	민무늬 티셔츠	
□ **striped**	줄무늬의	
a striped T-shirt	줄무늬 티셔츠	
□ **checked[checkered]**	체크무늬의	
a checked[checkered] tie	체크무늬 넥타이	
□ **wavy**	물결무늬의	
a wavy carpet	물결무늬 양탄자	
□ **polka-dot**	물방울무늬의	
a polka-dot swimsuit	물방울무늬 수영복	
□ **floral[flower-patterned]**	꽃무늬의	
a floral[flower-patterned] dress	꽃무늬 드레스	

05 할 일·부탁

□ **pack**	(짐을) 싸다, 포장하다		□ **fix**	수리하다
□ **prepare**	준비하다		□ **miss**	놓치다, 빠뜨리다
□ **assignment**	숙제		□ **exhibition**	전시회
□ **dish**	요리, 음식		□ **garage sale**	중고 물품 세일
□ **musical**	뮤지컬		□ **admission fee**	입장료
□ **agency**	대리점		□ **customer service**	고객 서비스
□ **repair shop**	수리점		□ **available**	시간이 되는
□ **volunteer work**	자원봉사 활동		□ **charity**	자선 (단체)
□ **owe**	빚을 지다			
owe you $5	너에게 5달러를 빚지다			

□ **drop off**	…을 내려주다
drop him off at his house	그를 집에 (차로) 내려주다
□ **pick up**	…을 마중 나가다
pick her up at the airport	(차로) 공항에 그녀를 마중 나가다
□ **stop by**	…에 들르다
stop by the grocery store	식료품점에 들르다
□ **ask a favor**	부탁하다
ask you a favor	너에게 부탁하다
□ **clean up**	청소하다
clean up the garbage	쓰레기를 청소하다[치우다]
□ **take a ride**	…을 타다
take a boat ride	보트를 타다
□ **take care of**	…을 돌보다
take care of the children	아이들을 돌보다
□ **put off**	연기하다, 미루다
put off my trip	여행을 연기하다
□ **take ... for a walk**	산책시키다
take my pet dog for a walk	내 반려견을 산책시키다
□ **be supposed to-v**	…하기로 되어 있다
I was supposed to do homework today.	나는 오늘 숙제를 하기로 되어 있었다.

1 할 일

□ **What should we do this weekend?**	우리 이번 주말에 뭘 할까요?
□ **What do you want to do?**	뭘 하고 싶나요?
□ **What do you have in mind?**	생각해 둔 것이 있습니까?
□ **Why don't you ...?**	…하는 게 어때요?
□ **Is there anything I can do for you?**	제가 당신을 위해 할 수 있는 일이 있나요?
□ **It's about time we began practicing.**	이제 연습을 시작할 시간이군요.

2 부탁

□ **I wonder if you ...**	당신이 …해 주실 수 있는지 궁금합니다.
□ **What would you like me to do?**	제가 무엇을 해드리길 원하세요?
□ **Would you mind if I ...? /** **May[Can] I ...?**	제가 …해도 될까요?

☐ Can[Could] you (please) ...? / Will[Would] you (please) ...?	⋯해 주시겠어요?		
☐ May I ask you a favor? / Can I ask you to do something for me?	부탁 좀 드려도 될까요?		

06 금액 계산

☐ charge	요금; 청구하다	☐ cost	비용이 들다
☐ lower	낮추다	☐ pricey	비싼
☐ extra charge	추가 요금	☐ extra fee	추가 수수료
☐ fare	표준 요금	☐ premium	할증금
☐ installment	할부	☐ additional	추가의
☐ membership card	회원 카드	☐ admission	입장(료)
☐ a quarter	4분의 1	☐ a half(one half)	2분의 1
☐ two thirds	3분의 2	☐ double[twice/two times]	2배
☐ three times	3배	☐ four times	4배
☐ 1 dollar = 100 cents		☐ 1 km = 1,000 m = 100,000 cm	
☐ place[take] an order		주문하다[주문을 받다]	
☐ take A off B		B에서 A를 깎다	
take 10% off the total price		전체 금액에서 10%를 깎다	
☐ get a discount		할인 받다	
get a 50% discount		50% 할인받다	
☐ change		거스름돈, 잔돈	
change for a $20 bill		20달러짜리 지폐의 거스름돈	
☐ How much will it cost?		그건 비용이 얼마나 들죠?	
☐ They are $7 each.		각각 7달러입니다.	

07 언급·내용 일치·도표

1 요금표

☐ bill	청구서	☐ warranty period	품질 보증 기간
☐ out of stock	품절[매진]된	☐ in stock	재고가 있는
☐ on installment	할부로	☐ reasonable	(가격이) 적정한

capacity	용량, 수용력	anniversary	기념일
catalog	(물품 등의) 목록	cost	값, 비용; (비용이) 들다

2 일정표

farewell party	송별회	housewarming party	집들이
volunteer	자원봉사하다	submit[hand in]	제출하다
cancel	취소하다	delay	미루다
contestant[participant]	참가자	bazaar	바자회
festival	축제	competition[contest]	대회, 시합

- sign up (for) — (…을) 신청하다
  sign up for a credit card — 신용카드를 신청하다
- register for — …에 등록하다
  register for a German class — 독일어 수업에 등록하다
- enroll in — …에 등록하다
  enroll in a course — 강좌에 등록하다
- participate[take part] in — …에 참가하다
  participate[take part] in a contest — 대회에 참가하다

3 학교

presentation	발표	take a lesson	수업을 받다
academic	학업의	subject[course]	과목
major	전공	session	수업 시간
project	과제	scholarship	장학금
evaluation	평가	organize	체계화하다
category	범주	credit	학점
lecture	강의	entrance exam	입학시험
final exam	기말고사	biology	생물학
physics	물리학	economics	경제학
architecture	건축학	computer science	컴퓨터 공학
politics	정치학	ethics	윤리학
literature	문학	statistics	통계학
chemistry	화학	dorm(dormitory)	기숙사
principal	교장	attendance	출석
cafeteria	간이 식당	semester	학기

□ absence	결석	□ join	가입하다

4 교통

□ accident	사고	□ heavy traffic	교통 체증
□ expressway	고속 도로	□ gas station	주유소
□ drunk(en) driving	음주 운전	□ traffic jam	교통 정체
□ rush hour	출퇴근 혼잡 시간	□ speed limit	제한 속도
□ dead end	막다른 길	□ driver's license	운전 면허증
□ crosswalk	교차로	□ traffic light	신호등
□ pedestrian	보행자	□ delay	연착
□ toll	통행료	□ stop	정거장
□ fare	운임	□ public transportation	대중교통
□ construction	공사	□ bound for	…행의, …을 향하는

□ access — 접근
have easy access to public transportation — 대중교통에 쉽게 접근하다

□ construction — 공사
inform someone of road construction — …에게 도로 공사를 알리다

□ transfer — 이동하다, 갈아타다
transfer to the next station — 다음 역에서 갈아타다

□ pull over — 차를 길가로 붙이다
pull the car over — 차를 길가에 세우다

□ clear off — …을 치워 버리다
clear snow off — 눈을 치우다

□ slow down — (속도 · 진행을) 낮추다
slow down at crosswalks — 교차로에서 속도를 줄이다

5 안내 · 공지

□ contribution	기부	□ committee	위원회
□ inconvenience	불편	□ direction	지시
□ notice	공지, 안내	□ emergency	비상사태
□ celebrate	축하하다, 기념하다	□ make an announcement	안내 방송을 하다

□ due to — …에 기인하는, …때문에
due to heavy rain — 폭우 때문에

□ according to — …에 따르면
according to the report — 보고서에 따르면

☐ be sure to-v		반드시 …하도록 하다	
be sure to get dressed up		차려입을 것을 명심하다	
☐ This is your captain speaking.		기장이 말씀드립니다.	

⁶ TV · 공연

☐ broadcasting station	방송국	☐ soap opera	연속극
☐ quiz show	퀴즈 프로그램	☐ cartoon	만화
☐ weather forecast	일기 예보	☐ headline	주요 뉴스
☐ genre	장르	☐ blockbuster	대 히트작, 성공작
☐ on the air	방송 중인	☐ cast	배역을 정하다[맡기다]
☐ stage	무대	☐ character	등장인물
☐ run	진행되다, 지속되다	☐ volume	용량; 음량
☐ breaking news	속보	☐ Stay tuned!	채널 고정하세요!

08 기타 (시간 · 날짜)

☐ a quarter	15분	☐ an hour and a half	1시간 30분
☐ half an hour	30분	☐ ten to seven	7시 (되기) 10분 전 (= 6시 50분)
☐ ten past five	5시 10분	☐ in 15 minutes	15분 후에
☐ weekday	평일	☐ weekend	주말
☐ the day before yesterday	그저께		
☐ the day after tomorrow	모레		
☐ three days from now	3일 뒤에		
☐ every other day[week]	이틀[2주]에 한 번		
☐ on February 17	2월 17일에		
☐ the end of this month	월말		
☐ It's two weeks from today.	오늘부터 2주일 후입니다.		

MEMO

만만한 수능영어

수능
만만

영어듣기
35회

혼자 하면 작심 3일,
함께 하면 작심 100일!

공부작당소모임

함께 공부하는 **재미**
학습 점검은 **철두철미**
다 같이 완북하는 찰떡궁합 **케미**
이번 학기 내 시험지는 전부 **동그라미!**

자신감을 키우는 진짜 스터디 그룹,
공작소에서 만나보세요.

공부작당소모임 APP 다운로드

앱스토어   구글 플레이

www.gongzakso.com

NE능률 교재 MAP

아래 교재 MAP을 참고하여 본인의 현재 혹은 목표 수준에 따라 교재를 선택하세요.
NE능률 교재들과 함께 영어실력을 쑥쑥~ 올려보세요!
MP3 등 교재 부가 학습 서비스 및 자세한 교재 정보는 www.nebooks.co.kr 에서 확인하세요.

수능

초1-2	초3	초3-4	초4-5	초5-6

초6-예비중	중1	중1-2	중2-3	중3
			첫 번째 수능 영어 기초편	첫 번째 수능 영어 유형편
				첫 번째 수능 영어 실전편

예비고-고1	고1	고1-2	고2-3, 수능 실전	수능, 학평 기출
기강잡고 독해 잡는 필수 문법	빠바 기초세우기	빠바 구문독해	빠바 유형독해	다빈출코드 영어영역 고1독해
기강잡고 기초 잡는 유형 독해	능률기본영어	The 상승 구문편	빠바 종합실전편	다빈출코드 영어영역 고2독해
The 상승 직독직해편	The 상승 문법독해편	맞수 수능듣기 실전편	The 상승 수능유형편	다빈출코드 영어영역 듣기
올클 수능 어법 start	수능만만 기본 영어듣기 20회	맞수 수능문법어법 실전편	수능만만 어법어휘 228제	다빈출코드 영어영역 어법·어휘
얇고 빠른 미니 모의고사	수능만만 기본 영어듣기 35+5회	맞수 구문독해 실전편	수능만만 영어듣기 20회	
10+2회 입문	수능만만 기본 문법·어법·어휘 150제	맞수 수능유형 실전편	수능만만 영어듣기 35회	
	수능만만 기본 영어독해 10+1회	맞수 빈칸추론	수능만만 영어독해 20회	
	맞수 수능듣기 기본편	특급 독해 유형별 모의고사	특급 듣기 실전 모의고사	
	맞수 수능문법어법 기본편	수능유형 PICK 독해 실력	특급 빈칸추론	
	맞수 구문독해 기본편	수능 구문 빅데이터 수능빈출편	특급 어법	
	맞수 수능유형 기본편	얇고 빠른 미니 모의고사	특급 수능·EBS 기출 VOCA	
	수능유형 PICK 독해 기본	10+2회 실전	올클 수능 어법 완성	
	수능유형 PICK 듣기 기본		능률 EBS 수능특강 변형 문제	
	수능 구문 빅데이터 기본편		영어(상), (하)	
	얇고 빠른 미니 모의고사		능률 EBS 수능특강 변형 문제	
	10+2회 기본		영어독해연습(상), (하)	

수능 이상/ 토플 80-89· 텝스 600-699점	수능 이상/ 토플 90-99· 텝스 700-799점	수능 이상/ 토플 100· 텝스 800점 이상		

만만한
수능영어

정답 및 해설

영어듣기
35회

정답 및 해설

01 ①	02 ③	03 ④	04 ⑤	05 ⑤	06 ③
07 ①	08 ②	09 ④	10 ②	11 ⑤	12 ③
13 ⑤	14 ④	15 ④	16 ③	17 ⑤	

01 ①

여: 먼저, 제 남자친구 제이드에게 중요한 질문을 할 기회를 주신 이 라디오 프로그램에 감사드리고 싶습니다. 지금 무척 떨리네요. 청취자 여러분, 모두 제게 행운을 빌어주세요. 제이드, 지금 당신이 정말 멀리 있잖아. 내가 그곳에서 당신에게 이 질문을 할 수 있다면 좋을 텐데. 당신이 떠나고 나서 난 당신이 내게 얼마나 중요한지, 그리고 내 인생에서 당신이 얼마나 필요한지 깨달았어. 그러니 제이드, 내 남편이 되어 날 이 세상에서 가장 행복한 여자로 만들어줄래? 만일 당신이 (청혼을) 승낙하고 이 반지를 받아준다면, 난 당신을 행복하게 해주기 위해 최선을 다할 거라고 약속할게.

어휘
nervous 불안한, 초조한

02 ③

여: 얘! 너 지금 땅에서 더러운 캔을 집었니?
남: 응, 그래. 난 사람들이 쓰레기를 버리는 게 싫어. 이걸 재활용 통에 넣을 거야.
여: 하지만 네가 한 것도 아니잖아. 왜 신경을 써?
남: 우리 행성은 하나뿐이니까 좋은 상태로 유지해야 해.
여: 그렇지만 한 사람이 영향을 줄 수는 없어.
남: 음, 모두가 조금씩 하면, 이런 공원은 깨끗하고 아름다워질 수 있어.
여: 그렇게 생각해본 적은 없는데.
남: 네가 자연을 돌보는 일에 관심을 더 많이 두게 되기를 바랄게.
여: 아마도, 그런데 난 너처럼 헌신적으로 되진 않을 것 같아.

어휘
litter (쓰레기를) 버리다 recycling bin 재활용 쓰레기통 make a difference 영향을 주다 be inspired to-v …하도록 마음이 내키다 dedicated 전념하는, 헌신적인

03 ④

남: 안녕하세요.
여: 안녕하세요. 찾으시는 걸 도와드릴까요?
남: 사실 전 온라인에 게시한 구인 광고를 보고 왔습니다.
여: 아, 그렇군요. 전에 가게에서 일해본 경험이 있으세요?
남: 아뇨. 없습니다. 하지만 저는 빨리 배우는 편이에요. 제가 하게 될 일은 어떤 것인가요?
여: 음, 당신은 선반에 충분한 스낵과 음료들이 있도록 해야 합니다.
남: 쉬울 것 같은데요. 할 수 있어요!
여: 그리고 때로는 손님들에게 돈을 받고 잔돈을 거슬러주게 될 거예요. 그건 높은 정확도를 필요로 하고 실수가 용납되지 않아요.

남: 알겠습니다. 그리고 근무시간은 어떻게 되나요?
여: 음, 저희는 24시간 영업합니다. 그래서 저는 자정부터 아침 6시까지 일할 사람이 필요해요.
남: 알겠습니다. 지원서를 작성해도 될까요?
여: 물론입니다. 여기 있어요. 와줘서 고마워요, 그리고 사흘 후에 제 결정을 알려드릴게요.

어휘
post 게시하다 change 잔돈 accuracy 정확도 midnight 자정

04 ⑤

남: 안녕하세요, 커 선생님. 전 토비의 아빠예요.
여: 안녕하세요, 윌리엄 씨. 토비의 유치원을 보러 와주셔서 감사해요.
남: 그 애가 여기 얘기를 많이 해요. 정말 이곳을 좋아하죠. 와! 벽에 있는 큰 나무가 정말 맘에 드네요.
여: 저도 그래요. 아이들이 이름을 잎사귀 위에 적어놨어요. 매우 열심이에요.
남: 놀랍네요! 그리고 다른 벽은 알파벳 글자들로 뒤덮여 있네요.
여: 네. 이런 식으로 아이들이 글자를 더 쉽게 배울 수 있어요.
남: 좋은 생각이네요. 오! 여기 토비가 말하던 선반이 있어요. 하나에는 책이, 다른 하나에는 장난감이 있네요.
여: 네, 아이들이 읽고 노는 걸 정말 좋아해요. 선반 옆에 곰 인형이 보이세요?
남: 네. 이미 저게 뭔지 알죠. 토비가 자신이 제일 좋아하는 장난감이라고 하더군요.
여: 맞아요. 그리고 아이들이 탁자 모서리에 다치지 않도록 방 가운데에 둥근 탁자를 두었어요.
남: 토비가 왜 이 장소를 좋아하는지 알겠네요.
여: 감사해요, 윌리엄 씨. 저희도 토비가 이곳에 있는 게 좋아요.

어휘
kindergarten 유치원 be covered with …로 덮이다 so that A won't A가 …하지 않도록

05 ⑤

남: 안녕, 신디. 너 괜찮니? 피곤해 보인다.
여: 음, 최근에 잠을 많이 못 잤거든.
남: 왜? 시험이나 뭐 공부하는 거야?
여: 아니. 새 룸메이트가 매일 밤 새벽 3시까지 자기 방에서 TV를 크게 틀어놓고 보거든.
남: 아, 이런. 그에게 그렇게 하지 말라고 부탁해야겠다.
여: 음, 그래. 하지만 문제가 그것만이 아니야.
남: 정말? 또 뭐가 잘못됐는데?
여: 음, 그 애는 더러운 접시들을 항상 싱크대 안에 그대로 두고, 절대로 쓰레기를 내다 버리지도 않아.
남: 이런. 넌 그 애랑 진지하게 얘기해야 할 것 같은데.
여: 알아. 하지만 너무 긴장돼서 직접 불만을 말할 수가 없어.
남: 그럼 대신에 그 애에게 짧은 편지를 쓰는 건 어때?
여: 그게 효과가 있겠다. 조언해줘서 고마워.

어휘
recently 최근에 sink 싱크대 trash 쓰레기 in person 직접

06 ③

남: 안녕, 키이라. 네가 나를 보고파 했다는 말을 들었어.

여: 맞아. 어제 연극 동아리 모임을 잊어버렸어?

남: 아니, 잊지 않았어. 나는 단지 참석할 수 없었어.

여: 어째서? 그건 매우 중요한 모임이었어. 우리는 연말에 할 연극에 대해서 논의했어.

남: 미안해. 미리 너에게 연락을 해야 했는데. 하지만 나는 인턴십을 위한 중요한 면접이 있었어.

여: 아, 그랬구나. 네가 겨울 방학 동안 인턴십을 하고 싶다고 말을 했지.

남: 맞아. 지난달에 지원했고 면접은 어제였어.

여: 오, 잘됐구나. 어떻게 됐어?

남: 너무 긴장해서 배가 아팠어. 하지만 전반적으로 잘된 것 같아.

여: 잘됐다. 어쨌든 나중에 회의에 대해서 전부 말해 줄게.

남: 고마워, 키이라. 다음 회의는 빠지지 않도록 할게.

어휘
year-end party 송년회, 연말 파티 beforehand 사전에, 미리 overall 전반적으로

07 ①

남: 안녕하세요. 오늘 무엇을 도와드릴까요?

여: 저는 머리를 좀 자르고 싶은데요. 비용이 얼마인가요?

남: 길이에 따라 다릅니다. 짧은 머리는 20달러, 중간 길이는 30달러이고 긴 머리는 40달러입니다. 고객님 머리는 짧은 기장으로 생각되네요.

여: 그렇군요. 괜찮은 것 같네요. 머리 감겨주시는 것도 포함이죠?

남: 네, 그렇습니다. 저는 파마를 하시는 것도 추천해 드립니다. 파마는 150달러예요.

여: 흠… 그건 좀 비싸네요.

남: 좀 더 말씀드릴게요. 파마하시면 머리 자르는 비용은 부과되지 않습니다.

여: 괜찮네요. 그러면 머리도 자르고 파마도 해주세요.

남: 좋습니다. 고객님의 모발은 약간 약해 보이네요. 모발 관리도 추가하는 건 어떠세요? 30달러면 되는데요.

여: 아뇨, 괜찮습니다. 저는 집에서 관리해요. 혼자서 할 수 있어요.

남: 알겠습니다. 신규 고객이시기 때문에, 제가 대신 무료로 두피 마사지를 해드릴게요.

여: 감사합니다! 훌륭하네요! 아, 잊을 뻔했네요. 당신에게 보여드릴 쿠폰 번호가 있어요. 여기 웹사이트에서 받았거든요.

남: 아, 네. 이것은 10퍼센트 할인으로 유효합니다. 이제 가운 입혀 드릴게요.

여: 좋아요.

어휘
rate 요금 depend on …에 달려 있다 get a perm 파마하다 Tell you what. 좋은 생각이 있다. 할말이 있다. charge 요금을 부과하다 scalp 두피

문제풀이
여자는 머리를 자르고 파마를 하기로 했는데, 파마를 하면 머리 자르는 비용은 들지 않으므로 총 150달러이다. 이 금액에서 10퍼센트 할인을 받으므로 여자가 지불할 금액은 135달러이다.

08 ②

여: 넌 누구에 관해 역사 보고서를 쓸 거니?

남: 조지 워싱턴 카버라는 사람으로 정했어.

여: 그는 미국의 초대 대통령이었지 않아?

남: 그건 조지 워싱턴이지. 조지 워싱턴 카버는 과학자이자 발명가였어.

여: 아, 그래? 뭘 발명했는데?

남: 가난한 농부를 도울 방법을 찾아냈어. 예를 들어, 농장의 토양을 향상시키는 법을 그들에게 제시해 주었어.

여: 어떻게 그걸 했는데?

남: 그들에게 항상 목화를 기르는 대신 땅콩과 콩을 심게 했어.

여: 땅콩과 콩? 그게 왜 도움이 되었는데?

남: 그것들은 토양에 질소를 보충해주는데 그것은 토양을 더 건강하게 만들거든.

여: 그렇구나. 그는 지금도 살아 있니?

남: 아니, 그는 1943년에 사망했어. 하지만 가장 인상적인 건 그가 노예로 태어난 아프리카계 미국인이라는 점이야.

여: 농담이겠지. 그건 그가 성공하기 힘들게 했을 게 분명해.

남: 분명히 그랬을 거야. 하지만 그는 그 점이 자신을 방해하게 두지 않았어.

어휘
inventor 발명가 soil 토양 encourage 권장[장려]하다 soybean 콩 cotton 목화 nitrogen 질소 pass away 사망하다 slave 노예

09 ④

남: 여러분은 경쟁에서 오는 전율을 좋아합니까? 여러분은 퍼즐 푸는 것을 즐깁니까? 그러면 여러분은 분명히 올해의 Great Astounding Race에 참가하고 싶을 겁니다! 이 행사에서 여러분은 도시 곳곳의 유명한 주요 명소를 방문하면서 퍼즐을 풀게 됩니다. 여러분은 팀으로 참가해야 하고, 각 팀은 2명으로 구성되어야 합니다. 가장 짧은 시간 안에 모든 도전 과제를 끝내는 팀은 하와이 여행을 상으로 받게 됩니다. 이 경주는 10월 5일에 열릴 것이며, 오전 11시에 Central Plaza에서 시작될 겁니다. 여러분은 경주가 시작하는 날에 오전 9시부터 등록할 수 있습니다. 등록비는 팀당 20달러이고 마련된 모든 기금은 아동 병원 재단에 기부될 것입니다.

어휘
definitely 분명히 participate in …에 참가하다 landmark 랜드마크, 역사적인 장소 be made up of …로 구성되다 challenge 도전 과제 register 등록하다 registration fee 등록비 fund 기금 foundation 재단

10 ②

여: 찾으시는 물건이 있나요?

남: 저는 어머니께 선물로 집에 가져갈 차를 찾는데요.

여: 멋지네요! 어머니께서 보통 어떤 종류의 차를 드시나요?

남: 어머니는 모든 종류의 차를 다 좋아하세요. 안타깝지만, 녹차 알레르기가 있으시고요.

여: 아, 알게 되어 다행이네요.

남: 네. 하지만 어머니께 근사한 무언가를 좀 드리고 싶어요. 크기는 어떤 것들이 있나요?

여: 저희는 세 가지 크기가 있습니다. 100그램, 200그램, 그리고 500그램이요.

남: 500그램은 너무 많네요. 아무래도 더 작은 크기 중 하나가 좋겠네요.

여: 네. 통에 든 차가 좋으세요, 아니면 지퍼백에 든 것이 좋으세요?

남: 지퍼백이요. 그게 제 짐에 넣고 다니기가 쉬울 것 같습니다. 제가 업무차 온 거라서요.

여: 알겠습니다. 그러면 이건 어떠세요? 이건 지금 재고가 없지만, 며칠 안에 구매할 수 있어요.

남: 사실, 제가 내일 떠날 거라서 기다릴 수가 없네요. 전 지금 재고가 있는 것이 필요합니다.

여: 그렇다면, 이걸 추천해 드립니다.

남: 좋아요! 그걸로 할게요.

어휘
luggage 짐, 수하물 on business 업무상 in stock 재고의 ship 수송하다; *구매할 수 있다 [문제] availability 이용 가능성

11 ⑤

남: 네가 들고 있는 큰 상자에는 뭐가 들어 있니?

여: 이건 과학 박람회를 위한 내 프로젝트야. 아직 끝마치지는 못했지만 말이야.

남: 그렇구나. 박람회는 언제 열리는데?

여: 12일에 개막해서 일주일 동안 계속될 거야.

어휘
fair 박람회 take place 열리다, 개최되다 [문제] last 마지막의; 바로 앞의, 지난; 지속되다

문제풀이
① 지난 주말에 열렸어.
② 박람회에서 2등 상을 탔어.
③ 아직 잘 모르겠어. 난 아직 결정 못 했어.
④ 난 항상 그 박람회에 가고 싶었어.

12 ③

여: 아빠, 용돈 좀 올려주실 수 있어요?

남: 글쎄다. 이미 매주 20달러를 주고 있잖니.

여: 그런데 방과 후에 친구들과 뭘 사 먹기엔 충분하지 않아요.

남: 그럼 돈 씀씀이에 더 신중해야겠구나.

어휘
allowance 용돈 [문제] careful with money 돈 씀씀이에 꼼꼼한 permission 허가, 허락

문제풀이
① 이번 주에 용돈은 벌써 줬잖니.
② 친구에게 돈 빌려달라고 부탁하지 말라고 했잖니.
④ 가능한 한 빨리 내 돈을 갚으면 좋겠구나.
⑤ 말도 안 돼! 친구들을 집에 초대하라고 허락하지 않았어.

13 ⑤

여: 우리는 겉모습만 보고 속을 판단하면 안 돼.

남: 무슨 말을 하려는 거야, 민영아?

여: 어떤 선생님들은 짧은 치마를 입거나 귀고리를 한 학생들이 문제가 있는 애라고 믿거든. 민수야, 너도 그렇게 생각하니?

남: 물론 아니지! 모든 사람은 원하는 대로 자신을 표현할 권리가 있어.

여: 나도 그렇게 생각해.

남: 하지만 나도 보라색 머리를 하거나 머리를 완전히 민 사람들을 다르게 대했다는 건 인정할 수밖에 없어.

여: 왜?

남: 단지 그 사람들이 문제를 일으킬 거라고 느껴서 그들을 피했거든.

여: 우린 우리와 달라 보이는 사람들에게 마음을 열 필요가 있어.

남: 맞아. 사람들은 있는 그대로 존중받아야 해.

어휘
judge 판단하다 troublemaker 골칫거리; *말썽꾸러기 open-minded 마음이 열린, 편견 없는 [문제] respect 존경하다; *존중하다

문제풀이
① 난 개인적으로 완전히 민 머리를 정말 좋아해.
② 아니, 나도 그런 사람들은 싫어.
③ 내가 겉모습만 보고 사람을 판단하지 말라고 했잖아.
④ 맞아. 모든 사람은 가능한 한 평범해 보여야 해.

14 ④

여: 제이크, 어딜 그렇게 급하게 가니?

남: 아, 안녕. 이동 통신 서비스 센터에 가는 중이야.

여: 전화기에 문제가 있니?

남: 아니. 오늘 오후에 통신 회사에서 전화를 받았는데, 내가 홍보 이벤트에서 스마트폰을 타게 됐다고 하더라고.

여: 와, 좋겠다! 그럼, 지금 그거 받으러 가는 거야?

남: 아니, 아니. 내가 당첨자라는 걸 확인해 주려고 그 사람들한테 내 이름과 주소를 말해줬거든. 그 뒤에 몇 가지 조사를 좀 해보고 나서 전부 사기라는 걸 알게 됐어.

여: 뭐? 무슨 사기를 말하는 거야?

남: 내가 그 전화기를 사용하면, 한 달에 통화료와 문자 메시지에 적어도 3만 원에서 4만 원을 쓰게 되어 있어. 그렇지 않을 때에는 휴대전화 가격을 지불해야 해. 전혀 경품이 아니야!

여: 어떻게 그런 식으로 사람을 속일 수 있지?

어휘
rush off 황급히 떠나다 promotional 홍보의, 판촉의 confirm 확인하다 scam 사기 [문제] deceive 속이다

문제풀이
① 어떤 종류의 (휴대)전화를 샀니?
② 네 (휴대)전화 좀 사용해도 될까?
③ 넌 정말 운이 좋구나. 난 뭔가 당첨된 적이 없어.
⑤ 나도 네 것 같은 스마트폰이 있었으면 좋겠어.

15 ④

남: 게리는 평생 자동차 경주를 하고 싶어했다. 어느 날 그는 유명한 포뮬러 원 경주용 자동차 운전자인 밥을 만났다. 게리는 밥에게 경주용 자동차를 운전하는 기술을 가르쳐 달라고 간청했다. 밥은 그러한 게리의 모습이 마음에 들어서 그를 데리고 있기로 했다. 곧 게리는 프로 선수처럼 운전했고 경주 트랙을 여러 번에 걸쳐 완주할 수 있게 되었다. 밥조차도 깊은 인상을 받았다. 밥은 자신의 매니저에게 가서 게리에 대해 말했다. 밥은 게리에게 차를 제공해 주도록 매니저와 재정 후원자들을 설득했다. 밥이 게리에게 그가 자신의 차를 갖게 될 거라고 말했을 때, 게리는 기쁨에 사로잡혔다. 그는 밥에게 항상 그의 가르침

을 기억하고 그를 스승으로 여기겠다고 말했다. 이런 상황에서, 밥이 게리에게 할 말로 가장 적절한 것은 무엇인가?

밥: 넌 잘 해낼 거야, 게리. 행운이 있길 바란다.

Formula One 포뮬러원(경주용 자동차) beg 간청[애원]하다 take A under one's wing A를 …의 보살핌 속에 두다 convince 확신시키다 financial backer 재정 후원자 fund 자금을 대다 be overcome with …에 사로잡히다, …이 넘쳐흐르다 mentor 멘토, 스승 [문제] ungrateful 은혜를 모르는 career 경력, 사회[직장] 생활

문제풀이
① 넌 정말 은혜를 모르는구나, 게리!
② 많이 가르쳐 줘서 고마워.
③ 언젠가 네 차를 갖길 바랄게.
⑤ 좋아. 내가 그를 네게 멘토로 소개해 줄게.

16 ③ 17 ⑤

남: 고대 이집트에 관한 저희의 특별 전시회에 오신 것을 환영합니다. 여러분이 여기서 보고 있는 벽은 파라오의 무덤에서 나온 것입니다. 여기를 보시면, 고대 이집트 문화에서 매우 흔했던 상징물을 보실 수 있습니다. 그것은 '앵크'라고 불렸습니다. 단지 이 벽면뿐만 아니라 이 전시회의 많은 인공물에도 많은 앵크가 있다는 것을 여러분은 알 수 있습니다. 앵크는 십자처럼 생겼지만, 기독교의 십자가와는 다르게 가장 윗부분에 고리가 있습니다. 앵크는 삶을 상징합니다. 여기, 여러분께서는 앵크를 들고 있는 내세의 여신을 보실 수 있는데 이집트인들은 그녀가 파라오를 부활시킬 수 있다고 믿었기 때문입니다. 이제 이쪽으로 저를 따라오시면, 앵크 모양의 거울을 보실 수 있습니다. 사실, 이집트어로 '거울'이라는 단어는 '앵크'였습니다. 이제, 전시회 주변을 자유롭게 다니시면서 많은 앵크들을 관람하시기 바랍니다.

어휘
exhibit 전시 tomb 무덤 pharaoh 파라오(고대 이집트의 왕) ankh 앵크 십자 artifact (역사적 · 문화적 의미가 있는) 인공물, 가공품 loop 고리 goddess 여신 afterlife 내세, 사후 세계 numerous 많은 [문제] mummy 미라 religion 종교

문제풀이
16 ① 파라오의 권력
② 미라의 제작 과정
③ 고대 이집트에서 삶의 상징
④ 고대 이집트인들의 종교
⑤ 고대 이집트인들이 상상했던 사후 세계

DICTATION Answers
본문 p.8

01 giving me a chance / how important you are / by becoming my husband
02 when people litter / make a difference / as dedicated as you
03 you posted online / a fast learner / mistakes are not allowed
04 wrote their names on the leaves / is covered with / so that the children won't hurt

05 haven't had much sleep / stop doing that / have a big talk / complain in person
06 How come / over the winter vacation / so nervous that
07 What are your rates / Tell you what / for free instead
08 how to improve / encouraged them to plant / makes it healthier / let it stop him
09 participate in / win a trip to Hawaii / will be donated to
10 to bring home / good to know / visiting on business
11 the fair take place
12 raise my allowance
13 judge a book / right to express themselves / open-minded toward people
14 rushing off / a promotional event / confirm / I'm supposed to
15 begged / under his wing / make it / overcome with joy
16~17 exhibit on ancient Egypt / at the top of / back to life / feel free to

02 영어듣기 모의고사
본문 ▲ p.12

01 ⑤	02 ②	03 ①	04 ⑤	05 ①	06 ③
07 ③	08 ④	09 ⑤	10 ⑤	11 ⑤	12 ④
13 ④	14 ③	15 ②	16 ②	17 ⑤	

01 ⑤

여: 안녕하세요, 여러분. 일 년 중 이맘때가 되면 우리 동아리는 다가오는 행사들을 위해 돈을 모금합니다. 전통적으로, 우리는 가족 구성원들에게 우리의 목적을 위해 돈을 좀 기부해 달라고 요청하기 위한 독창적인 방법을 찾고자 노력해 왔습니다. 비록 이것이 과거에는 성공을 거두었지만, 저는 우리가 이 동아리를 재정적으로 유지하기 위해서 부모님께 의존하는 것은 바람직하지 않다고 생각합니다. 우리는 우리 스스로 돈을 벌 수 있는 나이입니다. 우리 대부분은 방과 후나 주말에 몇몇 종류의 아르바이트를 하고 있습니다, 그렇죠? 그래서, 기부를 요청하기보다는, 뭔가 새로운 것을 시도하고 우리 스스로 동아리의 예산을 위해 직접 돈을 벌 수 있는 새로운 방법을 생각하는 게 어떨까요?

어휘
upcoming 다가오는 creative 창조적인; *독창적인 donate 기부[기증]하다 (n. donation) cause 원인; 이유; *대의, 목적 desirable 바람직한 financially 재정적으로 support 받치다; *지지[후원]하다 novel 새로운, 신기한 budget 예산

02 ②

남: 제임스 카메론이 우주 연구를 지원한다는 거 들었어?
여: 그래, 하지만 그게 가치 있다고 생각하지 않아.
남: 정말? 만약 지구에 똑바로 향하는 거대한 암석을 탐지한다면 어떨까?
여: 그런 일은 별로 가능성이 없는 것 같은데.
남: 전에 일어났었고, 과학자들은 그런 일이 다시 일어날 수도 있다고 해. 하지만 당장 우리 자신을 구할 기술이 우리에겐 없어.
여: 아주 무섭게 들리네. 그런데 과학자들이 위험을 벌써 알고 있다면 대비를 하고 있을 것 같아.
남: 그걸 대비하는 방법이 우주 연구를 하는 거야!
여: 응, 이제 알겠어. 결국, 우주 연구에는 가치가 있는 것 같네.
남: 맞아!

어휘
research 연구 value 가치 detect 발견[탐지]하다 head 향하게 하다 likely 가능성이 있는 be aware of …을 알다

03 ①

여: 안녕하세요, 존스 씨. 오늘 와주셔서 감사합니다. 밥에 관해 이야기를 나눌 필요가 있어요.
남: 안녕하세요, 피터스 씨. 네, 저도 동의합니다.
여: 저는 그 애에 대해 매우 걱정하고 있어요. 훌륭한 태도로 한 해를 시작했거든요.
남: 네, 밥은 당신의 수업들을 매우 즐거워했어요.
여: 하지만, 그 애는 일 년 동안 급격히 변한 것 같아요.
남: 제 아내와 저도 그 애의 성적이 떨어졌다는 것을 알고 걱정했습니다.
여: 밥은 숙제도 하지 않고 있고, 이미 25번이나 지각을 했어요.
남: 그랬군요. 그런 것들에 대해서는 몰랐네요. 제가 그에 대해 말해 볼게요.
여: 하지만 저는 그 애의 성적보다는 행복(안녕)이 더 걱정됩니다. 집에 무슨 일이라도 있나요?
남: 솔직히 말씀드리자면, 밥은 힘든 한 해를 보냈어요. 조부모님이 모두 돌아가셨거든요.
여: 그게 많은 것을 설명해 주네요. 상실감이 크시겠어요. 이제 알았으니, 밥을 도울 더 나은 방법을 찾아볼 수 있겠네요.
남: 신경 써 주셔서 감사합니다.

어휘
be concerned about …을 걱정하다 dramatically 극적으로, 급격하게 be aware of …을 알다 wellbeing 복지, 안녕, 행복 pass away 사망하다, 돌아가시다

04 ⑤

남: 새로운 직원인 브라운 씨 기억해요? 그녀는 탑승 수속 창구에서 일하고 있어요.
여: 머리를 묶은 여자분인가요?
남: 네. 그녀가 어떻게 하고 있는지 봅시다.
여: 좋아요. 아, 저 탑승 수속을 밟고 있는 남자는 하와이에 가려나 봐요.
남: 저도 그렇게 생각해요. 그는 짧은 소매 셔츠를 입고 선글라스를 머리에 얹었네요. 그는 하와이에 딱 맞겠어요.
여: 근데 줄을 선 저 소년은 어디로 가는지 궁금해요.
남: 네, 그 애는 스웨터를 입고 있어요. 아마도 추운 곳으로 가려나 봐요.

뒤에 있는 여자도 마찬가지 같네요.
여: 그녀는 심지어 팔에 코트도 들고 있네요.
남: 맞아요. 아, 벌써 10시 10분 전이네요. 우린 10시에 회의가 있잖아요. 서두릅시다.

어휘
check-in 【공항】 탑승 수속 (v. check in) pony tail 포니테일(머리를 하나로 묶은 형태) short-sleeved 짧은 소매의 fit in 맞다, 어울리다

05 ①

여: 실례합니다만, 부탁 좀 드려도 괜찮을까요?
남: 얼마든지요. 우린 이 줄에서 아주 오랫동안 기다리고 있었잖아요. 뭔가 해야 할 일이 있으신가 보군요.
여: 네. 사실은 사적인 통화를 해야 해서요.
남: 아, 알겠습니다.
여: 제가 화장실 근처로 가서 전화하는 동안 제 자리 좀 맡아주실 수 있나요?
남: 물론이죠, 하지만 너무 오래 지체하지는 마세요. 지금은 줄이 천천히 움직이고 있지만, 속도가 붙을지도 모르거든요.
여: 네, 저도 표를 구매할 기회를 놓치기는 싫어요.
남: 그건 정말 끔찍할 거예요. 전 이 연극을 이미 두 번이나 봤는데, 너무 좋아요. 이 연극을 놓치고 싶지 않으실 겁니다.
여: 저도 그렇게 들었어요. 어쨌든 절 도와주셔서 감사합니다.
남: 천만에요.

어휘
private 사적인, 개인적인 spot 점, 반점; *(특정한) 곳, 자리

문제풀이
① 그녀의 자리 맡아주기
② 그녀와 연극 표 교환하기
③ 좋은 연극 추천하기
④ 그녀가 그의 휴대전화를 빌려 쓰게 해주기
⑤ 화장실 가는 길 알려주기

06 ③

여: 안녕, 케빈. 너 어제 축구 시합을 했지, 그렇지 않아?
남: 맞아. 선수권 대회에서 가장 강력한 라이벌과 시합했어.
여: 정말? 그래서, 최종 점수가 어떻게 됐어?
남: 우리가 연장전에서 4대 3으로 그들을 이겼어.
여: 정말 잘 됐다. 그런데 넌 그것에 대해 많이 기뻐하는 거 같지 않은데.
남: 음, 시합이 끝날 무렵에 우리 팀 골키퍼가 부상을 당했어.
여: 오, 그렇구나. 심각한 게 아니었으면 좋겠다.
남: 그는 골대에 머리를 꽤 심하게 부딪혔어. 그는 아직 병원에 있어.
여: 정말 안됐다. 축구는 거친 운동일 수 있겠구나.
남: 맞아. 어쨌든 그가 괜찮길 정말 바라고 있어. 나중에 그를 방문할 거야.
여: 좋은 생각이야. 그는 분명히 고마워할 거야.

어휘
championship 선수권 대회 defeat 패배시키다, 이기다 overtime (경기) 연장전 injure 부상을 입다 goalpost 골대

07 ③

여: 안녕하세요! My Town 피트니스 센터에 오신 것을 환영합니다.
남: 안녕하세요. 저는 가입을 하려고 하는데요. 1개월 회원권이 얼마인지 알려주실래요?
여: 음, 2일권은 월 90달러이고 3일권은 월 120달러입니다.
남: 그건 제가 일주일에 두 번이나 세 번 운동할 수 있다는 건가요?
여: 맞습니다. 그리고 200달러짜리 무제한권도 있어요.
남: 주 3일권이 제게 충분할 것 같아요. 그리고 제가 빌릴 수 있는 운동복도 있나요?
여: 네, 하지만 그건 월 20달러를 추가로 내셔야 해요.
남: 괜찮아요. 일주일에 세 번 빨래하는 것보다 그게 더 편하죠.
여: 저도 그렇게 생각해요. 아, 이 근처 주민이세요?
남: 네, 맞아요. 저는 몇 블록 떨어진 곳에 살아요. 제가 할인받을 수 있나요?
여: 그렇습니다! 지역 주민들은 전체 가격에 10퍼센트 할인됩니다.
남: 멋진데요! 저를 설득하셨네요. 오늘 1개월 회원권을 등록할게요.

문제풀이
남자는 주 3일 회원권($120)을 1개월 등록하고 운동복을 대여($20)하며, 총액에 지역 주민 할인이 10퍼센트 적용되므로, 남자가 지불할 금액은 126달러이다.

08 ④

여: 저 그림 속의 남자는 누구니?
남: 그건 1480년에 태어난 유명한 탐험가인 퍼디낸드 마젤란이야.
여: 아, 그래? 그의 이름은 스페인어 같은데.
남: 사실, 그는 포르투갈 사람이야. 하지만 스페인 왕에 의해 후원받았지.
여: 그가 왜 그렇게 유명하니?
남: 그는 세계를 항해하는 첫 번째 원정을 이끌었어. 사실, 그는 향신료 무역을 위해 인도네시아에 가려고 했지. 하지만 그 전에 죽었어.
여: 무슨 일이 있었니?
남: 필리핀에서 원주민들과 싸우는 중에 죽임을 당했어.
여: 참 안됐구나. 그 원정 팀은 돌아왔어?
남: 아니. 남아 있는 선원들이 인도네시아에 가서 원하던 향신료를 얻었어. 그들은 원정을 시작한 지 3년 뒤인 1522년에 마침내 고향에 돌아갔어.
여: 인상적이다.
남: 응, 그리고 그게 그가 위대한 탐험가로 존경받는 이유야.

09 ⑤

남: 주목해 주시겠습니까? 잠시 시간을 내어 Breakwater 리조트의 내일 활동 일정을 알려 드리겠습니다. 우리는 해변에서 오전 10시에 가족 배드민턴 토너먼트로 하루를 시작하게 될 것입니다. 한 시간 후에는 고래 관찰 보트가 부두에서 출발할 것입니다. 그것은 여러분께서 정말로 놓쳐서는 안 될 경험입니다. 오후에는 3시에 마당에서 전통 무용 공연이 있을 것이며, 4시 30분에는 주 수영장에서 아이들을 위한 흥미진진한 게임이 이어질 것입니다. 마지막으로 우리는 저녁 식사 전에 특별한 칵테일 시간을 갖게 될 텐데요, 바텐더들이 특별 음료를 반값에 제공할 것입니다. 여러분께 재미난 하루를 약속드리며, 여러분 모두 오늘 밤 푹 주무시길 바랍니다.

10 ⑤

여: 이 Cool Star 선풍기 중 하나를 사기로 했는데, 어느 것을 주문해야 할지 모르겠어.
남: 그것들은 전부 비슷해 보인다. 서로 어떻게 달라?
여: 음, 일단은 작동 속도의 개수가 달라. 봐, 이건 두 개뿐이야.
남: 속도가 두 개뿐이야? 느리고 빠르고?
여: 응. 난 적어도 3개의 작동 속도를 가진 선풍기가 필요해. 그리고 이 중 일부는 안전 증서가 있지.
남: 그건 그것들이 더 안전하다는 걸 의미하는 거야?
여: 응, 맞아. 난 안전 증서가 없는 선풍기는 사지 않을래.
남: 알겠어. 리모컨은 어때?
여: 꼭 필요해. 그래야 내가 침대에서 벗어나지 않고서도 선풍기를 끌 수 있어서.
남: 품질 보증서도 필요해?
여: 응. 더 길수록 더 좋지. 전기 선풍기는 종종 고장이 나니까.
남: 음, 그러면 네가 살 건 이것인 것 같아.

11 ⑤

여: 그래서 그 쇼에 대해 어떻게 생각했어?
남: 댄서와 음악은 정말 좋았는데, 이야기는 지루했어.
여: 그 말을 들으니 뜻밖인데.
남: 다음에는 더 재미있는 거 보러 가자.

문제풀이
① 아마 네게 그걸 다시 보여줄 수 있을 거야.
② 오 이런, 댄스 수업에 늦었어!
③ 내 친구가 그걸 보고 싶다고 말했어.
④ 나는 언젠가 너와 함께 콘서트를 갈 수 있길 바라.

12 ④

남: 훗날 갖게 될 멋진 직업이 뭐가 될 거라고 생각하니?
여: 난 패션 디자이너가 되고 싶어.
남: 와! 멋지겠다.
여: 응, 하지만 그 산업에서 일하는 건 힘들 거라고 들었어.

어휘

someday 언젠가, 훗날 [문제] runway 런웨이, 패션쇼의 무대 industry 산업
Congratulations! 축하해!

문제풀이

① 아니, 그 패션 디자이너를 본 적이 없어.

② 응, 네가 런웨이에 등장할 수 있다면 좋을 텐데.

③ 응. 난 패션에 특별히 관심이 없어.

⑤ 축하해! 네가 패션 디자이너가 될 수 있을 줄 알았어.

13 ④

여: 여보, 저 왔어요. 당신 어디 있어요?

남: 침실에 있어요.

여: 늦어서 미안해요. 저녁은 먹었어요?

남: 아뇨, 배고프지 않아요. 점심을 많이 먹었거든요.

여: 그렇다고 저녁을 걸러야 하는 건 아니잖아요. 당신이 저녁을 거르면 내가 싫어한다는 걸 알잖아요.

남: 정말 난 배고프지 않아요.

여: 그럼, 내가 먹을 걸 해 줄게요. 뭘 먹고 싶어요?

남: 여보, 난 괜찮아요. 배고프면 내가 나중에 뭘 좀 먹을게요.

여: 나중이라뇨? 벌써 9시란 말이에요.

남: 알았어요. 가볍게 샌드위치나 한 개 먹을게요.

여: 좋아요. 10분 내로 준비할게요.

어휘

heavy 대량의 skip 거르다, 빠뜨리다 seriously 진심으로, 진지하게 fix 마련[준비]하다 [문제] would rather A than B B보다는 (차라리) A 하겠다

문제풀이

① 미안해요. 난 다이어트 중이에요.

② 나도 하나 만들어 줄래요?

③ 고마워요. 그건 내가 제일 좋아하는 음식이에요.

⑤ 아무것도 안 먹느니 간식이라도 먹어야겠어요.

14 ③

여: 와, 짐! 나 너무 감탄했어!

남: 무슨 뜻이야?

여: 음, 저 여자가 너한테 국립 민속 박물관에 가는 길을 물어봤을 때 네가 한국어로 자신 있게 얘기했잖아.

남: 아… 고마워. 아마 매일 연습해왔기 때문일 거야.

여: 정말? 넌 어떻게 공부하는데?

남: 난 그저 한국어를 배우는 라디오 프로그램을 따라서 말하기 연습을 할 뿐이야. 그리고 항상 한국인들과 말하려고 애쓰지.

여: 나도 너처럼 자신감이 있었으면 좋겠다. 난 낯선 사람들에게 영어로 얘기할 때마다 너무 쑥스러워.

남: 하지만 넌 미국에서 6개월간 영어를 공부했잖아, 그렇지?

여: 그래, 하지만 난 너무 수줍어서 교실 밖에선 연습을 못 했어. 난 주로 기숙사에서 지내면서 한국 영화를 봤지.

남: 음, (실력이) 많이 향상되지 않은 게 당연하구나.

어휘

impressed 감명받은 confidently 자신 있게 (n. confidence) direction 방향 embarrassed 창피한 dormitory 기숙사 [문제] nervous 불안해하는

문제풀이

① 사업상 영어를 구사하는 것은 중요해.

② 넌 네 영어 실력을 향상하기 위해서 많은 걸 하고 있어.

④ 난 저 낯선 사람에게 말하는 동안 긴장했어.

⑤ 누워서 영화 보는 것보다 더 좋은 건 없지.

15 ②

남: 캐런과 그녀의 남자친구는 영화를 보기 위해서 극장에 간다. 그들이 좌석을 찾아서 자리에 앉은 후, 캐런은 영화가 시작하기 전에 간식 판매대에서 먹을 걸 사고 싶어 한다. 그런데 그녀는 나갈 때 자신의 표를 좌석에 두고 온다. 그녀가 극장에 다시 들어가려고 하자, 그녀는 문에서 직원에 의해 제지당한다. 그는 그녀에게 입장권을 보여 줄 것을 요청하지만, 그녀는 표를 가지고 있지 않다. 직원은 그녀의 남자친구에게 전화해 볼 것을 제안하지만, 그녀의 휴대전화는 그녀가 역시나 좌석에 두고 온 재킷 안에 있다. 캐런은 밖에 나가서 그에게 전화하기 위해 공중전화를 이용할 수 있다고 말하지만, 이미 영화는 시작되고 직원은 그녀의 말을 믿기로 한다. 이런 상황에서, 극장 직원이 그녀에게 할 말로 가장 적절한 것은 무엇인가?

직원: 앞으로는 반드시 입장권을 소지하고 다니세요.

어휘

re-enter 다시 입장하다 payphone 유료[공중]전화 [문제] in advance 사전에, 미리

문제풀이

① 더 이상은 할인해 드릴 수 없습니다.

③ 손님께서는 미리 표를 예매하셨어야 합니다.

④ 걱정하지 마세요. 저희가 손님의 입장권을 훔쳐간 사람을 찾아낼 수 있을 겁니다.

⑤ 제안은 감사하지만, 영화가 끝난 이후에나 그에게 전화할 수 있어요.

16 ② 17 ⑤

여: 오늘 우리는 혜성에 대하여 배울 것입니다. 사람들이 혜성을 떠올릴 때, 종종 우주를 날아다니는 거대한 불덩어리의 암석들을 생각하지만, 이것은 매우 부정확한 이미지입니다. 혜성들은 불에 타고 있지도 않을뿐더러, 심지어는 암석도 아닙니다! 혜성은 대부분 먼지, 물, 얼음, 그리고 동결된 기체로 만들어집니다. 혜성에 약간의 암석이 있을 수는 있지만, 분명히 다수를 이루지는 않습니다. 그래서 불타고 있는 것과 거리가 먼 혜성은 사실 극도로 차갑습니다. 그것들은 일종의 더러운 눈덩이 같은 것입니다. 이제 여러분은 혜성의 뒤에 있는 긴 꼬리가 무엇인지 궁금해할지도 모릅니다. 음, 태양으로부터의 태양 복사가 혜성에 있는 기체를 증발시킵니다. 그런 다음에 태양풍은 혜성으로부터 기체를 가져가 버리고, 그렇게 꼬리가 만들어집니다. 그래서 혜성의 꼬리는 증발된 기체와 먼지로 이루어져 있습니다. 이 짧은 소개 후에 여러분이 더 이상은 혜성을 불타는 암석의 조각들로 떠올리지 않기를 바랍니다!

어휘

comet 혜성 inaccurate 부정확한 definitely 분명히, 틀림없이 majority 가장 많은 수, 다수 trail *(길게 연이은) 자국; 오솔길, 산길 solar radiation 태양복사 vaporize 증발하다

문제풀이

16 ① 혜성은 왜 불에 타고 있는가

② 혜성의 구성 성분
③ 어떻게 혜성이 지구에 영향을 미치는가
④ 혜성과 태양의 관계
⑤ 혜성과 불타는 암석의 차이점

01 raises money / earn money on our own / asking for donations

02 detect a huge rock / aware of the danger / value in space research

03 seems to have changed / worried about / To be honest / for your loss

04 with a ponytail / fit right in / I think the woman

05 asked you a favor / make a private call / keep my spot / helping me out

06 played our biggest rival / got injured / hope he's okay

07 a one-month membership / I can borrow / You've convinced me

08 led the first expedition / while fighting natives / is admired as

09 whale watching boat / a traditional dance performance / at half price

10 for one thing / without getting out of bed / The longer, the better

11 the story was boring

12 become a fashion designer

13 had a heavy lunch / fix you something / have a light sandwich

14 so confidently / try to speak / too shy to practice outside

15 get something to eat / leaves her ticket / use a payphone

16~17 Not only are / are made mostly of / you might wonder / no longer think of

03 영어듣기 모의고사
본문 ▲ p.18

01 ③	02 ③	03 ④	04 ③	05 ⑤	06 ①
07 ②	08 ③	09 ④	10 ⑤	11 ②	12 ⑤
13 ④	14 ⑤	15 ⑤	16 ①	17 ④	

01 ③

남: 모든 건물 입주민 여러분 주목해 주시겠습니까? 여러분도 아시다시

피, 어제의 심한 폭풍으로 인해 엄청난 양의 눈이 내렸습니다. 일반적인 상황에서는, 건물의 관리 부서원들이 그들의 평상시 일과의 일부로 눈과 얼음을 제거하는 작업을 처리합니다. 안타깝게도 이 눈 폭풍은 저희가 미처 알아차리지 못했으며, 저희는 건물의 진입로와 보도를 제때 치우지 못했습니다. 따라서 저희는 오늘 아침에 여유 시간이 있으신 주민 여러분의 도움을 요청합니다. 많은 학교와 사업체가 휴업 중이므로, 저희는 여러분들 중 일부가 시간이 된다면, 30분만 짬을 내어 저희를 도와주시기를 바랍니다. 어떤 도움이라도 진심으로 감사드리겠습니다.

어휘

resident 주민 circumstance (*pl.*) 환경, 상황 maintenance 보수 관리 removal 제거 routine 일상의 일 blizzard 눈보라 driveway 진입로 in a timely manner 제때 spare (시간·돈 등을) 할애하다 give ... a hand …을 돕다 sincerely 진심으로

02 ③

남: 점심 먹을 시간이야. 난 치즈버거와 감자튀김을 먹고 싶어.
여: 지방이 좀 더 적은 걸 먹는 게 어떨까? 그런 종류의 음식은 영양분은 많지 않으면서 살을 찌게 해.
남: 알아, 하지만 맛이 좋잖아.
여: 맛이 좋다고 지방 함량이 높은 음식들을 계속 먹으면, 고혈압 같은 병에 걸릴지도 몰라.
남: 음… 네 말이 일리가 있는 것 같아.
여: 맛도 좋으면서 조금 더 담백한 걸 먹으면 돼. 초밥은 어때?
남: 좋아. 지금부터 음식을 고를 땐 네가 한 말을 고려할게.

어휘

feel like v-ing …하고 싶다 fatty 지방이 많은 (*n.* fat) gain weight 체중이 늘다 nutrient 영양분 sushi 초밥

03 ④

여: 노동부에서 나왔습니다. 정기 안전 점검을 위해 저희가 이곳에 파견되었습니다.
남: 좋습니다. 공장을 보셔야 하나요?
여: 네, 그런데 먼저 몇 가지 질문에 답변을 해 주실 수 있을까요?
남: 그러죠. 어서 하세요.
여: 이 공장은 안전 절차가 문서로 되어 있나요?
남: 네, 그렇습니다.
여: 그럼 일하시면서 사고를 목격하신 적은 있나요?
남: 5년간은 없었습니다. 그 전에는, 있었지만요.
여: 경영진이 새로 바뀌기 전의 일이었나요?
남: 네, 그렇습니다.
여: 노동자들이 공평하게 대우받고 있다고 생각하시나요?
남: 네. 우린 불만 사항을 거의 받지 못했어요. 물론 돈을 더 벌고 싶어 합니다.
여: 네, 그렇겠죠. 고맙습니다.

어휘

the Ministry of Labor 노동부 routine *정기적인; 일상적인 safety check 안전 점검 facility (*pl.*) 시설, 설비 procedure 절차 management 경영진 take over (기업 등을) 인수하다 fairly 공평하게

04 ③

여: 실례합니다. 손잡이가 두 개 달린 바구니를 사용해도 될까요?
남: 그럼요. 여기 있습니다. 특별히 찾으시는 게 있나요? 가장 신선한 현지 과일이 있습니다.
여: 아, 그래요? 이 바나나는 얼마인가요?
남: 한 바구니에 10달러에요. 한 바구니에 네 다발이 들어 있어요.
여: 와, 괜찮은 가격이네요. 사과는 어때요?
남: 음, 한 상자에 보통 10달러인데 오늘 할인 중이에요. 그래서 9달러에 사실 수 있어요.
여: 그렇군요. 수박도 먹음직스러워 보이네요!
남: 세 개 모두 오늘 아침에 신선하게 들어왔어요.
여: 하나 사고 싶긴 한데 제가 들고 가기에는 너무 무거울 것 같아요.
남: 걱정하지 마세요. 저희가 배달해드려요.
여: 아, 저 현수막에 '무료 배달'이라고 쓰여 있네요! 그럼 두 개 살게요.

어휘
handle 손잡이 local 지역의, 현지의 bunch 다발, 송이 tempting 솔깃한, 구미가 당기는 deliver 배달하다 (*n.* delivery)

05 ⑤

남: 얘, 크리스틴. 너 시라도 쓰고 있는 거니?
여: 아니. 실은 파티를 위해 준비할 것들을 목록으로 작성하는 중이야.
남: 아, 그래? 무슨 파티인데?
여: 내 남자친구 생일을 위한 깜짝 파티지. 특별하게 해주고 싶거든.
남: 멋진 여자친구로구나! 무슨 계획을 세웠는데?
여: 음, 그의 친구들을 모두 초대할 거야. 그리고 특별한 음식도 좀 만들고 싶어.
남: 특별히 생각해 둔 거라도 있니?
여: 아니, 없어. 맛있는 것이어야 하는데 좋은 아이디어가 없네.
남: 난 파티를 열 때마다 인터넷에서 찾은 이 맛있는 타코 샐러드를 만들어.
여: 그거 아주 딱인데. 내 남자친구가 타코 샐러드를 좋아해! 어떻게 만드는지 보여줄 수 있어?
남: 상세한 요리법을 네게 이메일로 보내줄게. 네 남자친구가 아주 좋아할 거야.
여: 나도 그럴 거라 확신해. 고마워!

어휘
poem 시 step-by-step 한 걸음 한 걸음의, 단계적인 instruction 《*pl.*》*설명; 지시

06 ①

남: 안녕, 제인. 오랜만이야.
여: 안녕, 토니. 여기 은행에서 뭐 하고 있니?
남: 이 고지서들을 납부하려고 들렀어.
여: 아, 그렇구나. 은행이 오늘 아주 붐비네, 그렇지?
남: 응. 이 은행에서 한시적으로 새로운 저축 상품을 제공한다고 들었어. 그래서 많은 사람들이 새 계좌를 개설하려고 오나 봐.
여: 정말? 더 일찍 알았다면 나도 개설했을 텐데. 하지만 내일 휴가를 가기 때문에 오늘은 시간이 별로 없어.
남: 아, 어디로 가니?
여: 이탈리아.

남: 와, 진짜 멋지다. 유로화를 사려고 왔나 보구나.
여: 응. 은행 문 닫기 전에 할 수 있으면 좋겠어!
남: 음, 좋은 시간 보내. 그리고 돌아오면 전화 줘.

어휘
drop by 들르다 bill 고지서, 청구서 savings plan 저축 제도 limited 제한된, 한정된 account 계좌 purchase 구매하다

07 ②

남: 안녕하세요. 뭘 도와드릴까요?
여: 네, 이 양말을 교환하고 싶어서요. 어제 두 묶음을 샀거든요.
남: 알겠습니다. 양말에 문제가 있나요?
여: 음, 전 무릎까지 올라오는 양말이 필요한데 이 양말은 너무 짧아요.
남: 아, 저희 무릎 양말은 바로 이쪽에 있어요.
여: 이건 여섯 켤레짜리 묶음이네요. 어제 제가 구입한 묶음은 네 켤레씩 들어 있었는데.
남: 네, 하지만 저희가 가진 건 이것뿐이에요. 한 묶음에 15달러예요.
여: 알겠어요. 음… 양말이 너무 적은 것보다는 많은 게 나을 것 같군요. 열두 켤레 살게요.
남: 그렇죠. 하지만 어제 산 양말들이 총 24달러이기 때문에, 차액을 지불하셔야 합니다.
여: 그건 문제없어요.

어휘
exchange 교환하다 knee 무릎 difference 차이; *차액

문제풀이
여자는 한 묶음에 양말 여섯 켤레가 있는 15달러짜리 제품을 사려고 한다. 열두 켤레를 구입하려면 두 묶음($30)을 사야 하는데, 교환하려는 양말이 24달러어치이므로 차액으로 6달러를 더 지불해야 한다.

08 ③

남: 미셸, 너 과학 동아리에 들었다고 하더라.
여: 응, 그래. 가입하는 데 관심 있니?
남: 생각 중인데 우선 더 알아보고 싶었거든.
여: 음, 모든 학년의 회원들이 있어. 지금 11명이 있어.
남: 그렇구나. 얼마나 자주 모여?
여: 과학 실험실에서 매주 화요일 밤에 모여. 주로 과학 토론을 하지. 그런데 특별 행사도 많아.
남: 어떤 것들을 하는데?
여: 과학박물관에 가고 과학 박람회에도 참가해.
남: 멋진 것 같아. 그럼 누구나 가입할 수 있니?
여: 물론! 단지 네가 모든 과학 수업에서 최소한 C를 받기만 한다면 말이야.
남: 알겠어. 음, 질문에 답해 줘서 고마워. 오늘 밤에 더 생각해 볼게.
여: 좋아. 가입하기로 결정하길 바라!

어휘
lab 실험실 (= laboratory) stuff …것, 물건

09 ④

여: 모든 분들이 좋아하는 식용 버섯을 기념하는 사흘간의 행사인 Westchester 버섯 축제에 오신 걸 환영합니다. 아마도 여러분께서 아

시다시피, 이 지역은 수많은 버섯 농장으로 유명합니다. 하지만 여러분께서는 아마도 우리 숲에서 자라고 있는 야생 버섯 또한 풍부하다는 사실은 모르실 겁니다. 이 축제는 사람들에게 어떤 종류의 야생 버섯을 먹을 수 있는지, 그리고 어디서 찾을 수 있는지를 알려주는 데 목적이 있습니다. 온갖 종류의 맛있는 버섯을 이용한 음식을 제공하는 식품관뿐만 아니라 지역 전문가의 여러 발표회도 있을 것입니다. 입장료는 단지 4달러이며, 12세 이하의 어린이들은 반값에 입장할 수 있습니다. 그리고 마지막 날에는 모든 종류의 버섯을 주제로 하는 상품을 증정하는 특별한 경품 행사가 있으니 입장권을 가지고 계십시오.

어휘 edible 먹을 수 있는 fungus 균류; *버섯 wealth 재산; *풍부한 양 aim to-v …하는 것을 목표로 하다 presentation 발표 expert 전문가 admission 입장(료) giveaway 경품 theme …을 주제로 하다

10 ⑤

남: 안녕하세요. 제가 이 시간표를 보던 중이었는데, 도움이 좀 필요해요.
여: 네, 뭘 도와드릴까요?
남: 특정한 영어 실력을 향상시키는 과정을 수강할까 생각 중인데요.
여: 어떤 영어 실력을 향상시키길 원하시나요?
남: 전부 다요! 작문이 최우선이긴 하지만, 제 스케줄에 맞는 건 뭐든지 수강할 거예요.
여: 김 선생님은 원어민은 아니시지만, 선생님의 독해 수업은 학생들에게 정말 인기가 많답니다. 이날 밤에 시간이 괜찮으신가요?
남: 전 수요일을 제외하고는 매일 밤 시간이 있어요. 하지만 전 원어민 수업을 원해요.
여: 알겠습니다. 그럼 아침 시간은 어떠세요?
남: 너무 이르지만 않다면 아침도 괜찮아요. 9시 전에는 여기에 올 수가 없거든요.
여: 알겠습니다. 다른 필요한 사항은요?
남: 음, 저녁 8시까지는 끝났으면 해요. 그래야 여자친구를 만날 수 있거든요.
여: 그렇다면 이 수업이 당신에게 딱 맞는 것 같네요.

어휘 specific 구체적인; *특정한 fit …에 꼭 맞다 native speaker 원어민 requirement 《pl.》 필요(한 것) [문제] instructor 강사

11 ②

남: 엘렌, 커피 한 잔 마실래?
여: 고맙지만 사양할게. 난 커피를 그렇게 좋아하지 않아. 밤에 잠이 잘 안 오게 하거든.
남: 그렇구나. 대신 뭘 마시고 싶니?
여: 오렌지 주스를 마실게.

어휘 instead 대신에

문제풀이
① 마실 게 필요해.
③ 물론이야. 커피 한 잔 마실게.
④ 난 오늘 밤 푹 잘 거야.
⑤ 여기가 내가 제일 좋아하는 커피숍이야.

12 ⑤

여: 에단, 등산 가본 적 있니?
남: 응, 가족과 자주 등산해. 왜?
여: 이번 주말에 갈 계획이거든. 어떤 장비가 필요하니?
남: 야외용 옷, 신발, 지팡이가 필요해.

어휘 hike 하이킹[도보 여행]을 가다 equipment 장비 [문제] walking stick 지팡이

문제풀이
① 네가 지금 거기 가야 할 필요는 없어.
② 이번 토요일에 등산하러 가는 건 어때?
③ 이 등산용 신발은 편하지가 않아.
④ 스포츠용품점에서 장비를 살 수 있어.

13 ④

여: 잠시만요, 데니스? 방해해서 미안한데, 2시 10분이에요.
남: 벌써요? 전 이제 막 수업의 마지막 과를 끝내고 있었어요.
여: 네, 하지만 서둘러주세요. 수습 직원들이 이미 컴퓨터 수업에 늦었거든요.
남: 죄송해요. 15분만 주시면 수습 직원들을 보낼게요.
여: 15분이요? 그건 너무 긴데요.
남: 다른 반들은 벌써 끝났나요?
여: 네, 그리고 우리가 너무 오래 기다리면 컴퓨터 강사인 스미스 선생님은 그 사람들 없이 시작할 거예요.
남: 제가 전체 교육 프로그램을 방해하고 있다는 사실을 미처 몰랐어요!
여: 네, 안타깝게도, 우린 빡빡한 시간표에 따라 운영되고 있답니다.
남: 그렇다면 5분만 더 주시면 어떨까요?
여: 알겠어요. 스미스 선생님께 조금만 더 기다리시라고 얘기할게요.

어휘 interrupt 방해하다 trainee 수습 직원 hold up (…의 흐름을) 지연시키다[방해하다] run 진행하다 [문제] dedicated 전념하는, 헌신적인

문제풀이
① 당신은 정말 이 일에 대해서 좀 더 많은 시간을 할애해야 해요.
② 당신은 우리 회사에서 가장 헌신적인 수습 직원이에요.
③ 죄송합니다만, 컴퓨터 수업은 이미 끝났어요.
⑤ 물론이죠. 우린 3시까지는 다른 곳에 가지 않아도 돼요.

14 ⑤

[전화벨이 울린다.]
남: IT부서의 빈스입니다.
여: 안녕하세요, 빈스 씨. 전 제작부의 마리예요. 해리 조 씨 계신가요?
남: 아뇨. 아쉽게도 조 씨는 오늘 종일 외근입니다.
여: 오, 이럴 수가! 지금 당장 그의 도움이 필요한데요.
남: 유감이네요. 갑자기 우리 지사 중 한 곳으로 호출되어 나가셨어요.
여: 음, 그렇다면 거기 다른 누군가가 절 도와줄 수도 있겠네요.
남: 그럴 수도 있죠. 문제를 설명해 주시겠어요?
여: 제 컴퓨터에서 아무것도 출력할 수가 없어요. 지난번에 이런 일이 발생했을 때 조 씨가 고쳐 주셨는데.
남: 음. 만약 접속의 문제라면 프란신 씨가 해결할 수 있을 겁니다.

여: 맞아요! 조 씨가 접속을 재설정하는 것에 대해 뭔가 얘기하셨던 게 기억나네요.

남: 그럼 그녀를 바로 당신 자리로 보내드릴게요.

어휘

department (조직의) 부서 branch 나뭇가지; *지사 unexpectedly 예기치 않게 describe 서술[묘사]하다 connection 연결 recall 기억해 내다 reset 재설정하다

문제풀이

① 조 씨가 그 얘기를 듣는다면 기뻐할 거예요.
② 그가 실수한 게 틀림없는 것 같아요.
③ 당신이 문제를 해결했다고 그에게 말할게요.
④ 그녀는 그가 곧 돌아올 거로 생각했음에 틀림없어요.

15 ⑤

남: 일레인은 그녀의 친구인 제시카와 점심을 먹고 있는데, 제시카가 지난 주말에 대해서 불평하고 있다. 제시카와 그녀의 남자친구는 토요일에 교외에 위치한 놀이공원에서 시간을 보내기로 했던 것 같다. 그들은 아침 일찍 그곳에 도착해서, 각자 하루 종일 무제한으로 놀이기구를 탈 수 있는 특별 입장권을 구매했다. 그 입장권은 비쌌지만, 그들은 그것이 그만한 가치가 있을 거라고 생각했다. 그러나 20분 후에 비가 내리기 시작했고, 비는 종일 그치지 않았다. 이 때문에 그들은 놀이기구 대부분을 타지 못했다. 이런 상황에서, 일레인이 제시카에게 할 말로 가장 적절한 것은 무엇인가?

일레인: 정말 실망했었겠다.

어휘

complain 불평하다 amusement park 놀이공원 unlimited 무제한의 ride 탈것 entire 전체의 figure …라고 생각하다

문제풀이

① 우린 거기 같이 갔었어야 해.
② 부디 점심시간 후엔 날씨가 개면 좋겠다.
③ 너에게 같이 가자고 물어보지 않아서 미안해.
④ 우린 아직 롤러코스터를 탈 시간이 있어.

16 ①   17 ④

여: 세미나에 오신 것을 환영합니다. 저는 존슨 교수이고, 항공기에 관해 이야기하려고 합니다. 첫 번째 비행기의 비행은 1903년에 일어났습니다. 100년도 더 지난 후인 현재, 무인 항공기가 세상을 바꾸고 있습니다. 그것들은 드론으로 알려져 있는데, 원격으로 조종됩니다. 드론은 원래 안전하게 적군의 병력을 관찰하기 위해 군사적으로 사용되었습니다. 그러나 요즘, 그것들은 많은 다른 분야에서 사용되고 있습니다. 예를 들어, 기자들은 하늘 높이에서 중요한 일들을 촬영하려고 드론을 사용합니다. 이것은 시청자들에게 새롭고 유용한 시각을 제공합니다. 또한, 몇몇 큰 상점들은 배달 드론을 사용하기 시작했습니다. 가까운 미래에 소비자들은 온라인으로 상품을 주문해 그것을 집으로 빠르게 배송받을 수 있을 것입니다. 그리고, 농업에서 농부들은 이미 넓은 들판의 작물을 관찰하는 데에 드론을 사용하고 있습니다. 생각해보면, 가능성은 무궁무진합니다.

어휘

unmanned 무인(無人)의 aircraft 항공기 remotely 멀리서, 원격으로 originally 원래 observe 관찰하다 troop 부대, 군대 journalist 기자

film 촬영하다 perspective 시각, 시야 agriculture 농업 monitor 감시하다, 관찰하다 endless 끝없는 [문제] architecture 건축(학)

문제풀이

16 ① 드론의 각기 다른 쓰임들
   ② 비행기의 역사
   ③ 어떻게 최초의 드론이 발명되었는가
   ④ 현대 항공기의 유용한 특징들
   ⑤ 드론과 비행기의 차이점들

DICTATION Answers
본문 p.20

01 snow and ice removal / have some free time / give us a hand
02 makes you gain weight / get diseases / a little light
03 have written safety procedures / are treated fairly / earn more money
04 the freshest local fruit / on sale / look tempting
05 prepare for a party / make some special food / throw a party
06 dropped by / open a new account / purchase some euros
07 go up to my knees / $15 a pack / pay the difference
08 members from all grades / every Tuesday night / as long as
09 favorite edible fungus / can be eaten / get in for half price / a special giveaway
10 taking a course / whatever fits my schedule / There's no way
11 sleep at night
12 gone hiking
13 make it quick / other classes finished / on a tight schedule
14 out of the office / fix it for me / resetting the connection
15 bought a special pass / be worth it / unable to enjoy
16~17 took place in / film important events / using drones to monitor

04 영어듣기 모의고사
본문 ▲ p.24

01 ⑤	02 ①	03 ④	04 ④	05 ②	06 ①
07 ②	08 ②	09 ⑤	10 ③	11 ①	12 ④
13 ⑤	14 ③	15 ⑤	16 ③	17 ⑤	

01 ①

여: 여러분은 자신이 세상에서 가장 어려운 시간을 보내고 있다고 생각하실지 모릅니다. 하지만 제가 오늘날의 제가 되기 위해 견뎌야 했던 고통에 대해 여러분께 말씀드리겠습니다. 저는 건강한 소녀였으나, 사고가 제 다리를 앗아갔고 제 팔을 마비시켰습니다. 그런데도 저는 희망을 놓지 않았고, 많은 고투 끝에 손을 움직이는 법을 다시 익혔습니다. 그 후, 저는 다른 사람들을 돕기로 했고 의사가 되었습니다. 여러분도 저처럼 장애물을 이겨낼 수 있습니다. 절망하지 마십시오.

어휘 endure 참다, 견디다 take away 가져가다, 치우다 paralyze 마비시키다 hold onto …을 꼭 잡고 있다[놓지 않다] struggle 분투하다; *노력, 고투 overcome 극복하다, 이겨내다 obstacle 장애물 despair 절망; *절망하다

02 ①

남: 내일 비키네 집에 가는 거 잊지 마.

여: 아, 맞다. 그녀가 연휴를 축하하는 저녁 식사 파티를 열지.

남: 아주 재미있을 것 같아. 우리 친구들 모두 거기 올 거고, 그녀는 맛있는 저녁을 요리하겠지.

여: 몇 시에 시작하지?

남: 오후 7시에 시작하지만, 난 도중에 꽃을 사려고 조금 더 일찍 출발하고 싶어.

여: 꽃들이 무엇에 필요한데?

남: 음, 빈손으로 나타나는 건 무례할 것 같아. 파티 여주인에게 줄 뭔가가 있었으면 해.

여: 너 참 배려심이 깊구나.

어휘 celebrate 축하하다 pick up 사다 along the way 도중에 show up 나타나다 empty-handed 빈손으로 hostess 여주인 thoughtful 배려심 있는

03 ④

여: 안녕하세요. 뭘 좀 도와드릴까요?

남: 네, 다이앤 피터슨 씨께 온 소포가 있습니다.

여: 그건 전데요. 하지만 전 아무것도 받을 게 없는데요. 누구한테서 온 건지 말씀해주실 수 있나요?

남: 잠시만요… 영국의 데이비드 피터슨 씨가 보내셨습니다.

여: 아, 제 오빠예요. 정말 뜻밖이네요!

남: 여기 서명 좀 해주시겠어요?

여: 물론이죠. [잠시 후] 근데 옆집에 온 것은 없나요?

남: 사실은 있습니다. 왜 물어보시는 거죠?

여: 옆집 사람들이 몇 주 동안 휴가를 가면서 제게 배송 오는 걸 잘 봐달라고 부탁했어요.

남: 그렇군요. 그럼 이 물건에 대해서도 서명해주시겠어요?

여: 물론이죠, 기꺼이 할게요. [잠시 후] 여기요.

남: 대단히 감사합니다. 여기 있습니다.

어휘 package 소포 keep an eye out for …을 감시하다; *…을 지켜보다 delivery (물건·편지 등의) 배달

04 ④

남: 안녕, 새라. 이 사진 좀 봐. 내 남동생이 겨울 동화의 나라 전시회에서 찍은 거야.

여: 와! 사진 속의 모든 것이 얼음으로 만들어졌네!

남: 왼쪽에 있는 얼음 차가 마음에 든다. 사람들이 탈 수 있게 차 문이 열려 있네.

여: 응. 그걸 운전하면 재미있지 않을까?

남: 맞아. 그게 얼마나 빨리 갈 수 있을지 궁금하다. 얼음 조각 중에 네가 가장 좋아하는 건 뭐야?

여: 나는 가운데에 있는 게 마음에 들어. 그건 에펠탑처럼 보여, 그렇지?

남: 응, 그래. 꽤 인상적이다. 그리고 오른쪽에 있는 하프를 봐. 정말 섬세해 보여.

여: 거기에 있는 현들이 매우 가늘어. 연주도 할 수 있을 것처럼 보여.

남: 그걸 연주해보고 싶다. 그 뒤에는 얼음 백조도 있어.

여: 날개가 활짝 펼쳐져 있네.

남: 마치 날아가려고 하는 것 같아. 차 뒤에 있는 조각품은 뭐야?

여: 그건 꼭대기에 큰 별이 달린 크리스마스트리야.

남: 정말 멋진 전시회구나!

어휘 wonderland 동화의 나라 exhibition 전시(회) sculpture 조각품 delicate 섬세한 string 끈, 줄 swan 백조 spread 펼치다, 펴다 (spread-spread)

05 ②

남: 안녕, 낸시. 뭐 하고 있어?

여: 안녕, 마이크. 나는 커피를 끓이고 있어. 어제 선물로 새 커피 추출기를 받았어.

남: 향기가 정말 좋다. 커피 원두는 어디서 났어?

여: 시장에서 샀어. 인도네시아산이야.

남: 비싸지 않았어?

여: 비쌌어, 하지만 그만한 가치가 있어. 커피 맛이 정말 좋아.

남: 그렇구나. 나도 커피 추출기가 있으면 좋겠어. 그러면 매일 아침 커피를 만들 텐데.

여: 사실, 내 오래된 커피 추출기를 중고상에 팔 계획이었어. 원한다면 그걸 네게 줄게.

남: 정말? 내가 그 물건에 돈을 좀 줄게. 그냥 받을 순 없어.

여: 그럴 필요 없어. 어쨌든 네 생일이 곧 오잖아. 그냥 선물로 받아.

남: 오, 얼마나 고마운지 말로 표현을 못 하겠어.

어휘 brew (커피·차를) 끓이다 coffee bean 커피 원두 second-hand 중고의 appreciate 고마워하다

06 ①

여: 안녕, 앨버트. 뭘 도와줄까?

남: 안녕하세요, 해리스 선생님. 24일에 있는 시험에 관한 건데요. 제가 그날 여기 없을 거예요.

여: 왜 그렇지?

남: 제가 수영부거든요. 그날 저희는 도쿄에서 열리는 대회에 출전해요.

여: 아, 그렇구나. 우리 반에 다른 선수도 있니?

남: 아뇨, 저뿐이에요. 나머지 아이들은 모두 티즈데일 선생님 반인데, 그

선생님께서 시험을 다른 날로 옮기셨대요.

여: 그렇다면 넌 도쿄에 가기 전에 시험을 볼 수 있니?

남: 네, 그렇게 할 수 있어요. 내일 괜찮을까요?

여: 아니, 다른 아이들보다 네가 시험을 먼저 보니까 널 위해 다른 시험 문제를 출제해야 해. 방과 후 월요일 어떠니?

남: 괜찮아요.

여: 그리고 도쿄에서 행운을 빌어.

남: 고맙습니다.

어휘

competition 경쟁; *대회 schedule 일정; *일정을 잡다

07 ②

남: 안녕하세요. 아이스크림 좀 사려고요.

여: 그럼요! 오늘은 무엇을 드릴까요?

남: 전 아이스크림 컵 세 개가 필요해요, 하나는 초콜릿, 두 개는 딸기로요.

여: 알겠습니다. 저희는 (아이스크림) 크기가 두 가지 있습니다. 작은 스쿱은 2달러고요, 큰 스쿱은 4달러예요.

남: 그렇군요. 그럼 초콜릿은 큰 스쿱 하나랑 딸기는 작은 스쿱으로 두 개 할게요.

여: 네. 다른 건 필요하지 않으세요?

남: 괜찮아요, 그게 다예요.

여: 정말요? 토핑은 원하지 않으세요? 설탕 가루나 시럽을 1달러에 추가할 수 있어요.

남: 괜찮아요. 아, 5퍼센트 할인 쿠폰이 있는데요. 아직 사용할 수 있나요?

여: 잠깐 볼게요. [잠시 후] 네, 그건 이번 달 말까지 유효합니다.

남: 좋아요. 그걸 지금 사용하죠.

여: 알겠습니다. 여기 아이스크림 나왔습니다.

어휘

scoop 스쿱(한 숟갈의 양) sprinkle (디저트 장식용의) 설탕 가루 good 유효한

문제풀이

남자는 큰 스쿱 크기의 초콜릿 아이스크림 한 개($4)와 작은 스쿱 크기의 딸기 아이스크림 두 개($2×2)를 샀고 5퍼센트 할인 쿠폰이 있으므로, 남자가 지불할 금액은 7.6달러이다.

08 ②

여: 어젯밤에 무서운 영화를 봤어. 제목은 죽은 *태양*이었어.

남: 미래에 태양이 어두워지는 것에 관한 영화지?

여: 바로 그거야. 그건 진짜로 일어날 수 없겠지, 그렇지?

남: 사실상 그런 일은 일어날 거야. 태양은 행성이고 모든 행성은 죽으니까.

여: 농담이지!

남: 아니야. 지금 태양은 아주 안정적인 황색 행성이야.

여: 음, 그건 다행이야.

남: 그런데 결국에는 적색 거성이 되지. 훨씬 더 커지고 뜨거워질 거야.

여: 아, 그렇구나.

남: 시간이 지나면서 모든 에너지를 소진할 거야. 그리고 그다음에 죽게 될 거야.

여: 그 일이 생기는 데 얼마나 걸릴까?

남: 음, 태양은 나이가 약 50억 년이야. 그리고 아마도 50억 년쯤 남았을 거야. 그러니 우리에겐 시간이 많아!

어휘

scary 무서운 stable 안정적인 eventually 결국에는 use up 소진하다. 다 써 버리다 billion 10억

09 ⑤

남: 여러분 모두 지난주에 남아메리카에서 발생한 끔찍한 지진에 대해서 알고 계실 거라고 믿습니다. 생존자들을 돕기 위한 기금을 모으기 위해서, 우리 학교는 특별 록 콘서트를 주최할 것입니다. 그 콘서트는 3월 8일 토요일 오후 8시에 학교 강당에서 열릴 것입니다. 4개의 서로 다른 이 지역의 고등학교 밴드가 출연할 것입니다. 입장료는 10달러이며, 전액은 이 비극의 영향권 아래의 지역에서 일하는 자선 단체들로 직접 기증될 것입니다. 콘서트에 참석하는 분에게 또한 통조림 음식이나 중고 의류와 같은 기증품을 가져오시길 권장합니다. 입장권은 오늘부터 판매하며, 행정실에서 구매하실 수 있습니다.

어휘

earthquake 지진 raise funds 기금[자금]을 모으다 survivor 생존자 host (행사 등을) 주최하다 auditorium 강당 feature 특별히 포함하다, 특징으로 삼다 admission 입장(료) donate 기부[기증]하다 (n. donation) charity 자선 (단체) region 지역 affect …에 영향을 미치다 tragedy 비극(적인 사건) encourage 장려[권장]하다 canned 통조림으로 된 used 중고의 administrative 행정상의, 관리상의

10 ③

남: 도와드릴까요?

여: 네, 저는 새 손목시계를 찾는데요.

남: 이것들 좀 보시죠. 저희의 가장 인기 있는 모델들입니다.

여: 전부 다 아주 멋지네요, 이 삼각 모양만 빼고요. 그건 너무 이상해요.

남: 네, 하지만 청소년들에게는 인기가 있습니다. 시곗줄은 어떤 종류를 선호하시나요?

여: 저는 가죽끈이 있는 게 더 좋습니다.

남: 알겠습니다. 이건 어떠세요? 이건 굉장히 세련되고 유행하는 것입니다.

여: 아니에요, 괜찮아요. 사실, 저는 전자시계는 안 좋아해서요. 그것들은 저렴해 보여요.

남: 알겠습니다. 그러면 이 두 가지 제품을 더 자세히 보는 건 어떠세요?

여: 와! 이거 두 개 다 마음에 드네요. 그것들은 얼마인가요?

남: 여기에서 가격표를 보실 수 있습니다.

여: 아, 저건 너무 비싸네요. 덜 비싼 거로 할게요.

어휘

except (for) …을 제외하고 triangular 삼각형의 (n. triangle) leather 가죽 strap 끈 price tag 가격표 [문제] oval 타원형(의) square 정사각형(의) metal 금속 analog 아날로그식인

11 ①

여: 안녕 제임스, 여기서 뭐 하니?

남: 사촌이 이 학교에서 농구 선수로 뛰거든. 그 애를 응원하고 있어.

여: 그건 몰랐네. 어떤 선수니?

남: 경기장에서 가장 키가 큰 애야.

문제풀이
② 미안한데, 난 농구를 별로 좋아하지 않아.
③ 정말? 나도 네 농구팀에 들어가고 싶어.
④ 그는 스포츠 섹션에서 농구에 관한 책을 찾고 있어.
⑤ 슛을 전혀 막아내지 못해서 그 선수가 싫어.

12 ④

남: 네가 오늘 오후 과학 수업에서 발표한다고 들었어.
여: 응. 그래서 지금 엄청 스트레스를 받고 있어.
남: 음, 내가 널 도울 일이 있을까?
여: 응. 인쇄물을 복사해 줄 수 있겠니?

어휘
presentation 발표 stressed out 스트레스로 지친 [문제] make a copy
복사하다 handout 인쇄물, 수업 자료

문제풀이
① 물론이지, 내일 널 도울 수 있어.
② 아니, 네 발표는 완벽했다고 생각해.
③ 아니, 오늘 오후에는 시간이 날 것 같지 않아.
⑤ 그러고 싶어. 그런데 과학 선생님께 여쭤보는 게 좋을 거야.

13 ⑤

[전화벨이 울린다.]
남: 여보세요?
여: 여보세요, 패트릭. 저 선희예요.
남: 오. 안녕하세요, 선희 씨.
여: 너무 늦게 전화해서 죄송한데요, 내일 수업을 취소해야 해서요.
남: 오, 별일 아니었으면 좋겠네요.
여: 네, 별일 아니에요. 단지 내일 오후에 회의가 있어서 그래요.
남: 알았어요, 그럼 시간을 조정할까요?
여: 괜찮으시다면, 그렇게 하면 좋겠어요.
남: 언제가 좋으세요?
여: 금요일 오후 4시 어때요?
남: 괜찮아요. 같은 시간, 같은 장소에서요.
여: 고맙습니다. 갑작스럽게 말씀드린 일인데 정말 감사드려요.

어휘
cancel 취소하다 reschedule 일정을 재조정하다 [문제] short notice 갑작
스런 통보

문제풀이
① 회의에서 행운을 빌어요.
② 전 그때는 좀 곤란한데요.
③ 글쎄요, 그냥 취소해야 할 것 같아요.
④ 알아요. 그건 누구에게나 일어날 수 있는 일이었어요.

14 ③

여: 이건 너무 어려워. 난 포기할래!

남: 무슨 일이야, 힐러리?
여: 생일 선물로 이 스케이트보드를 받았는데, 탈 수가 없어.
남: 오, 전에 스케이트보드를 타 본 적이 없어?
여: 없어. 처음에는 정말 재미있어 보였는데, 너무 어려워. 이제는 그냥 좌
   절감을 느껴.
남: 이해해. 나도 어떻게 타는지 배우는 데 오랜 시간이 걸렸어.
여: 난 타려고 할 때마다 넘어져.
남: 너는 인내심을 가져야 해. 그냥 우선 며칠 동안은 직선으로 타려고 해
   봐.
여: 그건 지루해. 나는 점프나 묘기와 같은 재미있는 것을 하고 싶어.
남: 그런 것들은 나중에 배울 수 있어. 우선, 너는 기초적인 기술을 배워
   야 해.
여: 나도 알아. 하지만 내가 할 수 있을 것 같지 않아. 지금 난 그 위에 똑
   바로 서 있는 것조차 못하는걸.
남: 그냥 계속 연습하면 시간이 지나면서 나아질 거야.

어휘
frustrated 좌절감을 느끼는 trick 속임수; *묘기, 재주

문제풀이
① 내 스케이트보드를 빌릴래?
② 네 생일이었다고 나에게 말했어야지.
④ 대신 나에게 스케이트보드를 어떻게 타는지 가르쳐 줄 수 있어?
⑤ 너는 잘하고 있어. 이제 좀 더 어려운 묘기를 해 봐.

15 ⑤

여: 트레이시는 처음으로 뉴욕을 방문 중이다. 그녀는 모든 명소를 보면
   서 하루를 보내겠다고 결심한다. 모든 것이 순조로웠는데, 그녀는 관
   광 지도를 지하철에 두고 내린다. 타임스스퀘어를 찾을 수가 없어서,
   그녀는 한 지나가는 남자를 붙잡고 길을 묻는다. 그는 그곳에 가는
   가장 빠른 길을 설명해주지만, 그의 지시는 매우 복잡하다. 그는 설명
   을 몇 번이고 반복하지만, 트레이시는 이해하지 못한다. 결국, 그 남자
   는 그곳이 그다지 멀지 않으므로 트레이시가 길을 잃지 않도록 그녀
   와 걸어가겠다고 말한다. 이런 상황에서, 트레이시가 남자에게 할 말
   로 가장 적절한 것은 무엇인가?
트레이시: 저를 도우려고 이렇게 애를 써주셔서 감사해요.

어휘
sight 시력; *관광지 unable to-v …하지 못하는, …할 수 없는 direction 방향,
지시 complicated 복잡한 [문제] sightsee 관광 여행하다 go out of
one's way 특별히 애를 쓰다

문제풀이
① 그곳은 우리가 걸어가기엔 너무 멀어요.
② 사실, 전 오늘 관광할 계획이었어요.
③ 제가 도와드리고 싶지만, 제 지도를 잃어버렸어요.
④ 타임스스퀘어로 가는 길을 알려 주시겠어요?

16 ③ 17 ⑤

남: 생물학 수업에 오신 것을 환영합니다. 저는 여러분에게 질문이 있습
   니다. 먹고 마시는 것 중 여러분이 가장 좋아하는 것은 무엇입니까?
   자, 여러분들이 그것들을 즐길 수 있을 때 그렇게 하시는 편이 좋으실
   겁니다. 왜냐하면, 그것 중 일부는 곧 영원히 사라질 테니까요. 예를
   들어, 벌의 개체 수는 전 세계적으로 점점 감소하고 있습니다. 벌이 멸

종되면, 사람들이 먹을 꿀은 더 없겠죠. 바나나도 위험에 직면하고 있습니다. 바나나 나무들은 몇몇 지역에서 치명적인 질병을 앓고 있습니다. 이 질병은 뿌리를 공격해 결국 나무 전체를 죽이고 맙니다. 커피를 마시나요? 음, 커피도 매우 민감한 작물입니다. 지구 온난화와 다른 요인들이 커피 재배를 점점 더 어렵게 만들고 있고요. 기온 상승은 또한 연어도 위협하고 있습니다. 이 생선은 해산물 애호가들이 가장 좋아하는 것입니다. 하지만 그것들은 생존을 위해 차가운 물이 필요한데, 우리의 해양은 더 따뜻해지고 있습니다. 바라건대, 과학자들과 농부들이 이 문제들에 대한 해결책을 곧 찾아내면 좋겠습니다.

어휘
biology 생물학 population 인구, 개체 수 die out 멸종하다 deadly 치명적인 eventually 마침내 sensitive 민감한 crop 작물 factor 요인 threaten 위협하다 hopefully 바라건대 [문제] vulnerable …에 취약한[연약한]

문제풀이
16 ① 식품으로 유발된 질병들   ② 가장 영양이 풍부한 음식들
③ 멸종될 위기에 처한 음식들   ④ 기후 변화의 영향
⑤ 박테리아에 취약한 음식들

DICTATION Answers
본문 ▲ p.26

01 I had to endure / held onto my hope / Don't despair
02 celebrate the holidays / pick up some flowers / show up empty-handed
03 have a package / sign right here / keep an eye out for
04 for people to get in / spread wide open / What a cool exhibition
05 as a present / I wish I had / is coming soon
06 won't be here / take the test / make a different test
07 What can I get for you / Would you like anything else / Let me see
08 going dark in the future / get much bigger / use up all its energy
09 raise funds / It will feature bands / to bring donations
10 I'm looking for / modern and stylish / take the less expensive one
11 plays basketball
12 really stressed out
13 nothing is wrong / If you don't mind / Sounds good
14 I give up / how to ride one / stand up straight
15 spend the day seeing / asks him for directions / not very far
16~17 to eat and drink / soon disappear forever / for people to eat / find a solution

05 영어듣기 모의고사
본문 ▲ p.30

01 ⑤	02 ③	03 ②	04 ②	05 ④	06 ①
07 ⑤	08 ②	09 ④	10 ②	11 ④	12 ④
13 ④	14 ⑤	15 ④	16 ①	17 ②	

01 ⑤

남: 신사 숙녀 여러분, 주목해 주시겠습니까? 7살된 아이가 여기 접수대에 있습니다. 아이의 부모님은 판매 직원에게 가서 말씀하신 후 접수대로 오셔서 아이를 데려가시기 바랍니다. 여기에 오시면, 아이를 볼 수 있기 전에 아이의 인상착의를 자세히 설명하시고 이름을 말씀해 주셔야 합니다. 이는 단순히 유괴를 방지하기 위한 안전 예방책입니다. 경청해 주셔서 감사합니다.

어휘
front desk 접수대 kindly 부디 contact 연락하다 make one's way to …로 가다 describe 묘사하다 in detail 상세히 safety precaution 안전 예방책 prevent 막다 kidnap 납치[유괴]하다

02 ③

남: 이 소포를 보내러 우체국에 갈 거야.
여: 그냥 속달 서비스를 이용하는 게 어때?
남: 그게 우체국보다 더 비싸지 않을까?
여: 비용이 더 들지. 하지만 그걸 가져가는 사람을 보내주는 거잖아. 그럼 넌 이동하지 않아도 돼.
남: 그게 편리하겠구나.
여: 그리고 소포는 빠른 시간 내에 배달될 거야. 우체국은 최소한 며칠은 걸릴 거고.
남: 그런데 우체국은 속달 우편 서비스가 있잖아.
여: 응, 하지만 그게 비용이 더 들어. 그러니 너무 멀리 보내지 않는다면 큰 가격 차이는 없을 거야.
남: 알았어, 납득이 가네.

어휘
package 소포 courier 운반원, 배달원 save 아끼다; *피하게 하다, …하지 않아도 되다 trip 여행; *이동 convenient 편리한 express 속달의, 급행의 convinced 확신하는

03 ②

여: 안녕하세요! 오늘은 어떤 것을 도와드릴까요?
남: 음, 내일이 제 여자친구 생일이에요. 이미 그녀를 위해 예쁜 반지를 샀는데요. 이제 꽃을 좀 사주고 싶어서요.
여: 그거 근사하네요. 여자친구가 어떤 종류의 꽃을 좋아하나요?
남: 잘 모르겠어요. 아마도 당신이 제가 고르는 것을 도와주실 수 있을 것 같네요.
여: 그럼요. 말씀해 보세요, 여자친구는 어떤 분이신가요?
남: 음, 그녀는 재미있고 활동적이에요. 춤추는 것을 좋아하죠.
여: 저쪽에 있는 보라색 꽃들은 어떠세요? 특별해 보이잖아요.

남: 예쁘긴 하지만, 좀 더 낭만적인 것은 어떨까요?
여: 아, 알겠어요. 이쪽에 사랑스러운 장미들이 있어요. 빨간색은 로맨스를, 노란색은 우정을 뜻해요.
남: 그럼 붉은 장미 24송이로 할게요.
여: 여자친구가 분명 좋아할 거예요.
남: 고맙습니다. 도와주셔서 정말 감사해요.

어휘
describe 묘사하다, 말로 설명하다 energetic 활동적인 purple 보라색(의)
romantic 로맨틱한, 낭만적인 (n. romance) friendship 우정

04 ②

남: 창고를 치워야 할 것 같아.
여: 맞아. (치우는지) 꽤 됐어. 뭘 먼저 해야 할까?
남: 음, 천장에 큰 거미줄이 있어. 그런데 사다리가 너무 짧아서 그걸 떨어뜨릴 수가 없었어.
여: 알았어. 내가 이웃한테 더 긴 걸 빌릴 수 있을 것 같아.
남: 그래. 그리고 창문도 깨졌어. 수리해야 해.
여: 교체할 유리가 있니?
남: 내가 조금 주문했어. 오늘 오후에 올 것 같아. 아, 상자들을 버리는 것부터 시작할 수 있을 것 같아.
여: 겨우 세 개 있네. 아, 지난 여름에 썼던 큰 의자는 어떡할까?
남: 아, 그건 덮여 있네. 덮개만 세탁하면 될 거야.
여: 좋아.

어휘
storage 창고 spider web 거미줄 ceiling 천장 get down 내리다, 떨어뜨리다 ladder 사다리 replace 대체하다, 바꾸다 throw out 버리다

05 ④

여: 내일이 이 숙제 마감일이라니 믿을 수가 없어.
남: 응, 우린 주말에 컴퓨터 게임을 하지 말았어야 했나 봐.
여: 난 할 게 이렇게 많은 줄 몰랐어.
남: 그러게. 지금 보니까 우린 좀 더 일찍 숙제를 시작했어야 했어.
여: 오늘 밤에 급하게 하면 제대로 못할 텐데.
남: 어쩌면 마감일을 연장해 달라고 부탁해 볼 수도 있어.
여: 글쎄. 우린 구실이 없잖아.
남: 맞아. 늦게 내면 벌점이 얼마지?
여: 하루에 5점인 것 같아.
남: 음… 네 생각에 숙제를 늦게 내는 게 나을 것 같니?
여: 글쎄, 밤을 새워서 하면 제시간에 낼 수는 있는데.
남: 그렇긴 한데, 그다지 잘하지 못할 테니까 아마도 5점보다 더 많이 깎일 것 같아.
여: 네 말이 맞아. 벌점을 받더라도 시간을 충분히 갖고 하자.

어휘
project 과제, 숙제 involve 수반하다 rush through 급하게 처리하다
extension (기일 등의) 연장 deadline 마감일 doubt 의심하다 excuse 변명, 구실 penalty 벌점 worth …할 가치가 있는 stay up all night 밤새우다
get … in …을 모으다[거둬들이다]; *간신히 …을 하다

06 ①

여: 제이콥! 와! 이게 얼마만이야!
남: 안녕, 애슐리. 해외에서 살고 있었어. 이제 막 돌아왔어.
여: 네가 유럽을 여행하고 있다고 들었는데.
남: 음, 그게 내 계획이었지. 하지만 우선 돈을 좀 모아야 했어.
여: 무슨 일을 했어?
남: 한국에서 영어를 가르치는 일을 했어.
여: 와! 흥미로운데. 그래서 유럽을 가기 위한 돈을 좀 모았니?
남: 그랬지. 하지만 난 한국이 너무 좋아서 대신에 거기서 3년을 지냈어.
여: 3년이라니! 너 정말 거기 좋아했나 봐.
남: 응. 하지만 결국 집을 돌아오기로 결심했지.
여: 왜? 무슨 일이 있었어?
남: 아니, 내 생각엔 단지 집이 그리웠던 것 같아.

어휘
abroad 해외에(서) earn (돈을) 벌다 save (돈을) 모으다 stay 머무르다
homesick 향수에 잠긴

07 ⑤

남: 돌아온 걸 환영해! 카리브해 유람선 여행은 어땠어?
여: 아주 좋았어. 돈이 너무 많이 들긴 했지만.
남: 정말? 싸게 간 줄 알았는데.
여: 응, 유람선 여행 자체는 300달러밖에 안 들었는데, 여러 섬에서 많은 돈을 썼어.
남: 그럴 수밖에 없는 것 같아.
여: 맞아, 거기에 가는 데 하루가 걸렸고, 바하마 제도에서 이틀을 머물렀어. 거기서 200달러를 썼어.
남: 와! 또 어디 갔었어?
여: 자메이카에서도 하루 동안 있었지.
남: 거기서는 얼마를 썼는데?
여: 150달러쯤. 그리고 나서 바베이도스에서 이틀 있었고 거기서 250달러 정도를 썼어.
남: 거기가 마지막으로 머물렀던 곳이었기를 바라.
여: 응, 그런데 돌아오는 데 하루가 걸렸어.

어휘
Caribbean 카리브해의 cruise 유람선 여행 deal 거래 be bound to-v 반드시 …하다 stop 머무르다; 정거장; *머묾

문제풀이
여자는 유람선 여행에 300달러, 바하마 제도에서 200달러, 자메이카에서 150달러, 바베이도스에서 250달러를 지출했으므로, 여자가 지불한 여행 경비는 900달러이다.

08 ②

남: 이번 주말에 어디로 캠핑을 갈까?
여: Black Mountain 국립 공원에 가자.
남: 난 들어본 적이 없어.
여: 작년에 개장해서 많은 사람들이 알고 있진 않아.
남: 아, 공원이 근처에 있니?
여: Silver City의 동쪽이야. 거기까지 차로 3시간 정도 걸려.
남: 좋아. 공원이 크니?

여: 아주 크진 않지만, 강을 따라 보기 좋은 캠핑 구역이 있어.

남: 좋을 것 같아. 등산로는 어때?

여: 산을 오르는 세 개의 다른 등산로가 있어. 정상에서 보는 경치가 훌륭해.

남: 공원 주변에 관광지가 좀 있니?

여: 응, 차로 30분 거리 이내에 많은 역사 유적지가 있어. 할 게 아주 많을 거야!

어휘

nearby 가까운 곳의 hiking trail 등산로 view 경치 tourist attraction 관광지 historical site 역사 유적지 plenty of 많은

09 ④

여: 여러분도 아시다시피, 지난주에 졸업반 학생들 전원이 수학여행을 어디로 가고 싶은지에 대해 투표를 했습니다. 그 결과가 나왔습니다! 작년에 수도로 간 여행이 대성공을 거두었음에도, 올해는 213명의 학생만이 그곳에 가고 싶다고 투표했습니다. 나머지 표는 스키 여행과 Eagle Rock으로의 자연 체험 여행으로 나뉘었습니다. 표 차이가 근소해서 저희는 두 번 개표해야 했습니다. 학생 여러분, 올해는 Rocky Top 산으로 가게 되었으니 여러분의 스키 장비를 챙기십시오.

어휘

senior 연장자; *졸업반 학생 vote 투표하다; 투표수, 득표 school trip 수학여행 as for …에 관해서는 split 나누다 excursion 소풍, 짧은 여행 close (차이 등이) 근소한 pack up (짐 등을) 싸다 gear 장비 off 떠나, 출발하여

10 ②

여: 시에서 하는 연례 교육 학회가 다음 달에 있어. 너 나랑 같이 갈 거지?

남: 물론이야! 우린 매년 가잖아. 하지만 곧 참석할 워크숍을 골라야 해.

여: 난 일정표를 지금 가지고 있어. 넌 선호하는 게 있어?

남: 흠… 음, 나는 성인 교육 워크숍은 다시 가고 싶지 않아.

여: 응, 우리 작년에 거기 갔잖아. 그리고 난 점심 이후에 하는 것도 가기 싫어.

남: 오, 그래? 왜?

여: 점심을 먹고 나면 항상 너무 졸리더라고. 그렇지 않아?

남: 아니. 하지만 난 긴 워크숍 시간에 졸리더라. 그러니 두 시간보다 짧은 것을 고르자.

여: 나도 그게 좋아. 질의응답 세션은 어떤 것 같아?

남: 난 그게 워크숍에서 제일 중요한 부분 같아.

여: 동의해. 그러면 이 워크숍으로 선택하자. 이게 질의응답 시간이 있어.

남: 좋은 선택인 것 같아!

어휘

conference 회의, 학회 preference 선호 [문제] citywide 도시 전체의 serve (음식을) 제공하다

11 ④

남: 캐롤라이나, 우리 반에 새로 온 학생 봤니?

여: 새로 온 여자애? 응, 그 애랑 잠깐 얘기했어.

남: 그랬니? 그 애는 어때?

여: <u>그녀는 친절하고 다정한 것 같아.</u>

어휘

chat 수다를 떨다 [문제] by accident 우연히 warm-hearted 마음이 따뜻한

문제풀이

① 그녀는 우리 새 영어 선생님이야.

② 그녀는 내 가장 오랜 친구 중 하나야.

③ 그녀는 우연히 우리 반에 왔어.

⑤ 그녀는 자신의 엄마가 뉴욕 출신이라고 말했어.

12 ④

여: 리가 오디션을 통과했다는 거 들었니?

남: 정말? 음, 그녀가 노래와 춤에 뛰어난 건 알아.

여: 나도 그렇게 생각해. 아마 그녀는 언젠가 TV 스타가 될 거야!

남: <u>그녀의 사인을 미리 받아야겠어.</u>

어휘

audition 오디션 [문제] autograph 사인, 서명 in advance 미리 find out 알아내다, 알게 되다

문제풀이

① 그녀에게 그 소식을 말해줄게.

② 그녀는 시험에 합격하지 않을 거야.

③ 어떤 노래가 오디션에 좋을까?

⑤ 언제 그녀가 오디션의 결과를 알게 될까?

13 ④

남: 실례합니다, 뭔가에 대해서 항의하고 계신 건가요?

여: 아뇨. 저희는 이곳 시내에 새로운 어린이 병원을 세우기 위해서 모금 중이에요.

남: 정말 멋진 생각이네요!

여: 저희도 그렇게 생각해요. 저희는 이미 지역 사업체, 정치가, 그리고 많은 부모님들의 지지를 받고 있답니다.

남: 저도 왜 그런지 이해가 가요. 작년에 제 딸이 아팠을 때, 저흰 가장 가까운 병원에 가는 데 거의 한 시간 동안 차를 몰아야 했어요.

여: 저희들 중 다수가 비슷한 경험을 했답니다. 그게 바로 이 병원이 설립되도록 저희가 최선을 다하고 있는 이유죠.

남: 그렇다면, 어떻게 기금을 모으고 계세요?

여: 음, 당신과 같은 시민들로부터 기부를 받기도 하고, 지역 학교와 기업들에서 다양한 기금 마련 행사를 하고 있답니다.

남: 음, 저도 기부를 하고 싶지만 돈이 많지 않네요.

여: <u>기부하시는 금액이 얼마든 간에 감사히 받겠습니다.</u>

어휘

protest 항의[시위]하다; 항의, 시위 raise money 돈[기금]을 마련하다 support 지지; 지지하다 donation 기부, 기증 conduct *(특정한 활동을) 하다 fundraising 기금 모금 [문제] cause 원인; *대의, 주장 contribute 기부[기증]하다 facility 《pl.》 시설 outstanding 뛰어난

문제풀이

① 저희의 주장을 지지해주셔서 감사합니다.

② 당신의 딸이 곧 회복하길 바랍니다.

③ 저희 시위에 가담하셔서 언제든지 저희를 도와주실 수 있답니다.

⑤ 우리 도시의 건강 관리 시설은 뛰어나죠.

14 ⑤

여: 안녕하세요, 키팅 교수님. 지금 바쁘세요?
남: 아니 괜찮아. 들어오렴. 무슨 일 때문에 그러지?
여: 저, 보고서를 제때에 못 낼 것 같아요.
남: 규칙은 알고 있겠지. 하루 늦을 때마다 2점씩 감점이다.
여: 알아요. 안타깝게도 아버지께서 입원해 계세요.
남: 그거 유감이구나. 심각한 게 아니길 바란다.
여: 사실 아버지가 교통사고를 당하셔서 심하게 다치셨어요.
남: 그것 참 큰일이구나. 걱정이 많이 되겠구나.
여: 네, 맞아요. 아버지는 당분간 병원에 계셔야 할 거예요.
남: 음, 나도 몇 년 전에 사고를 당했지. 그 사고로 회복하는 데 몇 달 걸렸어.
여: 아버지가 쾌차하셨으면 좋겠어요. 어쨌든, 보고서는 다음 주 수요일까지 완성할 수 있을 것 같아요.
남: 보고서는 걱정하지 마라. 가서 아버지 곁에 있으렴.

어휘
injure 부상을 입다[입히다] badly 몹시, 심하게 recover 회복하다 [문제]
cautiously 조심스럽게 insurance 보험

문제풀이
① 좋은 생각이구나. 같이 가자.
② 응. 난 이제 좀 더 조심히 운전한단다.
③ 동의해. 그게 우리가 보험이 필요한 이유지.
④ 내일 내 책상에 과제물이 있어야 해.

15 ④

여: 션은 얼마 전에 운전면허를 땄다. 그의 어머니는 책임감 있게 운전하라고 장황하게 설명하신 후 그에게 차를 빌려주셨다. 지금 션은 사고를 내서 어머니 차에 흠집을 냈다. 새 차였으므로 그는 어머니에게 얘기할 일이 매우 걱정이다. 그리고 비록 그의 잘못으로 인한 사고가 아니었어도, 어머니가 자신을 믿고 차를 맡겼기 때문에 그는 속상하다. 그는 이 사고가 보험 처리가 된다는 걸 확신하고, 무슨 일이 일어났는지 어머니에게 솔직히 말하기로 결심한다. 그는 문 안으로 들어가, "엄마, 안 좋은 소식이 있어요. 작은 사고가 있었는데, 차에 흠집이 났어요."라고 말한다. 이 상황에서, 션의 어머니가 션에게 할 말로 가장 적절한 것은 무엇인가?
션의 어머니: 너 다쳤니? 너만 괜찮으면 자동차는 상관없단다.

어휘
driver's license 운전면허 loan 빌려주다 lecture 강의 responsibly 책임감 있게 dent 움푹 들어간 곳 fault 잘못 trust A with B A를 믿고 B를 맡기다 insurance 보험 cover 덮다; *(분실·상해 등을) 보장하다

문제풀이
① 축하한다! 드라이브하러 가자꾸나!
② 슈퍼마켓까지 날 좀 태워다 주겠니?
③ 운전하는 게 처음이니까 내가 같이 갈게.
⑤ 걱정하지 마라. 어차피 오래된 차였으니까 새 차를 사면 된단다.

16 ① 17 ②

남: 오늘은 여러분의 개의 감정을 읽는 방법에 대해 말하고자 합니다. 여러분이 개를 훈련시킬 때에는 개가 어떻게 느끼는지를 아는 것이 중

요합니다. 사람들은 종종 동물의 감정을 추측하고 싶어 하는데, 추측하는 것은 개에게는 필요하지 않습니다. 그 특정한 순간에 개가 어떻게 느끼는지 이해하기 위해서는 여러분은 단순히 개가 다양한 신체 부위를 움직이는 방식을 살펴보면 됩니다. 예를 들면, 행복한 개는 꼬리를 흔드는 경향이 있습니다. 그리고 만약 개가 놀고 싶어 한다면, 개가 앞발을 구부리고 귀를 세우며 꼬리를 든다는 것을 알아챌 수 있습니다. 반면에 꼬리를 다리 사이로 집어넣고 귀를 눕힌다면, 그것은 아마도 개가 겁을 먹은 것입니다. 가장 중요한 것은, 이빨을 드러내고 등의 털이 곤두서 있다면 멀리 떨어져 있어야 한다는 것입니다. 이것은 개가 화가 났다는 표시입니다.

어휘
emotion 감정 particular *특정한; 특별한 moment *순간; 잠깐, 잠시 wag 흔들다 tail 꼬리 tuck 밀어 넣다, 집어넣다 flatten 납작하게 하다 stay away from …에서 멀리 떨어져 있다 fur 털 stick up 튀어나오다, 곤두서다 [문제] behavioral 행동의 response 반응, 대응

문제풀이
16 ① 개의 신체 언어
   ② 효과적인 개 훈련 방법
   ③ 일반적인 개의 문제 행동
   ④ 개의 공격에 대한 적절한 대응
   ⑤ 개와 할 수 있는 다양한 게임들

DICTATION Answers
본문 p.32

01	make their way / describe her in detail / prevent kidnapping
02	more expensive than / pick it up / take at least a few days
03	get her some flowers / fun and energetic / yellow is for friendship
04	on the ceiling / replace it / throwing out the boxes
05	rush through it / no excuse / hand it in / take our time
06	living abroad / What did you do / got homesick
07	good deal / that's bound to / for a day
08	go camping / go up the mountain / within a half-hour drive
09	results are in / a huge success / count them twice
10	Do you have a preference / get very sleepy / That's fine with me
11	chatted with her
12	good at singing
13	protesting something / get this hospital built / to make a donation
14	hand my paper in / was injured badly / took me months to recover
15	got his driver's license / feels bad / insurance will cover
16~17	read your dog's emotions / moving its various body parts / its ears are flattened

01 ④	02 ⑤	03 ②	04 ④	05 ④	06 ④
07 ②	08 ④	09 ④	10 ②	11 ①	12 ④
13 ①	14 ⑤	15 ⑤	16 ②	17 ③	

01 ④

여: 안녕하세요, 학생 여러분. 이번 주 토요일은 '거국적인 변화를 만드는 날'입니다. 2004년 이날에는, 3백만 명의 사람들이 기꺼이 지역 사회에 봉사하겠다는 마음으로 수백 개의 마을에서 수천 가지의 프로젝트에 참여했습니다. Chelsea 고등학교 학생들은 저소득 가정의 생활 환경을 개선하기 위해 Vally View의 두 가정집에 페인트칠하는 일을 이미 신청했습니다. 여러분은 또한 보육원에서 시간을 보내거나, 거리에서 쓰레기를 주울 수도 있습니다. 여러분이 관심 있는 일을 선택하셔서 지역 사회를 위해 좋은 일을 하십시오. 여러분의 도움이 정말로 변화를 만들 수 있습니다.

어휘
care about …에 마음을 쓰다 community 지역 사회 volunteer 자원봉사로 하다 take part in …에 참가하다 sign up 계약하다; *신청하다 low-income 저소득의 orphanage 보육원 interest …의 관심을 끌다

02 ⑤

여: 우리 내일 치과에 가야 해.
남: 아, 그걸 잊고 있었네. 거기 몇 시에 가야 하지?
여: 우리 예약은 11시 40분으로 잡혀 있어.
남: 그럼 11시 20분쯤에 거기 도착하게 계획해야겠네.
여: 그렇게 생각해? 그건 여유 시간이 너무 많은데.
남: 음, 우린 새로운 환자잖아. 그러니까 서류를 작성해야 할 거야.
여: 그렇네.
남: 게다가 예정보다 앞서 끝나는 환자가 있으면 일찍 들어오라고 할 거야.
여: 아, 그러면 좋겠어. 우리 점심 약속에 더 많은 시간이 남을 테니까.
남: 네 말이 맞아. 박 박사님 진료실에 제시간에 도착하려면 11시에 집에서 출발해야 해.

어휘
appointment 약속, 예약 extra 추가의, 여분의 patient 환자 fill out 작성하다 paperwork 서류 ahead of …보다 빨리

03 ②

[전화벨이 울린다.]
여: 안녕하세요, ACD 컴퓨터 서비스입니다. 무엇을 도와드릴까요?
남: 음, 제 컴퓨터가 바이러스에 감염된 것 같아요.
여: 왜 그렇게 생각하시나요?
남: 음, 켤 수는 있는데, 자판을 치면 아무 반응이 없어요.
여: 연결선들은 점검해 보셨어요?
남: 네, 모두 잘 연결된 것 같아요.
여: 전원을 켜면 모니터에 뭔가 나타나나요?

남: 그냥 바탕화면이요. 아이콘들은 다 보여요.
여: 아이콘을 클릭하시면 뭔가 작동되나요?
남: 전혀 없어요. 바로 그게 문제랍니다.
여: 알겠습니다, 매장으로 컴퓨터를 가져오셔야 할 것 같네요.

어휘
key 열쇠; *(키보드 등의) 키 connection 연결 secure 안전한; *(매듭 등이) 튼튼한 desktop 바탕화면

04 ④

여: 토니, 연극을 즐길 준비 됐어?
남: 물론이야! 두 달 동안 모차르트 보기를 기대했어. 유명한 연극이잖아.
여: 맞아. 그리고 무대장치가 아주 멋져 보여. 조명이 매우 밝아!
남: 모차르트의 집처럼 보이는데. 왼쪽에 피아노가 있어. 그 위에 뭔가 쓰여진 종이도 있고.
여: 응. 그건 모차르트가 곡을 만들고 있는 것처럼 보여. 그리고 피아노 주변에 몇 장의 종이가 흩어져 있네.
남: 그러게. 탁자 위에는 와인 몇 병도 있어. 대부분은 비어 있다.
여: 아마 그가 전부 다 마셔버린 게 틀림없어. 안타깝게도 모차르트는 말년에 가난하고 병들었거든.
남: 그렇구나. 뭔가가 오른쪽에 스탠드를 덮고 있어. 저게 뭐지?
여: 모차르트의 가발이야. 난 그가 저 가발을 볼 때 그의 전성기를 떠올린다고 생각해.
남: 오, 조명이 이제 어두워진다! 연극이 곧 시작하려고 해!

어휘
play 연극 look forward to v-ing …하기를 고대하다 set 무대 장치 compose 작곡하다 scatter 흩어지게 하다 wig 가발 remind A of B A에게 B를 상기시키다

05 ④

남: 제인, 난 자동차 정비소에 갈게요. 차에 필요한 새 자동차 앞 유리 닦개를 사야 하거든요.
여: 알겠어요. 하지만 먼저 여기 와서 이것 좀 봐요.
남: 네. 무슨 일이죠?
여: 민디의 양말이 더는 발에 맞지 않아요. 우린 그 양말을 얼마 전에 샀는데 말이에요!
남: 믿을 수가 없네요! 우리 어린 딸이 이렇게 빨리 자라고 있다니.
여: 그러게 말이에요. 당신이 외출하는 동안 좀 더 큰 양말을 사다 줄 수 있어요?
남: 물론이죠. 자동차 정비소에서 도로 바로 아래쪽에 있는 옷 가게에 갈게요.
여: 고마워요. 하지만 반드시 양말에 동물무늬가 있어야 해요. 그렇지 않으면 그 애는 양말을 신으려고 하지 않을 거예요.
남: 알아요. 우리 아기 천사는 옷에 관해서는 아주 까다롭잖아요.
여: 동감이에요. 그건 날 닮은 것 같아요.

어휘
head (특정 방향으로) 가다 auto shop 자동차 정비소 windshield wiper 자동차 앞유리 와이퍼[닦개] fit (의복 등이) …에 맞다 unbelievable 믿기지 않는 otherwise 그렇지 않으면 picky 까다로운 when it comes to …에 관해서는 I'll say. 그럼요.

06 ④

여: 안녕하세요. 길을 잃은 것 같군요. 제가 도와드릴 수 있을 것 같아요.
남: 오 안녕하세요. 민속박물관을 찾고 있는데, 어디에서도 찾을 수가 없네요.
여: 길을 잘못 드신 것 같아요.
남: 제 여행 책자에 따르면 이 지하철역 7번 출구 밖인 바로 여기 위치해야 하거든요.
여: 그래요? 책을 잠시만 봐도 될까요?
남: 물론이죠.
여: [잠시 후] 아, 알겠어요. 이 책은 몇 년 된 것이네요. 지하철은 두어 해 전에 리모델링 되어서 출구 번호가 바뀌었어요.
남: 아, 그렇군요. 그래서 알 수 없었던 거군요.
여: 이제 9번 출구로 가셔야 해요. 이 길로 직진해서 100미터쯤 가면 있어요.
남: 잘됐네요. 도와주셔서 고마워요.

어휘
take a wrong turn 길을 잘못 들다 exit 출구 remodel 개조[리모델링]하다
figure out 파악하다

07 ②

남: 안녕하세요, 손님. 주차권을 보여주시겠어요?
여: 네, 여기요. 얼마 드려야 하죠?
남: 음, 요금은 시간당 3달러인데요. 여기 네 시간 동안 계셨네요.
여: 제가 쇼핑센터에서 몇 가지를 구매했어요.
남: 영수증을 보여주시겠어요? 100달러 이상 구매하셨으면, 주차비는 무료입니다.
여: 저는 약 70달러 정도 구매했어요. 그걸로 할인받을 수 있나요?
남: 사실, 50달러 이상 구매 시에는 두 시간 동안 주차가 무료입니다. 나머지 시간에 대해서는 여전히 요금을 지불하셔야 해요.
여: 아, 알겠어요. 잠깐만요, 경차 할인이 있는데, 맞나요? 제 차가 작아서요.
남: 어디 봅시다. [잠시 후] 그러네요. 그러면 20퍼센트 할인됩니다.
여: 좋아요. 현금이 없어서 신용카드로 지불해야겠어요.
남: 문제없습니다.

어휘
parking slip 주차권 owe 빚지다 receipt 영수증 remaining 잔여의, 남아 있는 small car 경차

문제풀이
여자는 시간당 3달러인 주차장에서 4시간을 머물렀고 50달러 이상 구매해서 2시간 동안 주차가 무료이다. 남은 2시간에 대한 요금을 내야 하는데, 20퍼센트의 경차 할인을 받으므로 여자가 지불할 금액은 4.8달러($3×2×0.8)이다.

08 ④

여: 왜 네 코트 곳곳에 흰 털이 있니?
남: 사실 그건 내 개의 털이야. 부모님께서 내 생일에 나에게 개를 주셨거든. 이름을 쇼팽이라고 지었지.
여: 아, 너 항상 개를 원했잖아. 그런데 왜 쇼팽이라고 불러?
남: 내가 가장 좋아하는 음악가의 이름을 따서 짓기로 정했거든.

여: 재미있네. 그런데 개를 돌보는 게 쉽지는 않지?
남: 응, 쉽지 않아. 하루에 세 번씩 산책을 시켜줘야 해.
여: 나도 개가 있었거든. 가끔 내 침대에서 잤어.
남: 내 개는 그러기엔 너무 커. 우리 집 마당에 있는 자기 집에서 자. [잠시 후] 아! 이제 가야겠어.
여: 어디 가는 거니?
남: 반려동물용품점에 가는 거야. 쇼팽에게 줄 간식을 사야 해.
여: 같이 가도 될까?
남: 물론이지. 가자.

어휘
name 이름 짓다 take care of …을 돌보다

09 ④

남: 대학 도서관에서 이번 학기에 몇 가지 새로운 정책을 도입할 것입니다. 첫째로, 3층 전체는 이제 그룹 스터디를 위해 배정됩니다. 혼자 공부하고자 하는 학생들은 1층과 2층을 이용해야 합니다. 컴퓨터는 지하로 옮겨졌으며 더 이상 예약이 되지 않습니다. 대신 컴퓨터는 선착순으로 이용 가능합니다. 우리는 또한 시간을 연장했습니다. 지금부터 우리는 주중에는 새벽 1시에 문을 닫고 주말에는 밤 11시에 문을 닫을 것입니다. 또한, 한 번에 대출 가능한 품목의 개수가 6개에서 8개로 늘어났습니다. 그렇지만, 연체료 또한 하루에 50센트에서 1달러로 인상되었습니다. 만약 질문이 있으시다면 (답변이) 가능한 직원에게 문의하시기 바랍니다.

어휘
institute 교육 기관, 협회; *(제도 등을) 도입[시작]하다 entire 전체의 reserve 예약하다; (자리 등을) 따로 남겨두다 basement 지하층 first come, first served 선착순의 basis 근거, 이유; *기준 (단위) extend 연장하다 check out (책 등을) 대출하다 late fee 연체료

10 ②

남: 제니퍼, 너 우쿨렐레 연주하지? 우쿨렐레 고르는 것 좀 도와줘!
여: 물론이야. 소프라노, 콘서트, 그리고 테너 이렇게 세 가지 크기가 있어.
남: 저기 있는 큰 건 무슨 사이즈야?
여: 저게 테너. 그게 제일 큰 거야.
남: 아, 그래. 난 저 사이즈는 별로야. 나한테는 너무 커서 쉽게 들고 다닐 수 없어.
여: 알겠어. 다음은 소재를 골라야 해. 원목과 합판이 있어.
남: 차이가 뭐야?
여: 원목이 더 소리가 좋은데, 합판은 더 튼튼해.
남: 난 합판으로 할래. 내가 떨어뜨렸을 때 깨지는 건 싫어.
여: 좋아. 그리고 이 가게에는 새 우쿨렐레와 중고 우쿨렐레 둘 다 있어.
남: 난 중고는 싫어. 새 걸로 할래.
여: 좋아. 그리고 마지막으로, 예산이 어떻게 되니?
남: 80달러 혹은 그보다 적어야 해. 그게 내가 쓸 수 있는 전부야.
여: 그러면 넌 이걸 사야겠다. 그리고 내가 오늘 밤에 무료로 강습을 해줄게!

어휘
carry 들고 다니다 material 물질; *소재 solid wood 원목 laminate wood 합판 drop 떨어뜨리다

11 ①

여: 안녕하세요. 이 스웨터를 반품하고 싶어요.
남: 알겠습니다. 그것에 무슨 문제가 있나요?
여: 네, 여기 어깨 부근에 구멍이 있어요.
남: <u>그렇군요. 영수증 있으시죠?</u>

어휘
return 반품하다 [문제] receipt 영수증 get A to-v A에게 …하게 만들다 massage 마사지하다

문제풀이
② 그것에 대해 의사에게 진찰을 받아야 합니다.
③ 스웨터는 지금 모두 품절이에요.
④ 하지만 오늘 스웨터를 입고 싶지 않아요.
⑤ 누군가를 시켜 당신의 어깨를 마사지하게 하는 게 좋을 거예요.

12 ④

남: 제니, 어디 있었니? 한참 동안 기다리고 있었는데.
여: 늦어서 정말 미안해. 예상보다 병원에서 더 오래 걸렸어.
남: 아, 병원에는 왜 가야 했어?
여: 아빠가 가벼운 사고를 당하셨는데, 지금은 괜찮으셔.

어휘
wait for a while 한참 동안 기다리다 [문제] minor 가벼운, 작은 drop by …에 들르다

문제풀이
① 병원은 여기서 별로 멀지 않았어.
② 병원에서 널 만나서 놀랐어.
③ 네가 병원으로 선생님 병문안을 갈 거로 생각해.
⑤ 내 여동생은 내가 유스호스텔에 들러주기를 원했어.

13 ①

남: 안녕, 멜리사. 너 뭐 하고 있니?
여: 여행안내 책자를 읽고 있어, 보이지?
남: 여행 가려고?
여: 이번 9월에 잰과 마크랑 유럽 배낭여행을 갈 생각이야. 너도 같이 갈래?
남: 물론이지. 재미있겠다! 어디로 갈 건데?
여: 글쎄, 그리스에 내려서 이탈리아를 여행하고, 남프랑스를 가로질러 스페인을 둘러본 다음에, 파리에서 비행기를 타고 오는 거야.
남: 정말 굉장하다! 얼마 동안 여행할 건데?
여: 5주 정도로 생각하고 있어.
남: 비행기 표는 벌써 샀니?
여: 응. 왕복에 약 백만 원 정도였어.
남: <u>좋아. 그 정도는 여유가 될 것 같아.</u>

어휘
travel guide 여행안내서 backpack 배낭여행을 가다 come along 함께 가다 land 착륙하다 loop around 고리 모양으로 이동하다 incredible 믿어지지 않는 round-trip 왕복 여행의 [문제] afford …할 여유가 있다 regard 《pl.》 안부(의 말), 인사

문제풀이
② 알았어. 전화해 줘서 고마워.

③ 잰하고 마크한테 안부 전해 줘.
④ 글쎄, 그건 전적으로 네 예산에 달렸어.
⑤ 여행 경비가 정말 비쌌구나, 그렇지?

14 ⑤

남: 지나야, 뭐 하고 있니? 음악 소리가 들렸는데.
여: 안녕 아빠. 새로운 곡을 피아노로 혼자 연습하는 중이에요.
남: 오. 난 네가 오늘 너의 방을 페인트칠할 줄 알았는데. 넌 몇 주 동안 그 일에 관해서 얘기했잖아.
여: 음, 전 오늘 아침에 페인트칠을 시작했는데, 벽 두 개만 끝냈어요. 나머지는 아마 내일 할 것 같아요.
남: 내일? 왜 오늘 하지 않고? 할머니를 위해서 네가 뜨려고 했던 스웨터는 기억나니? 지난달에 시작했는데, 넌 아직도 끝내지 못하고 있잖아.
여: 죄송해요, 아빠. 저한테 화내지 마세요. 단지 제가 하고 싶은 게 너무 많을 뿐이에요.
남: 화난 게 아니야. 네 관심사가 많은 건 좋다고 생각한단다. 하지만, 네가 시작한 일을 끝내는 게 중요해.
여: 아빠 말씀이 옳아요. 지금 제 방 페인트칠을 끝낼게요. 이 곡은 나중에도 배울 수 있어요.
남: 좋은 생각이구나.
여: <u>감사해요. 걱정과 조언에 정말 감사드려요.</u>

어휘
teach oneself 독학[자습]하다 knit (실로 옷 등을) 뜨다, 짜다 interest 관심, 흥미 [문제] on time 시간을 어기지 않고, 제시간에 appreciate 고마워하다 concern 우려, 걱정 advice 조언 Never mind. 걱정하지 마.

문제풀이
① 저도 마찬가지예요. 제 방은 페인트칠할 필요가 없어요.
② 걱정하지 마세요. 저는 새로 할 일을 찾아야 해요.
③ 좋아요! 지금 제가 그 노래를 연주하는 걸 듣고 싶으세요?
④ 걱정하지 마세요. 모든 게 제시간에 끝나도록 도와드릴게요.

15 ⑤

남: 2주 동안 블레이크는 반 친구 중의 한 명인 제인과 그룹 과제를 해 오고 있었다. 그는 혼자서만 과제를 했기 때문에 짜증이 났다. 마감일 이틀 전이었는데, 그가 혼자 과제를 하는 동안 제인은 피자를 먹으면서 TV를 보고 있었다. 그 과제는 미완성인 채로 바닥에 있었다. 그는 수집된 사진들과 만들어진 모형을 쳐다보았다. 어떻게 모레까지 다 끝낼 것인가? 그는 시끄러운 방에서 연구 보고서에 집중하려고 노력했다. 갑자기 그는 그녀에게 화가 났고 이에 대응하기로 했다. 이런 상황에서, 블레이크가 제인에게 할 말로 가장 적절한 것은 무엇인가?
블레이크: <u>넌 아무것도 하지 않는데 내가 모든 일을 하는 건 불공평해.</u>

어휘
annoyed 짜증이 난 due date 마감일 unfinished 미완성인 concentrate on …에 집중하다 research 연구, 조사 confront 맞서다 [문제] fair 공평한

문제풀이
① 나도 피자 좀 먹어도 될까?
② 도와줘서 고마워! 내일 보자!
③ 제발 조용히 얘기해줄래? 집중할 수가 없잖아.
④ 이 부분을 어떻게 하는지 확실히 모르겠어. 네가 좀 보여줄래?

16 ② 17 ③

여: 저는 채널 9 뉴스의 새라 킴입니다. 요즘, 이 지역의 많은 부모님이 유명한 패스트푸드 음식점인 Smile Burger로 아이들을 데리고 갑니다. 대부분 아이들이 거기서 파는 햄버거와 감자튀김을 즐겨 먹지만, 가장 큰 매력적인 요소는 Fun Food 세트입니다. 아이들이 이 세트를 주문하면 식사와 함께 재미있는 장난감을 받습니다. 그런데, 제공되고 있는 새 장난감 중 하나가 5세 미만의 아이에게 안전하지 않다는 사실이 최근에 주목을 끌게 되었습니다. 그것은 떼어낼 수 있는 작은 부속품이 달린, 말하는 기차인데요. 어제 이 지역의 세 살배기 아이가 이 부속품 중 하나를 입에 넣고 잘못해서 삼켰습니다. 이는 심각한 호흡 장애를 일으켰고 시립 병원에서 치료를 받아야만 했습니다. Smile Burger의 매니저에 따르면, 이 장난감은 더는 제공되지 않을 것이라고 합니다. 이 장난감이 포함된 Fun 세트를 이미 구매하셨다면 아이에게서 당장 그것을 치우셔야 합니다. 의문 사항이 있으시면 Smile Burger의 고객 서비스 센터로 전화하시면 됩니다.

어휘 remove 치우다, 없애다 accidentally 우연히, 잘못하여 swallow 삼키다 severe 심각한 breathing 호흡 take away 제거하다, 치우다

DICTATION Answers
본문 p.38

01 taking part in / improve their living conditions / pick up trash off

02 scheduled for / fill out paperwork / ahead of schedule

03 has a virus / checked the connections / click on an icon

04 looking forward to seeing / some bottles of wine / reminds him of better days

05 heading off / don't fit her feet / get some larger ones / very picky

06 took a wrong turn / was remodeled / straight down this road

07 How much do I owe you / get any discount / I don't have any cash

08 name him after / take him for a walk / slept in my bed

09 reserved for group study / extended our hours / late fees

10 easily carry it / What's the difference / if I drop it

11 return this sweater

12 It took longer

13 thinking of backpacking / travel up / That sounds incredible / round-trip

14 teaching myself / still haven't finished / finish what you start

15 felt annoyed / before the due date / concentrate on / confront her

16~17 the biggest attraction / under the age of five / caused severe breathing difficulties / containing this toy

---

07 영어듣기 모의고사
본문 ▲ p.42

01 ③	02 ②	03 ②	04 ⑤	05 ①	06 ⑤
07 ②	08 ①	09 ④	10 ④	11 ②	12 ④
13 ③	14 ③	15 ③	16 ②	17 ⑤	

01 ③

남: 안녕하십니까, 여러분. 이번 달 주민 회의를 시작하기에 앞서, 잠시 여러분께 건물 관리소에서 우리를 위해 온라인상에 특별한 공간을 마련했음을 상기시켜 드리고자 합니다. 그 사이트 주소는 제가 방금 배부한 카드에 쓰여 있습니다. 그곳은 건물에 관련된 최근 공지사항과 정보를 찾아보는 데 최적의 공간입니다. 또한, 여러분이 이웃들과 연락할 수 있는 토론회도 있습니다. 사람들은 이미 중고 물품을 팔거나 교환하려고 광고하기 위해서뿐만 아니라, 개인 소식이나 사진을 올리기 위해서 그것을 사용하고 있습니다. 그러므로 한번 둘러보시고, 만약 질문이 있다면 저에게 알려 주시기 바랍니다.

어휘 remind 상기시키다 set up 세우다 hand out …을 나눠주다 up-to-the-minute 최신 정보를 가진 notice 공지사항 concerning …에 관한 forum 포럼, 토론회 post 게시하다 trade 거래하다

02 ②

남: 남미 여행은 어땠어?
여: 즐겁게 보냈어! 몇 주 동안 머물러서 지역을 탐험할 시간이 많았어.
남: 흥미로운 음식은 먹었니?
여: 물론이지! 특별한 양념이 된 새로운 요리를 몇 가지 먹어봤어.
남: 해외여행을 한 게 이번이 처음이었니?
여: 아니. 난 가능한 한 자주 해외여행을 하려고 해. 할 때마다 새로운 걸 배우거든.
남: 그게 네 세계관을 정말로 변화시키는 것 같구나.
여: 그래. 다른 문화를 체험하는 것은 모든 것에 관한 시각을 넓혀주는 것 같아.
남: 그렇다면, 나도 여행을 계획해야겠어.

어휘 dish 음식 season 계절; *양념하다 spice 양념, 향신료 overseas 해외로 broaden 넓히다 perspective 관점, 시각

03 ②

여: 안녕하세요. 도움이 좀 필요한데요.
남: 물론이야. 뭘 도와줄까?
여: 반납할 물건이 세 가지 있어요. 유감스럽게도 4일이나 늦었네요.
남: 그래. 연체료를 내야 한다는 점을 이해해주길 바란다.
여: 알고 있어요. 책 한 권당 하루에 50센트라고 들었어요.
남: 안타깝지만 그렇단다. 그럼 총 연체료는 6달러라는 얘기지.
여: 근데 전 현금이 그렇게 많지 않아요. 그리고 전 수업을 위해 이 DVD를 대출해야 해요.

남: 음, 연체료를 지불하기 전까지는 새로운 품목을 대출할 수 없어. 직불 카드를 갖고 있니?

여: 네, 있어요. 여기서 사용할 수 있나요?

남: 물론이지. 그 카드를 이 기계에 통과시키고 비밀번호만 누르면 돼.

여: 와, 정말 편리하네요. 감사합니다.

어휘

late fee 연체료 check out (책 등을) 대출하다 type 입력하다 passcode 암호

04 ⑤

여: 피터, 이 사진 좀 봐. 이건 새 동물원에서 온 거야.

남: 오! 그 동물원 이제 막 개장했지?

여: 응, 그리고 거긴 많은 시설과 전시물들이 있어.

남: 왼쪽에 있는 저 피크닉 구역이 멋져 보인다. 파라솔, 탁자, 그리고 나무가 있네.

여: 동의해. 먹고 쉬기에 완벽한 장소일 거야.

남: 가운데에 있는 폭포랑 호수도 좋아 보인다. 호수에 동물도 있을까?

여: 응, 하마 두 마리가 그 안에서 수영을 하고 있어. 그것들은 정말 행복해 보인다.

남: 그럴 만도 하지. 그리고 오른쪽에 재미있는 팝콘 판매대를 봐. 거대한 팝콘 상자 같아 보여.

여: 기발하다! 저 학생 두 명은 뭘 하는 거지?

남: 그들은 울타리 안으로 토끼를 쓰다듬고 있네. 아마 아이들이 동물을 만질 수 있는 동물원에 있는 게 틀림없어.

여: 그 동물원은 방문하기에 정말 멋진 곳처럼 보여.

남: 응, 언젠가 곧 가야겠다.

여: 좋아!

어휘

facility (pl.) 시설 exhibit 전시(品) waterfall 폭포 hippo 하마 I don't blame you. 그럴만도 해. stand 가판대 clever 영리한 pet 쓰다듬다 petting zoo 동물을 만질 수 있는 동물원 terrific 정말 멋진

05 ①

[전화벨이 울린다.]

여: 여보세요?

남: 티나 패터슨 씨인가요?

여: 네, 제가 티나인데요.

남: 안녕하세요. 제가 언젠가 공원에서 지갑을 하나 찾았는데요, 그게 당신 것 같아서요.

여: 오, 제 지갑을 찾으셨다고요! 근데 제 전화번호를 어떻게 아셨죠?

남: 그 안에 당신 전화번호가 적힌 오래된 휴대전화 요금 청구서가 있었어요.

여: 그거 다행이네요. 그 지갑 안에 제가 잃어버리고 싶지 않았던 귀중한 사진이 몇 장 들어 있거든요.

남: 그럼 제 사무실로 오셔서 언제라도 찾아가시면 돼요.

여: 안타깝게도 전 지금 출장 중이에요.

남: 그러시다면, 저한테 주소를 알려주시고 제가 오늘 오후에 그걸 우편으로 보낼게요.

여: 그러면 정말 좋죠. 매우 친절하시네요.

남: 도움이 되었다니 그저 기쁠 따름이에요.

어휘

purse 지갑 belong to …의 소유이다 precious 귀중한 come by 잠깐 들르다

06 ⑤

여: 이봐, 저스틴. 나랑 영화 보러 갈래?

남: 미안하지만, 해야 할 숙제가 많아.

여: 그래? 우리가 해야 할 과제가 있어?

남: 기억 안 나? 우린 내일 아침까지인 과학 과제가 있잖아.

여: 정말? 그게 뭐였지?

남: 책 보고서잖아. 우린 생태계에 관한 책을 읽어야 해.

여: 아, 이제 기억나! 까맣게 잊고 있었어. 어떡하지?

남: 음… 당장 그걸 시작하는 게 좋을 거야. 읽을 페이지가 많거든.

여: 알려줘서 고마워. 그렇지 않았다면 내일 정말 곤란해졌을 거야.

남: 천만에.

어휘

assignment 과제 due 기한인 ecosystem 생태계 completely 완전히 otherwise 그렇지 않으면

07 ②

남: 린다, 추천해줄 만한 적당한 아르바이트 자리 있니?

여: 너 일자리를 찾고 있니? 돈이 왜 필요한데?

남: 올여름에 스페인에서 연구 강의를 수강할 거거든.

여: 아, 나도 지난해에 그런 강의 중 하나를 들었어. 약 2,000달러의 비용이 들지, 맞지?

남: 사실, 전체 비용은 2,500달러야. 근데 난 학교를 통해서 예약했기 때문에, 10퍼센트 할인을 받을 수 있어.

여: 그렇다면 급여가 꽤 많은 일이 필요하겠구나.

남: 음, 실은 부모님이 1,000달러를 내주실 거야. 난 그냥 나머지만 부담하면 돼.

여: 그런 상황이라면, 카페에서 서빙하는 일자리를 구해 봐. 한 달에 300달러 정도 벌 수 있어.

남: 그거 정말 딱이다. 조언해줘서 고마워.

여: 별거 아니야. 일자리 찾는 데 행운을 빌어.

어휘

suitable 적합한 book 예약하다 cover (돈을) 대다 rest 나머지 wait tables 식당에서 서빙하다

문제풀이

원래 2,500달러인 학비는 10퍼센트 할인 후 2,250달러인데, 부모님께서 1,000달러를 내주실 것이므로 남자가 부담할 액수는 1,250달러이다.

08 ①

남: 우즈 씨, 이 일에 지원해 주셔서 고맙습니다. 이력서가 매우 인상 깊었어요.

여: 그렇게 말씀해 주시니 기뻐요.

남: 6월 15일 10시에 와서 컴퓨터 능력 시험에 응시하길 바랍니다.

여: 알겠어요. 시험 시간은 얼마나 걸리나요?

남: 음, 4개의 다른 부문이 있는데요. 대부분 사람들은 약 2시간 후에 끝냅니다.

여: 몇 명이 그 시험을 보나요?

남: 다른 30명이 있을 거예요. 여러분 중에서 5명이 이 일자리에 뽑히게 될 거예요.

여: 알겠습니다. 그럼 시험을 보러 이곳에 다시 와야 하나요?

남: 실은 4층에 있는 회의실에서 시험에 응시하게 될 거예요.

여: 좋아요. 결과는 언제 발표되죠?

남: 시험 후 일주일 뒤에 결과를 발표할 예정입니다.

여: 알려주셔서 감사합니다.

어휘

apply for …에 지원하다 impressed 인상 깊게 생각하는 résumé 이력서 select 선택[선발]하다 conference room 회의실 announce 발표[공지]하다

09 ④

여: 이 메시지는 전 직원에게 해외 영업부에 곧 생길 공석에 대해서 공지하려는 것입니다. 앤 비즐리의 퇴직으로, 우리는 일본 시장의 새로운 영업 담당자를 구하게 될 것입니다. 자격을 갖춘 모든 지원자들은 학사 학위와 자사 혹은 타사에서 최소한 2년간의 영업 경력이 있어야 합니다. 지원자들은 또한 일상 대화 수준으로 일본어를 구사할 수 있어야 합니다. 이 직위의 급여는 경력에 따라 연간 6만 달러에서 7만 5천 달러입니다. 현 직원들이 이 직위에 우선 고려대상이지만, 만약 적합한 사람이 구해지지 않는다면 우리는 3월 1일부터 외부에서 지원을 받을 것입니다.

어휘

notify A of B A에게 B를 알리다 opening 구멍; *공석 retirement 은퇴, 퇴직 seek 찾다 coordinator 조정자, 책임자 qualified 자격 있는 applicant 지원자 degree 정도; *학위 conversational (일상) 대화의 salary 급여 consideration 고려

10 ④

남: 안녕하세요. 무엇을 도와드릴까요?

여: 안녕하세요. 무선 스피커 좀 보여주실래요?

남: 네, 물론입니다. 다섯 가지 모델이 있고, 특징에 따라 가격이 다릅니다.

여: 그렇군요. 저는 250달러 이상은 쓸 수 없을 것 같아요.

남: 그렇다면 이것들이 손님의 예산에 맞겠네요.

여: 각각 무게가 얼마나 나가나요? 저는 그걸 많이 들고 다닐 예정이라서, 가급적이면 1킬로그램 미만의 가벼운 것이 필요해요.

남: 알겠습니다. 여기 1킬로그램 미만의 모델들입니다.

여: 충전하는데 얼마나 걸리는지도 궁금해요. 1시간 이내로 스피커가 완전히 충전되면 좋겠어요.

남: 손님의 예산에 맞고 급속 충전되는 두 가지 모델이 있습니다. 이 타원형 모양의 제품은 어떠세요?

여: 잘 모르겠어요. 정육면체가 나을 것 같네요. 무척 깔끔해 보이네요.

남: 훌륭한 선택이에요! 현금으로 결제하시겠어요, 아니면 신용카드로 하시겠어요?

여: 현금으로 낼게요.

어휘

wireless 무선의 feature 특징 preferably 오히려, 가급적이면 charge 충전하다 oval 타원형(의) cube 정육면체 neat 단정한, 깔끔한

11 ②

남: 애나, 이 웹사이트는 무엇에 관한 거니?

여: 자선 기부를 위한 홈페이지야. 편지를 보내서 다른 나라의 가난한 아이들을 도울 수 있어.

남: 훌륭한데! 처음에 어떻게 그것에 대해 알게 됐어?

여: 친구 중 한 명이 나한테 추천했어.

어휘

charitable 자선의 donation 기부 [문제] sign up for …에 등록하다 volunteer 자원봉사하다

문제풀이

① 이 사이트에서 많은 활동을 할 수 있어.

③ 우선, 이 프로그램에 등록해야 해.

④ 그들에게 편지를 쓰면 기분이 좋아져.

⑤ 내 친구들은 다른 나라로 자원봉사하러 갔어.

12 ④

여: 알렉스, 오늘 우리 축구를 할 수 있을 것 같니?

남: 아니, 밖엔 비가 오고 있어. 젖은 잔디 위에서는 하기 힘들지.

여: 맞아. 그럼 언제 경기를 해야 할까?

남: 모르겠네. 비가 그칠 때까지 기다려야겠지.

어휘

grass 풀, 잔디 match 경기 [문제] take place 개최되다 wash up 세수하다

문제풀이

① 다음에는 최선을 다할게.

② 그 경기는 보기에 지루했어.

③ 그 경기는 그 경기장에서 열릴 거야.

⑤ 우리 이제 세수하고 시합을 준비해야겠어.

13 ③

여: 안녕하세요. 제가 여기서 이 치마를 샀는데, 환불하고 싶어요.

남: 알겠습니다. 그 치마에 무슨 문제라도 있나요?

여: 그냥 저한테 어울리지 않아서요.

남: 알겠습니다. 영수증이 있으시면, 전액 환불받으시거나 다른 품목으로 교환하실 수 있습니다.

여: 여기 있어요. 그리고 전 그냥 환불받을래요.

남: 그러세요. [잠시 후] 아, 잠시만요. 그 치마를 3주 전에 사셨네요?

여: 네, 내일이면 3주가 될 거예요.

남: 죄송합니다만, 저희 환불 정책은 구매 후 15일간만 유효하답니다.

여: 아, 그래요? 예외로 해주시면 안 될까요?

남: 죄송하지만, 그렇게 해 드릴 수는 없습니다. 그렇지만, 저희 교환 정책은 6주간 유효합니다.

여: 그럼 그걸 교환할 수밖에 없겠네요.

어휘

suit …에게 맞다[어울리다] receipt 영수증 refund 환불 exchange 교환하다 policy 정책 extend 연장하다; *포괄하다 exception 예외 [문제] be[get] stuck with (싫은 일을) 강요당하다

문제풀이

① 그럼 저보고 이 이상한 치마를 계속 입으라고요?

② 그걸 알게 돼서 좋긴 한데, 그건 당신 잘못이에요.

④ 그럼 영수증을 갖고 다시 올게요.
⑤ 이 경우를 예외로 처리해주셔서 감사합니다.

14 ③

여: 안녕, 제이콥. 어떻게 지내요?
남: 나쁘지 않아요. 지금 막 월간 재무보고서를 마쳤어요.
여: 잘됐네요. 있죠, 우리 집에서 송년회를 할 거라는 사실을 알려 주고
   싶었어요.
남: 아, 정말요? 그거 멋지네요. 언제죠?
여: 12월 20일 토요일 오후 2시에 시작할 거예요.
남: 20일이요? 아, 미안하지만, 전 참석하지 못할 거예요.
여: 오, 그거 정말 안됐네요. 부서 사람들 전부 거기에 올 텐데요.
남: 저도 가고 싶지만, 그날 컴퓨터 연수가 있어요.
여: 아. 잠깐만요, 새로운 급여 지급 시스템을 위한 연수 말이에요?
남: 네, 맞아요.
여: 그건 다음 달로 연기되었다고 들었어요.
남: 그렇다면 어쨌든 저도 갈 수 있겠네요.

어휘

financial report 재무보고서 year-end party 송년회 department 부서
payroll 임금 대장, 급여 지급 (총액) postpone 연기하다

문제풀이

① 안타깝지만, 저도 갈 수 없어요.
② 네, 그 말을 듣게 되어 유감이네요.
④ 그럼 워크숍에서 만나요!
⑤ 하지만 그때쯤이면 휴일이 다 끝날 텐데요.

15 ③

여: 대니얼은 고등학교 학생이다. 다음 주에 중요한 과학 시험이 있기 때
   문에, 그는 주말 내내 그 시험공부를 했다. 월요일 아침, 대니얼은 몹
   시 피곤하지만 시험에 대해서 자신이 있다. 하지만 그는 수학 숙제하
   는 것을 잊었다는 사실을 갑자기 깨닫는다. 수학 시간에 자리에 앉아
   서 선생님이 도착하시길 기다리는 동안, 그는 그럴듯한 변명을 생각
   해내려고 필사적으로 애쓴다. 아무런 생각이 나질 않자, 친구인 줄리
   에게 돌아서서 문제를 설명한다. 줄리는 대니얼에게 선생님이 편찮으
   셔서 오늘 수업에 오지 않으실 거라고 말해준다. 이런 상황에서, 대니
   얼이 줄리에게 할 말로 가장 적절한 것은 무엇인가?
대니얼: 이번에는 내가 운이 좋았던 것 같아.

어휘

exhausted 지친 confident 자신감 있는 desperately 필사적으로 come
up with …을 생각해 내다 excuse 변명 [문제] cheating 컨닝, 부정행위

문제풀이

① 네가 얼른 나으면 좋겠다.
② 부정행위는 결코 좋은 생각이 아니야.
④ 그게 내가 걱정했던 거야.
⑤ 내가 공부했던 게 헛된 일이라는 뜻이니?

16 ② 17 ⑤

남: 안녕하세요, 교사 여러분. 오늘 세미나에서 저는 여러분의 학생들에
   관하여 이야기하고 싶습니다. 여러분들은 아마 이미 그들이 모두 다

르다는 것을 알아차리셨을 것입니다. 하지만 이 차이점들은 그들의
외모나 성격 이상입니다. 각 학생은 독특한 방식으로 학습합니다. 그
들 중 일부는 그들의 시각에 의존합니다. 그들은 도표나 사진 같은 그
래픽을 빨리 이해할 수 있습니다. 다른 학생들은 그들의 청각을 사용
하는 것을 선호합니다. 그들은 내용을 들을 때 더 잘 학습합니다. 짧
고 간단한 노래는 이 유형의 학습자들을 가르치기 위한 좋은 기술입
니다. 물론, 읽기와 쓰기를 통해 가장 잘 학습하는 몇몇 아이들도 있
습니다. 그들에게는 답을 작성하도록 하는 많은 퀴즈가 주어져야 합
니다. 그리고 몸을 쓰는 것이 필요한 아이들도 있습니다. 이들이 더 효
율적으로 학습하도록 돕기 위해서는 그들에게 교실 주변을 돌아다니
도록 많은 기회를 주십시오. 역할극은 이러한 학생들이 학습할 수 있
는 재미있고 활동적인 방식입니다. 따라서 여러분이 훌륭한 교사가
되고 싶다면, 여러분들의 학생을 아셔야만 합니다.

어휘

notice 주목하다; *알아채다 go beyond …을 넘어서다 appearance (겉)모
습, 외모 rely on …에 의지하다 material 자료 chant 찬트(단순하고 반복적인
노래) technique 기법, 기술 plenty of 많은

문제풀이

16 ① 가장 효율적인 학습법
   ② 학생들의 각각 다른 학습 형태들
   ③ 학생들의 나쁜 행동 다루기
   ④ 학생들이 가장 좋아하는 교사가 되는 법
   ⑤ 교실에서 흔히 저지르는 교사들의 실수

DICTATION Answers

본문 p.44

01 a special place online / notices and information / advertise
    used items

02 explore the area / traveling overseas / broadens my
    perspective

03 three items to return / pay a late fee / check out

04 to eat and relax / I don't blame them / a terrific place to
    visit

05 found a purse / pick it up anytime / drop it in the mail

06 a lot of homework / due tomorrow morning / would have
    been in big trouble

07 a suitable part-time job / the total cost / cover the rest

08 impressed by your résumé / Five of you / on the fourth
    floor

09 a new sales coordinator / two years of sales experience /
    depending on experience

10 differ according to features / within your budget / that
    charge quickly

11 help poor children

12 play on wet grass

13 doesn't suit me / extends for 15 days / make an exception

14 a year-end party / a computer training workshop / it was
    postponed

15 spent all weekend studying / do his math homework /
    come up with

08 영어듣기 모의고사
본문 ▲ p.48

01 ①	02 ①	03 ①	04 ③	05 ③	06 ④
07 ③	08 ③	09 ②	10 ④	11 ③	12 ①
13 ①	14 ⑤	15 ④	16 ②	17 ⑤	

01 ①

여: 안녕하세요, 저는 Delstar 상사의 로라 블레인입니다. 오늘 귀하의 사무실이 이미 문을 닫은 것 같네요. 제가 어제 아침 귀사에 주문한 레이저 프린터 용지 12상자 2세트와 관련하여 전화 드렸습니다. 제가 종이의 수량을 잘못 파악했고 비품실에 저희가 생각했던 것보다 더 많은 종이를 보유한 것으로 확인되었습니다. 주문한 게 아직 배송되지 않았으면 12상자 2세트를 1세트로 줄여주실 수 있나요? 이 요청에 문제가 있으면 아침에 제일 먼저 제게 전화해주세요. 제 번호는 862-8088에 내선 43번입니다. 감사합니다.

어휘
in reference to ⋯에 관하여 carton 큰 상자 place an order 주문하다 turn out ⋯인 것으로 드러나다 ship 실어 나르다, 운송하다 reduce 줄이다, 축소하다 request 요청 extension 확장; *구내전화, 내선

02 ①

남: 와, 또 공과금이 많이 나왔어! 내가 매달 에너지에 얼마나 많은 돈을 쓰는지 믿기질 않는군.
여: 그걸 낮추기 위해 네가 할 수 있는 일이 많아.
남: 예를 들면?
여: 우선, 전기 제품을 더 효율적인 것으로 바꾸면 돼.
남: 지금 당장은 새 전기 제품을 살 여유가 없는 것 같아.
여: 알겠어, 그럼 에어컨을 켜는 대신에 선풍기를 이용하는 건 어때? 선풍기는 전기를 훨씬 더 적게 쓰지.
남: 그건 해볼 수 있을 거야.
여: 또, 밖이 정말 더울 때는 커튼을 닫아 둬. 비쳐 들어오는 햇빛은 방을 더 덥게 만들 뿐이야.
남: 그건 전에 생각해 본 적이 없어. 그게 정말로 에너지 비용을 아끼는 데 도움이 된다고 생각해?
여: 틀림없이. 때론 작은 변화가 큰 차이를 만들 수 있어.

어휘
utility bill 공과금 replace 대체하다 appliance 전기 제품 efficient 효율적인 afford to-v ⋯할 여유가 있다 fan 선풍기 definitely 틀림없이

03 ①

남: 지난주에 이 아파트를 봤어요. 꽤 괜찮아요.
여: 넓어 보이네요.
남: 나도 그렇게 생각했어요. 그리고 옷장들 좀 봐요, 정말 커요!
여: 잘됐네요. 우린 옷이 많잖아요.
남: 맞아요. 우리에게 그 공간은 정말 필요할 거예요.
여: 그렇죠. 특히 태어날 아기랑 있으려면요.
남: 중개인 말이 세탁기가 거의 새 거래요.
여: 그건 중요하죠.
남: 그리고 우리가 이곳으로 이사하면 당신 부모님과 가까이 살게 돼요. 우린 그분들을 더 자주 볼 수 있는 거죠.
여: 네, 그게 또 다른 장점이네요. 이 집이 우리한테 좋을 것 같아요.
남: 좋아요! 내가 중개인에게 말할게요. 그녀가 아래층에서 기다리고 있어요.
여: 잠시만요. 생각해 볼 시간을 좀 가져요.
남: 알겠어요. 그녀에게 우리가 내일 전화하겠다고 말할게요.

어휘
closet 벽장, 옷장 on the way 가는 중인; *(아기가) 아직 태어나지 않은 agent 중개인 brand-new 신상품의, 아주 새로운 advantage 장점, 이점

04 ③

남: 마당 세일을 했어. 자, 그 사진 중 몇 장을 좀 보여줄게.
여: 좋아. 아, 바닥에 있는 거울이 낯이 익어.
남: 사각형 모양의 것? 우리 거실에 있었던 거야.
여: 맞아. 아, 아기를 데리고 온 여자가 저 원피스를 좋아하는 것 같은데. 그녀가 그걸 샀니?
남: 응, 그녀의 딸에게 리본이 달린 저 작은 원피스가 잘 어울릴 거야.
여: 아, 옷걸이에 있는 옷깃이 있는 셔츠는 네 아들 것이니?
남: 응. 지금 그 애는 저 옷을 입기엔 너무 커버렸지.
여: 아, 2달러짜리 배낭이 있네. 가격 괜찮다.
남: 맞아. 그건 빨리 팔렸어.
여: 그렇구나. 그래서 내놓은 걸 모두 팔았니?
남: 탁자 아래의 꽃무늬 상자 안에 있는 것만 빼고 전부 다 팔았어.

어휘
yard sale 마당 세일(개인 주택의 마당에서 사용하던 물건을 파는 것) ground 바닥, 땅 collared 옷깃이 달린 hanger 옷걸이 put out 내놓다

05 ③

여: 와, 너희들 정말 바쁜 것 같구나.
남: 응. 우리는 지역 독립 영화제를 준비하고 있어.
여: 너희들 정말 그걸 할 거니? 난 그냥 생각만 하는 줄 알았어.
남: 음, 그냥 생각했던 건데, 현실이 되고 있어. 5월 첫째 주에 할 거야.
여: 정말 멋지다. 자원봉사자가 좀 필요하니?
남: 물론 필요하지. 네가 우리를 좀 도와주겠니?
여: 물론이야. 내가 무엇을 할 수 있는지만 말해줘.
남: 생각해볼게. [잠시 후] 얘! 너 학교 축제 때 그 셔츠들 디자인 했었지, 그렇지?
여: 응, 그랬지. 그것들은 꽤 인기가 있었어.
남: 맞아. 음… 우리 영화제를 위해서도 비슷한 일을 할 수 있을 것 같니?
여: 하고 싶어. 그게 표를 팔거나 의자를 놓는 것보다 확실히 낫겠지.

정답 및 해설 • 27

남: 잘됐다! 도와줘서 정말 고마워.

어휘

independent film 독립 영화 reality 현실 (상황) volunteer 자원봉사를 하다; *자원봉사자 fair 공평한; *축제, 박람회 beat 치다; *능가하다 [문제] recruit (사원·회원을) 모집하다

문제풀이

① 의자 설치하기 ② 자원봉사자 모집하기
③ 셔츠 디자인하기 ④ 축제 표 판매하기
⑤ 영화제 참석하기

06 ④

남: 나는 이번 주말이 기대돼!
여: 왜? 이번 주말에 무슨 일이 있는데?
남: 친구들하고 돌고래 관광을 하러 갈 거야.
여: 와! 부럽다. 나는 작년 여름에 가족과 돌고래 관광을 갔지만, 우리는 돌고래를 전혀 보지 못했어.
남: 부러워할 필요 없어. 너도 이번 주말에 우리와 함께 가면 돼.
여: 그러고 싶지만 너희와 함께 갈 수 없어.
남: 왜? 부모님이 허락을 안 해 주실까?
여: 우리 부모님은 개의치 않으실 거지만 내가 시간이 없어.
남: 다음 달에 웅변대회에 참가하는 것 때문에 그러니?
여: 아니, 하지만 학교 연극이 다음 주에 있어. 그래서 이번 주말이 예행연습을 할 마지막 기회가 될 거야.
남: 오, 알겠어. 음, 너를 위해서 사진을 좀 찍어 올게!
여: 좋았어. 네가 돌고래를 좀 볼 수 있기를 바라. 여행 즐겁게 해!

어휘

jealous 질투하는 permission 허락 rehearse 예행연습을 하다

07 ③

남: 안녕하세요. 어머니께 드릴 꽃을 좀 주문하려고요. 장미가 얼마죠?
여: 12송이에 20달러입니다.
남: 6송이면 10달러인가요?
여: 아뇨, 6송이는 12달러입니다.
남: 네, 그거 괜찮네요. 그럼 장미 6송이 주세요, 빨간 장미로요.
여: 알겠습니다. 다른 것은요?
남: 카네이션 12송이요… 아뇨, 그것도 6송이요.
여: 알겠습니다. 카네이션은 10달러입니다.
남: 거기에 또 뭐가 어울릴까요?
여: 안개꽃이요, 작은 흰 꽃 종류예요. 아주 예쁘고, 2달러만 더 내시면 됩니다.
남: 음, 아니에요. 그냥 장미를 두 송이 더할래요.
여: 장미를 추가하시면 한 송이에 2달러씩이에요. 하지만 전 안개꽃을 정말 추천합니다. 장미와 카네이션을 더 돋보이게 해주거든요.
남: 알겠어요, 그럼 추가하려던 장미는 빼고 안개꽃으로 할게요.
여: 좋은 선택입니다!

어휘

dozen 12개 go with …에 어울리다 baby's breath 안개꽃 throw in 곁들이다, 끼워 넣다 complement 보완하다

문제풀이

남자는 장미 6송이($12)와 카네이션 6송이($10), 그리고 안개꽃($2)을 구입했으므로, 남자가 지불할 금액은 24달러이다.

08 ③

남: 안녕, 앨리슨! 어떻게 지냈니?
여: 안녕, 마이크! 난 내 가게를 열었어.
남: 잘됐다. 어떤 가게야?
여: 장신구 점이야. 손으로 만든 목걸이와 팔찌, 반지, 헤어 액세서리를 팔아.
남: 정말? 네가 직접 만들어?
여: 아니, 내가 아는 지역 예술가들에게 사는 거야.
남: 언제 한번 가서 보고 싶다.
여: 잘됐네. 가게는 5번가 Townville Mall에 있어. 할인 행사가 있으니 수요일에 들러 줘.
남: 그래야겠다. 할인을 많이 해줄 거니?
여: 그럼! 오전 9시부터 오후 1시까지 매장의 모든 목걸이와 반지를 20퍼센트 할인할 거야.
남: 꼭 일찍 가야겠네. 그리고 너랑 같이 점심을 먹어도 되고.
여: 그게 좋겠네. 곧 또 보자.

어휘

jewelry 보석, 장신구 handmade 손으로 만든 necklace 목걸이 bracelet 팔찌

09 ②

남: 안녕하세요, 여러분. 수업에 오신 것을 환영합니다. 물론 이것은 컴퓨터 입문 과정이고, 우리는 일주일에 두 번씩 저녁 8시에 만나게 될 것입니다. 여러분 모두 이 수업을 위해 교재를 구매하셔야 합니다. 그것은 여러분의 강의 계획표에 나와 있고 대부분 서점에서 구매하실 수 있습니다. 자, 이 수업은 컴퓨터의 업무적인 측면에 초점을 맞출 것입니다. 이 수업은 직장으로 돌아가 컴퓨터에 대한 기초 지식을 증진할 필요가 있는 성인들을 위해 특별히 만들어졌습니다. 여러분이 바쁘신 것을 알기 때문에 과제는 많지 않을 것입니다. 하지만 전 여러분이 수업 시간에 배운 것을 연습하시길 바랍니다. 만약 여러분에게 개인 컴퓨터가 없다면, 저희 컴퓨터실에 있는 것을 사용하실 수 있습니다.

어휘

purchase 구매하다 textbook 교과서 syllabus 요강, 강의 계획표 specifically 분명하게, 명확히

10 ④

남: 안녕하세요. 오늘은 무엇을 찾으시나요?
여: 예전 것을 대체할 새 스마트폰을 사려고요.
남: 알겠습니다. 용량이 얼마나 필요하세요?
여: 제가 지금 쓰는 전화기가 16기가바이트인데요, 충분치가 않아요.
남: 알겠습니다. 카메라는 어떠세요? 1,300만 화소면 될까요?
여: 아뇨. 저는 풍경 사진을 찍는 걸 좋아해서요. 적어도 1,600만 화소는 되어야 해요.
남: 그건 문제가 안 될 것 같네요. 가격대는 어떠세요?
여: 1,000달러 넘게 돈을 지불할 순 없어요.
남: 그런 경우라면, 이 두 가지 전화기가 손님의 요구에 맞습니다. 왼쪽 것

이 더 저렴하고요.

여: 오, 알겠습니다. 꽤 멋져 보이네요.

남: 하지만, 오른쪽 것이 더 높은 고객 평가를 받았죠.

여: 정말요? 그러면 오른쪽의 것으로 할게요. 가격보다는 품질이 더 중요해요.

어휘
replace 대신[대체]하다 storage 저장(량) GB 기가바이트 (= gigabyte)
megapixel 100만 화소 landscape 풍경 fit 들어맞다 rating 평가, 순위
quality 품질

11 ③

여: 엄마는 내일 집에 오실 거야. 우리는 부엌을 청소해야 해.

남: 알았어. 내가 설거지를 다 할게. 네가 바닥을 닦아 줄래?

여: 그래. 내가 청소 용품을 가져올게. 우리는 욕실도 청소해야 해.

남: 좋은 생각이야. 함께 청소하자.

어휘
mop (대)걸레로 닦다 supply 《pl.》 용품

문제풀이
① 아니, 엄마는 다음 주에 여기 오실 거야.
② 내가 이미 바닥을 닦았다고 말했잖아.
④ 나 대신 설거지를 해줘서 고마워.
⑤ 먼저 노크를 해! 엄마가 욕실에 계실지도 몰라.

12 ①

남: 수지에게 새 강아지가 생겼다는 거 알고 있었니?

여: 아니, 하지만 그녀가 동물을 좋아하는 건 알아. 그건 어떻게 생겼니?

남: 작고 갈색 점이 있는 흰색 강아지야. 아주 귀여워!

여: 그녀는 틀림없이 무척 신나겠구나.

어휘
puppy 강아지 spot 점 [문제] vet 수의사 (= veterinarian)

문제풀이
② 그녀가 강아지를 수의사에게 데리고 간 것 같아.
③ 그녀는 나에게 자신의 문제에 대해 말한 적이 없어.
④ 와. 그녀가 식물에 대해 잘 알고 있겠구나.
⑤ 음, 그 개를 본 적이 있는데, 전혀 귀엽지 않았어.

13 ①

남: 안녕, 몰리. 나는 오늘 아침에 흥미로운 기사를 읽었어.

여: 오, 정말? 그것에 대해 말해 줘.

남: 우리는 끊임없이 전자파에 노출되어 있다고 해.

여: 그게 어떻게 가능하지?

남: 휴대전화나 전자레인지와 같이 우리가 일상생활에서 사용하는 것들 대다수가 전자파를 만들어 낸대.

여: 위험할 것 같아!

남: 적은 양은 해롭지 않지만 시간이 흐르면서, 정기적인 노출이 우리의 몸에 영향을 미치고 심지어 암을 초래할지도 몰라.

여: 끔찍하다. 나 자신을 어떻게 보호하지?

남: 음, 네가 아파트에서 특정한 종류의 화초를 키우면, 그것들이 전자파의 일부분을 흡수할 거야.

여: 알겠어. 또 다른 건 없어?

남: 있지. 네가 휴대전화로 이야기할 때 얼굴에서 멀리 떨어지게 잡아.

여: 그걸 사용할 때마다 명심할게.

어휘
article 기사 constantly 끊임없이 be exposed to …에 노출되다
electromagnetic wave 전자파 microwave 전자레인지 exposure 노출
absorb 흡수하다 [문제] symptom 증상

문제풀이
② 네 증상이 지속되면 의사에게 가 봐.
③ 전화기를 고치게 수리점에 가지고 가.
④ 우리는 새 전자레인지를 사려고 쇼핑을 하고 있었어.
⑤ 나는 식물이 전자파를 흡수하는지 몰랐어.

14 ⑤

남: 안녕하세요, 대출 신청을 하고 싶은데요.

여: 음, 저희 대출 담당자인 더글라스 씨가 오늘 안 계십니다만, 제가 도와드리도록 할게요. 어떤 대출 상품을 찾고 계시나요?

남: 새집 마련을 위한 것으로요.

여: 알겠습니다. 집을 사시는 게 이번이 처음이세요?

남: 아뇨, 두 번째입니다.

여: 알겠습니다. 고객님께서는 일부 특별 상품에 대해서는 자격이 없을 수 있음을 미리 말씀드릴게요.

남: 괜찮습니다. 기대하진 않았어요.

여: 저희가 제공하는 대출 상품들을 살펴보실 수 있도록 책자를 좀 드릴게요.

남: 알겠습니다. 대출받는 게 얼마나 어려울까요?

여: 그에 대한 답변은 정확하게 드릴 수가 없을 것 같네요. 그에 대해선 더글라스 씨가 더 자세히 알려드릴 수 있어요.

남: 제가 언제 더글라스 씨와 상담할 수 있을까요?

어휘
apply for …을 신청하다 loan 융자, 대출 be eligible for …에 자격이 있다

문제풀이
① 제가 지금 당장 그를 데려오겠습니다.
② 유감이지만, 그렇게 못하실 것 같습니다. 그는 그만뒀거든요.
③ 이건 제가 생각했던 것보다 훨씬 수월하네요.
④ 분명히 그가 이해할 거예요. 행운을 빌어요!

15 ④

여: 캐시는 남편이 승진되었음을 알리는 전화를 받았다. 이를 축하하려고 그녀는 와인과 스테이크로 특별한 저녁 식사를 준비하기로 결심했다. 그래서 그녀는 슈퍼마켓에 가서 와인과 스테이크를 조금 구입했다. 심지어 꽃가게에 들러 장미 한 다발도 샀다. 하지만 요리를 시작하려고 했을 때, 그녀는 스테이크의 유통 기한이 하루 전까지였음을 알게 되었다. 짜증이 난 그녀는 슈퍼마켓으로 다시 갔다. 이런 상황에서, 캐시가 직원에게 할 말로 가장 적절한 것은 무엇인가?

캐시: 이건 기한이 지났어요. 다른 걸로 교환하고 싶어요.

어휘
promote 승진시키다 florist 꽃집 bouquet 꽃다발 be about to-v …하려고 하던 참이다 expiration date 만료 날짜, 유통 기한 irritated 짜증이 난
[문제] withered 시든 expired 만료된, 기한이 지난 exchange A for B A

를 B로 교환하다

문제풀이
① 죄송하지만, 이 스테이크는 너무 비싸네요.
② 이걸 요리하는 방법을 설명해 주시겠어요?
③ 이 장미들은 전부 시들었어요. 환불받고 싶어요.
⑤ 전 레드 와인은 좋아하지 않아요. 화이트 와인을 좀 보여주시겠어요?

16 ② 17 ⑤

남: 안녕하세요, 여러분. 저는 지역 경력 센터의 스캇 페르난데즈입니다. 여러분은 로봇이 여러분의 직업을 대체할 것을 걱정하고 있나요? 염려하시는 게 당연할지도 모릅니다. 저는 오늘 이곳에 곧 사라질 것 같은 몇 가지 직업에 관해 이야기하려고 왔습니다. 먼저, 텔레마케터는 실직할 것으로 예상됩니다. 컴퓨터들이 이미 그들의 일에 월등히 잘 할 수 있습니다. 세무 대리인도 위기에 처해 있습니다. 통상적인 계산 업무는 기계에 의해 쉽고 효율적으로 처리될 수 있습니다. 법무 보조원도 마찬가지입니다. 기계들은 반복적인 작업의 많은 부분을 빠르게 처리할 수 있죠. 게다가, 사람들은 더 이상 패스트푸드 요리사로서 일할 것이라고 기대하지 않습니다. 패스트푸드 산업의 점점 더 많은 직업이 자동화되고 있습니다. 향후 20년 동안 기술에 의해 사라져버릴 다른 직업들이 있을까요? 그에 관해 몇 분 동안 생각해 봅시다.

어휘
take over 대체하다 concerned 걱정[염려]하는 disappear 사라지다 excel 뛰어나다, 탁월하다 tax preparer 세무 대리인 routine 일상적인, 지루한 calculation 계산 legal assistant 법무 보조원 repetitive 반복적인 automate 자동화시키다 wipe out 없애버리다 [문제] healthcare 건강 관리(의); 의료(의)

문제풀이
16 ① 미래 직업에 지원하는 법
   ② 미래에 가장 안전하지 않은 직업들
   ③ 인간을 대체할 훈련 로봇들
   ④ 특별한 기술을 필요로 하는 새로운 직업들
   ⑤ 기계들은 절대 할 수 없는 업무들

DICTATION Answers 본문 p.50

01	in reference to an order / made a mistake / can you reduce it
02	replace your appliances with / running the air conditioner / make a big difference
03	lots of clothes / almost brand-new / another advantage
04	looks familiar to / collared shirt on the hanger / That's a great deal
05	need any volunteers / You designed those shirts / do something similar
06	I'm excited about / go with you / take part in
07	order some flowers / throw in / complements
08	handmade necklaces / having a sale / will be 20% off
09	purchase a textbook / the business aspect of computers / use one in our computer lab

10	to replace my old one / take photos of landscapes / more important than cost
11	mop the floor
12	got a new puppy
13	being exposed to / have an effect on / hold it away from
14	apply for a loan / be eligible for / give you some brochures
15	got promoted / stopped by the florist / the expiration date
16~17	taking over your job / right to be concerned / excel at their job / will be wiped out

09 영어듣기 모의고사
본문 ▲ p.54

01 ⑤	02 ③	03 ⑤	04 ④	05 ③	06 ②
07 ④	08 ⑤	09 ④	10 ④	11 ⑤	12 ④
13 ②	14 ⑤	15 ①	16 ③	17 ③	

01 ⑤

여: 안녕하세요, 학생 여러분. 여러분 모두가 아시다시피 오늘 밤에 졸업 무도회가 있습니다. 여러분 모두가 아주 즐겁고 안전한 밤을 보내시길 바랍니다. 이와 관련하여 여러분 모두에게 다시 한번 당부할 내용은 무도회장에 주류는 허용되지 않는다는 것과 술에 취한 것처럼 보이는 사람은 입장이 안 된다는 것입니다. 졸업 무도회는 여러분의 고교 시절에서 가장 기억에 남을 만한 추억 중 하나가 되어야 하며, 이 특별한 밤이 술로 인해 가장 불행한 밤이 될 수도 있습니다. 감사합니다, 멋진 밤 보내십시오.

어휘
prom 졸업 무도회 note 주의, 주목 remind 상기시키다 alcoholic 알코올성의, 술의 (n. alcohol) beverage 음료 permit 허용[허락]하다 intoxicated (술 등에) 취한 memorable 기억에 남을 만한 unfortunate 불행한

02 ③

여: 목요일까지는 『제인 에어』를 다 읽지 못할 거야.
남: 그럼 할 수 있을 때 끝내도록 해.
여: 하지만 시험은 목요일이야!
남: 알아. 하지만 책을 단숨에 읽으면 많은 걸 얻지 못할 거야.
여: 하지만 시험을 더 잘 볼지도 모르잖아.
남: 그게 정말로 중요한 거니?
여: 그래! 난 좋은 성적을 원해.
남: 넌 성적에 대해 너무 신경을 많이 쓰고 있어. 교육의 요점은 배우는 거야. 아무것도 배우지 못한다면 성적이 뭐가 중요해?
여: 지금 내가 받는 성적이 미래의 내 기회에 영향을 미치겠지.

남: 있잖아… 배우는 데 집중하고 시험에는 덜 신경 쓰렴. 좋은 성적은 따라올 거야.

어휘
rush through 서둘러 처리하다 point 요점, 가장 중요한 것 care 관심을 가지다 matter 중요하다 opportunity 기회 focus on …에 초점을 두다

03 ⑤

여: 실례합니다, 저를 좀 도와주실 수 있나 해서요. 제가 생물학 수업을 들어야 하는데요. 그리핀 교수님의 수업이 9월에 있나요?

남: 네, 다음 학기에 생물학을 가르치세요. 등록하실 건가요?

여: 네, 등록해야 해요. 12월에 졸업할 계획이거든요.

남: 아, 이런… 안타깝게도, 그 수업은 이미 정원이 꽉 찼네요. 그래도 대기자 명단이 있는데요. 누군가가 수업을 그만두면, 그 자리에 들어갈 수 있어요.

여: 네, 명단에 제 이름 좀 올려주세요. 명단에 몇 명이 있나요?

남: 확인해 볼게요. [잠시 후] 이런, 이미 명단에 14명이나 있네요. 어쨌든 당신의 이름을 적어 놓을게요.

여: 제가 그 수업을 듣기는 어려울 것 같네요. 졸업을 위한 필수 과목이라 문제네요.

남: 네, 이해해요. 근데 그리핀 교수님이 과거에는 예외를 두시기는 했어요.

여: 무슨 말씀이세요?

남: 특별한 상황이라면, 때때로 한 명 정도는 더 수업을 들을 수 있게 해주시지요.

여: 아. 그럼 교수님께 직접 말씀을 드려봐야겠네요.

남: 네. 그럼 약속을 잡아드릴게요.

어휘
biology 생물학 requirement 필요(한 것) exception 예외 circumstance 환경; *상황, 형편 appointment 약속

04 ④

여: 아이스크림 먹을래?

남: 아, 저쪽에 가게가 보이네.

여: 간판에 거대한 아이스크림콘이 있어.

남: 아, 이미 모든 자리에 사람들이 있어. 어디 앉아야 하나?

여: 봐! 저 가로등 옆에 벤치가 있어. 저기 앉자.

남: 그래. 뭘 먹고 싶니?

여: 음. 난 바닐라 아이스크림 하나를 먹을래. 넌?

남: 저 꼬마 남자애가 먹고 있는 거랑 같은 아이스크림콘을 먹을래.

여: 줄무늬 반바지를 입은 남자애 말이니?

남: 응. 그 애의 아이스크림이 맛있어 보이거든.

여: 좋아. 콧수염 있는 남자가 주문을 받고 있어. 그는 힘이 세 보이고 아이스크림을 누구보다도 더 잘 퍼낼 거라고 확신해.

어휘
sign 표지판, 간판 streetlight 가로등 scoop 숟갈, 스쿠프; (숟갈로) 뜨다, 파다 striped 줄무늬가 있는 shorts 반바지 yummy 아주 맛있는 mustache 콧수염 take an order 주문을 받다

05 ③

남: 퇴근 후에 뭐 하실 거예요?

여: 때에 따라 다르죠. 비가 안 오면 공원에 갈까 해요.

남: 저 구름을 보아하니 (공원에 갈) 운이 없으시겠는데요.

여: 네, 맞는 말씀 같군요. 그냥 집에 가서 독서나 해야겠어요.

남: 저기, 우리 몇 명이서 영화를 보러 갈까 하는데 관심 있으면 같이 가요.

여: 정말요? 무슨 영화를 하는데요?

남: 링이요. 공포 영화예요.

여: 오, 전 공포 영화 안 좋아해요.

남: 정말요? 코미디 영화는 어때요? 다른 사람들은 그걸 보러 갈 거예요.

여: 그렇다면 저도 갈게요. 책은 집에서 아무 때나 읽을 수 있으니까요.

어휘
That depends. 그것은 때와 형편에 달렸다. by the look(s) of …의 모양으로 보건대, 아마도 out of luck 운이 나빠서 come along 함께 가다 scary 무서운 in that case 그런 경우에는

06 ②

여: 안녕, 닉. 기운이 없어 보여. 무슨 일이니?

남: 이번 학기 시간표 때문에 문제가 있어.

여: 오 정말? 왜 그런 거니?

남: 음, 그리스어 수업에 등록했는데 방금 폐강되었어.

여: 그렇구나. 유감이네. 무슨 일이 있었던 거야?

남: 그 수업을 신청한 학생들이 많지 않았겠지.

여: 있잖아, 대신 독일어를 등록할 수 있을지도 몰라. 난 이번 학기에 그 수업을 들을 거야.

남: 하지만 난 다음 여름에 그리스에 갈 계획이라서 그 언어를 조금은 말할 수 있길 정말 원했거든.

여: 그리스 문화 수업을 듣고 여유 시간이 나면 그 언어를 공부하는 게 어때?

남: 괜찮은 생각이네. 그렇게 해야겠어.

어휘
look down 의기소침해 보이다 register for …에 등록하다 That's a shame. 그거 유감이다. sign up for …을 신청하다

07 ④

여: 안녕하세요. 저는 Marple 고등학교 졸업 앨범을 주문하고 싶은데요.

남: 안녕하세요. 저희에게는 두 가지 선택사항이 있어요. 일반 졸업 앨범은 50달러고, 학년 졸업 앨범은 60달러예요.

여: 학년 앨범은 일반 앨범과 어떻게 다르죠?

남: 일반 앨범은 모든 학생의 사진이 다 포함되어 있고요. 학년 앨범은 손님 학년의 학생들만 들어있습니다.

여: 좋네요. 저는 저희 학년 것만 있는 걸로 할게요. 저는 10학년입니다.

남: 알겠어요. 사진들의 디지털 파일도 원하세요?

여: 제가 사진을 몇 장이나 갖게 되는 거죠? 그건 얼마인가요?

남: 10달러에 50장 분량의 파일을 가질 수 있어요. 그리고 고객님이 원하시는 걸 고르실 수 있어요.

여: 흠… 좋은데요. 150장을 살게요. 여기 제 신용카드입니다.

남: 네. 여기 졸업 앨범과 사진들을 요청하는 양식을 작성해주세요.

어휘
yearbook 졸업 앨범 cost 비용이 들다 select 고르다 fill out 작성하다

request 요청하다

여자는 자신의 학년만 담긴 졸업 앨범($60)과 50장에 10달러인 디지털 사진 파일 150장($10×3)을 산다고 했으므로, 여자가 지불할 금액은 90 달러이다.

08 ⑤

여: 자, 나는 네가 자동차 운전을 처음 한다는 걸 알아. 우리가 출발하기 전에 너는 몇 가지 해야 할 것들이 있어.
남: 알았어. 나 정말 긴장돼.
여: 그러지 마. 맨 먼저 네가 해야 할 것은 안전벨트를 매는 거야.
남: 아, 그래. 잊어버릴 뻔했네.
여: 좋아. 이제 편안해졌으면 백미러를 조정해 볼래?
남: 그래. 좋아.
여: 그럼 이제 시동을 걸 준비가 된 거야. 브레이크 밟았어?
남: 응.
여: 됐어, 해봐. [잠시 후] 자, 뒤에 아무것도 없으면 후진 기어를 넣고, 브레이크에서 발을 천천히 떼면 돼.
남: 알았어, 후진해서… 이런.
여: 왜 그래?
남: 움직이질 않아. 기름이 거의 떨어진 것 같아.

nervous 긴장되는, 떨리는 put on …을 착용하다 seat belt 안전벨트 adjust 조정[조절]하다 rearview mirror (자동차의) 백미러 start 시작하다; *(차에) 시동을 걸다 in reverse 역방향으로, 후진으로 ease A off B A를 B에서 천천히 떼다 out of (물건이) 바닥난, 다 떨어진

09 ④

남: Norris 강당에 오신 것을 환영합니다. 다큐멘터리를 시작하기 전에, 몇 가지 안내 말씀을 드리겠습니다. 여러분 주변의 학생들이 다큐멘터리를 즐길 수 있도록 모든 휴대전화의 전원을 꺼주십시오. 다큐멘터리가 상영되는 동안 방해가 되는 어떠한 행동도 용인되지 않을 것이며, 그런 학생들은 보안 요원에 의해 퇴실 조치를 받을 것입니다. 오늘 또한 몇몇 학생들이 복도를 돌아다니며 Jenny 펀드를 위한 기금을 모을 것입니다. Jenny 펀드는 모금된 기부금 전액을 암 치료법을 찾도록 돕는 데 사용할 것입니다. 여러분이 낸 어떠한 기부금도 대단히 감사할 것입니다. 들어주셔서 감사드리며 즐겁게 영화를 감상하십시오.

documentary 다큐멘터리 announcement *공고; 발표 disruptive 파괴적인, 지장을 주는 tolerate 관대하게 다루다, 묵인[허용]하다 security personnel 보안 직원 up and down 아래위로; *왔다갔다 aisle 통로, 복도 donation 기부(금)

10 ④

남: 안녕하세요. 저는 내일 토론토에 가는 기차표를 한 장 사고 싶어요.
여: 물론입니다. 몇 시에 출발하려고 하세요?
남: 제 일정이 유동적이지만, 정오 이전 즈음에 출발했으면 합니다.
여: 알겠습니다. 저희 열차 중 일부는 토론토로 직행하고, 나머지는 한두

번 갈아타야 합니다.
남: 한 번 갈아타는 것은 괜찮지만, 두 번은 하고 싶지 않네요.
여: 그리고 선호하시는 좌석이 있나요?
남: 네, 저는 창가 자리가 좋아요. 지나가는 풍경을 보는 게 즐겁거든요.
여: 그렇다면 타실 수 있는 두 개의 열차가 있네요. 하지만 더 이른 것은 약간 더 비쌉니다.
남: 오 그래요? 왜 그런가요?
여: 그건 (출퇴근) 혼잡 시간대에 출발해서 통근자들에게 인기가 많아요.
남: 알겠습니다. 그러면 저는 더 저렴한 열차로 표를 살게요.
여: 알겠습니다. [잠시 후] 여기 있습니다.

flexible 유연한 transfer 이동, 환승; 갈아타다, 환승하다 preference 선호도 scenery 풍경 rush hour (출퇴근) 혼잡 시간대 commuter 통근자

11 ⑤

여: 저기, 이 넥타이 중에 어떤 게 더 좋은 선물이라고 생각해?
남: 왼쪽에 있는 게 난 더 좋아. 오른쪽에 있는 건 유행이 지난 거야.
여: 정말? 난 반대로 생각했는데.
남: 음, 네가 좋아하는 색깔인 걸 고르는 게 낫겠다.

out of style 유행이 지난 [문제] suit 정장 trend 유행, 동향

① 그것들은 분명히 멋진 선물이 될 거야.
② 개인적으로 난 넥타이를 거의 매지 않아.
③ 헤어스타일 동향이 요즘 매우 빠르게 변해.
④ 그건 내가 입고 있는 정장에 어울리지 않는 것 같아.

12 ④

남: 캠핑 여행이 너무 기다려져.
여: 나도 그래. 난 캠핑을 가본 적이 없어. 거기서 낚시할 수 있니?
남: 응, 하지만 난 새 낚싯대를 사야 해.
여: 백화점에서 할인 중인 걸 좀 봤어.

fishing pole 낚싯대 [문제] aquarium 수족관

① 네 텐트를 어디에서도 본 적이 없어.
② 여기서 캠핑하는 게 안전하다고 생각하니?
③ 낚시에 관한 TV쇼를 본 적이 없어.
⑤ 내 아들도 수족관에서 물고기를 보는 걸 좋아할 거야.

13 ②

여: 이번 주말에 산으로 캠핑 가자.
남: 그거 정말 좋은 생각이네! 주말 동안 맑고 화창할 거야.
여: 응, 이번 주말이 캠핑하기에 좋을 거라고 들었어. (캠핑을) 준비해 보자. 텐트가 어디 있는지 아니?
남: 아, 스미스 씨 가족이 빌려 갔는데 아직 돌려주지 않은 것 같아.
여: 2주 전에 그들의 바비큐 파티에 갔을 때 우리가 가져온 것 같은데.
남: 음, 난 기억이 안 나. [잠시 후] 아, 이제 생각난다! 우리가 가져오려고 했는데, 잊어버렸잖아.

여: 차고에 가서 한번 확인 좀 해 볼래? 난 침낭하고 카메라를 챙길 테니까. *[잠시 후]* 찾았어?

남: 아니, 거기에 없어.

여: 알았어, 그럼 <u>내가 스미스 씨 가족에게 전화해 볼게.</u>

문제풀이

① 넌 스웨터를 입어야 해.

③ 쓰레기 내놓는 것 잊지 마.

④ 그는 그걸 거기에 둔 걸 잊었어.

⑤ 우린 정말 차고를 치워야 해.

14 ⑤

여: 콘서트 어땠어?

남: 아주 좋았어. 그들은 정말 대단한 밴드야.

여: 난 그들의 최근 앨범은 마음에 안 들더라.

남: 그래, 나도 이전 앨범들이 훨씬 더 좋아.

여: 옛날 곡도 많이 연주했니?

남: 응, 전곡을 조금씩 연주했어. 그들은 활기도 넘쳤어.

여: 누구랑 같이 갔었어?

남: 옛날 학교 친구들이랑. 우린 예전에 그 밴드를 엄청나게 좋아했었거든.

여: 그들을 전에도 본 적 있어?

남: 아니, 그래서 그렇게 재미있었지.

여: 음, 너희는 정말 좋은 시간을 보낸 것 같구나.

남: 그랬어. <u>너도 기회가 있으면 꼭 보라고 추천하고 싶어.</u>

어휘

stuff 작품 **[문제] look forward to** …을 기대하다 **sort** 종류

문제풀이

① 넌 거기에 가면 안 될 것 같아.

② 응, 난 정말 기대돼.

③ 응. 날 초대해 줘서 고마워.

④ 글쎄. 난 그런 거 정말 안 좋아해.

15 ①

여: 어제 샘은 아버지께 친구들과 심야 영화를 보러 가도 되는지 여쭤 보았다. 아버지는 그건 좋지 않다고 생각하셨고, 샘에게 집에서 시험 준비를 하라고 말씀하셨다. 나중에 아버지가 잠자리에 든 후, 샘은 조용히 문을 열고 집을 나섰다. 그와 그의 친구들은 영화관에서 좋은 시간을 보냈다. 그들은 팝콘도 먹고 영화를 보고 나서 게임도 했다. 샘이 집에 도착했을 때, 그는 아버지가 깨어나 거실에서 기다리고 계신 것을 발견하고는 매우 놀랐다. 이런 상황에서, 그의 아버지가 샘에게 할 말로 가장 적절한 것은 무엇인가?

샘의 아버지: 샘, <u>내가 나가지 말라고 했잖니!</u>

어휘

shock 충격을 주다, 놀라게 하다 **awake** 깨어 있는 **[문제] fail** 낙제하다

문제풀이

② 배고프니? 저녁은 식탁 위에 있어.

③ 네 친구들이 낸 소리가 날 깨웠단다.

④ 시험에서 낙제하다니 너한테 정말 화가 난다.

⑤ 네 친구들은 어디에 있니? 친구들이 여기서 잘 거 아니니?

16 ③ 17 ③

남: 오늘날, 많은 사람이 전 세계를 여행하고 외국 동전들을 가지고 집으로 돌아옵니다. 떠나기 전에 모든 동전을 사용하는 것은 거의 불가능하고, 환전소는 동전을 거의 받아주지 않습니다. 비록 동전은 무겁고 더럽지만, 당신은 돈을 버리고 싶지는 않을 것입니다. 여기 남은 동전으로 무엇을 할지에 대해 유용한 팁(조언)이 좀 있습니다. 첫째로, 공항에서 가능한 많은 동전을 다 쓰도록 노력해 보세요. 예를 들어, 선물 가게에서 친구들을 위한 작은 기념품을 살 수 있습니다. 둘째로, 동전을 온라인에서 팔아보세요. 동전의 가치를 전부 받을 수는 없겠지만, 아무 것도 안 받는 것보다는 낫습니다. 셋째로, 동전들에 자그마한 자석들을 붙여 그걸 냉장고를 꾸미는 데 사용할 수 있습니다. 또는 그것들을 자선 단체에 기부할 수 있습니다. 많은 자선 단체들이 외국 동전들을 받아줄 것이고, 그걸 제일 필요로 하는 사람들에게 보내 줄 것입니다. 이건 다른 사람들을 돕는 쉬운 방법입니다! 그러니 앞으로는 남은 외국 동전들을 현명하게 사용할 수 있도록 해 보십시오.

문제풀이

16 ① 가치 있는 외국 동전들의 오용

　② 동전을 세척하는 최고의 방법

　③ 남은 동전들을 사용하는 여러 가지 방법들

　④ 외국 동전 수집의 인기

　⑤ 전 세계의 다양한 종류의 동전들

DICTATION Answers　　본문 p.56

01	enjoyable and safe night / alcoholic beverages / your most memorable experiences
02	rush through the book / What do grades matter / Focus on learning
03	I wonder if / add me to the list / put your name down / talk to him directly
04	next to the streetlight / in striped shorts / taking orders
05	That depends / out of luck / In that case
06	having some trouble with / signed up for that class / take the Greek culture class
07	We have two options / How much does that cost / fill out this form
08	put your seat belt on / adjust the rearview mirror / in reverse
09	turn off all cell phones / Any disruptive behavior / help find a cure
10	one train ticket to Toronto / require a transfer or two / It leaves during rush hour

11	is out of style
12	buy a new fishing pole
13	great for camping / returned it / We meant to
14	a lot better / used to love / had a really good time
15	if he could go / left the house / was shocked to find
16~17	foreign coins / throw money out / what to do with / be sure to

10 영어듣기 모의고사

본문 ▲ p.60

01 ①	02 ④	03 ②	04 ⑤	05 ⑤	06 ③
07 ④	08 ⑤	09 ④	10 ①	11 ①	12 ③
13 ④	14 ①	15 ⑤	16 ①	17 ③	

01 ①

여: 뮤지컬을 보고 싶은데 어디서 언제 봐야 할지 모르세요? Whitman 고등학교의 학생들과 직원들이 뮤지컬 *신데렐라의 유리 구두*를 마련하였습니다. 학생들은 3월 이후로 모두 숙제와 다른 활동들을 계속해 나가는 한편, 음악을 익히고, 대사를 외우고, 리허설에 참여하는 등 준비를 해오고 있습니다. 이 극은 약간의 희극적인 반전이 있는 고전 신데렐라 이야기의 개작입니다. 그러므로 5월 4일 또는 5일 저녁 7시에 학교 강당으로 *신데렐라의 유리 구두*를 보러 오십시오. 입장료는 현장에서 5달러입니다. 이것은 가족끼리 보기 좋은 오락물이니, 아이들을 데려오세요!

어휘
put together 구성하다, 만들다 rehearsal 리허설 keep up with …에 뒤떨어지지 않다 remake 개작, 개조 comical 재미있는, 웃긴 twist (이야기의 예상 밖의) 전환, 전개 auditorium 강당 admission 입장료 bring along …을 데리고 오다

02 ④

여: 거기 뭐가 있니, 에릭?
남: 연꽃이야. 모퉁이에 있는 꽃집에서 샀어.
여: 바로 그건 줄 알았어. 연꽃은 예쁘기만 한 게 아니라 의미도 있어. 이 꽃이 몇몇 고대 문화에서 (특별한) 의미가 있다는 걸 알고 있었니?
남: 물론이지! 고대 이집트에서는 그게 부활의 상징이었다고 읽었어.
여: 불교에서도 의미가 있지.
남: 오, 정말?
여: 응. 그 종교에서 연꽃은 순수와 완전함을 상징해.
남: 힌두교에서의 의미와 비슷하네.
여: 그럼, 힌두교에서도 순수를 나타내는 데 그걸 쓴단 말이니?

남: 응, 하지만 그게 다가 아니야. 힌두교 신자들은 그것을 미와 젊음의 상징으로도 여겨. 하나의 꽃이 그렇게 많은 여러 사람에게 중요할 수 있다는 걸 누가 알았겠어?

어휘
lotus 연(꽃) meaningful 의미가 있는 significant 중요한 ancient 고대의 symbol 상징 (*v.* symbolize) rebirth 부활 Buddhism 불교 religion 종교 purity 순수 perfection 완전 Hinduism 힌두교 represent 나타내다 Hindu 힌두교 신자 youth 젊음, 청춘

03 ②

여: 오늘 무슨 문제로 오셨나요?
남: 몸이 계속 안 좋아요. 목도 아프고요.
여: 알겠습니다. 뭐 복용하신 것 있나요?
남: 아스피린만요. 그리고 약사가 추천해준 다른 약하고요.
여: 도움이 되나요?
남: 별로요. 목 아픈 게 더 심해지고 있는 것 같아요.
여: 감염이 됐을 수도 있겠네요. 체온을 재볼게요.
남: 열이 있는 것 같지는 않아요.
여: 음, 약간 높아요. 아마 의사 선생님이 검사를 더 하실 겁니다.
남: 알겠습니다. 심각한가요?
여: 제가 뭐라고 말씀드릴 수는 없네요. 금방 진료하러 들어오실 거예요.
남: 고맙습니다.
여: 여기 기다리시는 동안 볼 수 있는 잡지가 있어요.
남: 감사합니다.

어휘
sore throat 인후염, 목이 아픈 증상 pharmacist 약사 infection 감염 fever 열 shortly 곧, 이내

04 ⑤

여: 데이비드, 해변에 있어서 정말 행복하다. 멋지지 않아?
남: 응, 정말 멋진걸. 날씨가 완벽해.
여: 저 높은 야자나무 근처에 우리 수건을 두자. 야자나무들 덕분에 열대섬에 있는 것 같은 기분이 들어.
남: 좋은 생각이다! 거기에 해먹도 걸려 있네.
여: 응. 그건 정말 평화로워 보여. 어! 바다에서 저 두 사람은 뭘 하는 거야?
남: 그들은 패들보드를 타고 있어. 그건 재미있는 스포츠지.
여: 오! 나도 할 수 있을 거로 생각해?
남: 물론이야. 우린 이 근처 서핑 상점에서 강습을 들을 수 있어.
여: 좋다! 점심 먹고 나서 같이 가보자.
남: 그래. 저기를 봐. 한 소년과 소녀가 모래성을 같이 만들고 있네. 너무 귀엽다.
여: 그러게, 그 옆에는 여자 몇 명이 비치 발리볼을 하고 있어. 굉장히 신나는 게임이야!
남: 우리가 오늘 여기 온 게 정말 기뻐.
여: 나도! 아무튼, 우리 이제 점심 도시락을 꺼내자!

어휘
terrific 정말 멋진 tropical 열대의 hammock 해먹(나무 등에 달아매는 그물·천으로 된 침대) paddle (카누용의) 작은 보트, 노 volleyball 배구

05 ⑤

여: 와! 너에게 아주 중요한 밤이야! 넌 정말 흥분되겠다!
남: 실은 아주 많이 떨려.
여: 오, 이런! 왜?
남: 난 이렇게 많은 사람 앞에서 공연해 본 적이 없거든.
여: 걱정하지 마. 우리가 응원해 줄게. 아주 잘 될 거야.
남: 부탁 좀 들어줄 수 있어?
여: 물론이지.
남: 내 카메라 좀 들고 있어 줄래? 무대 뒤에 두기가 좀 그래서.
여: 응, 문제없어. 내가 사진을 찍어줄게.
남: 그럴 필요는 없어. 어떤 사진과 학생이 공짜로 찍어 준다고 했거든.
여: 더 찍어도 되잖아.
남: 알아, 하지만 배터리가 꺼졌어.
여: 아, 알았어. 그럼 걱정하지 마. 내가 잘 보관할게.
남: 고마워. 이따 봐.

어휘
cheer 응원하다 do ... a favor …의 부탁을 들어주다 backstage 무대 뒤에서 dead 죽은; *방전된, 작동을 안 하는 [문제] charge (요금·값을) 청구하다; *충전하다

문제풀이
① 연극에 참여하기　　　② 배터리를 충전하기
③ 사진사를 고용하기　　　④ 그의 사진을 찍기
⑤ 카메라를 가져가기

06 ③

남: 안녕, 나라, 학교 연극에 대해서 들었어? 잃어버린 공주의 공연이 있을 거래.
여: 오, 정말? 그건 내가 가장 좋아하는 연극이야!
남: 나도 역시 좋아해. 너 거기 나가 볼 거니?
여: 아니, 안 나갈 거야.
남: 왜? 나는 네가 주연을 하면 굉장할 거라고 확신해.
여: 그런데 공주 역은 대사가 많아.
남: 연습하고 외울 시간이 많을 거야.
여: 그게 문제가 아니야. 나는 그 모든 사람 앞에서 연기하는 게 두려워.
남: 자신감을 가져. 넌 할 수 있어.
여: 그렇지 않아. 너는 어때? 너는 나가 볼 거니?
남: 나도 안 할 거야. 나는 축구 연습 때문에 충분한 시간이 없을 거야.
여: 우리 둘 다 연극에 서지는 못할 것 같네. 그냥 그 공연이 상영될 때 같이 보러 가자.

어휘
try out (선발 시험에) 나가 보다 leading role 주연 line 대사 memorize 암기하다 confidence 자신감 put on (연극을) 상연하다

07 ④

남: 믿어지니? 나한테 5달러밖에 없어.
여: 그럼 학생 식당에서 먹어야겠네.
남: 좋아. 여기서 뭘 살 수 있지?
여: 음, 생선이 3달러야. 그리고 샐러드가 1달러이고. 충분하네.
남: 응, 생선은 좋아. 하지만 난 샐러드는 내키지 않아. 근데 목이 말라.

여: 물 마셔.
남: 싫어, 차를 마시고 싶어. 50센트밖에 안 하잖아.
여: 그래.
남: 그리고 감자튀김도 먹을래!
여: 그건 몸에 별로 안 좋아!
남: 상관없어. 얼마지?
여: 1달러 25센트.
남: 좋아. 충분해.

어휘
cafeteria 구내식당 be into …에 관심이 있다 thirsty 목이 마른

문제풀이
남자는 생선($3), 차($0.5), 감자튀김($1.25)을 먹을 것이므로, 총 4달러 75센트를 지불해야 한다.

08 ⑤

여: 다음 달에 내 생일이 있어요, 토드.
남: 5월에 태어났어요? 그럼 당신의 탄생석은 에메랄드군요.
여: 그건 몰랐어요. 에메랄드는 어떻게 생겼어요?
남: 다이아몬드랑 비슷한데 색이 진한 녹색이에요.
여: 아 그래요, 전에 본 적이 있어요. 다이아몬드만큼 단단해요?
남: 아니, 그렇지 않아요. 강하지만, 다이아몬드보다는 더 쉽게 부서져요.
여: 그렇군요. 어디에서 발견되나요?
남: 전 세계에서 발견되지만, 대부분은 남아메리카와 아프리카에요.
여: 가치는 있나요?
남: 네. 금이 안 간 에메랄드는 다이아몬드보다 더 가치가 있을 수 있어요.
여: 와, 하나 갖고 싶네요.
남: 미국에서 에메랄드는 55번째 결혼기념일의 전통적인 선물이에요. 그러니 기다려야겠죠!

어휘
birthstone 탄생석 similar to …와 비슷한 valuable 가치가 큰, 값비싼 crack 금, 틈 anniversary 기념일

09 ④

남: Goldberg 초등학교가 8월 3일 일요일에 하계 도서전을 개최합니다. 이 도서전은 시내 도서관 앞 보도에서 열립니다. 그러나 날씨가 안 좋을 경우에는 도서관 로비로 옮겨집니다. 도서전은 오전 10시에 시작해서 오후 4시까지 계속됩니다. 성인 도서와 모든 연령대의 아동도서가 특별 할인 가격으로 판매됩니다. 도서전에서 마련된 돈은 학교에서 새 교과서를 구매하는 데 사용됩니다. 모든 Goldberg 학생들이 참여할 것을 권장합니다. 우리는 또한 책을 판매하는 데 기꺼이 시간을 내서 자원봉사해 주실 부모님을 모집하고 있습니다. 무료 음식과 음료수, 그리고 지역 극단에서 하는 인형극 공연도 있을 예정입니다!

어휘
fair 박람회 sidewalk 보도 lobby 로비 be willing to-v 기꺼이 …하다 puppet 인형, 꼭두각시 performance 공연

10 ①

남: 음악 축제 기간이 곧 시작될 거야. 내가 일정표를 내려받았어. 올해 우리 하나 가보자.

여: 멋지다! 넌 어떤 종류의 음악을 좋아해?

남: 난 록과 재즈. 클래식도 좋아하는데, 클래식 축제는 너무 지루할 거야.

여: 동의해. 어느 달이 가장 좋을까?

남: 8월과 10월 둘 다 괜찮아. 안타깝게도 우린 9월에 시험이 있잖아.

여: 그러네. 우리 부산 축제 중의 한 곳을 가자. 거긴 아름다운 도시야.

남: 나 부산 정말 좋아하는데, 너무 멀어. 서울에 있는 걸 고르자.

여: 좋아. 넌 입장권에 얼마나 (돈을) 쓸 수 있어?

남: 얼마든지 괜찮아. 아르바이트로 돈을 많이 저축해뒀거든.

여: 대단하다. 하지만 난 1일권에 50달러만 쓸 수 있어.

남: 그러면 이곳에 가자.

여: 좋아! 내가 오늘 온라인으로 입장권 두 장 주문할게.

어휘

unfortunately 안타깝게도, 불행히도 afford …할 여유가 있다 a pair of 한 쌍의

11 ①

여: 이봐 팀, 연필 좀 빌려줄 수 있니?

남: 쉿! 시험 중에는 말하면 안 되는 거잖아.

여: 알아, 하지만 내 연필이 부러졌어. 달리 뭘 해야 할지 모르겠어.

남: <u>알겠어, 이걸 가져가.</u>

어휘

lend 빌려주다 be supposed to-v …하기로 되어 있다

문제풀이

② 아마 넌 잘 했을 거야.

③ 아니, 그건 내 연필이 아니야.

④ 나중에 그 시험공부를 할게.

⑤ 그걸 내게 빌려줘서 고마워.

12 ③

남: 미안하지만, 클로이, 나쁜 소식이 있어.

여: 저런, 무슨 일이야?

남: 네 친구 애비가 넘어져서 팔이 부러진 것 같아.

여: <u>우리 내일 그녀에게 가 봐야겠어.</u>

어휘

break one's arm …의 팔이 부러지다

문제풀이

① 아, 내가 네 팔을 좀 볼게.

② 애비는 그렇게 말하지 않을 거야.

④ 아니, 난 뉴스 보는 것을 좋아하지 않아.

⑤ 나쁜 일이 일어나지 않았다니 다행이야.

13 ④

여: 파티 재미있어?

남: 정말 재미있어. 날 초대해줘서 고마워.

여: 너 없이는 나도 올 수 없었을 거야.

남: 근데 여기 있는 많은 사람을 모르겠어.

여: 음, 파티 주최자들인 비키와 테오는 만나 봤겠지. 난 테오랑 일해.

남: 알아.

여: 그리고 나머지 사람들 대부분은 테오나 비키랑 함께 일하는 사람들이야. 나도 다 알진 못해.

남: 비키는 어디서 일해?

여: 어느 광고 회사인 것 같아. 음식은 어때?

남: 저 새우를 잔뜩 먹었어. 더는 못 먹을 것 같아.

여: 음, 그럼 춤출 시간이네.

남: 아직 안 돼, 배가 너무 불러. 조금 있다가.

여: 그럼 뭐 하고 싶어?

남: <u>신선한 공기 좀 마시게 잠깐 나가자.</u>

어휘

host 주인, 주최자 advertising 광고 stuff oneself on[with] …을 잔뜩 먹다 shrimp 새우 [문제] step outside 외출하다. 자리를 뜨다

문제풀이

① 테오랑 춤추자.

② 댄스 수업을 듣자.

③ 새우를 좀 더 먹고 싶어.

⑤ 춤추는 곳이 어딘지 가르쳐 줘.

14 ①

남: 얘! 너 아직도 공부해? 금요일 밤이야.

여: 영어 수업에 대비해서 공부할 게 많아.

남: 나중에 해도 되잖아. 나가자.

여: 월요일까지 허클베리 핀을 읽어야 해. 얼마나 두꺼운지 한번 봐.

남: 걱정하지 마. 누구나 그 이야기를 알잖아.

여: 하지만 우린 시험을 봐. 중요한 시험이란 말이야.

남: 공부만 하고 놀지 않으면… 그 속담 무슨 뜻인지 알지.

여: 네 말이 맞는 것 같아. 주말 내내 시간이 있으니까.

남: 들어봐. 릴리와 조가 그 카페에 간대. 가서 그 아이들을 만나러 가자.

여: 그거 솔깃하다. 그런데 좀 더 있다가 가야 할 것 같아. 한 30분쯤?

남: 좋아. 네가 가고 싶어 할 줄 알았어.

여: <u>네 말에 넘어간 거야.</u>

어휘

tempting 유혹적인. 사람의 마음을 끄는 [문제] talk A into B A가 B하도록 설득하다 babysit 아이를 봐주다 grab 움켜쥐다

문제풀이

② 안 돼, 난 못 가. 공부해야 해.

③ 난 내 여동생을 돌봐야 해.

④ 그래, 그 책을 탁자 위에 올려놓을게.

⑤ 그래, 지갑만 챙겨서 나가자.

15 ⑤

남: 스테파니는 회사 사장과 회의를 준비하고 있다. 어제, 그녀는 동료인 존에게 중요한 매출 보고서 한 부를 인쇄해 달라고 부탁했다. 하지만 그가 그것을 가져다줄 때, 그녀는 그가 잘못된 버전을 인쇄했다는 것을 깨닫는다. 그녀는 그에게 올바른 버전으로 인쇄해 달라고 부탁하지만 그는 그녀에게 인쇄기가 작동을 멈췄다고 말한다. 그녀는 2층에 다른 인쇄기가 있다는 것을 아는데, 그것은 단지 흑백으로만 인쇄가 된다. 그녀는 컬러로 된 보고서가 필요하다. 그때 스테파니는 길 건너 인터넷 카페에 컬러 인쇄기가 있다는 것을 기억해 낸다. 회의는 30분도 채 남지 않았고, 그녀는 정확한 보고서가 정말로 필요하다. 이런

상황에서, 스테파니는 존에게 뭐라고 말을 할까?
스테파니: 인터넷 카페로 달려가서 보고서를 인쇄해 주세요.

어휘
coworker 동료 copy 복사(본); 한 부 sales report 판매 보고서 print 인쇄하다

문제풀이
① 보고서는 색이 있을 때 더 나아 보인다고 생각해요.
② 대신 카페에서 회의합시다.
③ 미안하지만 내가 잘못된 보고서를 부탁했음이 틀림없네요.
④ 유감스럽게도, 사장님은 회의에 참석하지 않을 거예요.

16 ①   17 ③

여: 안녕하십니까? 오늘 밤의 플라멩코 공연에 와 주셔서 감사합니다. 공연이 시작되기 전에, 여러분께 이 열정적인 예술에 대한 배경지식을 좀 드리고 싶습니다. 플라멩코는 로마니 족(집시족), 이슬람교도, 그리고 유대인들이 스페인 정부를 피하려고 도망친 스페인의 산악지대에서 생겨났습니다. 실제로, 당시 스페인 정부는 이 소수 민족을 거부하고 그들을 함부로 대했습니다. 그 결과, 이 세 문화가 뒤섞여 플라멩코가 탄생하였습니다. 스페인 정부가 (이들에게) 점점 더 수용적으로 되자, 산에 살던 사람들은 도시로 돌아갔고, 그곳에서 플라멩코를 전파했습니다. 거기서 스페인 사람들은 그들만의 아이디어를 이 예술에 더하였습니다. 오늘날, 플라멩코는 다양한 장소에서 공연되지만, 종종 스페인의 길거리에서도 볼 수 있습니다. 이런 즉흥적인 공연들이 때로는 가장 멋집니다. 오늘 밤, 여러분들은 열정적인 무용수와 음악가들이 이 오래된—하지만 여전히 매우 인기 있는—예술을 공연하는 것을 보게 될 것입니다. 그럼 지금부터 무대가 시작되겠습니다!

어휘
background information 배경지식 passionate 열정적인 Romani 로마니, 집시 Muslim 이슬람교도, 회교도 Jewish 유대(교)인의 flee 달아나다, 도망치다 (fled-fled) reject 거부하다 minority 소수; 소수집단 (a. minor) poorly 형편없이, 좋지 못하게 blend 섞다, 혼합하다 accepting 받아들이는 venue 장소 spontaneous 즉흥적인 popular 인기 있는, 대중적인 (n. popularity) [문제] representation 표현 South American 남미사람, 남미인

문제풀이
16 ① 플라멩코의 기원
　　② 스페인 소수 부족의 역사
　　③ 스페인에서 플라멩코의 인기
　　④ 플라멩코의 많은 표현 방식들
　　⑤ 스페인 전통춤의 다양성

DICTATION Answers
본문 p.62

01 put together / keeping up with / fourth or fifth
02 that's what it was / was significant to / consider it a symbol of beauty
03 have a sore throat / check your temperature / while you wait
04 The weather is perfect / I can try it too / playing beach

---

volleyball
05 cheering for you / take some photos / take good care of
06 you'd be great / Have some confidence / put on the performance
07 What can I buy / I'm not into salads / get French fries
08 as hard as diamonds / found all over the world / without any cracks
09 In case of / children of all ages / are encouraged to attend
10 What kind of music / Any price is fine / Sounds good
11 lend me a pencil
12 fell and broke her arm
13 met the hosts / I stuffed myself / time to dance
14 have a lot to study / how thick it is / All work and no play / It's tempting
15 getting ready for / has stopped working / in less than 30 minutes
16~17 was born in / treated them poorly / get to see

---

11 영어듣기 모의고사

본문 ▲ p.66

01 ①	02 ⑤	03 ②	04 ⑤	05 ②	06 ⑤
07 ②	08 ②	09 ③	10 ④	11 ②	12 ①
13 ⑤	14 ⑤	15 ④	16 ③	17 ④	

01 ①

남: 신사 숙녀 여러분, Rich's 스포츠 용품 매장을 이용해 주셔서 감사합니다. 지금 시각은 8시 50분이며, 저희 매장은 10분 후에 폐점할 예정입니다. 구매할 상품이 있으시면 지금 계산대로 가 주시기 바랍니다. 5번과 6번 계산대가 아직 열려 있습니다. 쇼핑을 마치지 못하셨다면, 저희 매장은 내일 오전 10시에 영업을 시작합니다. 이번 주말에는 특별 세일이 있을 예정입니다. 오늘 저녁에 나가시는 길에 전단지를 한 부 가져가시거나, 내일 아침 신문을 확인하시기 바랍니다. 곧 다시 뵙기를 바랍니다. Rich's를 이용해 주셔서 감사합니다.

어휘
goods 상품 merchandise 상품 cashier 계산원, 출납원 circular 순환의; *(광고용) 전단지

02 ⑤

여: 다음 주 생물학 시험을 볼 준비가 되었니?
남: 아니 별로. 도서관에서 일주일 내내 공부했는데, 아직도 준비가 안 된 것처럼 느껴져.

여: 함께 공부하는 건 어떨까?

남: 그건 너무 정신이 산만할 것 같아.

여: 아니, 그게 좋을 거야. 다양한 주제에 관해 서로 퀴즈를 낼 수 있어. 또, 우리 중 한 명이 질문하면 다른 한 명이 대답해줄 수 있을 거야.

남: 네 말이 맞는 것 같구나. 때때로 약간의 도움이 필요하거든.

여: 그리고 우리 둘이니 그렇게 지루하지 않을 거야.

남: 좋아. 그럼 몇 시에 만날까?

여: 내일 아침 10시는 어때?

어휘

biology 생물학 feel like …한 느낌이 있다 distracting 마음을 산란케 하는
quiz 퀴즈; *질문하다

03 ②

여: 새로운 소설을 내신 걸 축하해요. 또 한 권의 베스트셀러군요.

남: 감사합니다. 제 모든 독자들에게 감사할 따름입니다.

여: 당신의 성공 비결이 무엇이라고 생각하십니까?

남: 정말로 비결이 있다고는 생각하지 않아요. 전 제 경험을 토대로 글을 쓰는 평범한 사람일 뿐입니다.

여: 그럼 소설 속의 모든 이야기가 개인적인 경험에서 나온 건가요?

남: 그렇다고 할 수 있죠. 전 새로운 걸 창조하지는 않습니다. 전 제가 아는 것을 씁니다.

여: 놀라워요. 이 책을 쓰기 위해 전 세계를 여행하셨던 게 틀림없군요.

남: 음, 모든 곳에 가 본 것은 아니지만, 많은 장소에 갔었어요. 글을 쓰지 않으면 여행을 다니죠.

여: 그래서 당신의 책이 그렇게 재미있군요. 아무튼, 시간을 내주셔서 정말 감사합니다.

남: 아뇨, 이렇게 이야기를 나눌 수 있어서 제가 감사하죠.

어휘

storyline 줄거리 personal 개인의 on the road 여행 중인

04 ⑤

남: 여기가 내가 가장 좋아하는 카페야. 여기서 공부하자.

여: 어, 저 유리문 뒤에 있는 책들은 뭐야?

남: 여기는 북카페야. 읽을 책을 고를 수 있지.

여: 와, 좋다. 어디 앉고 싶니?

남: 저 소파는 어때?

여: 음, 사람들이 주문한 걸 찾아가는 곳 바로 옆이잖아. 우리를 방해할 수도 있어.

남: 알겠어. 그럼 중앙에 있는 긴 직사각형 탁자는 어때? 저기서는 방해받지 않을 거야.

여: 좋을 것 같아. 그리고 탁자 위에 귀여운 꽃 모양 전등이 있어! 그걸 켜 봐야겠어.

남: 좋아. 계산대에 주문하는 사람이 아무도 없네. 내가 가서 우리 음료를 주문할게.

여: 응, 주문하기 좋을 때야.

어휘

pick up 찾아가다 order 주문한 음식; 주문하다 disturb 방해하다
rectangular 직사각형의 counter 계산대, 판매대

05 ②

남: 킴, 부동산 중개인과 벌써 이야기해 봤어요?

여: 아뇨. 나는 먼저 우리가 어디서 살고 싶은지에 대해서 이야기해야 한다고 생각했어요.

남: 음, 나는 편리한 곳에 살고 싶어요. 그래서 지하철역과 가까운 장소를 찾고 싶어요.

여: 내가 오로지 신경 쓰는 것은 크기예요. 우리 아이들은 이제 자기만의 방이 필요하잖아요.

남: 하지만 큰 아파트는 비용이 훨씬 더 많이 들지 않을까요?

여: 아마도요. 하지만 지하철역에서 가까운 것도 역시 비싸요.

남: 우선 인터넷으로 검색해 보는 게 어때요? 서로 다른 아파트 가격을 비교할 수 있는 웹사이트가 있다고 들었어요. 나는 지금은 검색할 시간이 충분하지 않아요.

여: 알았어요, 내가 할게요. 부동산 중개인을 방문하기 전에 온라인으로 가격을 비교하는 게 좋은 생각인 것 같아요.

남: 나도 그렇게 생각해요. 당신이 준비되면 언제든지 중개인에게 전화를 할게요.

여: 좋아요.

어휘

real estate agent 부동산 중개인 convenient 편리한 compare 비교하다
search 찾아보다, 검색하다

06 ⑤

남: 리사, 나랑 내일 전자 제품 시장에 갈래?

여: 물론이야. 무엇을 사야 하는데?

남: 나는 새 태블릿 PC를 사고 싶어.

여: 하지만 너는 바로 작년에 하나를 샀잖아. 그걸 벌써 잃어버렸어?

남: 아니, 아직 가지고 있어. 그렇지만 그것에 만족을 못 하겠어.

여: 그래? 왜? 너무 느려?

남: 아니, 꽤 빨라. 그리고 화면도 크고, 그래서 정말 좋아.

여: 그러면 이유가 뭐야? 배터리가 너무 빨리 닳니?

남: 아주 오래가지는 않지만 개의치 않아. 문제는 무게야.

여: 오, 알겠다. 너는 항상 그걸 가지고 다니지, 그렇지?

남: 그래, 그런데 그건 너무 무거워. 그래서 매우 불편해.

여: 그러면 내일 너에게 더 적절한 것을 찾아보자.

어휘

electronics 전자 제품 be satisfied with …에 만족하다 all the time 내내, 항상 inconvenient 불편한 suitable 적절한

07 ②

[전화벨이 울린다.]

남: Z Tix입니다. 무엇을 도와드릴까요?

여: 안녕하세요. 저는 Yay Sound 콘서트 표를 세 장 구매하고 싶어요.

남: 그러시군요. 저희는 세 종류의 좌석이 있어요. 측면, 발코니, 그리고 오케스트라예요. 각각 80달러, 60달러, 100달러입니다.

여: 어느 좌석이 무대에서 가장 가까운가요?

남: 오케스트라 좌석이 제일 가까운데, 가까울수록 가격이 비싸요. 측면 좌석들도 여전히 꽤 가깝고 합리적인 가격입니다.

여: 알겠습니다. 그럼 측면 좌석 세 장을 살게요. 그런데 요즘 비수기 아닌가요?

남: 네. 성수기 동안에는 같은 좌석에 120달러를 지불하셔야 해요. 싸게 파는 거랍니다!

여: 오, 좋아요! 저는 조기 예매 할인도 받을 수 있다고 온라인에서 읽었는데요. 그건 너무 늦었나요?

남: 할인받기에 늦지 않으셨지만, 회원이셔야 해요.

여: 저 회원이에요! 제 회원 번호는 QX665입니다.

남: 어디 봅시다… *[타자치는 소리]* 그렇네요. 10퍼센트 할인됩니다.

여: 좋아요. 전화로 결제할 수 있나요?

남: 네. 신용카드 번호만 알려주세요.

어휘
purchase 구매하다 respectively 각각 pricey 값비싼 reasonably 합리적으로 price 가격을 매기다 slow season 비수기 (↔ peak season) bargain (정상가보다) 싸게 사는 물건 early bird discount 조기 구매 할인

문제풀이
여자는 80달러짜리 측면 좌석을 세 장 사고 10퍼센트 할인을 받으므로, 여자가 지불할 금액은 216달러이다.

08 ②

남: 붉은 행성에 관해 보고서를 쓰고 있어요, 엄마.

여: 붉은 행성? 화성을 말하는 거겠네.

남: 맞아요. 그 색깔이 토양에 있는 철에서 나온 걸 알고 계셨나요?

여: 아니, 몰랐어. 또 어떤 걸 배웠니?

남: 음, 화성의 이름은 로마 전쟁의 신의 이름을 따서 지어졌어요. 또 태양으로부터 네 번째 행성이죠.

여: 그렇구나. 지구처럼 위성이 있니?

남: 사실, 두 개가 있어요. 그런데 우리 위성보다는 훨씬 더 작아요.

여: 화성 자체가 지구보다 더 작지, 그렇지?

남: 네, 그래요. 크기가 약 절반이죠. 하지만 몇 가지 면에서는 비슷하기도 해요.

여: 예를 들어줄 수 있니?

남: 그럼요. 화성의 하루는 지구의 하루보다 겨우 30분 더 길어요.

여: 흥미롭구나. 아마도 언젠가 거기에 갈 수 있을 거야.

어휘
planet 행성 iron 철 be named for …의 이름을 따서 짓다 moon 달; *위성 last 지속되다 someday 언젠가, 훗날

09 ③

여: 지난주에, Seaview 마을은 최초의 Slow Food 축제를 개최했습니다. 그 행사는 토요일과 일요일에 Elm 공원에서 열렸으며, 5천 명 이상의 방문객을 유치했습니다. 그 축제의 하이라이트는 농부들의 장터였는데, 거기에서는 지역에서 재배된 농산물이 할인 가격에 판매되었습니다. 또한, 그 지역의 음식점들이 가장 건강에 좋은 메뉴들을 시식용으로 제공한 많은 음식 부스도 있었습니다. 유기농법과 건강식 요리에 관한 개별적인 발표가 있었는데, 그것들은 교육적이고 재미있었습니다. 그리고 이틀에 걸쳐서 마을 최고의 밴드들이 공연했으며, 많은 사람들이 해가 질 때까지 춤을 추었습니다.

어휘
attract 끌다, 유인하다 highlight 가장 중요한[두드러진] 부분 produce 농산물 bargain 싸게 사는 물건, 특가품 booth 작은 공간, 부스 tasting 시음, 시식 organic 유기농의 educational 교육의, 교육적인 entertaining 재미있는

throughout …동안 내내 performance 공연, 연주 crowd 군중, 무리

10 ④

남: 있잖아! 나 교환학생 프로그램에 합격했어!

여: 잘됐다! 어느 프로그램을 고를지 결정했어?

남: 아직. 이게 좋아 보이는데, 수업을 중국어로 하네.

여: 넌 중국어를 충분히 잘하지 않잖아. 넌 영어로 진행되는 수업이 있는 프로그램이 필요해.

남: 네 말이 맞아. 그리고 난 1년 동안 지속되는 걸로 골라야 해.

여: 응. 한 학기는 문화에 관해 배우기엔 좀 충분하지가 않지.

남: 하지만 내 예산에도 신경을 써야만 해.

여: 맞는 말이야. 얼마나 쓸 수 있어?

남: 부모님께선 최대 만 5천 달러라고 말씀하셨어.

여: 나쁘지 않네. 넌 교외 주택에 머무는 프로그램을 골라. 그게 더 재미있을 거야.

남: 아냐, 난 그렇게 생각하지 않아. 학교 안에서 사는 게 더 안전하다고 느낄 것 같거든.

여: 그렇구나. 음, 그러면 이게 너한테 가장 잘 맞는 것처럼 보인다.

어휘
exchange 교환 whole 전체(의) semester 학기 maximum 최대(의) off-campus housing 교외 주택

11 ②

남: 어떻게 해야 할지 모르겠어요. 우리 아들을 어디서도 찾을 수가 없어요!

여: 장난감 매장은 둘러봤어요? 그 애가 숨어있을 수도 있잖아요.

남: 네, 그럼요. 아, 미아 보호소에 가봐야겠어요.

여: 저기 봐요! 저 진열품 바로 옆에 그 애가 있어요.

어휘
hide 숨다 [문제] display 진열(품) hide-and-seek 숨바꼭질

문제풀이
① 이 매장은 오늘 아주 붐비지는 않네요.
③ 귀여워라! 전에 당신의 아들을 만났던 것 같아요.
④ 집에서 숨바꼭질하는 게 어때요?
⑤ 당신은 다른 매장에서 새 직장을 구해야 할 것 같아요.

12 ①

여: 아빠, 토요일 밤에 차 빌려주실 수 있으세요?

남: 글쎄다. 어디 가고 싶은데?

여: 쇼핑센터에 친구들을 태워 가려고 해요.

남: 조심해서 운전하겠다고 약속한다면.

어휘
lend 빌려주다 [문제] promise to-v …하기로 약속하다

문제풀이
② 미안하지만 난 토요일 밤엔 바빠.
③ 우리 차는 캠핑을 하기에는 충분히 크지 않아.
④ 알았어. 오늘 밤 저녁 식사 후에 널 쇼핑하는 데 데리고 갈 수 있어.
⑤ 그리고 싶지만, 난 쇼핑센터에서 아무것도 살 필요가 없어.

13 ⑤

남: 티나! 오랜만이야!

여: 만나서 반가워, 재민아. 어떻게 지냈어?

남: 잘 지냈지. 그냥 평소처럼 매일 그냥 일하고 운동하면서 보냈어.

여: 그렇구나. 아직도 같은 컴퓨터 회사에서 일하니?

남: 응, 그래. 너는 어때? 일은 어떻게 되고 있니?

여: 음… 사실 난 좀 힘든 시간을 보냈어. 직장을 잃었거든.

남: 오, 정말? 무슨 일 있었어?

여: 우리 회사의 재정 상태가 좋지 않아서 나를 다른 몇 명의 직원과 함께 해고했어.

남: 끔찍하네.

여: 그래, 아주 충격적이었어. 하지만 운이 좋게도 일이 내게 유리하게 모두 해결되었어.

남: 오, 잘됐다. 그런데 어떻게? 새로운 일자리를 찾았어?

여: 응. 내가 교사 채용 시험을 봐서 합격했어. 다음 달부터 학교에서 일할 거야.

남: 와! 네 불운이 좋은 기회로 바뀌었구나.

어휘

go through …을 겪다 financial 재정적인 fire 해고하다 employee 직원 work out …을 해결하다 recruitment 채용, 신규 모집 examination 시험 (= exam) [문제] misfortune 불운

문제풀이

① 곧바로 운동을 시작하는 것이 어때?

② 네가 다음 달에 시험에 합격할 거라고 확신해.

③ 네가 예전 일자리를 다시 찾았다는 소식을 들으니 좋다.

④ 너는 왜 컴퓨터 회사에서 일하고 싶니?

14 ⑤

여: 안녕, 마크. 뭘 읽고 있어?

남: 안녕, 수잔. 이건 도서관에서 빌려 온 정말로 흥미진진한 소설이야. 사실, 방금 다 읽었어.

여: 무엇에 대한 건데?

남: 어떤 의사에 관한 이야기인데, 그 의사의 환자들이 불가사의한 질병에 걸렸어.

여: 불가사의한 질병? 결국에는 그 의사가 그들을 치료할 수 있니?

남: 응. 그는 처음에는 어려움을 겪지만, 결국은 그들을 다 치료해.

여: 정말 재미있을 것 같네. 작가가 대단한 상상력을 가지고 있는 게 틀림없어.

남: 그는 확실히 멋진 작가지만, 나는 그 책이 완전히 허구적이라고 생각하지 않아.

여: 정말? 왜 그런데?

남: 그 작가는 30년간 의사였어.

여: 그의 경험이 그가 그 책을 쓰는 데 도움을 준 게 틀림없구나.

어휘

check out (도서관에서) 책을 빌리다 in the end 결국 eventually 마침내 cure 낫게 하다, 치유하다 fictional 허구적인

문제풀이

① 별로 사실적인 책처럼 들리진 않는다.

② 도서관에 한 부 더 있는 것 같아.

③ 그가 그의 모든 환자를 치료할 수 있었다니 기쁘다.

④ 그 이야기들은 아마도 그의 상상력에서 나왔을 거야.

15 ④

남: 켈리는 사만다의 가장 친한 친구이다. 어느 날, 사만다는 켈리가 점심으로 늘 말라빠진 빵을 가져온다는 것을 알아차렸다. 켈리의 가족은 아버지가 해고된 후부터 재정적인 어려움을 겪고 있었다. 켈리 가족의 상황을 알고 난 사만다는 어머니에게 점심을 두 개 싸달라고 부탁했다. 사만다는 점심 도시락을 두 개 가져와서 켈리에게 한 개를 주곤 했다. 켈리는 사만다의 배려가 정말 고마웠다. 이런 상황에서, 사만다는 켈리가 그녀에게 고맙게 여길 때 뭐라고 말하겠는가?

사만다: 친구 좋다는 게 뭐니?

어휘

notice 알아채다 financial 재정적인 get laid off 해고당하다, 일자리를 잃다 consideration 배려 [문제] be on a diet 다이어트 중이다

문제풀이

① 난 다이어트 중이야.

② 넌 요리를 잘하는구나.

③ 그가 좋아할 거라고 확신해.

⑤ 샌드위치는 한 개당 1달러야.

16 ③ 17 ④

여: 안녕하세요, 학생 여러분. 지난 시간에 우리는 겨울에 동면하는 몇몇 동물들에 대해 이야기했었죠. 그런데 동면이란 무엇일까요? 오늘, 우리는 동물들이 동면할 때 그들에게 어떤 일이 벌어지는지를 이야기해 보려고 합니다. 동면하는 동물들이 하는 첫 번째 일은, 그들의 체온을 낮추는 것입니다. 때때로 그들의 체온은 섭씨 18도까지 떨어질 수 있습니다. 동면하는 동물들은 또한 그들의 호흡을 느리게 만듭니다. 그들은 분당 한 번의 호흡만 할 수도 있습니다. 그들의 심장박동도 마찬가지입니다. 어떤 동면하는 동물들은 심장박동을 75퍼센트 수준으로 낮춥니다. 게다가, 동면하는 동물들은 신진대사 활동도 줄입니다. 이것은 그들이 에너지를 많이 사용하지 않음을 의미합니다. 결과적으로 그들은 아주 오랜 시간 동안 음식이나 물을 먹지 않은 채로 생존할 수 있죠. 동면이란 놀랍지 않습니까? 여러분이 동면에 관해 알고 싶은 다른 것들이 있나요? 이제, 우리는 어디서 동물들이 동면하는지에 대해 이야기해 봅시다.

어휘

hibernate 동면하다 (n. hibernation) temperature 온도 up to …까지 breath 호흡하다 (n. breath) heart rate 심박동 수 metabolism 신진대사 consume 소모하다; *먹다, 마시다

문제풀이

16 ① 동물들이 동면하는 장소들

② 동물들이 어떻게 겨울에 살아남는가

③ 동면이 어떻게 동물에게 영향을 끼치는가

④ 어떤 동물이 겨울에 동면하는가

⑤ 신진대사를 줄이는 방법들

01   be closing / still open / some special sales / on your way out

02   Are you ready for / that might be too distracting / it won't be so boring

03   thankful to all my readers / writes from his own / personal experience / on the road

04   it is right next to / be disturbed there / ordering at the counter

05   where we want to live / look online / whenever you're ready

06   I'm not satisfied with / The problem is / a more suitable one

07   purchase three tickets / the closer, the pricier / not too late to get

08   comes from the iron / the fourth planet / much smaller than

09   took place in / sold at bargain prices / organic farming and healthy cooking / put on performances

10   Not yet / that lasts a whole year / I'd feel safer

11   look around the toy store

12   lend me the car

13   How have you been / went through / all worked out

14   checked out from the library / in the end / Why is that

15   financial problems / got laid off / pack two lunches / very grateful

16~17   hibernate in winter / lower their body temperature / drop up to / without consuming any food

12 영어듣기 모의고사
본문 ▲ p.72

01 ①	02 ③	03 ③	04 ③	05 ④	06 ③
07 ⑤	08 ③	09 ②	10 ⑤	11 ①	12 ④
13 ①	14 ①	15 ④	16 ②	17 ⑤	

01   ①

여: 우리 몸의 98퍼센트가 물로 구성되어 있다는 것은 잘 알려져 있습니다. 우리 몸이 충분한 양의 물을 가지고 있지 않으면, 우리는 심각한 건강 문제를 겪을 수 있습니다. 심지어 죽음에 이를 수도 있습니다. 충분한 물을 섭취함으로써, 우리는 에너지를 얻을 수 있고 정상적으로 활동할 수 있습니다. 이것은 어떤 비타민이나 영양소보다 더 중요

합니다. 우리 몸뿐만 아니라 우리가 사는 세상도 대부분 물로 이루어져 있습니다. 우리 인간은 이 귀중한 물질 없이는 살 수 없다고 해도 틀린 말이 아닙니다.

어휘
be composed of …로 구성되다   **consume** 소비하다; *먹다, 마시다 **sufficient** 충분한   **function** 기능[작용]하다; *활동하다   **nutrient** 영양소 **mainly** 주로   **valuable** 귀중한   **substance** 물질

02   ③

여: 난 책을 다 읽었어.
남: 세종대왕에 관한 책? 어떻게 생각해?
여: 한동안 읽은 최고의 책 중의 하나야. 빌리고 싶니?
남: 아니, 괜찮아. 난 소설 읽는 걸 선호해.
여: 아, 하지만 역사적인 실화를 읽으면 배울 수 있는 게 정말 많아.
남: 그래, 하지만 독서의 목적은 상상력을 훈련하는 거라고 생각해. 그래서 난 소설을 읽는 거야.
여: 하지만 세종대왕 시대의 삶이 어땠는지 상상할 수 있잖아.
남: 그건 여전히 실제 삶에 기반을 두지. 허구 소설에서는 인간이 본 적 없는 것들에 대해 읽거든.
여: 외계인이나 용 같은 것 말이야?
남: 그래. 그게 상상력을 자극하고, 또 모두들 그게 더 많이 필요해.

어휘
fiction 소설, 허구   **nonfiction** 논픽션, 실화   **exercise** 운동시키다, 훈련하다 **imagination** 상상(력)   **alien** 외계인   **challenge** 도전하다; *(흥미·상상력 등을) 자극하다

03   ③

[전화벨이 울린다.]
여: 톰슨 씨 사무실입니다. 무엇을 도와드릴까요?
남: 저는 프레더릭 존스입니다. 톰슨 씨 계십니까?
여: 지금 통화 중이십니다. 메모를 남기시겠습니까, 아니면 기다리시겠습니까?
남: 얼마나 오래 통화하실 것 같습니까?
여: 얼마나 길어질지 모르겠는데요. 벌써 10여분 동안 통화하고 계세요.
남: 그럼 메모를 남길게요.
여: 네. 말씀하세요.
남: 제 이름은 말씀드렸죠. 저는 전에 그에게 일을 의뢰했던 옛 친구인데, 다음 주에 시내로 돌아올 예정입니다. 돌아오는 길에 그를 만나보고 싶어서요.
여: 톰슨 씨께서 귀하의 전화번호를 아시나요?
남: 분명히 그럴 겁니다만, 다시 불러 드릴게요.

어휘
be on the phone (전화로) 통화 중이다   **hold** (전화를 끊지 않고) 기다리다 **business** 일, 업무

04   ③

남: 가족들과 나는 지난주에 아주 편안한 캠프장에 갔어. 이 사진에서 볼 수 있어.
여: 아, 큰 바위들로 둘러싸여 있구나.

남: 응, 그게 가장자리를 분명하게 하는 데 도움이 됐지.

여: 아, 저 커다란 나무가 그늘이 되어 좋았겠다.

남: 맞아. 그래서 우리가 텐트를 그 아래에 세웠지.

여: 팔걸이가 있는 두 개의 편안한 의자도 보이네. 책 읽기에 좋았을 것 같아.

남: 응, 그랬어.

여: 의자 옆에 통나무는 왜 쌓아둔 거지?

남: 우리는 매일 밤 불을 피웠거든. 정말 흥미진진했지.

여: 저기서 정말 재미있었겠구나.

남: 응. 심지어 두 나무 사이에 해먹을 걸었어. 사진 속의 여동생이 하는 것처럼 그곳에서 낮잠을 잤지.

> **어휘**
>
> campsite 캠프장, 야영지 surround 둘러싸다 edge 가장자리 shade 그늘 armrest 팔걸이 log 통나무 pile up 쌓다 hammock 해먹(나무 등에 달아매는 그물침대) take a nap 낮잠을 자다

05 ④

여: 여보, 내일 아침에 언제 일어날 거예요?

남: 모르겠는데요. 난 몇 시에 잘지도 모르겠어요. 왜요?

여: 내일 정말 일찍 일어나야 해서요. 당신이 날 깨워 줄 수 있을까요?

남: 몇 시에 일어나야 해요?

여: 준비를 다 하고 새벽 5시에는 집에서 출발해야 해요.

남: 정말 이른 시간이군요. 자명종을 맞춰 놓지 그래요?

여: 맞춰놨는데, 확실하게 일어나고 싶어서요.

남: 난 일하느라 늦게까지 깨어 있을 거라서 약속은 할 수 없겠는데요.

여: 그럼, 당신이 잠자리에 들 때 날 깨우면 어때요?

남: 정말 그랬으면 좋겠어요?

여: 물론이죠. 내가 아침에 일어나기만 한다면 괜찮아요.

남: 알았어요. 이제 가서 자요, 그래야 충분한 휴식을 취할 수 있으니까요.

> **어휘**
>
> sweetheart 여보, 당신 set the alarm clock 자명종 시계를 맞추다 be up late 늦게까지 깨어 있다 promise 약속하다

06 ③

남: 좋습니다. 거의 다 되었어요. 지금 손님 초상화의 마지막 부분을 그리고 있어요.

여: 정말 기대되네요. 빨리 보고 싶어요.

남: 자… 끝냈어요! 여기 있습니다. 어떻게 생각하세요?

여: 오 이런! 이게 정말 저예요?

남: 네. 잘못된 게 있나요? 마음에 안 드세요?

여: 아니요, 정말 맘에 들어요. 하지만 제가 이렇게 아름답나요?

남: 네, 당신은 아름다워요. 제가 당신의 친절함과 생기 넘치는 성격을 담아낸 것 같아요.

여: 그걸 오늘 밤에 집에 가져가고 싶군요!

남: 사실, 그럴 수 없어요. 물감이 마르려면 적어도 24시간이 걸리거든요.

여: 오. 안타깝지만 이해해요. 며칠 뒤에 다시 올게요.

> **어휘**
>
> portrait 초상화 capture 정확히 포착하다[담아내다]

07 ⑤

남: 수잔, 엄마 선물 샀니?

여: 응, 방금 귀금속 상점에 선물을 가지러 갔었어. 여기 있어.

남: 진주 목걸이구나! 얼마야?

여: 그렇게 비싸지는 않았어. 나한테 150달러만 주면 돼.

남: 150달러? 너무 과한데, 수잔! 무슨 생각을 하고 있었던 거야?

여: 음, 네 예산이 200달러라고 했잖아. 우리가 반반씩 낼 거니까 그렇게 비싼 건 아니야.

남: 난 우리 둘이 합해서 200달러를 생각한 거였어. 아무튼, 괜찮아. 근사한 목걸이긴 하네.

여: 응. 아주 멋있어. 엄마가 정말 좋아하실 거야.

남: 응, 나도 그렇게 생각해. 내가 내일 내 몫의 돈을 줄게.

여: 알겠어.

> **어휘**
>
> pearl 진주 necklace 목걸이 budget 예산, 경비 gorgeous 화려한, 멋진 share 몫, 지분

> **문제 풀이**
>
> 여자는 엄마의 선물을 사는데 남자와 돈을 절반씩 나누어 내기로 하였고 남자에게 150달러를 달라고 했으므로, 여자가 귀금속 상점에 지불한 금액은 300달러이다.

08 ③

여: 어디 가니, 폴?

남: 스터디 모임에 가. 매주 화요일과 목요일마다 일주일에 두 번씩 모여.

여: 아! 네가 스터디 모임의 일원인 줄 몰랐어.

남: 응. 내 친구 킴이 두어 달 전에 시작했어.

여: 특별히 공부하는 게 있니?

남: 응, 수학과 과학이야. 내게 가장 어려운 두 가지 수업이거든.

여: 그렇구나. 솔직히, 난 도서관에서 혼자 공부하는 걸 더 좋아해.

남: 정말? 왜 그런 거니?

여: 너무 많은 사람이 함께 공부하면, 난 쉽게 산만해져.

남: 개인적으로, 나한테는 도움이 돼. 우린 6명인데 서로의 질문에 답해 주거든.

여: 음, 모두 자신만의 공부 스타일이 있는 거지.

남: 맞아. 어쨌든, 난 뛰어가야겠어. 나중에 얘기할게.

> **어휘**
>
> get together 모이다 in particular 특별히 to be honest 솔직히 말하면 get distracted 정신이 산만해지다 personally 개인적으로

09 ②

남: 최신 라디오 프로그램인 북부의 아침을 들어주셔서 감사합니다. 저희는 평일 아침 6시부터 8시까지 방송되며, 여러분께 뉴스를 알려 드리고 최신 히트곡을 들려 드리며, 바라건대 여러분을 웃음 짓게 만들어 드릴 것입니다. 제 이름은 찰스 리이며 제가 사회자가 될 것입니다. 여러분은 국제 정치에 관한 제 이전 프로그램인 찰스와의 주말 덕분에 제 목소리를 알아채셨는지도 모르겠네요. 저는 세 명으로 이루어진 캐나다의 최고 라디오 기자팀과 함께, 토론토 지역의 지역 기사를 보도할 것이며, 또한 여러분께 국제 뉴스에 대해서도 때맞춰 보도해 드릴 것입니다. 그리고 만일 프로그램을 놓치신다면, 여러분은 언제

든지 저희 웹사이트에 로그인하셔서 온라인으로 다시 들으실 수 있습니다.

어휘
on the air 방송 중에 latest 최신의 host 진행자, 사회자 recognize 알아 보다 politics 정치 cover (뉴스 등을) 보도[방송]하다 timely 시기적절한, 때맞춘 replay 재경기; *다시 보기[듣기]

10 ⑤

[전화벨이 울린다.]
여: '여보세요.
남: 안녕, 레이첼. 지금 집에 가는 길이야. 뭘 하고 있니?
여: 집에 일찍 오는구나! 아, 벌써 오후 7시네. 난 다큐멘터리 보고 있었어.
남: 정말? 웬일이야? 넌 다큐멘터리 안 좋아하잖아!
여: 맞아, 그런데 숙제 때문에 봐야 했어. 이제 그 내용을 바탕으로 에세이를 쓰기 시작해야 해.
남: 아… 언제 끝날 것 같아?
여: 음… 두 시간 뒤에?
남: 그다음에 TV 볼래?
여: 음, 나는 그 시간대에 하는 TV 쇼를 안 좋아해. 그런 종류의 쇼는 너무 지루해.
남: 그렇다면, 어떤 종류의 쇼를 보고 싶은데?
여: 내 에세이랑 다큐멘터리에 대한 생각을 떨쳐버리기 위해서라면 리얼리티 쇼를 봐도 괜찮아.
남: 좋아! 음, 난 어떤 것이든 좋아. 단지 스페인어로 된 것만 아니었으면 해. 자막을 읽고 싶지 않거든.

어휘
essay 과제물, 에세이 get one's mind off …에 신경을 쓰지 않다 subtitle 자막

11 ①

여: 안녕하세요, 손님. 표와 여권을 볼 수 있을까요?
남: 네, 여기 있습니다.
여: 감사합니다. 손님의 비행기는 7번 게이트에서 출발할 겁니다. 가방을 부치시겠어요?
남: 네, 이 검은 가방만요.

어휘
passport 여권 check 확인하다; *(수하물을) 부치다

문제풀이
② 아니요, 전 가방이 이미 많아요.
③ 아니요, 그 가방을 본 적이 없어요.
④ 네, 그 큰 갈색 가방이 좋겠네요.
⑤ 잘 모르겠어요. 제 여권을 확인해 볼게요.

12 ④

남: 실례합니다만, 이게 베를린으로 가는 기차인가요?
여: 네, 그래요. 몇 분 뒤에 출발할 거예요.
남: 고마워요. 그런데 기차에 음식이 있나요?
여: 네, 저쪽에 식당칸이 있어요.

어휘
by the way 그런데 [문제] dining car 식당칸 express 급행열차

문제풀이
① 그건 건강에 매우 좋지 않다고 들었어요.
② 거리 건너편에 카페가 있어요.
③ 네, 베를린에는 음식점이 많아요.
⑤ 네, 우리는 급행열차로 갈아탈 수 있어요.

13 ①

남: 오늘이 12월 29일이야. 새해가 다가오고 있다는 걸 믿을 수가 없어.
여: 나도 그래. 나는 언제나 새해를 시작하기 전에 흥분이 돼.
남: 나도 그래. 너의 새해 결심이 뭐니?
여: 오, 난 새해 결심을 하지 않아.
남: 왜?
여: 나는 사람들이 비현실적인 목표를 세운다고 생각해. 그들은 오랫동안 그 목표를 유지할 수 없고, 그럼 그저 자기 자신에 대해서 실망을 하게 돼.
남: 네가 무슨 말을 하는지 알겠지만, 내 결심은 작년에 잘 풀렸어. 나는 8월까지 5킬로그램을 감량하려고 결심했지.
여: 정말? 성공적이었어?
남: 내 목표를 달성하지는 못했어. 나는 겨우 3킬로그램 감량했지만, 여전히 내가 무언가를 성취한 것 같은 기분이 들어.
여: 오, 네가 네 목표를 달성하지 못해도, 결국에는 결심이 이로울 수 있을 거 같아.
남: 그게 내가 생각하는 바야.
여: 결심을 하는 것이 정말 가치가 있는 것 같아.

어휘
resolution 결심 unrealistic 비현실적인 work out (일이) 잘 풀리다 accomplish 성취하다 beneficial 유익한, 이로운 [문제] be worth v-ing …할 가치가 있다 meet a goal 목표를 달성하다

문제풀이
② 아마도 넌 몇 가지 체중 감량에 관한 조언을 내게 해 줄 수 있을 거야.
③ 아쉽지만, 작년에 내 목표를 달성하지 못했어.
④ 알겠어, 그렇지만 나를 다시 실망시키지 말아 줘.
⑤ 너는 작년에 좀 더 어려운 목표를 세웠어야 했어.

14 ①

남: 넌 이번 크리스마스에 뭘 할 거니?
여: 실은, 몇몇 친구들과 작은 콘서트를 열까 계획 중이야.
남: 난 네가 악기를 연주하는지 몰랐어.
여: 난 피아노를 꽤 쳐. 로비도 남편 스티브와 함께 와서 클래식 기타를 연주할 거야.
남: 많은 사람이 참여하는 것 같구나. 멋질 것 같아.
여: 분명 재미있을 거야. 난 크리스마스 캐롤 연주를 위해 옛날 학교 밴드도 다시 모이게 할 거야.
남: 모두가 음악가 같다. 노래 부르는 사람은 없니?
여: 노래를 잘 부를 수 있는 사람이 한 명도 떠오르지 않네. 누구 아는 사람 없니?
남: 난 매주 교회 성가대에서 노래하거든.
여: 와! 잘됐다! 콘서트에서 우리와 함께 노래해 줄래?

남: 그래, 기꺼이 도와줄게.

어휘
instrument 악기 involved 관여하는, 연관된 choir (교회의) 성가대 [문제]
spring break 봄 방학

문제풀이
② 성가대는 몇 시에 시작하니?
③ 유감스럽게도 이번 봄 방학에 다른 계획이 있어.
④ 미안하지만, 난 피아노를 칠 줄 몰라.
⑤ 내가 피아노 칠 수 있는 사람을 찾아볼까?

15 ④

남: 로리 커밍스는 학급에서 수석으로 졸업할 예정이다. 그녀는 휠체어에 있을 수밖에 없기 때문에 힘든 학교 생활을 했다. 그녀는 그냥 모든 걸 포기하고 싶었던 순간들이 많이 있었다. 그러나 그녀가 그런 난관에 직면할 때마다, 그녀의 친구들과 반 친구들이 그녀 곁에 있어 주었다. 그들은 그녀가 자신 스스로를 믿도록 도왔다. 로리는 그들의 지지가 그녀가 성공할 수 있게 해 주었다고 느낀다. 그녀는 학교를 졸업하게 되어 기쁘지만, 반 친구들에게도 고마움을 느낀다. 오늘은 그녀의 졸업식 날이고, 그녀는 졸업반 친구들에게 고마움을 표현하고 싶다. 이런 상황에서, 로리는 그녀의 반 친구들에게 뭐라고 말하겠는가?
로리: 그동안 저를 도와주신 여러분들에게 감사의 인사를 드리고 싶습니다.

어휘
be confined to …에 갇히다 face 직면하다 dilemma 딜레마, 진퇴양난
support 지지, 지원 grateful 고마워하는 graduation 졸업 (v. graduate)
gratitude 감사 [문제] failure 실패 challenge 도전, 어려운 일

문제풀이
① 우리는 모두 우리의 실패로부터 배울 것이라고 확신합니다.
② 여러분들이 같은 어려움에 직면했었다는 것을 압니다.
③ 선생님들의 노고에 감사드립니다.
⑤ 좋은 성적으로 졸업하는 여러분이 정말 자랑스럽습니다.

16 ② 17 ⑤

여: 신사 숙녀 여러분, 최초로 열리는 국제 음악 축제에 오신 것을 환영합니다. 이 행사가 가능할 수 있게 정말 열심히 일한 모든 자원봉사자 여러분께 감사드립니다. 저희 개막식은 Whirlwind 발레단의 춤과 함께 오전 10시에 시작될 것입니다. 그다음으로는 11시에 중앙 무대에서 노래 대회가 이어질 것입니다. 누구나 자유롭게 대회에 참가할 수 있고 우승자는 상금을 받게 됩니다. Neilson 재즈 트리오는 오후 2시에 공연을 할 예정입니다. 그들은 국내 최고 밴드 중 하나이니 놓치고 싶지 않으실 것입니다. 공연에 이어 무대 옆 부스에서 그 밴드의 CD를 구매할 수 있다는 점에 주목하세요. 오후 4시에 여러분만의 나무로 된 플루트를 만드는 법을 배우시려면 소풍 구역으로 오세요. 모든 재료는 소액의 수수료를 받아 제공됩니다. 또한, 오후 6시에는 중앙 무대에서 다양한 록 밴드의 음악을 들으실 수 있습니다. 모든 분이 오늘 멋진 시간 보내시길 바랍니다.

어휘
volunteer 자원봉사자 ceremony 의식 be followed by …이 이어지다
cash prize 상금 note 주목하다 purchase 구매하다 booth 작은 공간, 부스
wooden 나무로 된, 목재의 material 재료 fee 수수료; *요금, 회비

---

01 composed of water / lead to death / have energy and function / this valuable substance

02 the best books I've read / to exercise your imagination / With fiction novels

03 leave a message or hold / gave him some business / have your number

04 it's surrounded by / must have been / the logs piled up

05 wake me up / set your alarm clock / get enough rest

06 It's almost done / Is something wrong / It takes

07 pick it up / for both of us / It's gorgeous

08 get together / a couple of months ago / prefer to study alone

09 be on the air / recognize my voice / covering local stories / listen to replays online

10 on the way home / In two hours / that's on at that time / get my mind off

11 checking any bags

12 We'll be leaving

13 New Year's resolutions / become disappointed in themselves / I still feel like

14 played an instrument / seems to be a musician / my church choir

15 confined to a wheelchair / what allowed her to succeed / share her gratitude with

16~17 will be followed by / is free to participate in / won't want to miss / for a small fee

13 영어듣기 모의고사
본문 ▲ p.78

01 ③	02 ③	03 ④	04 ④	05 ③	06 ②
07 ②	08 ③	09 ③	10 ③	11 ①	12 ④
13 ④	14 ⑤	15 ③	16 ①	17 ④	

01 ③

남: 오늘 오후에 와주신 모든 부모님을 환영합니다. 이렇게 시간을 내어 참석해 주셔서 감사합니다. 저는 졸업 앨범 위원회의 레이니스 교사입니다. 올해의 졸업 앨범 발행을 앞두고, 모든 부모님께 이번이 자녀들에게 특별 축하 메시지를 제출하실 수 있는 마지막 기회라는 것을 상기시켜 드리고 싶습니다. 이것은 여러분의 사랑하는 자녀들에게 그들이 얼마나 특별한지 알려줄 수 있는 좋은 방법입니다. 모든 메시지는

졸업 앨범에 인쇄될 것이고, 그들 중 일부는 졸업 앨범 파티에서 낭독될 것입니다. 출판 마감 때문에, 모든 메시지는 늦어도 2월 15일까지 제출되어야 합니다. 경청해 주셔서 감사드리고, 여러분의 연락을 기다리겠습니다.

어휘
yearbook 졸업 앨범 committee 위원회 upcoming 다가오는, 곧 나올
release 발표, 발매 submit 제출하다 publishing 출판(업) deadline 기한,
마감 일자 look forward to v-ing …하기를 고대하다

02 ③

남: DNP 디자인 회사의 리랑 전화 통화를 막 끝냈어.
여: 그런데?
남: 좋은 소식과 나쁜 소식이 있어. 내가 그들의 디자인 팀에 오길 바란대…
여: 너에게 잘 된 일이네! 그런데?
남: 그런데 자원봉사직이야.
여: 뭐? 네가 무료로 그들을 위해 일해줄 거라고 생각할 리 없어.
남: 어떻게 해야 할지 모르겠어. 아마 몇 달 동안 자원봉사를 해주면 그들이 나를 고용해서 급여를 주겠지.
여: 샘, 넌 경력 있는 디자이너야. 비급여 인턴으로 일하기에는 너무 솜씨가 좋아.
남: 알아, 하지만 이건 좋은 기회야, DNP 디자인은 높이 평가되는 회사잖아.
여: 그게 그들이 인턴에 의존할 필요가 없는 이유인 거야. 직원에게 돈을 줄 여유가 있잖아.

어휘
volunteer 자원봉사자; 자원봉사로 하다 experienced 경력 있는 well-respected 존경을 받는, 높이 평가되는 rely on …에 의존하다 afford to-v …할 여유가 있다

03 ④

남: 점심시간이 거의 다 되었어. 갈 준비됐니?
여: 아직이야. 나는 우선 이 사진들을 봐야 해.
남: 이 사진들이 잡지 표지에 사용할 것들이니?
여: 그래. 그중에 잘 어울리는 사진 두 장을 골라야 해.
남: 흠… 내 생각에는 이 두 개가 완벽한 짝이야. 그것들은 심지어 이번 달 주제와도 어울리잖아.
여: 나도 그렇게 생각해. 그 사진들을 써야겠어.
남: 좋아. 표지 결정이 끝났으니까, 뭐 좀 먹으러 갈래?
여: 미안하지만 지금 또 다른 걸 끝내야 해. 한 시간 안에 레이아웃을 디자인해서 편집장에게 가져다줘야 하거든.
남: 그렇다면 내가 샌드위치를 사서 너에게 가져다줄게.
여: 그러면 좋겠다. 정말 고마워!
남: 천만에. 너를 도울 수 있어서 기뻐.

어휘
go well 잘 어울리다 match 잘 어울리는 것[사람]; 어울리다 theme 주제
layout 레이아웃, 배치

04 ④

남: 우리 잡지 표지를 위해 이 축제에 있는 누군가의 사진이 필요해.
여: 난 록 밴드에서 드럼을 연주하는 남자애가 마음에 들어.
남: 안 돼, 우리는 이미 작년에 그 밴드의 사진을 썼어.
여: 그럼 학교 이름이 써 있는 커다란 표지판을 들고 있는 학생은 어때?
남: 난 슈퍼히어로로 복장을 한 남자애가 더 흥미로워 보이는데.
여: 아, 하늘을 나는 것처럼 한쪽 팔을 내밀고 있는 남자애?
남: 응. 아, 개와 고양이처럼 옷을 입고 손을 잡고 있는 저 둘도 마음에 들어.
여: 사랑스러워 보이네. 그런데 난 저 남자애 사진을 사용하기로 정했어.
남: 누구?
여: 아이들에게 사탕을 나눠주고 있는 애 말이야. 그게 최고의 사진이라고 생각해.

어휘
costume 복장 dress up 변장하다 adorable 사랑스러운 hand out 나눠주다 shot 발사; *사진

05 ③

여: 미치겠어. 토요일 파티 준비를 할 수 있을지 모르겠어.
남: 내가 뭘 도와줄까?
여: 글쎄, 이제 막 이 소파를 옮길 참이었어…
남: 문제없어. 내가 도와줄게.
여: 그런데 아직 모든 일을 어떻게 진행해야 할지 모르겠어.
남: 초대는? 내가 몇몇 사람들한테 전화해 줄까?
여: 지금 당장은 말고. [잠시 후] 알겠다. 간식이 좀 필요해.
남: 감자 칩, 프레첼, 그리고 음료수 말이야?
여: 그렇지. 내가 당장은 가게에 갈 수가 없어.
남: 음, 내가 지금 바로 갈게. 나한테 목록만 적어줘.
여: 그래주면 큰 도움이 될 거야.
남: 전혀 문제 될 거 없어. 아주 멋진 파티가 될 거야.

어휘
go crazy 미치다 be about to-v 막 …하려고 하다 invitation 초대(장)
pretzel 프레첼(매듭·막대 모양의 비스킷)

06 ②

남: 안녕 케이트. 잘 지내니?
여: 안녕 빌리. 사실을 말하자면, 난 꽤 기분이 언짢아.
남: 무슨 일이니?
여: 음, 엄마를 위해 새로운 손목시계를 주문했는데 그 귀금속 가게가 주문을 엉망으로 만들었어.
남: 무슨 일이 있었는데?
여: 정확히는 모르겠지만, 오늘 준비가 안 된다고 말했어.
남: 끔찍한데. 어떻게 할지 결정은 했니?
여: 아직 못 했어. 다른 손목시계를 살 수 있겠지만, 그 위에 특별한 메시지를 새긴 걸 원했거든.
남: 내 사촌이 귀금속 가게에서 일해. 그가 너를 도와줄 수 있을지도 몰라. 네가 좋다면 그에게 지금 전화해 볼게.
여: 아, 그럼 너무 좋지. 정말 고마워.

어휘
fairly 상당히 annoyed 화가 난, 짜증이 난 mess up 망치다 engrave (돌·

쇠붙이 등에) 새기다

07 ②

[전화벨이 울린다.]

여: Special Cake 서비스입니다. 무엇을 도와드릴까요?

남: 안녕하세요. 저는 제 아내 생일을 위해 케이크를 주문하고 싶어요.

여: 물론이죠. 어떤 종류의 케이크를 원하세요?

남: 웹사이트에서 장미 케이크를 봤어요. 그건 뭐죠?

여: 저희는 장미꽃으로 크림을 만듭니다. 그 케이크는 장미로 장식되기도 하고요. 장미처럼 보이고 장미 향이 나죠.

남: 흥미롭네요. 얼마인가요?

여: 30달러입니다. 하지만 장식을 더 하고 싶으시다면 비용을 좀 더 내셔야 해요.

남: 알겠습니다. 케이크 위에 '생일 축하해, 엘리자베스!'라고 써주시고 가운데에 여왕 조각품을 더해주세요.

여: 그러면 글자 쓰는 게 4달러이고 조각이 6달러입니다.

남: 흠… 할인받을 수 있는 방법이 있나요?

여: 회원이 되시면 20퍼센트 할인됩니다.

남: 그래요! 가입할게요.

여: 좋습니다. 손님의 이름과 주소를 말씀해 주세요.

어휘

blossom 꽃 sculpture 조각(품)

문제풀이

남자는 케이크($30)를 사고 글씨 서비스($4)와 조각품($6)을 추가로 구입한 후 20퍼센트 할인을 받으므로, 남자가 지불할 금액은 32달러이다.

08 ③

[전화벨이 울린다.]

남: 여보세요. 무엇을 도와드릴까요?

여: 네, 제 아들을 위한 여름 캠프를 찾고 있는데 귀사의 캠프에 대한 정보를 얻고 싶어서요.

남: 네. 저희 캠프를 강력히 추천합니다. 여러 가지 활동이 많거든요.

여: 예를 들면요?

남: 저희 캠프에는 호수가 있어서, 낚시와 같은 많은 수상 스포츠들이 있답니다.

여: 좀 더 활동적인 것은요?

남: 카약도 있습니다. 그리고 카누 같은 다른 종류의 배들도 있고요.

여: 승마는요?

남: 근처에 있는 마구간으로 한두 차례 아이들을 데리고 갑니다. 이것은 선택 사항이고요.

여: 그렇군요.

남: 또한, 이곳에서는 하이킹하기도 좋아요. 아름다운 지역이죠.

여: 좋은 것 같네요. 남편과 상의해 볼게요.

남: 아드님이 이 캠프에 참가하면 좋겠습니다. 전화해주셔서 감사합니다.

어휘

highly 매우 recommend 추천하다 kayak 카약(에스키모인의 수렵용 소형 가죽배); *카약을 타다 canoe 카누(노로 젓는 작은 배) horseback riding 승마 stable 차분한, 안정된; *마구간 optional 선택 사항의 hiking 도보 여행 discuss 의논하다

09 ③

여: Kew 왕립식물원은 런던에 위치하고 있으며 유네스코 세계 문화유산입니다. 이곳은 식물에 관한 연구와 교육의 중요한 중심지입니다. Kew 왕립식물원은 세계 최대 규모로 현존하는 식물을 보유하고 있습니다. 수집은 1700년대에 처음 시작되었고, Kew는 1840년에 영국의 국립식물원이 되었습니다. 오늘날에는 희귀 식물의 종자를 보존하고 식물학을 연구하는 중요한 장소입니다. 이곳은 또한 세계 최대 규모로 식물에 관한 책을 소장하고 있는 도서관을 보유하고 있습니다. Kew 왕립식물원은 거대한 빅토리아식 온실로 유명한데, 그 온실들은 유럽에서 동종 최초의 건축물에 속했습니다. 이 모든 유명한 매력적인 요소들을 갖춘 Kew 왕립식물원은 매년 2백만 명의 방문객을 맞이합니다.

어휘

heritage 유산 collection 수집, 소장 preserve 보존하다 seed 종자, 씨앗 rare 희귀한 Victorian 빅토리아 시대 양식의 greenhouse 온실 structure 구조물, 건축물 attraction *매력(적인 요소); 명소 annually 매년

10 ③

남: 요가 센터입니다. 무엇을 도와드릴까요?

여: 네, 귀사의 워크숍에 참가하고 싶어서요.

남: 좋습니다. 정확히 언제 오고 싶으세요?

여: 4월 첫째 주 이후가 가장 좋을 것 같아요.

남: 훌륭한 워크숍 프로그램이 많습니다. 강사 훈련 과정으로 하실 건가요?

여: 네, 그럼요. 곧 강사 일을 시작할 거라서요.

남: 그럼 4월 9일과 10일에 있는 '전문가 지도' 워크숍은 어떠세요?

여: 전문가요? 아뇨, 전 아직 그럴 준비는 안 됐어요.

남: 알겠습니다. 그럼, 그달 말에 상트페테르부르크에서 '일반인 지도' 워크숍이 있어요. 그거라면 딱 맞을 겁니다.

여: 좋은 것 같네요. 그런데 상트페테르부르크까지는 너무 멀어요.

남: 음, 똑같은 워크숍이 같은 달에 좀 더 일찍 다른 지역에서 있어요. 그리고 기립 자세를 가르치는 것에 대한 워크숍도 있고요.

여: 음… 그렇다면 저는 더 저렴한 워크숍으로 시작할게요. 그 워크숍이 괜찮으면, '일반인 지도' 워크숍도 들을게요!

어휘

expert 전문가 location 위치, 장소 [문제] tuition (소규모) 수업, 교습

11 ①

남: 내일 내 친구 집에 가는 데 같이 갈래?

여: 상황에 따라 달라. 친구 어디에 살고 있니?

남: 그는 부산에 살아. 우린 해변에도 갈 수 있어.

여: 아주 재미있을 것 같아.

어휘

That depends. 상황에 따라 달라. [문제] crowded 붐비는 must have v-ed …했음이 틀림없다 should have v-ed …해야 했는데 (하지 못했다)

문제풀이

② 난 네 여자친구를 만난 적 없어.

③ 오늘 해변은 너무 붐비네.

④ 바다는 여전히 추웠던 게 틀림없어.

⑤ 너랑 그 해변에 갔어야 했는데.

12 ④

여: 안녕, 제이크. 너 괜찮아? 몸이 별로 안 좋아 보이는데.
남: 사실, 일주일 내내 독감을 앓았어.
여: 안됐구나. 내가 해줄 수 있는 게 있을까?
남: <u>뜨거운 수프를 가져다줄 수 있니?</u>

flu 독감 [문제] medicine 약 catch a cold 감기에 걸리다

문제풀이

① 응, 그게 내게 도움이 될 거야.
② 이 약은 너에게 도움이 될 거야.
③ 아니, 널 위해 아무것도 할 수 없어.
⑤ 아니, 난 올해에는 감기 걸린 적 없어.

13 ④

남: 일이 끝나서 기분이 좋아.
여: 그래, 마침내 금요일이야.
남: 아, 이런, 날씨 좀 봐.
여: 비가 퍼붓네.
남: 흘딱 젖겠어.
여: 넌 오늘 차 안 갖고 왔어?
남: 응, 걸어왔어. 오늘 아침에는 날씨가 좋았잖아.
여: 일기 예보 안 봤어?
남: 응, 그런 것 같아.
여: 주말 내내 비가 온다고 했거든.
남: 아, 이런. 중요한 계획이 있었는데.
여: 정말? 뭔데?
남: 새로 산 자전거를 타볼 겸 장거리 여행을 하려고 했지.
여: <u>아마 그걸 미뤄야 할 거야.</u>

어휘

be over 끝나다 pour 쏟다, 붓다 get soaked 젖다 weather forecast 일기 예보 try out …을 시험해 보다 [문제] give ... a ride …을 태워주다 postpone 미루다, 연기하다

문제풀이

① 집까지 나 좀 태워다줄 수 있어?
② 네 차를 어디에 주차했어?
③ 자전거 타기는 정말 좋았어, 그렇지?
⑤ 내가 일기 예보를 확인했어야 했어.

14 ⑤

남: 안녕, 재스민.
여: 안녕, 트로이. 뭘 보고 있는 거니?
남: 야구 게임이야. 내 생각에 넌 스포츠 열성 팬은 아닌 것 같아.
여: 왜 그렇게 말하는데? 나 스포츠 아주 좋아해.
남: 아, 미안해. 내가 야구 얘기를 했을 때, 네가 별로 신이 난 것 같지 않았거든.
여: 음, 솔직히 말해서, 난 야구의 모든 규칙을 아는 건 아니야.
남: 그래, 상당히 복잡하지. 음, 채널 13번에서는 축구 경기도 하고 있어.

여: 정말? 사실 축구는 내가 잘 아는 스포츠야. 내가 고등학교 팀에서 경기를 했었거든.
남: 네가 운동선수였는지 몰랐는걸. 채널을 변경할까?
여: 아니, 그럴 필요는 없어. 네가 나한테 규칙을 설명해주면 되잖아, 그렇지?
남: 물론이지. 일단 네가 규칙을 이해하고 나면, 경기가 더 재미있을 거야.

어휘

mention 말[언급]하다 complicated 복잡한 [문제] would rather A than B B보다 (차라리) A 하겠다 once (접속사로) 일단 …하면

문제풀이

① 아니. 난 네가 야구를 하는 것보다는 차라리 축구하는 걸 보고 싶어.
② 미안해. 다음에는 네가 좋아하는 드라마를 보자.
③ 물론 아니지. 스포츠를 직접 하는 게 TV를 보는 것보다 더 재미있어.
④ 문제없어. 내가 오늘 오후에 내일 경기 입장권을 살게.

15 ③

여: 어느 젊은 역도 선수가 아주 열심히 훈련했는데, 대회에서 여전히 좋은 성적을 내지 못하고 있다. 그는 완벽하게 짠 식이요법을 준수하려고 애쓰며, 어느 경쟁 선수들보다도 훈련하는 데 더 많은 시간을 들인다. 강도 높은 훈련에도 불구하고, 그는 여전히 5, 6위밖에 못한다. 낙심했지만 포기하지 않으려는 그는 언젠가는 우승을 하리라고 맹세한다. 그리고 마침내 그는 해낸다. 그가 1위를 차지한 것이다! 이런 상황에서, 그의 코치가 그에게 할 말로 가장 적절한 것은 무엇인가?
코치: <u>힘들었던 네 모든 노력이 성과를 거두었구나.</u>

어휘

weightlifter 역도 선수 competition 경쟁; *대회 follow 따라가다, 따르다 rival 경쟁자 despite …에도 불구하고 intense 극심한, 강렬한 manage to-v 간신히 …하다 come in (순위상) …등을 하다 frustrated 낙심한 vow 맹세하다 [문제] pay off 성과를 거두다

문제풀이

① 다음번에는 운이 따를 거야.
② 행동하는 것보다 말하는 게 더 쉬워.
④ 언젠가 너에게도 기회가 올 거야.
⑤ 기운 내! 나는 네가 낙담하지 않으면 좋겠어.

16 ① 17 ④

남: 안녕하세요, 학생 여러분. 여름방학은 재미있게 놀고 푹 쉬기에 좋은 시간이지만, 여러분은 또한 조심해야 합니다. 여름철에 흔한 심각한 건강 문제들이 많이 있습니다. 이러한 문제 중 하나는 식중독입니다. 이것은 여러분을 아주 괴롭게 할 것입니다! 사람들은 해로운 병균으로 오염된 음식이나 물을 섭취했을 때 이 질병에 걸립니다. 또 다른 여름철 건강 문제는 열사병입니다. 그것은 직사광선에서 너무 많은 시간을 보내는 것에 의해 발생합니다. 그것은 여러분을 어지럽고 메슥거리게 할 것입니다. 수두 또한 여름철에 흔합니다. 수두에 걸린 사람들은 가렵고 빨간 혹이 피부에 생깁니다. 재채기나 기침을 하는 감염자들이 주변에 있을 때 이 병에 걸리게 됩니다. 많은 사람이 여름철에 또한 피부 발진을 겪게 됩니다. 이것은 사람들이 땀을 많이 흘릴 때 나타납니다. 땀은 사람들의 피부를 자극하고, 가렵고 붉게 만듭니다. 이번 여름에 이러한 건강 문제를 어떻게 피할 수 있을까요? 저는 여러분의 생각들을 들어보고 싶습니다.

어휘

food poisoning 식중독 awful 끔찍한 consume 소비하다; *섭취하다
contaminated 오염된 heat stroke 열사병 dizzy 어지러운 nauseous
메슥거리는 chicken pox 수두 itchy 가려운 bump 혹 infected 감염된
cough 기침하다 sneeze 재채기하다 irritate 성가시게 하다; *자극하다

DICTATION Answers

본문 p.80

01 the yearbook committee / submit a special message / the publishing deadline

02 want me to join / give me a salary / too skilled to work

03 go well together / you're finished with / help you out

04 in a rock band / holding out one arm / handing out candy to children

05 I'm going crazy / call some people / make me a list

06 tell you the truth / messed up the order / engraved with a special message

07 order a cake / decorated with roses / become a member

08 highly recommended / lots of water sports / It's optional

09 contains the world's largest collection / preserving the seeds of rare plants / which were among

10 participating in / a number of / perfect for you

11 to visit my friend

12 had the flu

13 It's pouring rain / see the weather forecast / go for a long ride

14 a big sports fan / pretty complicated / explain the rules

15 follow a perfect diet / manages to come in / in first place

16~17 that has been contaminated / feel dizzy and nauseous / who cough or sneeze / makes it itchy and red

14 영어듣기 모의고사

본문 ▲ p.84

01 ①	02 ④	03 ②	04 ⑤	05 ②	06 ⑤
07 ④	08 ③	09 ④	10 ③	11 ②	12 ①
13 ①	14 ③	15 ⑤	16 ⑤	17 ②	

01 ①

여: 제가 여기서 일한 지 15년이 넘는 동안, 우리 회사가 많은 위대한 것을 성취하는 것을 보게 되어 영광이었습니다. 그리고 전 특히 우리가 작년에 개발한 상품이 현재 동종의 상품 중에서 가장 많이 팔린다는

사실 덕분에, 이곳에서 대표였던 것이 무척 자랑스럽습니다. 그래서 이것은 쉬운 결정이 아니었지만, 저는 더 이상 ACN Industrial 주식회사의 최고경영자로서 근무하지 않을 것을 발표해야겠습니다. 현 부사장인 아이반 슬론이 제 일을 인계받을 겁니다. 그의 열정과 경험을 고려해 볼 때, 저는 ACN Industrial 주식회사가 슬론 씨의 지도력 아래 장래가 밝다고 확신합니다. 수년간 여러분 모두의 노고와 지지에 감사를 전하고 싶습니다.

어휘

be honored to-v …하게 되어 영광이다 particularly 특히 due to …때문에 serve 근무하다 CEO 최고경영자 (= chief executive officer) Inc. 주식회사 (= Incorporated) current 현재의 vice president 부사장 take over …을 인계받다

02 ④

여: 이봐, 마크. 나 새 평면 TV를 살까 생각 중이야. 어떻게 생각해?
남: 나라면 지금 사지 않을 거야.
여: 왜 그렇게 말해?
남: 음, TV는 보통 모든 연말 세일이 시작되는 11월에 최저가가 되거든.
여: 맞아. 좀 더 기다릴 가치가 있겠어.
남: 그런데, 새 디지털카메라를 살 생각이라고 하지 않았니?
여: 그것 역시도 11월이 사기 좋을 때야?
남: 사실, 가게에서는 보통 1월과 2월에 카메라 가격이 가장 저렴해.
여: 그렇지만 그땐 휴가철이 아니잖아.
남: 아니지. 하지만, 그때는 모든 최신 모델이 나오는 때야. 그래서, 그때 구형 모델을 사면 아주 좋은 가격에 구할 수 있어.

어휘

flat-screen 평면 스크린의 worth …의 가치가 있는 in the meantime 그런데 as well 역시도, 또한 latest 최신의 at that time 그때

03 ②

남: 여러분, 안녕하세요. 전 닥터 쿨이고 여러분은 KSPR을 듣고 계십니다. 오늘 밤 우리의 특별 손님은 미셸 화이트입니다. 환영합니다, 미셸.
여: 안녕하세요, 닥터 씨. 여기에 나오게 돼서 정말 기쁩니다. 청취자들을 위해 몇 곡을 부를 수 있어서 신이 나네요.
남: 훌륭해요! 그런데 시작하기 전에, 질문이 몇 가지 있는데요.
여: 좋아요.
남: 유명 배우 조 피처와 당신과 관련된 소문에 대해 좀 더 말씀해 주실 수 있나요?
여: 아, 그거요? 아무것도 아니에요. 우리는 파티에서 한 번 만난 것뿐이에요.
남: 그렇군요, 그리고 건강은 어떠세요? 마지막 콘서트에서 발목을 삐지 않았었나요?
여: 네. 정말 아팠어요! 하지만 이제 훨씬 좋아졌어요.
남: 그거 잘됐군요. 마지막으로, 작업 중인 새 앨범은 무엇인가요?
여: 팝송 모음이에요. 그건 8월 15일에 나올 거예요.
남: 우리 모두 그 새 앨범이 나오길 고대하고 있죠! 자, 이제 그 새 앨범 중에서 몇 곡을 불러줄 수 있나요?
여: 물론이죠. 기꺼이 불러드릴게요.
남: 청취자 여러분들, 광고 후에 돌풍을 일으키고 있는 팝 스타의 라이브 무대로 돌아오겠습니다. 채널을 고정해주세요!

어휘
sprain (팔목·발목 등을) 삐다 ankle 발목 hurt 다치게 하다; *아프다
sensation 느낌, 감각; *선풍적인 인기를 끄는 것 commercial 상업의, 상업적
인; *광고 Stay tuned. 채널을 고정해라.

04 ⑤

여: 현대 음악실로 갈까요? 표지판에 왼쪽에 있다고 쓰여 있어요.
남: 아직요. 우리 딸이 아직 큐레이터가 저 전시관에 무엇이 들어있는지
    설명하는 걸 듣고 있어요.
여: 그래요, 1700년대에 사람들이 이러한 악기들을 만들었다는 사실에
    감명을 받은 것 같군요.
남: 네, 그녀는 악기를 만드는 것에 매우 흥미 있어 해요. 와, 이 바이올린
    을 봐요! 정말 오래되어 보이네요.
여: 네, 그렇네요. 오른쪽 표지판에 500년이나 되었다고 쓰여 있네요.
남: 놀랍네요. 어, 저기 동그란 테두리의 액자 안에 있는 사람은 누구죠?
여: 제 추측엔 저 악기를 실제로 연주했던 사람인 것 같네요.
남: 그는 정말 음악가처럼 생겼네요. [잠시 후] 아, 봐요! 저 남자가 하프의
    사진을 찍고 있어요.
여: 악기들의 사진을 찍는 것이 허락되는지 잘 모르겠네요.
남: 아마도 여기서 일하는 누군가에게 물어봐야겠어요.

어휘
curator 큐레이터(박물관·미술관 등의 전시 책임자) display case 진열장
instrument 악기 sign 표지판, 간판 harp 【악기】 하프

05 ②

여: 데이비드, 오늘 저녁에 시간 좀 있니?
남: 그럼. 난 아무 계획도 없어. 무슨 일인데?
여: 역사 숙제로 10쪽짜리 에세이를 써야 해. 내일이 제출 마감일인데 절
    반도 못했어.
남: 음, 난 항상 역사를 잘했어. 기꺼이 도와줄게.
여: 사실 에세이는 도와줄 필요 없어. 단지 에세이를 쓸 수 있는 조용한
    시간이 필요할 뿐이야.
남: 그래서 내가 해줄 게 뭔데?
여: 있지, 오늘 밤에 내가 알렉시스에게 공항으로 마중을 나간다고 약속
    했어. 그 애가 멕시코 여행에서 돌아올 거거든.
남: 알렉시스가 멕시코에 간 줄 몰랐네.
여: 응. 멕시코 시티에 있는 그녀의 가족에게 가 있었어. 그녀의 비행기 도
    착 시각이 9시야.
남: 아. 그럼, 기꺼이 그녀를 데리러 갈게. 네 차 좀 써도 될까?
여: 물론이지. 정말 고마워.

어휘
due (제출·지불) 기일이 된 halfway 중간에; *부분적으로, 불완전하게 flight 비
행; *항공편 appreciate 고마워하다 [문제] browse (정보를) 열람[검색]하다

문제풀이
① 여행 사이트 검색하기　　　② 그녀의 친구 마중 가기
③ 그녀에게 자동차 빌려주기　④ 그녀를 공항에 데려다주기
⑤ 그녀의 숙제를 도와주기

06 ⑤

[문 두드리는 소리]
여: 오. 안녕하세요, 앤디.
남: 안녕하세요, 웬디. 이렇게 일찍 찾아와서 미안해요.
여: 별말씀을요. 일어난 지 몇 시간 됐어요. 무슨 일이시죠?
남: 당신의 아드님 샘에 관해 드릴 말씀이 있어서요.
여: 어머. 그 애가 무슨 짓을 했나요?
남: 저, 불평하고 싶진 않은데, 오늘 아침에 샘과 샘 친구들이 새로 산 제
    자동차 유리창을 깼어요.
여: 설마요. 어쩌다 그렇게 됐죠?
남: 야구공을 주고받으면서 학교에 걸어가고 있었나 봐요. 공이 제 차에
    맞았는데 앞 유리에 금이 갔어요.
여: 분명히 샘이었나요?
남: 그 애가 도망가기 전에 아주 똑똑히 봤어요.
여: 정말 죄송합니다. 수리비를 보상해 드리고, 아들 녀석을 꼭 혼을 낼게
    요.

어휘
come by 들르다 complain 불평하다 crack 금이 가게 하다 windshield
(자동차의) 앞 유리 repair 수리, 수선 punish 벌하다

07 ④

남: 여보! 커튼에 어떤 종류의 천을 사용할지 정했어요?
여: 네, 나는 이 흰색 천이 좋아 보일 것 같아요. 그리고 식탁보로는 저 체
    크무늬 천을 사요.
남: 좋은 생각이네요. 난 커튼을 만들려면 8야드의 천이 필요할 거라고
    생각해요.
여: 확실해요? 나는 우리 방에는 적어도 10야드가 필요할 거라고 생각하
    는데요.
남: 알았어요. 식탁보는요?
여: 흠… 2야드면 충분할 거예요, 그렇죠?
남: 그래요, 그만하면 될 거예요.
여: 음, 흰색 천은 야드당 3달러이고, 체크무늬 천은 5달러예요.
남: 판매세를 더하는 걸 잊지 말아요.
여: 맞아요. 그게 얼마인가요?
남: 10퍼센트요.
여: 알겠어요. 그만하면 모든 천 값으로 나쁘지 않네요.
남: 맞아요. 내가 계산대에 가서 돈을 낼게요.

어휘
fabric 천 checkered 체크무늬의 tablecloth 식탁보 sales tax 판매세
counter 계산대

문제풀이
남자는 야드당 3달러의 천을 10야드, 5달러의 천을 2야드 구매하고 10퍼
센트의 판매세를 내야 하므로, 남자가 지불할 금액은 44달러이다.

08 ③

남: Blueport 마을에 오신 걸 환영합니다.
여: 고마워요, 존슨 시장님. 마을에 대해 좀 말씀해주세요.
남: 물론이죠. Blueport는 인구 3천 명이 안 되는 다소 작은 마을입니다.
여: 네, 그런데 산과 바다 사이에 아름답게 위치해 있어서 최근 국내 최고

관광지로 선정되기도 했죠.

남: 네, 방문하는 관광객의 수는 지난 15년 동안 꾸준히 증가했어요.

여: 그럼 관광업이 주요 산업인가요?

남: 네. 대부분의 마을 주민은 현지 호텔과 식당에서 근무하고 있죠.

여: 그렇군요. 또한, 현재 Blueport는 자동차가 없는 지역이라고 들었어요.

남: 네. 이것은 환경에 도움이 되고 거리를 더 안전하게 만들죠.

여: 사람들은 어떻게 돌아다니죠?

남: 많은 사람이 자전거를 이용하지만 이용할 수 있는 공공 버스가 있습니다.

여: 정말 흥미롭네요. 저희와 이야기 나눠 주셔서 감사합니다.

어휘
tourist attraction 관광지 steadily 꾸준히 townspeople 도시[읍] 주민
employ 고용하다 car-free 자동차 없는

09 ④

남: Mountain Scouts는 사람과 야생 동물 모두를 위해 우리 마을 주변에 있는 숲을 깨끗하고 안전하게 지키고자 열심히 일하는 자원봉사자들입니다. 올해의 부모-자녀 프로젝트로 우리는 Rocky 강 근처에서 3일간 야영을 할 것입니다. 그곳에 있는 동안 우리는 강가에 떨어져 있는 모든 쓰레기들을 주우며 시간을 보낼 것입니다. 깨끗한 강은 우리 모두에게, 특히 물속에 사는 물고기와 동물들에게 더 좋을 것입니다. 이번 여행에 저희와 함께하고 싶으신 분들은 반드시 목요일 오후까지 신청해 주시기 바랍니다.

어휘
volunteer 자원봉사자 forest 숲 wildlife 야생 동물 trash 쓰레기 edge 가장자리 sign up 신청하다

10 ③

남: City 렌탈소입니다. 무엇을 도와드릴까요?

여: 이번 주 금요일부터 일주일 동안 차를 한 대 빌리려고 합니다.

남: 알겠습니다. 탑승자가 몇 명이나 되시죠?

여: 네 명이요. 저희 남편과 아이 둘이 저랑 함께 탈 겁니다.

남: 그렇군요. 그리고 짐을 넣을 공간이 얼마나 필요하시나요?

여: 저희는 두 개의 가방을 가져갈 겁니다. 차를 어디서 가져가면 되나요?

남: 저희는 사무실이 두 개인데요, 하나는 시내에 있고 다른 하나는 공항에 위치해 있습니다.

여: 좋네요. 저는 공항에서 차를 가져갈게요. 차에 에어컨은 다 있죠?

남: 전부 다는 아닙니다. 하지만 올해 이맘때는 날씨가 시원해서요, 필요하지 않으실 겁니다.

여: 실은 제가 더위를 쉽게 잘 타요. 그래서 저는 에어컨이 있는 차를 선호합니다.

남: 문제없습니다. 공항에 도착하시면 적합한 차가 손님을 기다리고 있을 거예요.

여: 훌륭하네요. 도와주셔서 감사합니다.

어휘
rental 임대, 대여 passenger 승객, 탑승자 downtown 시내 located …에 위치한 pick up 찾다, 찾아오다 suitable 적합한

11 ②

여: 나는 다음 달에 필리핀에 방문할 때 스쿠버 다이빙을 하러 갈 거야.

남: 좋은 생각이야. 예전에 다이빙하러 가 봤니?

여: 아니, 이번이 처음이라서 나는 약간 걱정돼. 위험할 것 같아.

남: 안전 수칙을 따르면 괜찮을 거야.

어휘
scuba diving 스쿠버 다이빙 [문제] experienced 숙련된

문제풀이
① 언젠가 필리핀에 다이빙하러 가렴.
② 나에 대해서 걱정하지 마. 나는 숙련된 다이빙 선수야.
③ 모든 사람들이 해외에 가기 전에 초조해해.
④ 나는 전에 그곳에 휴가를 여러 번 갔었어.

12 ①

남: 케이트, 프레젠테이션 강당을 예약할 수 있었나요?

여: 나쁜 소식이에요. 주말에는 벌써 다 예약이 찼어요.

남: 알겠어요. 그럼 우리가 어떻게 해야 할까요?

여: 다른 장소를 찾아봐야죠.

어휘
presentation 발표, 프레젠테이션 [문제] venue 장소 right person 적임자 crowded 붐비는

문제풀이
② 프레젠테이션을 정말 보고 싶었어요.
③ 강당은 2층에 있어요.
④ 당신이 프레젠테이션의 적임자예요.
⑤ 강당은 주말에 아주 붐빌 거예요.

13 ①

남: 에리카, 너 어디 가니?

여: 안녕, 댄. 조깅하러 갈 준비하고 있어.

남: 난 네가 조깅하는 줄 몰랐어.

여: 아, 최근에 시작했어. 지난달에 병원에 갔었는데, 의사 말이 내가 운동을 좀 더 자주 해야 한대.

남: 그래, 무슨 말인지 알아. 난 책상에 앉아서 보내는 시간이 너무 많거든. 살이 찌기 시작했어.

여: 원한다면 나랑 같이 조깅하러 가도 되는데.

남: 넌 얼마나 자주 해?

여: 일주일에 세 번. 매주 월요일, 수요일, 금요일 오전 10시에 해.

남: 음. 할 수 있을 것 같아. 같이 해도 괜찮겠어?

여: 물론이지. 같이 할 사람이 있으면 좋을 것 같아.

어휘
jogger 조깅하는 사람 come along 함께 가다 [문제] company 동료; *동행인

문제풀이
② 고맙지만 사양할게. 난 조깅하는 걸 좋아하지 않아.
③ 물론이지. 난 병원에 혼자 가는 게 싫거든.
④ 문제없어. 화요일 아침에 봐.
⑤ 미안하지만 난 그 날 약속이 있어.

14 ③

여: 안녕하세요. 찾으시는 물건 있으세요?
남: 네. 파인애플 있나요?
여: 있어요. 바로 여기 바나나 옆에요.
남: 아, 보이네요. 와, 파인애플 하나에 5달러요? 좀 비싸지 않나요?
여: 비싼 거 아니에요. 멀리 하와이에서 수입해 온 거예요.
남: 정말 좋아 보이기는 하네요. 제가 이번 주말에 여는 파티에 쓸 것으로 여러 개가 필요하거든요.
여: 음, 오늘 필요한 게 아니시라면, 내일 다시 들르시는 게 더 나을 거예요. 그때 과일과 채소를 세일할 거예요.
남: 파인애플도 세일하나요?
여: 네. 할인 가격은 파인애플 3개에 10달러가 될 거예요.
남: 정보 고마워요. 내일 다시 올게요.

어휘
sort of 조금, 약간 import 수입하다 on sale 할인[세일] 중인 [문제] tip 조언, 힌트

문제풀이
① 너무 싼데요. 15달러는 어떨까요?
② 모르겠네요. 하와이는 꽤 멀잖아요.
④ 좋아요. 오늘 파인애플 3개를 살게요.
⑤ 파인애플 고마워요. 맛있네요.

15 ⑤

남: 피오나는 고등학생이다. 그녀는 매일 자전거를 타고 등하교를 한다. 어느 날 방과 후, 그녀는 진입로에 자전거를 세워 둔다. 그날 저녁에 그녀의 오빠 팀이 일을 끝내고 집에 도착하니 날이 어둡다. 그는 주차 도중 잘못하여 피오나의 자전거를 친다. 피오나가 집 밖으로 뛰어나와 자신의 찌그러진 자전거를 본다. 그녀는 팀에게 무슨 일이 일어났냐고 묻는다. 이런 상황에서, 팀이 피오나에게 할 말로 가장 적절한 것은 무엇인가?
팀: 미안해. 어두워서 네 자전거를 못 봤어.

어휘
driveway (도로에서 집까지의) 진입로 accidentally 우연히, 잘못하여 twisted (형태가) 뒤틀린, 일그러진 [문제] run over (차가 사람·사물을) 치다

문제풀이
① 누군가가 내 자전거를 쳤어.
② 어두울 때는 자전거를 타면 안 돼.
③ 내 자전거를 세우다가 자동차에 치였어.
④ 그 고양이를 피하려고 했지만, 내가 너무 빠르게 가고 있었어.

16 ⑤ 17 ②

여: 저는 크리스틴 후버 박사이고, 오늘 여러분에게 에너지에 관해 이야기하고자 합니다. 우리는 모두 화석 연료와 같은 전통적인 에너지원을 알고 있습니다. 그리고 여러분은 아마 바람이나 태양으로부터 오는 지속 가능한 에너지에 대해서도 많이 들어보셨을 겁니다. 하지만 여러분은 바이오매스 에너지에 관해서는 잘 모르실 겁니다. 그것들은 잎과 같은 식물의 일부나 식물을 이용함으로써 만들어집니다. 식물들은 끊임없이 재성장하기 때문에, 바이오매스 에너지는 재생 가능한 에너지의 형태로 간주됩니다. 또한, 풍부한 식물 공급이 가능합니다.

우리는 화석 연료처럼 그것을 찾으려고 할 필요가 없습니다. 이 때문에, 바이오매스는 또한 생산하는 것이 더 저렴합니다. 땅에 구멍을 뚫거나 배관을 건설할 필요가 없습니다. 마지막으로, 바이오매스는 유연한 에너지원입니다. 그것은 많은 다양한 형태로 변환될 수 있고, 그래서 그것은 다양한 종류의 유기재로 만들어질 수 있습니다. 그것은 미래의 청정 에너지원이 될 것입니다.

어휘
fossil fuel 화석 연료 sustainable 지속 가능한 biomass 바이오매스(에너지원으로서 이용되는 생물 자원) constantly 끊임없이 regrow (간격을 두고) 재성장하다 renewable 재생 가능한 abundant 풍부한 drill 구멍을 뚫다 pipelines 배관 flexible 유연한, 융통성 있는 convert 변환하다 organic material 유기재 [문제] conserve 보존하다 resource (pl.) 자원 alternative 대체 가능한, 대안의 carbon-free 탄소가 없는

문제풀이
16 ① 다른 에너지원들
② 대체 에너지의 이점들
③ 화석연료 사용의 단점들
④ 에너지원을 보존하는 방법들
⑤ 미래의 대체 에너지원

DICTATION Answers
본문 p.86

01 due to the fact that / take over my duties / over the years

02 hit their lowest prices / getting a new digital camera / all of the latest models

03 sprain your ankle / looking forward to / after some commercials

04 on the left / impressed by the fact / allowed to take pictures

05 It's due tomorrow / time to work / pick her up / flight comes in

06 broke a window / throwing a baseball / pay for the repairs

07 what kind of fabric / at least ten yards / Don't forget to add

08 less than 3,000 / makes our streets safer / have public buses

09 volunteers who work hard / picking up all the trash / be sure to sign up

10 passengers will there be / is located in / get hot really easily

11 the first time

12 fully booked

13 getting ready to / what you mean / if I come along

14 They're imported / We're having a sale / The sale price

15 leaves her bicycle parked / accidentally hits / her twisted bicycle

16~17 traditional energy sources / Since plants constantly regrow / be converted into

01 ①	02 ④	03 ①	04 ⑤	05 ⑤	06 ③
07 ③	08 ④	09 ③	10 ④	11 ①	12 ⑤
13 ⑤	14 ②	15 ④	16 ③	17 ③	

01 ①

남: 고객 여러분, 주목해 주시기 바랍니다. 미아를 찾기 위한 안내 방송입니다. 올랜도 밀러라는 이름의 어린 소년이 행방불명 되었습니다. 그 소년은 10분 전에 어린이 코너에 있었습니다. 여섯 살이며 평균 키와 몸무게의 소년입니다. 갈색 머리카락과 갈색 눈을 가졌습니다. 파란색과 흰색의 줄무늬 셔츠를 입고 뉴욕 양키스 야구 모자를 쓰고 있습니다. 현재 고객 서비스 센터에서 소년의 아버지가 기다리고 계십니다. 만약 이 아이를 보게 된다면, 아이와 아버지가 다시 만날 수 있도록 아이를 고객 서비스 센터로 인도해 주십시오. 감사합니다.

[어휘]
announcement 안내 방송 missing 없어진, 실종된 average 평균의 assist 돕다, 거들다 reunite 재회시키다

02 ④

여: 이 키보드가 또 말을 안 듣네.
남: 무슨 일이야?
여: (글자를) 입력할 때 글자 몇 개가 계속 끼어서 움직이질 않아. 키보드가 너무 낡은 것 같아.
남: 음, 그걸 청소하는 건 어떨까? 키보드에 공기를 뿌리면 먼지가 제거되어서 문제가 해결될 거야.
여: 그건 벌써 해봤지. 효과가 없는 것 같았어.
남: 그럼 내가 그걸 가져가서 수리를 받을게.
여: 상점에서는 그런 서비스를 제공하지도 않을 거야. 키보드는 요즘 비싸지 않으니까 그냥 교체하는 게 더 쉬워.
남: 그러니까, 넌 새 걸 더 좋아하는 거구나.
여: 그게 가장 쉬운 해결 방법인 것 같아.

[어휘]
act up 제 기능을 못 하다 get stuck 꼼짝 못 하게 되다 spray 뿌리다, 살포하다 dust 먼지 do good 도움이 되다 inexpensive 비싸지 않은 replace 대체[교체]하다

03 ①

남: 실례합니다, 크리스틴? 잠시 시간 있으세요?
여: 네, 하지만 빨리 해주세요. 제가 아주 바쁘거든요.
남: 그러죠, 잠깐이면 될 거예요. 지난주 호의 제 기사에 관해 질문이 있어서요.
여: 당신 기사요? 아, 네. 대기 오염의 영향에 대한 기사 말이군요. 그거 아주 좋았어요.
남: 고마워요. 전 단지 당신이 왜 마지막 문장을 바꿨는지 궁금해서요. 그 문장에 잘못된 게 있었나요?

여: 제 기억이 정확하다면, 그 문장의 문법에는 아무런 문제가 없었어요. 전 단지 어색하다고 생각했어요.
남: 그래요? 알겠습니다, 당신 말이 맞을 거예요. 기사의 나머지 부분은 어땠나요?
여: 말씀드린 바와 같이 아주 좋았습니다. 철자 오류가 몇 개 있긴 했지만, 전 당신의 문체가 마음에 들어요. 계속 잘해 주시기 바랍니다.
남: 고맙습니다. 이제 다시 일하도록 당신을 보내드려야겠네요.
여: 고마워요. 오늘 정말 할 일이 많아요.

[어휘]
article 기사 issue 발행 부수, …판[쇄] effect 영향 pollution 오염 correctly 정확하게 grammar 문법 awkward 어색한 spelling 철자 style 문체 keep up …을 유지하다

04 ⑤

여: 안녕, 코리! 너랑 화상 채팅하게 돼서 좋구나! 그게 너의 새로운 방이니?
남: 네, 엄마. 저의 홈스테이 가족들이랑 막 방 꾸미기를 끝냈어요.
여: 네 침대 위에 있는 저건 사진이니?
남: 네, 우리 가족과 제 친구들의 사진들이에요. 제가 해외에서 공부하는 동안 그들이 제 기운을 북돋아 줄 거예요.
여: 좋구나! 책상은 어디에 뒀니?
남: 이 큰 창문 아래에 있어요.
여: 잘했다. 책을 둘 장소는 있어?
남: 책상 옆에 네 칸짜리 선반이 있는 책장이 있어요. 맨 아래 선반이 비어 있거든요. 거기에 제 새 책들을 채울 거예요.
여: 좋은 생각이야. 소파에 있는 건 책가방이니?
남: 네! 두 개의 쿠션 바로 사이에 있네요.
여: 바닥 대부분이 저 깔개로 덮여 있네. 방이 춥니?
남: 아뇨, 그냥 제 방을 더 아늑하게 만들고 싶었어요. 그래서 깔개를 샀어요. 어때 보여요?
여: 줄무늬가 멋져 보이는구나! 학교에서 즐겁게 지내렴!
남: 고마워요, 엄마! 곧 또 얘기해요!

[어휘]
decorate 장식하다 cheer 응원하다, 힘을 북돋우다 bookcase 서가, 책장 couch 소파 rug 깔개 striped 줄무늬의

05 ⑤

여: 마크, 너 내가 부엌 페인트칠 도와준 거 기억해?
남: 물론 기억하지. 정말 큰 도움이 되었는걸. 아직도 너한테 진 신세를 못 갚고 있네.
여: 맞아. 그래서 말인데, 네가 내 새 컴퓨터에 관해 좀 도와줄 수 있나 해서.
남: 네가 컴퓨터를 새로 산 줄 몰랐어.
여: 응, 부모님이 크리스마스 선물로 새 노트북 컴퓨터를 사주셨어. 문제는 내가 그것을 인터넷에 연결하지 못한다는 거야.
남: 나도 널 도와주고는 싶은데, 난 인터넷에 대해 아는 게 없어.
여: 그렇다는 거 알아. 하지만 네 사촌 대니는 잘 알잖아. 컴퓨터 천재잖니.
남: 응, 정말 그래. 그 애한테 전화해 볼까?
여: 그럼 정말 고맙지.

남: 알았어. 그 애가 오늘 밤에 와서 손봐줄 수 있는지 알아볼게.

owe A B A에게 B를 빚지다   laptop 휴대용 컴퓨터   connect 연결되다
genius 천재

06 ③

여: 안녕, 톰. 이번 주말에 뭘 할 거니?

남: 가족들과 제주도로 여행을 갈 거야.

여: 와. 아주 신이 나겠구나.

남: 물론 그렇지. 그런데 내 부탁 좀 들어줄 수 있니?

여: 그래. 뭔데?

남: 우리 가족이 떠나 있는 이틀 동안 내 개를 돌봐줄 사람이 필요해.

여: 음… 내가 해주면 좋은데, 할 수가 없어.

남: 왜?

여: 사실은, 주말에 아르바이트해야 하거든.

남: 네가 아르바이트하는지 몰랐어. 어디서 일해?

여: 집 근처 이탈리아 음식점에서 서빙을 해. 어쨌든, 다른 사람에게 부탁
하는 게 어때?

남: 그래야겠구나.

do ... a favor …의 부탁을 들어주다   part-time job 시간제 일, 아르바이트

07 ③

남: 안녕하세요! Donut World에 오신 걸 환영합니다. 오늘은 뭘 드릴까
요?

여: 제가 초등학교 학부모회에 가는데, 도넛을 좀 가져가려고요. 도넛 20
개가 얼마죠?

남: 어떤 종류의 도넛으로 하느냐에 따라 다릅니다.

여: 음. 저 초콜릿 도넛이 좋아 보이네요. 얼마죠?

남: 저건 하나에 50센트입니다.

여: 알겠습니다. 그럼 젤리가 들어간 도넛은요?

남: 젤리 도넛은 약간 더 비쌉니다. 하나에 1달러입니다.

여: 음, 저것도 좋아 보이네요. 초콜릿 도넛 10개와 젤리 도넛 10개로 할
게요.

남: 알겠습니다. 더 필요한 것 있으세요?

여: 아뇨, 그거면 됐어요. 도와줘서 고마워요.

depend on …에 달려 있다   apiece 각각, 하나에

초콜릿 도넛 10개는 5달러($0.5×10), 젤리 도넛 10개는 10달러($1×10)
이므로, 여자가 지불할 금액은 15달러이다.

08 ④

여: 이건 거대한 미국삼나무야. 세계에서 가장 큰 나무지.

남: 얼마나 크게 자랄 수 있니?

여: 음, 세계 기록은 115미터가 넘어.

남: 그렇게 크게 되려면 시간이 오래 걸리겠구나.

여: 응, 그래. 그런데 미국삼나무는 2천 년을 넘게 살 수 있어.

남: 여기 캘리포니아에서만 자라니?

여: 여기가 대부분의 미국삼나무가 발견되는 곳이지만, 중국에서 자라는
종도 있어.

남: 정말? 그건 몰랐어.

여: 불행히도, 모든 미국삼나무 종은 멸종 위기에 처해 있어. 1850년 이
래로 미국삼나무 숲의 95퍼센트가 사라졌어.

남: 그것 참 유감이네. 아름다운 나무인데.

여: 그 이상이지. 대기로부터 많은 탄소를 없애주기도 하거든.

남: 응. 그리고 그건 지구 온난화를 늦추는 데 도움이 되는 거잖아. 우리
는 그것들을 정말로 보호해야겠네.

redwood (미국)삼나무   species (생물) 종   unfortunately 불행히도
endangered 멸종 위기에 처한   remove 제거하다   carbon 탄소
atmosphere 대기   slow down 늦추다   global warming 지구 온난화

09 ③

여: 공식적으로는 X 상이라고 알려진 Ansari X 상은 1996년 봄에 X 상
재단 덕분에 만들어졌습니다. 재단은 성공적인 우주선을 발사할 수
있는 첫 번째 단체에 천만 달러를 제안했습니다. 우주선은 2주라는
기간 내에 2번의 발사를 할 수 있어야 했습니다. 게다가, 우주선은 사
람이 타고 있어야 하고 정부 기금을 써서 제작될 수 없었습니다. 수년
간의 시도 끝에, 상은 마침내 2004년에 수여되었습니다. 그 상은 민
간 기업과 단체가 우주 수송 분야에 기여할 수 있다는 것을 입증하기
위해 창설되었습니다. 민간 우주 개발 경쟁에 대한 관심을 불러일으
킴으로써, Ansari X 상은 차세대 탐험가를 창출하는 데 도움이 되고
있습니다.

foundation 재단   launch 발사하다; 발사   spacecraft 우주선   manned 사
람을 실은, 유인(有人)의   fund 자금   attempt 시도   demonstrate 입증하다
corporation 기업   contribute to …에 기여하다   transport 수송
generate 생성하다   space race 우주 개발 경쟁   explorer 탐험가

10 ④

여: 우리 회사가 올해 산업 회의를 주최할 예정이에요.

남: 맞아요. 우리는 그걸 개최할 호텔을 선택해야 하죠.

여: 제 상사는 현대적인 장소를 원한다고 하셨어요.

남: 현대적인 장소요? 그게 무슨 말인가요?

여: 그녀는 저희가 2012년이나 혹은 그 이후에 수리한 호텔을 선택하길
바라시더라고요.

남: 알겠습니다. 음, 회의실이 얼마나 많이 필요할까요?

여: 적어도 10개의 회의실을 가진 곳이 필요해요. 8개는 충분하지 않아요.

남: 알겠습니다. 객실은 어떤가요? 200개는 너무 적은가요?

여: 아니에요, 200개면 괜찮을 겁니다. 하지만 그것보다 적어서는 안됩니
다.

남: 저희한테 몇 개의 선택권이 있는 것 같네요. 예산은 어떻게 되죠?

여: 꽤 적어요. 더 저렴한 것으로 합시다.

남: 좋습니다. 지금 바로 호텔에 전화할게요.

host 주최하다   conference 회의   hold 개최하다, 열다   renovate 개조[수리]
하다   a couple of 몇 개의   budget 예산   [문제] venue (행사의) 장소

11 ①

남: 점심으로 뭘 먹을 거야?
여: 피시 앤 칩스를 먹을까 봐.
남: 정말? 넌 생선 먹는 거 안 좋아하는 줄 알았는데!
여: 아니, 난 그저 날생선을 안 먹는 거야.

어휘
fish and chips 피시 앤 칩스(생선튀김에 감자튀김을 곁들인 요리) [문제] raw fish 날생선 lunch box 도시락

문제풀이
② 괜찮아. 아침을 많이 먹었어.
③ 시도해봐야 해. 그리 나쁘지 않아.
④ 아니, 오늘 점심 도시락을 싸 왔어.
⑤ 응, 생선 먹는 것에 관해 많이 알고 있어.

12 ⑤

여: 발표를 잘하는 것에 관한 수업을 듣기 시작했어. 아주 중요한 기술이지.
남: 그래? 작년에 그 수업을 들었는데.
여: 아, 그건 몰랐어. 더 나은 공개 연설가가 되는 데 도움이 된 것 같니?
남: 물론이지. 그 수업에서 자료를 정리하는 법도 가르쳐줬어.

어휘
presentation 발표 public speaker 공개 연설가 [문제] deliver a speech 연설을 하다 definitely 틀림없이 application form 지원서 양식 available 구할 수 있는 organize 정리하다 material 자료

문제풀이
① 미안한데, 난 그 수업을 더 이상 듣지 않아.
② 고마워. 내일 수업에서 만나자.
③ 연설자는 인상적인 연설을 했어.
④ 틀림없어. 지원서는 온라인에서 구할 수 있어.

13 ⑤

[전화벨이 울린다.]
남: 여보세요?
여: 안녕, 척. 나 오빠 동생이야.
남: 안녕, 비키. 잘 지내고 있어?
여: 별일 없어. 지난주에 오빠가 내 잔디 깎는 기계를 빌려 간 거 기억해?
남: 물론이지. 내건 작동이 잘 안 되고, 우리 집 마당의 잔디는 꽤 길게 자라 있었거든.
여: 알아. 하지만 이제는 오빠네 잔디가 짧고 말끔하지?
남: 물론이지. 네 잔디 깎는 기계가 성능이 아주 좋던데.
여: 그럼 이제 돌려줘야 한다고 생각하지 않아?
남: 미안해. 네가 그게 필요한지 몰랐어. 언제든 와서 가져가.
여: 어제 내 차가 고장 났어. 오빠가 우리 집으로 갖다 줄래?

어휘
lawn mower 잔디 깎는 기계 yard 마당 neat 깔끔한 [문제] afford …할 여유가 있다 break down 고장 나다

문제풀이
① 내가 그걸 빌려 가도 괜찮은 거 확실해?
② 맞아. 내 잔디 깎는 기계가 오빠 것보다 더 좋아.
③ 우리가 새 잔디 깎는 기계를 살 여유가 있는지 모르겠네.
④ 그걸 빌려야 했다면 나한테 진작 물어봤어야지.

14 ②

여: 안녕하세요. 오늘은 무엇을 도와드릴까요?
남: 정말 멋진 도서관이네요! 전 며칠 전에 가족과 함께 이곳으로 이사왔어요.
여: 오, Northville에 오신 것을 환영합니다.
남: 감사합니다. 오늘 도서 대출 카드를 신청할 수 있나 해서요.
여: 물론이죠. 신분증 아무거나 보여 주시면 됩니다.
남: 음, 전 아직 운전면허증이 없어요. 하지만 전 대학생이에요.
여: 학생증을 갖고 계세요?
남: 아뇨, 다음 주 월요일에나 받아요.
여: 오. 죄송합니다만, 신분증이 있어야 도서 대출 카드를 발급해 드릴 수 있습니다.
남: 알겠어요, 다음 주에 다시 올게요.

어휘
apply for …을 신청하다 library card 도서 대출 카드 ID 신분증 (= identification) driver's license 운전면허증 [문제] due 지불[제출] 기일이 된 register for …에 등록하다

문제풀이
① 이 책들은 반납 기한이 언제죠?
③ 알겠어요, 여기 제 운전면허증이 있어요.
④ 제가 언제 수강 신청을 할 수 있죠?
⑤ 하지만 전 수년째 이 마을에 살고 있어요.

15 ④

여: 알렉스와 모니카는 마케팅 부서에서 함께 일을 하고 있다. 그들은 새 마케팅 계획을 세우는 팀 프로젝트를 맡았다. 그들은 작업을 반으로 나누기로 했다. 알렉스의 일은 정보를 수집하는 것이고, 모니카의 일은 발표를 하는 것이다. 알렉스는 이미 인터넷에서 모든 필요한 정보를 수집했다. 게다가 그는 발표를 위한 개요까지 적었다. 하지만 모니카는 아직 발표 파일을 만드는 것을 시작하지 않았다. 그녀는 그저 다른 것을 하느라 너무 바쁘다고 말을 할 뿐이다. 알렉스는 발표 파일 없이는 그의 일을 계속할 수 없다고 생각한다. 그는 또한 이것이 팀 프로젝트인데도, 그가 대부분의 일을 하고 있다고 느낀다. 그는 그녀가 자기 일을 제대로 하기를 원한다. 이런 상황에서, 알렉스는 모니카에게 뭐라고 말을 할까?
알렉스: 이 팀 프로젝트에 대해서 좀 더 책임감을 가지세요.

어휘
presentation 발표 outline 개요

문제풀이
① 시작하기 전에 개요를 만들어 주세요.
② 언제 정보 수집을 끝낼 수 있나요?
③ 언제든 당신이 하고 싶을 때 발표 파일을 만드세요.
⑤ 바쁘시다면, 제가 발표 파일을 만들겠습니다.

16 ③ 17 ③

남: 안녕하세요, 여러분! 미국의 지진 희생자들을 돕기 위해 기부하는 것에 관심이 있는데, 누구에게 돈을 보내야 할지 모르겠다고요? 제가

당신을 위한 해결책을 가지고 있습니다. 당신의 돈을 미국 적십자사에 기부하세요. 당신은 이것이 국제 적십자사와 똑같다고 생각할지도 모르겠습니다만, 그렇지 않습니다. 사실, 국제 적십자사 위원회는 1863년에 설립되었습니다. 미국 적십자사는 약 20년 정도 뒤에 클라라 바튼이라는 이름의 미국 여성에 의해 시작되었습니다. 그녀는 국제 기구에서 아이디어를 얻었습니다. 바튼은 교사로서 그녀의 경력을 시작했습니다. 그러나 미국 남북전쟁이 일어나는 동안 그녀는 간호사로 일했습니다. 전쟁이 끝난 뒤, 그녀는 유럽으로 휴가를 떠났습니다. 그곳이 바로 그녀가 국제 적십자사에 대해 배운 곳이었습니다. 그녀가 미국으로 돌아왔을 때, 그녀는 미국 정부에 (적십자사의) 미국 지부의 필요성을 설득시켰습니다. 이것은 뉴욕주에서 단지 몇몇 지사로 시작했습니다. 이제 미국 전역에 650개 이상의 소재지를 가지고 있습니다. 미국 적십자사와 국제 적십자사는 자선 사업을 위한 세계적인 리더가 되었습니다.

어휘
earthquake 지진 victim 희생자 solution 해법, 해결책 found 설립하다 convince A of B A에게 B를 이해시키다[설득하다] branch 나뭇가지; *지사, 분점 charity 자선 (단체)

문제풀이
16 ① 미국 적십자사가 하는 일
② 적십자 기호가 의미하는 것
③ 미국 적십자사는 어떻게 설립되었나
④ 누가 국제 적십자사 위원회를 설립했는가
⑤ 각국의 적십자 지사 사이의 관계

DICTATION Answers

본문 p.92

01 a lost child / average height and weight / waiting for him / be reunited

02 acting up / spraying some air / get repaired

03 make it quick / article in last week's issue / sounded awkward / a couple of spelling errors

04 cheer me up / fill it up with / How does it look

05 owe you a favor / get it to connect / computer genius / come over and fix it

06 going on a trip / take care of / a part-time job

07 20 donuts cost / That depends on / $1 apiece

08 more than 2,000 years / 95% of redwood forests / slow down global warming

09 perform two launches / using government funds / By generating interest

10 to hold it at / that was renovated / no less than that

11 have for lunch

12 how to make good presentations

13 borrowed my lawn mower / return it now / come over

14 may I help you / apply for / driver's license / get you

15 make the presentation / too busy with / do her work properly

16~17 help earthquake victims / was founded in / worked as a nurse / where she learned about

---

16 영어듣기 모의고사
본문 ▲ p.96

01 ②	02 ①	03 ③	04 ⑤	05 ④	06 ②
07 ④	08 ④	09 ⑤	10 ③	11 ⑤	12 ⑤
13 ⑤	14 ⑤	15 ⑤	16 ②	17 ①	

01 ②

여: 안녕하십니까, 여러분. 오늘의 취업 박람회는 앞으로 한 시간 후에 막을 내릴 것입니다. 참석해주셔서 대단히 감사드리며, 폐막 시각 전에 여러분이 관심을 두고 계신 모든 부스에 방문하실 것을 상기시켜 드리고자 합니다. 만약 여러분이 시간이 없어 방문하지 못한 부스가 있다면, 박람회는 이곳에서 금요일까지 열릴 것이며, 매일 오전 10시부터 오후 6시까지 운영합니다. 다양한 분야의 산업체와 전문 직종의 대표들과 함께 하는 저희 취업 박람회는 여러분이 졸업 이후에 첫 직장을 찾을 최고의 장소입니다. 따라서 남은 시간을 최대한 활용하시기 바랍니다. 대단히 감사합니다.

어휘
career fair 취업 박람회 attendance 참석 booth (전시장의) 부스 representative 대표 industry 산업(계) professional 전문적인 graduation 졸업 make the most of …을 최대한 활용하다 remaining 남아 있는, 남은

02 ①

여: 어, 저 스마트폰 봐. 아주 저렴하네.
남: 응, 그런데 상표가 뭔지 모르겠어.
여: 그래도 좋아 보이고 케이스가 포함되어 있네.
남: 음, 품질이 그렇게 좋지 않을지도 모르잖아. 그걸 사기 전에 상표를 조사해 봐야 해.
여: 네 말이 맞는 것 같은데, 그게 어렵지 않을까?
남: 그렇지 않아. 인터넷에서 구매 후기를 살펴보거나 휴대폰 가게의 판매원한테 얘기하면 돼.
여: 괜찮은 생각인 것 같아. 금방 고장이 나는 걸 원하진 않아.
남: 바로 그거야. 잘 알고 있는 게 최상의 선택을 하는 데 도움이 되지.
여: 그래, 네 말이 맞는 것 같아.

어휘
recognize 알아보다 quality 질, 품질 break 고장 나다 informed 잘 아는

03 ③

남: 안녕하세요, 리드 씨. 18일에 관한 추가 정보들이 있습니다.
여: 아, Plaza 호텔을 예약했나요?
남: 네, 했습니다. 그리고 뷔페 식사를 야외에서 제공하는 건 어떨까요?
여: 야외에서요? 그건 생각해보지 않았는데요.
남: 음, 점점 더 많은 고객분께서 야외에서 파티를 여는 것을 선택하고 계십니다. 특히 여름철에는 말이죠.
여: 알겠어요. 좋은 것 같네요. 뷔페에는 뭔가 진척이 있나요?
남: 네. 요청하신 대로 유기농 음식을 제공하는 출장 뷔페 업체를 찾았습

니다.

여: 잘했어요. 초대 손님들 대부분이 환경 운동가라서, 우린 정말 다른 방법으로는 할 수가 없어요.

남: 그렇죠. 좋습니다, 뷔페 선택 사항을 오늘 밤에 이메일로 보내드릴게요.

여: 고마워요, 토비. 당신이 최고예요.

어휘
booking 예약 serve (음식을) 제공하다 host (행사 등을) 주최하다 progress 진척, 진행 catering 음식 공급(업) deal in (상품을) 취급하다 organic 유기농의 request 요청하다 environmental campaigner 환경 운동가 option 선택 (사항)

04 ⑤

[휴대전화벨이 울린다.]

여: 여보세요?

남: 젠킨스 씨, 제 사무실에 콘서트 표를 두고 왔어요. 찾아봐 주실래요?

여: 그럼요. [문 여는 소리] 오! 누군가 화이트보드에 당신의 회의 스케줄을 적어둔 것 같아요.

남: 네, 다음 주에 회의가 몇 차례 있거든요.

여: 그럼 표를 어디에 두었어요?

남: 아마도 제 책상 위에 노트북과 태블릿 PC 사이에 있을 거예요.

여: 아니, 없어요. 그리고 책상 의자 위에도 없는데요.

남: 창문 옆의 식물 근처도 확인해 줄래요?

여: 여전히 보이지 않아요.

남: 아, 미안해요. 지갑 안에서 표를 막 찾았어요. 제 실수예요. 하지만 당신이 거기 있으니, 창문이 닫혀 있는지도 확인해 줄래요?

여: 네, 닫혀 있네요. 또 필요한 건 없으세요?

남: 아뇨. 당신을 귀찮게 해서 미안해요. 내일 봐요!

어휘
whiteboard 화이트보드 wallet 지갑 make sure 확실히 …하다 trouble 괴롭히다; *귀찮게 하다

05 ④

[전화벨이 울린다.]

여: 여보세요, 줄리아입니다.

남: 안녕, 줄리아! 나 마이크야. 태국에서 전화하는 거야!

여: 마이크! 어떻게 지내니?

남: 아주 잘 지내. 휴가를 멋지게 보내고 있지. 넌 어떠니?

여: 잘 지내, 하지만 네가 부럽다. 여긴 폭설이 내렸어.

남: 알아. 인터넷에서 읽었거든.

여: 그래, 무슨 일 있니? 걱정하지 마. 네가 부탁한 대로 네 고양이에게 먹이를 줬어.

남: 오, 고마워! 근데 너에게 부탁할 일이 또 있어.

여: 그게 뭔데?

남: 너도 알다시피 내 신발 가게 앞에 버스 정류장이 있잖아, 내가 그곳에 없으니 가게에 있는 눈을 치울 수가 없어.

여: 아, 알겠다. 사람들이 눈 위에서 미끄러져 넘어질까봐 걱정하는 거구나?

남: 맞아. 그것 좀 관리해 줘. 그렇게 해준다면, 내가 멋진 선물을 사다 줄게!

여: 알겠어. 내가 대신 해 줄게.

어휘
feed 먹이를 주다 (fed-fed) shovel 삽질하다. 삽으로 파다[옮기다] slip 미끄러지다 fall 넘어지다

06 ②

여: 안녕, 제임스. 네 새 전화기는 어떠니?

남: 전화기는 괜찮은데, 솔직히 휴대전화 통신 업체를 바꿀까 생각 중이야.

여: 정말? 다른 회사의 서비스 제도가 지금 것보다 더 좋은 거니?

남: 꼭 그런 건 아니야. 현재 내가 이용하고 있는 통신 업체에 짜증이 나서.

여: 문제가 뭔데?

남: 음, 지난주에 통신망 유지 관리 작업 중이어서 (전화) 수신이 전혀 안 됐어.

여: 안 좋구나.

남: 맞아. 또, 통신망이 매우 느려서 때때로 인터넷을 쓰기가 어려워.

여: 아주 불만스럽겠네.

남: 응. 더 나은 서비스를 제공하는 통신 업체로 바꾸고 싶어.

어휘
to be honest 솔직히 말해서 provider 제공자, 제공 기관 maintenance 유지 관리 reception 접수처; *수신 상태 frustrating 불만스러운

07 ④

남: 와, 여기 예쁜 기념품들이 많네요.

여: 네. 둘러보세요.

남: 이 인상파 작품 달력은 얼마죠?

여: 12달러입니다.

남: 저희 어머니에게 딱이에요. 이걸로 살게요.

여: 탁월한 선택이에요. 그리고 저쪽에 다양한 머그잔들도 보셨나요?

남: 네, 예쁘네요. 반 고흐의 작품이 (그려져) 있는 게 마음에 들어요.

여: 저도 그렇답니다. 이 머그잔들은 하나에 10달러예요.

남: 좀 비싸네요. 박물관 로고가 있는 더 작은 크기의 이 컵은 얼마예요?

여: 그건 6달러입니다.

남: 좋아요, 그걸로 할게요.

여: 네. 그게 다인가요?

남: 아, 이 펜도 두 개 살게요. 개당 1달러네요.

여: 알겠습니다.

어휘
souvenir 기념품 take a look around 주위를 둘러보다 Impressionist 인상파의 range *다양성; 범위 mug 머그잔 artwork 미술품

문제풀이
남자는 달력 한 개($12)와 작은 컵($6), 두 개의 펜($1×2)을 산다고 하였으므로, 20달러를 지불해야 한다.

08 ④

여: 오늘 밤 초대손님은 론 그린입니다. 그는 이제 막 국회의원으로 선출되었습니다. 환영합니다, 론.

남: 고마워요, 새라. 초대해 주셔서 기쁩니다.

여: 자, 지난주 선거에 몇 명이 참여했나요?

남: 선거구에서 대략 만 6천 명이 투표했어요.
여: 그렇군요. 그럼 그 표 중에서 몇 표를 받으셨죠?
남: 거의 만 3천 표를 받았는데, 그건 전체의 70퍼센트가 넘어요.
여: 선거 운동이 왜 그렇게 성공적이었다고 생각하세요?
남: 사람들이 세금을 낮추고 고속도로 체계를 개선하겠다고 약속한 사실을 마음에 들어하신 것 같아요.
여: 선거 운동에 비용이 얼마나 들었는지 여쭤봐도 될까요?
남: 물론이죠. 약 4백만 달러를 썼어요. 많은 돈이지만 그럴 가치가 있었죠.
여: 네, 행운을 빌겠습니다.
남: 저를 쇼에 초대해 주셔서 감사합니다.

elect 선출하다 the National Assembly 국회 election 선거 district 지구, 지역 vote 투표하다; (선거의) 표 worth …의 가치가 있는

09 ⑤

남: 바쁜 스케줄과 패스트푸드의 세상에서 몇몇 사람들은 속도를 늦출 방법을 찾고 있습니다. 슬로우푸드 운동은 다른 이들과 음식을 나눠 먹는 기쁨을 회복하는 데 목표를 두고 있습니다. 이 운동은 1980년대에 이탈리아에서 패스트푸드 음식점 개업에 대응하여 시작되었습니다. 이 운동의 참가자들은 식사할 때 음식과 대화를 함께 나누는 즐거움에 중점을 둡니다. 설립 이래로 이 운동은 150개가 넘는 나라에 퍼졌고, 현재 약 10만여 명의 회원들이 있습니다. 사교적 활동으로서의 음식에 중점을 두는 이 운동은 음식이 영양분의 근원뿐만 아니라 사교적 즐거움이 될 수 있다는 것을 우리에게 상기시켜 줍니다.

slow down 속도를 늦추다 be aimed at …을 목표로 하다 restore 회복하다 get started 시작되다 in response to …에 대응하여 found 설립하다 spread 퍼지다 social 사회의, 사회적인 remind 상기시키다 nutrient 영양분

10 ③

남: 이게 네가 졸업 무도회를 위해 고려하고 있는 드레스들이니?
여: 응. 어느 것을 주문할지 결정할 수가 없어.
남: 음, 어떤 길이를 원해?
여: 긴 것만 빼고 아무거나. 긴 드레스를 입고서는 춤추기 너무 어려워.
남: 그건 그래. 그리고 두 개의 다른 스타일이 있는 것 같은데.
여: 응. 단순한 건 A자형 드레스들이고 화려한 건 무도회용 드레스야.
남: 무도회용 드레스들이 예쁘지만, 졸업 무도회에는 너무 과해. 난 단순한 게 좋아.
여: 나도 그래. 이제 색을 골라야 해. 솔직히 난 다 마음에 들어.
남: 초록색은 사지 말자. 난 네 머리카락 색이랑 초록색은 안 어울린다고 생각해.
여: 사실, 네 말이 맞아. 그리고 난 드레스에 100달러 넘게 쓸 수가 없어.
남: 그러면 여기 있는 이게 가장 좋은 선택이다.
여: 응! 지금 주문할게. 나한테 잘 어울렸으면 좋겠다!

consider 고려하다 prom 졸업 무도회 length 길이 make sense 의미가 통하다, 이해가 되다 fancy 화려한 ball gown 무도회용 드레스 to be honest 솔직히 말하면

11 ⑤

여: 이봐, 조슈아, 너 괜찮니? 오늘 안 좋아 보여.
남: 지난 주말부터 두통이 있었어.
여: 왜 병원에 가보지 않았니?
남: 갔지. 약을 먹었는데 도움이 안 됐어.

headache 두통 since …부터, …이후 [문제] clinic 병원 medicine 약

① 알아. 전에 그를 만났던 것 같아.
② 너는 주중에 그를 만나야 해.
③ 병원은 모퉁이를 돌면 바로 있어.
④ 난 네게 마을에서 최고의 병원이 어디에 있는지 알려줄 수 있어.

12 ⑤

남: 안녕, 메리. 오늘 공항에 갈 거지, 그렇지?
여: 응, 이번이 내 첫 해외여행이야. 오랫동안 이 순간을 기다려왔어.
남: 몇 시에 비행기가 이륙하니?
여: 세 시간 정도 후에. 거기 가려면 난 서둘러야 해.

abroad 해외로 take off 이륙하다 [문제] available 구할[이용할] 수 있는

① 기차로 30분이 걸려.
② 난 지난달부터 기다려왔어.
③ 지금 6시 30분이야. 제시간에 거기로 쉽게 갈 수 있어.
④ 괜찮아. 표는 내일 구할 수 있을 거야.

13 ⑤

여: 안녕하세요, 선생님. 영어 수업 성적 때문에 상의 좀 드리려고요.
남: 그래. 들어와라. 네가 마리아구나, 맞지?
여: 네, 맞아요. 안타깝게도 선생님의 수업에서 낙제했어요.
남: 맞아. 기말고사에서 성적이 안 좋았던 것 같더구나.
여: 네. 전 수업 내용은 이해했는데, 초조해지고 당황했어요. 시험에 응시할 때 스트레스를 많이 받아요.
남: 유감이구나. 다른 성적도 한번 보자.
여: 감사합니다. 쪽지 시험에서는 성적이 더 좋았어요. 선생님의 수업을 정말 통과하고 싶어요.
남: 음, 네 말이 맞아. 네 쪽지 시험 성적은 기말고사 성적보다 훨씬 좋구나.
여: 선생님 수업에서 정말 열심히 노력했어요. 제 성적을 바꿔주실 수 있나요?
남: 미안하지만 그럴 순 없어. 지금 네 출석 기록을 보고 있는데, 넌 8번이나 수업에 결석했구나.
여: 네, 하지만…
남: 학교 규칙상 결석은 최대 5번까지 할 수 있어.
여: 그럼 선생님께서 해주실 수 있는 건 없는 건가요?
남: 미안하다. 넌 이 수업을 재수강해야 할 거야.

discuss 상의하다 fail 실패하다; *낙제하다 (↔ pass) stressed out 스트레스를 받는 attendance 출석 policy 정책 absent 결석한 maximum 최대

(의) [문제] Give it a shot. 한번 시도해 봐. admit 인정하다 repeat 되풀이하다

문제풀이
① 네 최종 점수는 A가 될 거야.
② 한번 해보렴. 그가 받아들여 줄지도 몰라.
③ 네 출석률이 좋았던 건 인정한다.
④ 그럴 필요 없어. 넌 성적이 아주 우수했잖니.

14 ⑤

남: 안녕하세요, 전 리 박사입니다. 우리가 이 비행기에서 서로 옆에 앉아서 가는 것 같네요.
여: 만나서 반갑습니다.
남: 잠시만요, 낯이 익네요! 유명한 가수시죠, 그렇죠?
여: 절 알아보시다니 놀라운데요. 대체로 제 팬들은 박사님보다 훨씬 더 어려서요.
남: 아, 전 어린이 병원의 암 전문의인데 제 환자 중 다수가 당신의 팬이죠.
여: 아, 이제 알겠네요.
남: 실은… 뭐 좀 여쭤봐도 될까요?
여: 물론이죠.
남: 자원봉사로 시간을 내서 병원에서 작은 콘서트를 열어 주시면 어떨까요? 아이들이 아주 기뻐할 겁니다.
여: 오, 정말 영광이에요.
남: 고맙습니다. 정말 감사드려요.
여: 천만에요. 도울 기회를 주신 것에 대해 제가 박사님께 감사드려야죠.

어휘
familiar 친숙한 recognize 알아보다 cancer 암 specialist 전문의(醫) volunteer 자원하다 delighted 아주 기뻐하는 honor 명예, 영광 [문제] autograph 서명 celebrity 유명인사 opportunity 기회

문제풀이
① 사인해 드려서 기뻐요.
② 전 유명인사 옆에 앉을 거라고는 생각도 못 했어요.
③ 죄송하지만, 전 자원봉사 일을 할 수가 없어요.
④ 박사님께서는 방금 제가 했던 콘서트에 참석하셨어야 했어요.

15 ⑤

남: 제인은 고등학생이고 학교에서 멀리 떨어져 살고 있다. 그녀의 동네에는 지하철이 없고 버스와 택시도 거의 없다. 그녀의 부모님이 매일 그녀를 학교에 태워다 주시곤 했는데, 그녀의 오빠가 최근에 운전 면허증을 취득했다. 이제 그는 종종 그녀를 학교에 태워다 주고 그녀가 쇼핑을 가야 할 때 쇼핑몰에 태워다 준다. 그는 언제나 천천히, 조심스럽게 운전을 하지만, 한 가지가 제인을 걱정하게 한다. 그녀의 오빠는 운전 중에 전화기에 문자메시지를 받으면, 보통 즉시 그것들을 읽는다. 때때로 그는 심지어 그 메시지에 답을 하기도 한다. 그녀는 그것이 결국은 사고를 낼까봐 두렵다. 이런 상황에서, 제인은 그녀의 오빠에게 뭐라고 말하겠는가?
제인: 운전 중에는 도로에 집중해 줘.

어휘
driver's license 운전 면허증 eventually 결국은, 마침내 [문제] option 선택 (사항) concentrate on …에 집중하다

문제풀이
① 버스를 타고 학교에 가는 것이 네 최고의 선택이야.
② 나는 서둘러야 해. 학교로 태워다 줄 수 있어?
③ 우린 무슨 일이 일어났는지 엄마와 아빠에게 이야기해야 해.
④ 내가 보낸 메시지에 왜 답을 안 했어?

16 ② 17 ①

여: 암에 관한 한, 예방과 조기 발견이 당신이 취할 수 있는 최상의 수단입니다. 대부분의 암 형태를 식별하는 데는 임상 검사가 필요하지만, 피부암은 단지 육안만으로도 종종 발견할 수 있습니다. 개개인은 그들의 피부에서 의심스러운 부분을 스스로 점검해 볼 수 있습니다. 정기적으로 몸 전체를 점검해야 합니다. 이렇게 하면 어떤 변화가 생겼는지 당신이 알 수 있게 됩니다. 모양이 고르지 않아서, 양쪽이 서로 대칭의 형상이 아닌 점을 찾아보세요. 점 주변의 불규칙한 가장자리도 문제의 징후일 수 있습니다. 점의 색깔에도 유의해야 합니다. 동일한 점에서 다른 색들이 나타나면, 의사에게 보고해야 합니다. 또한, 시간이 지나면서 크기나 모양이 변화하는 점도 고려할 또 다른 이유가 됩니다. 위의 증상 중 어떤 것이라도 즉시 의사에게 보고해야 합니다. 그렇게 하면, 추가적인 검사가 시행될 수 있습니다. 이 조언들을 부지런히 따름으로써 통제할 수 없게 되기 전에 피부암을 막을 수 있다는 걸 기억하십시오.

어휘
prevention 예방 detection 발견 (v. detect) laboratory 실험실 identify 식별하다 sight 시력; 보기, 봄 examine 조사하다; *검사[진찰]하다 (n. examination) suspicious 의심스러운 inspect 점검하다 mole (피부의) 점 uneven (무늬나 형태가) 고르지 않은 mirror image 거울상(像), 상호 대칭성 구조 border 경계, 가장자리 concern 우려, 걱정 symptom 증상 diligently 부지런히, 열심히 [문제] breakthrough 돌파구

문제풀이
16 ① 점들의 다른 특성
② 피부암의 자가 진단
③ 피부암 예방의 새로운 돌파구
④ 피부암 임상 검사의 장점
⑤ 피부암 증상을 보고하는 방법에 대한 조언

DICTATION Answers
본문 p.98

01 your attendance / closing time / the remaining hour
02 recognize the brand / look at reviews / make the best choice
03 make the booking / make any progress / the buffet options
04 left the concert tickets / check near the plant / the window is closed
05 had heavy snow / shovel the snow away / slip and fall
06 annoyed with my current company / hard to use / better service
07 lots of beautiful souvenirs / I'll take it / a dollar each
08 took part in / nearly 13,000 votes / lower taxes
09 in response to / sharing food and conversations / as a

        social activity

10   which one to order / goes well with / looks good on me

11   had a headache

12   your flight take off

13   discuss my grade / poorly on the final test / did much better / a maximum of

14   you look familiar / volunteering your time / be an honor

15   far away from / got his driver's license / cause an accident

16~17 any suspicious areas / pay attention to / further testing / out of control

17 영어듣기 모의고사

본문 ▲ p.102

01 ③	02 ⑤	03 ①	04 ⑤	05 ③	06 ④
07 ④	08 ④	09 ③	10 ①	11 ⑤	12 ②
13 ④	14 ⑤	15 ②	16 ①	17 ①	

01 ③

남: 지난번 공연을 놓쳤던 모든 팬 여러분에게 좋은 소식이 있습니다! *맘마미아*가 돌아왔습니다! Abba의 노래를 기반으로 한 이 세계적인 인기 뮤지컬은 지난 3개월간의 서울 공연에서 전석 매진이었습니다. 이번 맘마미아 앙코르 공연에는 세계적으로 유명한 배우들이 출연할 것이며, 이번 기회를 놓치신다면 다시는 그들 모두를 한 무대에서 보기 어려울지도 모릅니다. 앙코르 공연은 한국 음악 홀에서 1월 14일부터 1월 31일까지 공연될 것입니다. 표가 금방 매진될 것으로 예상되므로, 오늘 당신의 표를 구입하세요!

어휘
global 세계적인 sold out 매진된 run 장기 공연; (연극 등이) 계속 공연되다 cast (연극·영화의) 출연진, 배역들 encore 앙코르, 재청 season 공연[상영] 기간 feature (배우를) 주연시키다

02 ⑤

남: 내일 시험이 또 하나 있어. 시험 보는 게 진저리가 나.
여: 아, 나도 그래. 우리는 시험을 너무 많이 보잖아. 선생님이 우리의 (실력) 향상을 점검하는 더 나은 방법이 있는 데 말이야.
남: 어떤 종류의 평가를 생각하는 거야?
여: 연구 프로젝트지. 단지 사실을 기억하는 것 대신에 수업에서 배운 것을 연구할 수 있잖아.
남: 그게 더 흥미로울 것 같아.
여: 전에 그런 역사 프로젝트를 했었거든. 역사 사건들에 대해 더 깊게 이해할 수 있었지.

남: 수업 시간에 발표할 수도 있겠는데.
여: 그래, 학교에서 우리가 평가받을 수 있는 더 나은 방법들이 확실히 있어.

어휘
be sick of …에 싫증 나다 progress 진보, 향상 assessment 평가 (v. assess)

03 ①

남: 좋아, 오늘 수업은 여기까지야.
여: 감사합니다, 앤더슨 선생님. 저는 피아노 치는 게 너무 좋아서 집에서도 연습하고 싶어요.
남: 그거 좋지. 집에 피아노가 있니?
여: 아뇨, 하지만 곧 하나 사려고요.
남: 어쿠스틱 피아노를 살 거니, 아니면 디지털 피아노를 살 거니?
여: 실은 그것에 대해 선생님의 조언을 구하고 싶었어요.
남: 넌 주택에 사니, 아파트에 사니?
여: 가족들과 아파트에서 살아요.
남: 그럼 디지털로 된 것을 사는 게 좋겠구나.
여: 왜요?
남: 디지털 피아노는 밤에 사람들을 방해하지 않도록 네가 헤드폰을 이용해서 연습하게 해주거든.
여: 그거 좋은 점이네요.

어휘
acoustic 음향의, 청각의; *전자 장치를 쓰지 않는 disturb 방해하다

04 ⑤

여: 우리가 빅 벤을 보고 있다니 믿기지 않아.
남: 뭐? 탑 꼭대기에 있는 저 시계? 나는 전에 본 적이 있어.
여: 하지만 사진에서만이잖아. 대부분 사람들은 빅 벤을 직접 보고 싶어 해.
남: 그게 저 관광객들이 그 앞에서 사진을 찍는 이유겠지.
여: 물론이야. 템스 강에 있는 저 배를 봐.
남: 응, 많은 관광객들을 나르고 있어. 저 여자가 타고 있는 버스가 보이니?
여: 그건 이층버스야. 런던에서는 흔한 거야.
남: 2층으로 된 버스야? 낯설어 보여.
여: 그럴 거야. 탑 옆에 두 마리 새들이 보이니?
남: 난 새 한 마리만 보이는데. 그건 탑 주위를 날고 있어.
여: 탑 위에 하나가 더 앉아 있어. 저기에 사는 것 같아.

어휘
in person 직접 get on …에 타다 double-decker bus 이층버스 story 이야기; *(건물의) 층

05 ③

남: 제인, 당신이 하는 일이라곤 TV를 보는 것뿐이군요!
여: 음, 직장에서 긴 하루를 보내고 집에 오면 피곤해요.
남: 나도 마찬가지예요. 나만 항상 집안일을 하는 게 공평한가요?
여: 글쎄요, 아니겠죠. 하지만 난 어떻게 하는지 몰라요.
남: 식기 세척기를 사용해 본 적이 있어요?

여: 아뇨, 없어요.

남: 그럼, 세탁기는 사용해 본 적 있어요?

여: 한번 해 봤어요. 근데 내가 옷을 다 망쳐놨잖아요. 당신은 정말 화를 냈고요, 기억나요?

남: 그럼, 그냥 이 걸레를 들고 방마다 돌아다니면서 탁자와 선반의 먼지를 닦아요.

여: 네. 그건 할 수 있어요.

어휘
fair 공평한 housework 집안일 dishwasher 식기 세척기 washing machine 세탁기 ruin 망치다 cloth 천; *걸레 wipe A off B B에서 A를 닦아내다 dust 먼지

06 ④

여: 오랜만이야, 토마스!

남: 안녕, 잘 지내니?

여: 난 좋아. 요즘 어떻게 지내니?

남: 주중에 아르바이트를 두 개 하느라 바쁜 것 빼곤 꽤 괜찮아.

여: 와, 피곤하겠구나! 공부와 일을 어떻게 동시에 해낼 수 있니?

남: 사실 상당히 힘들어. 일을 그만둘까 생각 중이야.

여: 그래? 하지만 일을 하는 것도 이점이 있잖아. 예를 들어, 돈을 벌고 유용한 경험을 얻을 수 있어.

남: 맞는 말이야. 그런데 일이 공부하는 시간을 빼앗아가잖아. 그래서 이번 학기에 성적이 떨어졌어.

여: 그런 경우라면 아무래도 관두는 게 옳은 결정인 것 같아.

남: 동의해. 어쨌든, 나는 일하러 가야겠어. 나중에 보자!

어휘
tiring 피곤한 manage to-v (가까스로) 해내다 quit 그만두다 earn 벌다 gain 얻다 practical 실용적인, 유용한

07 ④

여: 안녕하십니까, 손님. PC World에 오신 것을 환영합니다.

남: 감사해요. 제 딸에게 줄 컴퓨터를 사고 싶습니다. 도와주시겠어요?

여: 물론이죠. 어떤 모델에 관심이 있나요?

남: 이 17인치 LCD 모니터가 있는 데스크탑은 얼마죠?

여: 75만 원입니다. 같은 컴퓨터지만 더 큰 19인치 모니터를 구입하시면 5만 원이 추가됩니다.

남: 그렇군요. 딸애는 프린터도 필요한 것 같아요. 좀 있나요?

여: 두 가지 모델이 있습니다. 잉크젯 프린터는 10만 원이고 레이저 프린터는 25만 원입니다.

남: 알겠습니다. 컴퓨터와 19인치 모니터를 살게요.

여: 프린터도 사시겠습니까?

남: 네. 레이저 프린터로 할게요.

문제풀이
17인치 LCD 모니터가 있는 데스크탑은 75만 원인데 남자는 이 사양에 19인치 모니터를 산다고 했으므로 5만 원이 추가되었고, 여기에 25만 원인 레이저 프린터도 구매했으므로, 남자가 지불할 금액은 105만 원이다.

08 ④

여: 오늘 밤 계획 있니?

남: 아니, 없어.

여: 음, 새 영화가 지난 목요일에 개봉했어. 보러 가자.

남: 제목이 뭐야?

여: 제목은 *해적 공격*이야. 내가 좋아하는 영화 웹사이트에서 좋은 평을 받았어. 본 사람들이 10개 별 중에 9.5개 별을 줬어.

남: 액션 영화니?

여: 응. 멋진 특수 효과가 있을 걸로 예상돼. 그리고 3D야!

남: 신날 것 같아. 몇 시에 상영하니?

여: 세 번 상영해. 첫 번째가 6시고 다음이 8시와 10시에 있어. 저녁을 먹은 다음 오후 8시 영화를 보러 가는 게 어떨까?

남: 음, 얼마나 오래 하는지 아니? 부모님께서 내가 너무 늦게까지 나가 있는 걸 허락하지 않으시거든.

여: 90분 동안 상영해.

남: 그렇다면 괜찮아.

어휘
release 개봉하다 pirate 해적 review 논평, 비평 be supposed to-v …인 것으로 여겨진다, …로 예상된다 special effect 특수 효과

09 ③

여: 새로 오신 윈터 선생님을 환영해 주세요. 여러분께 간단히 선생님을 소개하겠습니다. 선생님은 캘리포니아 대학을 졸업하시고 계속해서 그곳에서 대학원 과정까지 마치셨습니다. Harper에 오시기 전, 선생님은 North Sacramento의 Norwood 중학교에서 교편을 잡으셨습니다. 선생님은 이곳 Harper 고등학교에서 영어를 가르치실 겁니다. 여가 시간에, 윈터 선생님은 아마추어 시립 합창단에서 노래하고 계시며, 시립 대회의장에서의 음악회에서 공연을 해오고 계십니다. 선생님은 두루 여행하셨고, 5명의 자녀가 있으며, 시험지 채점하기를 좋아하신답니다!

어휘
briefly 간단히, 잠깐 graduate 졸업하다; 대학원의 choir 합창단 extensively 널리, 광범위하게 grade 등급; *채점하다

10 ①

남: 신혼여행을 어디로 가야 할지 찾으려면 이 안내 책자를 봅시다.

여: 아, 난 항상 시드니에 가는 것을 꿈꿔왔어요.

남: 나도 그랬는데, 시드니는 너무 비싸요. 우리의 예산은 2천 달러로 제한되어 있어요.

여: 알아요. 뭐, 괜찮아요. 당신은 휴가를 며칠 낼 수 있어요?

남: 바쁜 시기라서 난 5일 넘게 (휴가를) 낼 수 없어요.

여: 그렇군요. 음, 좋아요, 하지만 여행 가이드는 필요해요.

남: 음… 내 생각엔, 뭘 하고 뭘 먹을지 우리 스스로 결정하는 게 재미있을 것 같아요.

여: 하지만 이건 우리 신혼여행이에요. 아무것도 신경 쓰지 않고 쉴 수 있으면 좋겠어요.

남: 그래요, 좋아요. 그럼 우리가 갈 수 있는 곳은 딱 두 곳인 것 같군요. 어떤 장소가 더 좋아요?

여: 아, 사은품을 볼까요. 음… 저는 확실히 와인보다는 마사지가 좋아요.

남: 나도 그래요. 내가 예약을 할게요.

어휘
honeymoon 신혼여행 take off …(동안)을 쉬다 [문제] destination 목적지.

도착지 complimentary 무료의

11 ⑤

남: 린지, 내 지갑을 본 적 있니?
여: 아니, 못 봤어. 점심시간 전에는 가지고 있었니?
남: 그런 것 같아. 교실을 떠날 때 가지고 가는 걸 잊었나 봐.
여: 걱정하지 마. 우린 확실히 찾을 수 있을 거야.

어휘
wallet 지갑 forget to-v …하는 것을 잊다

문제풀이
① 넌 지갑을 잃어버릴 뻔했어.
② 내일 네게 점심을 사줄 수 있어.
③ 넌 제시간에 교실에 들어와야 해.
④ 점심시간에 날 방문하는 걸 잊지 마.

12 ②

여: 안녕, 재키! 아직도 밴드에서 연주하니?
남: 응, 우린 금요일마다 방과 후에 연습해. 와서 우리가 연주하는 거 들어볼래?
여: 좋을 것 같아. 너 아직도 드럼을 치지, 그렇지?
남: 아니, 난 요즘 전기 기타를 연주해.

어휘
[문제] electric guitar 전자 기타 price 가격을 매기다 reasonably 상당히;
*합리적으로

문제풀이
① 아니, 우리 밴드는 금요일에만 연주해.
③ 응, 밴드는 그만두었지만 아직 드럼을 쳐.
④ 응, 그 드럼들은 가격이 적정한 것 같아.
⑤ 유감이지만 우린 악기가 하나도 없어.

13 ④

여: 뭘 읽고 있니, 마틴?
남: 이번 토요일에 부모님을 모시고 갈 만한 좋은 식당을 찾는 중이야. 이 잡지에서 괜찮은 곳을 몇 군데 추천하고 있거든. 봐, 너 여기 가 본 적 있어?
여: 어디?
남: 여기 말이야. 서울에서 제일 맛있는 이탈리안 레스토랑 중의 하나라는데.
여: The Sweet Garden? 아, 말도 안 돼. 거기 가 봤는데, 내가 가 본 이탈리안 레스토랑 중의 최악이야.
남: 어떤 면에서 그러니?
여: 글쎄, 잡지에서 말한 것처럼 분위기는 괜찮았어. 근데 음식은 비싼 가격을 생각해 봐도 그냥 그랬어. 하지만 가장 끔찍했던 건 서비스였어!
남: 뭐가 문제였는데?
여: 웨이터들이 손님들에게는 신경도 안 쓰고 계산대에 모여서 자기들끼리 수다를 떨고 있더라고. 음식은 식은 채로 나왔고. 난 저녁 식사 후에 불만을 얘기했어.
남: 아무래도 다른 식당을 찾아봐야겠다.

어휘
spot 장소, 지점 ridiculous 우스꽝스러운, 터무니없는 atmosphere 대기;
*분위기 average 평균의, 보통의 consider 고려[생각]하다 chat 잡담하다

문제풀이
① 그밖에 또 필요한 게 있니?
② 나랑 같이 또 거기에서 식사할래?
③ 네 말을 주의 깊게 들었어야 했는데.
⑤ 그럼 빨리 예약하는 게 낫겠다.

14 ⑤

남: 뭘 하고 있니, 새라?
여: 30분 넘게 제 새 스카프를 찾고 있어요. 그걸 보셨어요, 아빠?
남: 저쪽에 있는 저것 아니니?
여: 어디요?
남: 네 코트 옆에 벽에 걸려 있는 것 말이야.
여: 아뇨, 그건 예전 스카프예요. 제 새 스카프는 빨간색에 검은 줄무늬가 있어요.
남: 아. 오늘 아침에 켈리가 두르고 있는 걸 본 것 같구나.
여: 정말요? 켈리가 새로 산 제 스카프를 하고 있었다고요?
남: 응. 그 애는 그게 자신의 새 코트와 잘 어울린다고 했어.
여: 믿을 수가 없어요! 그 스카프 바로 어제 산 건데.
남: 이런. 그 애에게 너무 화내지 말거라.
여: 전 그저 제 물건을 빌리기 전에 그 애가 저한테 먼저 물어보면 좋겠어요.

어휘
hang 걸다, 걸리다 [문제] prefer A to B B보다 A가 더 좋다 beyond …의 범위를 넘어서 stuff 물건

문제풀이
① 전 새 스카프보다 예전 스카프가 더 좋아요.
② 그것은 제 예산을 초과하네요. 살 수가 없어요.
③ 그 애가 새 코트를 어디서 샀는지 알면 좋겠어요.
④ 새로 산 제 스카프를 못 찾겠어요. 어디 있는지 아세요?

15 ②

남: 마이크와 그의 할머니는 매우 각별한 사이다. 그분은 그를 무척 사랑하신다. 다음 주에 마이크는 대학을 졸업하게 된다. 그곳(대학교)에 가기 위해 장거리를 여행해야 하는 할머니를 포함해서 그의 가족 모두가 참석할 계획이다. 안타깝게도, 그의 졸업식을 이틀 앞두고 마이크의 할머니가 편찮으시다. 심각한 것은 아니지만, 졸업식 참석을 위해 거동하시기는 힘들 정도이다. 할머니는 마이크에게 전화를 해서 자신의 상황을 설명하고 졸업식에 가지 못하는 것에 대해 사과를 하신다. 이런 상황에서, 마이크가 그의 할머니께 할 말로 가장 적절한 것은 무엇인가?
마이크: 걱정하지 마세요. 어서 쾌차하시기만 하세요.

어휘
prevent A from v-ing A가 …하는 것을 막다 graduation ceremony 졸업식 apologize 사과하다

문제풀이
① 여행이 즐거우셨다니 기뻐요.
③ 엄청난 실수를 저지른 데 대해 사과드리고 싶어요.

④ 제게 이 선물을 보내주시다니 정말 친절하시네요.
⑤ 그러고 싶지만 안 되겠어요. 다음에 해요.

16 ① 17 ①

여: 4천 년이 넘는 시간 동안, 깃발은 사람들이나 조직이 그들 자신을 나타내고, 메시지를 보내고, 사상을 표현하는 것을 도왔습니다. 중세 시대에는, 기사들이 무거운 갑옷을 입고서 전투에 나갔고, 그래서 누가 누구인지를 구분하는 것은 어려웠습니다. 그래서 그들은 적들로부터 그들 자신을 구분하기 위해서 깃발을 들고 다니기 시작했습니다. 오늘날, 거의 모든 조직들은 깃발을 가지고 있습니다. 예를 들어, 걸스카우트는 세계의 어린이들 위로 빛나는 태양을 상징하기 위해 푸른 바탕에 금빛의 세잎 식물을 사용하는데, 이것은 젊은 여성과 소녀들이 세계를 통틀어 능동적인 지도자가 되도록 격려합니다. 미국의 각 주 정부들도 그 주의 핵심적인 가치를 나타내는 그들 고유의 깃발을 가지고 있습니다. 올림픽 깃발에서는, 다섯 개의 원이 다섯 개의 대륙에 상응합니다. 유엔의 깃발에는, 지구를 떠받치는 올리브 가지들이 세계 평화를 나타냅니다. 심지어 개인들도 깃발을 가지고 있습니다. 왕이나 여왕들은 종종 그들이 현재 머무는 건물에 개인적인 깃발이 휘날리도록 합니다.

어휘
organization 조직, 단체 represent 표현하다 battle 전쟁 armor 갑옷 distinguish A from B B에서 A를 구별하다 trefoil 세잎 식물 symbolize 상징하다 core 핵심의, 가장 중요한 correspond to …에 상응[일치]하다 signify 의미하다, 나타내다 fly 날다, 날리다 (flew-flied[flown]) [문제] function 기능 evolution 진화

문제풀이
16 ① 깃발들의 목적과 기능
   ② 역사를 통틀어 본 깃발의 진화
   ③ 자기표현을 위한 깃발의 중요성
   ④ 현대에 가장 흥미로운 깃발들
   ⑤ 중세 시대의 깃발 제작 과정

DICTATION Answers
본문 ▲ p.104

01 this global hit musical / miss this opportunity / sell out

02 sick of taking tests / Instead of just remembering facts / give presentations

03 practice at home / ask your advice / disturb people at night

04 in person / carrying lots of tourists / flying around the tower

05 Is it fair / ruined all the clothes / wiping the dust off

06 at the same time / quitting my jobs / my grades have dropped

07 this desktop / need a printer / take the computer / the laser printer

08 got good reviews / have great special effects / stay out too late

09 introduce her to you briefly / In her free time / has traveled extensively

10 go for our honeymoon / take off / make the reservations

11 forgot to take it

12 playing in a band

13 recommends some great spots / I've been to / chatted with each other

14 Hanging on the wall / black stripes / looked good with

15 travel a long distance / prevent her from traveling / apologize for

16~17 to distinguish themselves / become active leaders / is flown above the building

18 영어듣기 모의고사
본문 ▲ p.108

01 ③	02 ⑤	03 ③	04 ⑤	05 ①	06 ②
07 ④	08 ④	09 ③	10 ③	11 ①	12 ④
13 ①	14 ⑤	15 ⑤	16 ②	17 ④	

01 ③

여: 전 학교에 다닐 때 공부하면서 음악 듣는 것을 좋아했습니다. 부모님은 그렇게 하면 집중이 안 된다고 생각하셔서 허락하지 않으시곤 했죠. 부모님이 모르셨던 건 음악을 듣는 것이 제가 공부하는 데 도움이 된다는 점이었어요. 많은 사람들이 공부에는 정도(正道)가 없다는 것을 깨닫지 못합니다. 사람들에게는 저마다의 다른 학습 방법이 있지요. 전 제 아들이 공부하는 모습을 보면 긴장이 되는데, 그 애가 계속 돌아다니기 때문입니다. 제가 볼 때 그렇게 하는 것은 시간 낭비처럼 보이지만, 그건 그의 학습 방법입니다.

어휘
distracting 마음을 산란케 하는 correct 옳은 tense 긴장한 keep v-ing 계속 …하다 waste 낭비

02 ⑤

남: 그 신발 상자 버리지 마.
여: 왜? 신발 넣는 데 더 이상 필요 없는데.
남: 알아. 그런데 다른 데 쓸 수 있어. 사무용품을 정리하는데 좋을 거야.
여: 너 정말 물건을 버리는 걸 싫어하는구나?
남: 맞아. 항상 오래된 물품의 또 다른 용도를 찾으려고 하지.
여: 돈을 절약하는 걸 좋아하기 때문이니?
남: 그건 일부야. 하지만 쓰레기 매립지를 채우는 걸 더 걱정하는 거야. 모두가 더 적은 물품을 버릴 수 있게 물건을 최대한 활용하려고 노력해야 해.
여: 그럼 넌 플라스틱 포크나 숟가락을 좋아하지 않겠구나.

남: 사실 난 그걸 씻어서 다시 사용해!

어휘
throw out 버리다 (= throw away) use 사용하다; 사용, 이용 organize 정리하다 supply 《pl.》 용품, 비품 fill up 채우다 landfill 쓰레기 매립지

03 ③

여: 중요한 경주가 한 달도 안 남았어.

남: 알아요. 전 더 많이 훈련할 거예요.

여: 지금은 너무 많이 하면 안 돼. 오전에는 2마일을 천천히 달리는 게 좋겠어.

남: 알겠어요.

여: 그리고 저녁에 선선할 때 속도 조절을 할 거야.

남: 좋은 계획 같네요.

여: 어제 넌 전보다 더 빨리 달렸어.

남: 네. 또 그렇게 할 수 있다면, 전 우승할 거예요.

여: 그렇고말고. 이젠 운동이 다가 아니야. 앞으로 몇 주 동안은 식단도 살펴봐야 해.

남: 알아요, 단백질과 탄수화물을 많이 먹어야 한다는 걸요.

여: 맞아. 내가 널 위해서 적어온 특별 식단표가 있어. 이건 네가 훈련 효과를 극대화하고 큰 경주를 준비하는 데 도움이 될 거야.

남: 고맙습니다. 자, 시작해요.

어휘
overdo 지나치게 하다 jog 천천히 달리다, 조깅하다 work on …에 애쓰다[공들이다] protein 단백질 carbohydrate 탄수화물 write out 상세히 쓰다 make the most of …을 최대한 활용하다

04 ⑤

남: 너와 함께 공원에서 산책하니 너무 좋다.

여: 응, 그래. 우린 여길 좀 더 자주 와야겠어.

남: 오, 태권도 유니폼을 입고 코치와 같은 자세를 취하는 학생들이 세 명 있어.

여: 코치도 같은 유니폼을 입었는데, 그의 유니폼에는 소매에 태극기가 있어.

남: 그래, 우리 저기 벤치에 앉아서 그들을 구경해도 되겠다.

여: 나무 옆의 벤치 말이니?

남: 응, 그늘 안이잖아.

여: 그렇긴 한데, 우린 거기 못 앉을 것 같아.

남: 왜 안 돼? 나무에서 사과도 따면 재미있을 것 같은데.

여: 그래, 하지만 개를 데리고 온 저 여자가 벤치 쪽으로 걸어가고 있잖아.

남: 아, 그렇구나. 그 여자는 아마도 여기에 앉을 거야.

어휘
pose 자세 flag 기, 깃발 sleeve 소매 shade 그늘 towards …쪽으로

05 ①

남: 저 학생들은 오늘 밤 댄스파티 때문에 정말 신나 있군요.

여: 그러게요, 드레스와 양복을 입으면 다들 정말 멋져 보일 거예요.

남: 당신도 거기 일하러 가시나요?

여: 아뇨. 시험 채점 때문에 너무 바빠서 일을 도우러 자원하지 못했어요.

남: 아, 그렇군요.

여: 하지만 지금 채점 일이 끝나서 들러보려고요.

남: 저기, 아서가 아프다는 얘기 들었어요?

여: 아뇨. 그가 봉사 일을 하기로 하지 않았었나요?

남: 네, 이제 우린 한 명이 부족해요. 당신이 입구에서 표 좀 받아 줄 수 있을까요?

여: 오 그럼요. 재미있겠네요. 근데 뭘 입어야 할지 모르겠어요.

남: 음, 차려입을 필요 없어요. 전 평소에 근무할 때 입는 옷을 입을 거예요.

여: 좋아요. 그럼 저도 그래야겠네요.

어휘
grade 채점하다 check out …을 확인하다 ill 아픈 short 부족한 dress up 옷을 갖춰 입다

06 ②

여: 오늘 신용카드를 쓰려고 했는데, 계산원이 결제 한도에 도달했다고 말하더라고요. 어떻게 된 건지 알아요?

남: 오, 미안해요. 제 잘못이에요. 지난 주말에 예기치 못하게 신용카드를 썼어야 했어요. 내가 내일 해결할게요.

여: 지난 주말이요? 당신 친구들과 캠핑 갔던 때요? 무슨 일 있었어요?

남: 음, 제가 당신한테 주말 동안 비가 왔다고 말한 거 기억나요?

여: 네. 캠프장이 물에 잠겼다면서요.

남: 맞아요. 그래서 우린 호텔 방 몇 개를 얻어야 했어요.

여: 네, 당신이 이미 저한테 다 말해줬어요. 모든 걸 당신이 계산했다는 얘기예요?

남: 음, 아무도 돈을 많이 가져오지 않아서요. 우린 캠핑할 동안 돈이 많이 필요할 거라고 생각 못 했죠.

여: 그럴 수 있죠.

남: 그래서 제가 신용카드로 전부 결제했어요. 내일 친구들을 만나면 저한테 다시 갚을 거예요.

여: 그래요. 이제 이해가 가네요. 돈을 받으면 꼭 신용카드 대금을 결제하도록 해요!

어휘
cashier 출납원, 계산원 credit limit 신용 한도 unexpectedly 뜻밖에, 예상외로 straighten out 해결하다, 바르게 하다 campsite 야영지, 캠프장 flood 물에 잠기다, 침수되다 pay … back …에게 (빌린 돈을) 갚다 bill 고지서, 청구서

07 ④

남: 안녕하세요, 도와드릴까요?

여: 예, 저는 내일 있을 연극 티켓 두 장을 사고 싶습니다. 얼마인가요?

남: 좌석에 따라 다릅니다. 일반 티켓은 장당 20달러이지만, 앞 세 줄에 있는 좌석은 비용이 10달러 더 들어요.

여: 앞에 있는 좌석을 원합니다. 추가로 지불하는 건 상관없어요.

남: 알겠습니다. 학생이세요? 학생은 20퍼센트의 할인을 받아요.

여: 아니요. 하지만 저는 극장 회원 카드를 가지고 있어요.

남: 알겠습니다. 그 카드는 각 티켓에 대해서 10퍼센트 할인을 해 줍니다.

여: 좋아요. 여기 카드가 있어요.

남: 감사합니다. [잠시 후] 흠… 지금은 카드로 할인을 받을 수 있습니다만, 다음 달에 만료가 되는 것 같네요.

여: 그런가요? 제 회원 자격을 갱신하려면 비용이 얼마나 들죠?

남: 6개월에 10달러이고 일 년에 15달러입니다.

여: 알겠어요. 그 티켓 두 장하고 회원 자격을 일 년 갱신하겠습니다.

어휘

row 열, 줄 front 앞면, 앞부분; *앞쪽 expire 만기가 되다. 만료되다 renew 갱신하다

문제풀이

여자는 30달러짜리 티켓 두 장에 대해 각각 회원 할인(10%)을 받고, 회원 자격을 일 년 갱신하기 위해 15달러를 별도로 지불하므로, 여자가 지불할 금액은 69달러이다.

08 ④

여: 지난 화요일에 대단히 흥미로운 강연에 갔었어, 매튜. 너도 좋아했을 거야.

남: 정말? 누가 강연을 했는데?

여: 그는 독일에 있는 대학의 교수였어. 그는 우주를 연구해.

남: 아, 그렇구나. 그럼 우주여행에 관해 얘기했니?

여: 사실 그의 강연은 시간을 통한 여행이 가능한가에 관한 것이었어.

남: 공상 과학 영화 같은데.

여: 바로 그거야. 그의 생각은 아주 독창적이었어.

남: 네가 좋았다니 놀라워. 과학이 네 장기는 아니잖아.

여: 그래, 하지만 그가 아주 명쾌하게 설명해주었거든. 거의 세 시간짜리였는데 지루하지 않았어.

남: 나도 갔더라면 좋았을 텐데.

여: 재미있었어. 그 뒤에, 우리 무리는 도서관에 가서 토론했지.

남: 다음에 흥미로운 강연에 갈 때, 내게 알려줘!

어휘

fascinating 대단히 흥미로운 universe 우주 whether (접속사로) …인지 아닌지 science fiction movie 공상 과학 영화 strong point 장점, 장기 afterwards 이후에

09 ③

남: 세계에서 가장 큰 종교 사원인 앙코르 와트에 오신 걸 환영합니다. 이 사원은 12세기 초에 힌두교 사원으로 건설되었습니다. 그러나 나중에 불교 사원으로 바뀌었습니다. 사원의 외벽은 길이가 3.5킬로미터가 넘습니다. 내부에는 십자가 모양으로 배열된 다섯 개의 탑뿐만이 아니라 세 개의 긴 통로가 있습니다. 원래 이 사원은 Preah Pisnulok으로 명명되었지만, 나중에 이는 '사원의 도시'를 의미하는 앙코르 와트로 바뀌었습니다. 이 사원은 캄보디아의 국가적 상징이 되었고 심지어 캄보디아 공식 국기에도 등장합니다. 이제 안으로 가서 둘러봅시다.

어휘

religious 종교의 temple 사원 passageway 통로 arrange 배열하다 cross 십자가 feature 특별히 등장하다 official 공식적인

10 ③

여: 원두 공장에 오신 것을 환영합니다. 어떻게 도와드릴까요?

남: 안녕하세요. 전 커피 0.5킬로그램을 사고 싶어요.

여: 네. 제 뒤의 차트에 오늘의 선택이 나와 있어요.

남: 아! 전 진한 맛 커피가 더 좋아요. Arabica 원두가 좋은 선택일까요?

여: 사실, Robusta 커피가 더 진한 맛이에요. 꽤 쓰답니다.

남: 정말요? 그럼 그걸로 할게요. 그리고 커피 가는 기계가 없어서, 통원두는 원하지 않아요.

여: 네. 선호하는 신선도가 있으신가요?

남: 네. 볶은 지 일주일보다 덜 된 것이 좋아요.

여: 문제없어요. 손님의 기호에 맞는 선택지가 두 개 있어요.

남: 네, 저도 알겠네요. 더 저렴한 거로 살게요.

여: 아무렴요. 여기 있습니다.

어휘

half a kilogram 0.5킬로그램 flavor 맛, 풍미 grinder 그라인더, 가는[빻는] 기구 freshness 신선함 preference 선호 roast (콩·땅콩 등을) 굽다, 볶다 [문제] grind 갈다 (ground-ground)

11 ①

여: 반려동물을 원하니?

남: 물론이야, 고양이나 개를 정말 갖고 싶어.

여: 잘됐다. 내 개가 강아지를 낳았는데, 그중 하나를 네가 데려갈 수 있을까 생각했어.

남: 그러고 싶어. 강아지들이 몇 살이니?

어휘

puppy 강아지 [문제] fantastic 환상적인 get along (well) with …와 잘 지내다

문제풀이

② 넌 강아지를 잘 돌봐야 해.

③ 정말? 강아지를 얼마 주고 샀니?

④ 환상적이야. 내 고양이들은 나와 노는 걸 좋아해.

⑤ 맞아. 개는 고양이와 잘 지내지 못해.

12 ④

남: 루시, 저 학생들 좀 봐! 열람실에 음료수를 가지고 가잖아.

여: 정말? 도서관 규정에는 어떤 음식과 음료수도 허용되지 않는다고 명시되어 있는데.

남: 사서가 뭔가 해야 하는 거 아니야?

여: 그래. 그들이 학생들에게 규칙을 일깨워줘야지.

어휘

reading room (도서관의) 열람실 librarian 사서 [문제] vending machine 자동판매기 check out (책 등을) 대출하다 strict 엄격한 violate 위반하다 remind A of B A에게 B를 상기시키다

문제풀이

① 이 건물은 모든 학생들에게 개방되어 있어.

② 규칙이 엄격해서 아무도 어기지 않아.

③ 그들은 로비에 있는 자동판매기를 사용할 수 있잖아.

⑤ 물론이지. 그들은 학생들이 책을 대출하는 걸 도와줄 거야.

13 ①

여: 나 정말 떨려.

남: 왜?

여: 오늘 밤에 처음으로 남자친구 부모님을 만나거든.

남: 그의 부모님을 만난다고? 그것참 부담스럽겠다.

여: 그러게. 제발 그분들이 날 마음에 들어 하시면 좋겠어.

남: 당연히 널 좋아하실 거야.

여: 내가 바보 같은 말을 하면 어쩌지? 넌 내가 긴장하면 어떻게 되는지 알잖아.

남: 네가 바보 같은 소리 하는 거 난 한 번도 들어본 적 없어. 긴장 풀어.

여: 잘못될 것 같아서 말이야.

남: 있지, 그런 태도를 갖고 있으면 그렇게 돼. 그냥 네 여자인 친구의 부모님들 중 한 분을 만난다고 생각해.

여: <u>좋은 생각이야. 시도해 볼게.</u>

14 ⑤

남: 우린 교수님께 그룹 과제에 대해 말씀드릴 거야. 너도 갈 수 있니?

여: 언제 갈 건데?

남: 수요일 교수님 면담 시간 중에.

여: 교수님 면담 시간이 언제야?

남: 오후 2시에서 4시까지야.

여: 그래, 갈게. 난 이 과제에 대해 질문할 게 많거든.

남: 응, 우리도 그래. 교수님과 얘기하고 난 후에 도서관에 가서 과제 하자.

여: 좋은 생각이야. 근데 너만 편하다면 우리 집에 가서 하자.

남: 사실 그것도 아주 좋지. 내가 다른 애들한테 물어볼게.

여: 그럼 같이 커피도 마시고 간식도 먹을 수 있잖아.

남: <u>도서관보다 훨씬 좋다.</u>

15 ⑤

남: 조쉬와 그의 아들인 라이언은 지붕에서 구멍을 수리하고 있다. 조쉬는 라이언이 이전에 지붕 위에서 일해본 적이 없기 때문에 그가 다칠까봐 조금 걱정이 된다. 얼마 후, 조쉬는 라이언에게 반대쪽 지붕에 있는 그의 연장통에서 망치를 가져다 달라고 말한다. 아버지를 기쁘게 해드리고 싶어서, 라이언은 일어나서 지붕을 건너가 연장통 쪽으로 간다. 조쉬는 그를 걱정스럽게 쳐다본다. 그가 망치를 찾아 일어나서 돌아오려고 할 때, 조쉬는 그에게 무언가 소리치기로 한다. 이러한 상황에서, 조쉬가 라이언에게 할 말로 가장 적절한 것은 무엇인가?

조쉬: <u>천천히 그리고 조심히 걸어오렴!</u>

16 ② 17 ④

여: 오늘날 세계에서 제 2언어를 구사하는 것은 성공을 위한 필수 기술입니다. 중국은 빠르게 국제 사업 강대국이 되어가고 있습니다. 그러니 여유 시간에 비즈니스 중국어를 배우는 게 어떨까요? Little's 어학원은 최근에 개원했고 표준 중국어와 광둥어 강좌를 모두 제공합니다. 막 시작하셨다면, 언어의 기본을 가르치는 초보자 강좌가 몇 개 있습니다. 또 중국어를 이미 하시는 분들을 위해 비즈니스 강좌와 일대일 지도를 제공합니다. 저희 모든 수업은 면접과 프레젠테이션, 협상을 포함한, 비즈니스 환경에서 쓰이는 특정 어휘와 구조에 중점을 둡니다. 일부 학교는 전통적인 교실에서 비즈니스 언어 기술을 가르치는 반면, 저희 수업은 사무실 환경에서 이루어집니다. 그리고 바쁜 일정으로 가끔 수업을 빠질 수밖에 없다면, 추가 비용 없이 세 번의 보충 수업을 수강할 수 있습니다. 게다가, 한 달에 2개 이상의 수업에 등록하면 20퍼센트 할인을 받게 됩니다. 더 알고 싶으시다면 저희 수업 중 하나를 청강하러 저녁에 언제든지 들러 주세요.

11   if you'd take one

12   drinks are allowed

13   That's pretty serious / say something stupid / with that attitude

14   group assignment / so do we / go to my place / have snacks

15   fixing a hole / Eager to please / to yell something

16~17 essential skill for success / one-on-one tutoring / occasionally miss a class / get a 20% discount

19 영어듣기 모의고사

본문 ▲ p.114

01 ②	02 ①	03 ④	04 ④	05 ③	06 ①
07 ③	08 ④	09 ③	10 ④	11 ②	12 ①
13 ②	14 ⑤	15 ⑤	16 ④	17 ⑤	

01 ②

남: 사람들은 종종 다른 사람들과 함께 있을 때도 외로움을 느낍니다. 그리고 자녀들이 집을 떠나거나, 부모나 배우자가 세상을 떠나면 흔히 심한 공허함을 느낍니다. 반려동물들은 여러분의 공허함을 채워줄 뿐만 아니라 여러분을 행복하게 해줄 것입니다. 어떤 이들은 반려동물이 지저분하고 그들이 필요로 하는 것들을 돌보는 데 너무 많은 시간이 들기 때문에 반려동물을 좋아하지 않는다고 말합니다. 하지만 여러분이 반려동물에게 먹이를 주고, 목욕을 시키고, 그들을 산책시켜야 하기 때문에 반려동물을 소유하는 것은 여러분을 좀 더 활동적이고 부지런하게 만듭니다. 어쩌면 여러분은 반려동물을 기르는 이웃들과 친구가 될 수도 있습니다. 이것이 제가 사람들이 반려동물을 길러야 한다고 생각하는 이유입니다.

어휘
empty 빈, 공허한 (n. emptiness) active 활동적인 diligent 부지런한
bathe 목욕시키다

02 ①

여: 최근에 주차장의 상태를 본 적 있어요?

남: 네. 구덩이가 잔뜩 있더라고요.

여: 오늘 아침에 내 타이어가 구덩이에 끼어 꼼짝도 안 했어요.

남: 정말 불편하죠.

여: 그게 다가 아니에요. 그것은 고객들에게 나쁜 첫인상을 주잖아요. 우리 건물에 차를 몰고 들어오면서 고객들이 처음으로 보는 게 낡아빠진 주차장이거든요.

남: 그래서 그걸 어떻게 할 거죠?

여: 이건 허용될 수 없다는 걸 건물 관리인에게 알리려고 해요. 그는 가능한 빨리 구덩이를 보수해야 해요.

남: 그가 당신의 말을 들으면 좋겠네요.

여: 저도요. 당장 그에게 전화해야겠어요.

어휘
state 상태 hole 구멍 stuck (…에 빠져) 꼼짝 못 하는 inconvenient 불편한
run-down 황폐한 acceptable 받아들일 수 있는

03 ④

여: 무엇을 도와드릴까요?

남: 침실 한 개짜리 아파트를 찾고 있어요. 광고를 봤거든요.

여: 아직 남아 있는지 살펴볼게요. [잠시 후] 없군요, 죄송합니다. 침실 하나짜리는 다 나갔습니다.

남: 벌써요?

여: 음, 적어도 학기가 시작하기 두 달 전에 대부분의 계약이 이루어지죠.

남: 다른 것들은 없을까요?

여: 아직 원룸은 여러 개 있습니다.

남: 원룸도 괜찮아요. 한 달에 얼마죠?

여: 한 달 임대료는 500달러이고, 보증금 500달러를 오늘 내셔야 합니다.

남: 알겠어요. 계약은 1년 단위인가요, 아니면 6개월 단위인가요?

여: 1년 계약만 제공합니다.

어휘
advertisement 광고 contract 계약 option 선택 (사항) studio 원룸 (아파트) monthly 매달의 rental fee 임대료 security deposit 보증금

04 ④

여: 여기 스키 리조트 멋지다! 꼭대기에서 리조트 주위의 모든 산을 볼 수 있을 거야.

남: 응, 근데 스키 리프트를 타려면 오래 기다려야 할 것 같아.

여: 그러게. 스키 리프트를 기다리는 줄이 상당히 기네.

남: 그런데 벤과 그의 아들은 어디 있어?

여: 저기에. 그는 아들이 스키 신는 걸 도와주고 있어.

남: 오, 아들이 참 귀엽지 않니? 벤 옆에 있는 저 스노보드는 새것 같네. 보드 위의 세로줄 무늬가 마음에 들어.

여: 벤이 그의 아내 캐리가 그 애의 생일 선물로 저것을 줬다고 했어. 어, 그녀가 이쪽으로 오네.

남: 아, 스키 스틱을 들고 있는 여자분?

여: 응. 가서 인사하자.

어휘
awesome 굉장한, 아주 멋진 ski lift 스키 리프트 put on 입다, 착용하다
vertical stripe 세로줄 무늬 ski pole 스키 스틱

05 ③

여: 에단, 어버이날이 다가오고 있는 거 알아?

남: 응, 알아. 엄마 아빠께 뭘 드릴 거야?

여: 그게, 사실은 난 선물 살 돈이 전혀 없어.

남: 넌 정말 돈을 못 모으더라!

여: 알아. 우리 둘이 합하면 부모님께 선물을 드릴 수 있을 것 같은데. 내

가 나중에 너한테 돈을 갚으면 되잖아.

남: 미안하지만 안 돼. 네게 계속 돈을 주고 싶진 않아! 아르바이트를 구하는 게 어때?

여: 알겠어, 좋아. 근데 난 일을 잘 못 구해. 어디서부터 시작해야 할지 모르겠어.

남: 내가 도와줄게. 이력서 쓰는 것부터 시작해보자.

여: 알았어.

06 ①

[휴대전화벨이 울린다.]

여: 여보세요, 블레이크.

남: 여보세요, 크리스틴. 소풍 갈 준비 됐니?

여: 미안한데, 난 못 갈 것 같아.

남: 왜? 무슨 일이니?

여: 집의 지붕 한쪽이 무너져서 수리해야 해.

남: 정말 충격적이구나. 아무도 안 다쳤으면 좋겠는데.

여: 다행히 모두 괜찮아. 그런데 최대한 빨리 수리해야 해.

남: 그래야겠네. 사실, 나는 어젯밤에 우리가 먹을 음식을 만들었어. 그걸 같이 못 먹다니 아쉽다.

여: 그랬어? 안타깝다. 네 친구들하고 나눠 먹어야겠네.

남: 그러려고. 그래서 수리공에게는 연락했니?

여: 응, 했어. 지금 오는 중이야.

07 ③

남: 안녕하세요, 수업에 쓸 공책이 좀 필요한데요.

여: 네. 여기 공책들이 많이 있어요.

남: 저 빨간 건 얼마죠?

여: 3달러이고, 100페이지로 되어 있습니다.

남: 전 더 두꺼운 게 필요해요. 200페이지로 된 공책이 있나요?

여: 그럼요. 저 주황색 공책이 200페이지로 되어 있고 5달러입니다. 안에 선이 있는 것과 없는 것 두 종류가 있어요.

남: 그렇군요. 안에 선이 있는 300페이지짜리 공책이 있나요?

여: 네. 저 검은색 공책이 300페이지이고 7달러입니다.

남: 네. 선이 있는 주황색 공책 두 권과 선이 있는 300페이지짜리 공책 한 권을 살게요.

여: 네, 그럼 공책 세 권이군요. 여기 있습니다.

남: 고맙습니다.

08 ④

남: 콘서트는 어땠어?

여: 좋았어. *피터와 늑대*는 정말 재미있었어.

남: *피터와 늑대*라고? 꼭 동화 같은데.

여: 맞아. 해설자가 늑대를 잡는 한 소년과 새에 대한 이야기를 들려 줘. 해설자가 이야기를 하는 동안 오케스트라는 이야기와 어울리는 곡을 연주해.

남: 음악은 어때? 아이들을 즐겁게 하려면 분명 신이 나야 할 텐데.

여: 응, 맞아. 오케스트라의 다양한 악기들이 모두 이야기 속 등장인물처럼 소리를 내거든.

남: 정말? 어떤 악기가 늑대처럼 무서운 소리를 낼 수 있지?

여: 그건 프렌치 호른이야. 플루트는 새처럼 소리를 내고, 오보에는 오리처럼, 그리고 클라리넷은 고양이처럼 소리를 내지.

남: 피터는?

여: 아, 현악기들이 피터를 위해 행복한 선율을 연주해. 난 그가 가장 좋았어. 그리고 바순은 피터의 할아버지를 연주하고. 그는 매우 늙고 성격이 나쁜 것처럼 소리가 나지.

남: 무척 재미있을 것 같아.

여: 응, 아이들을 위한 명작으로 여겨져.

09 ③

여: 우주에서 온 것처럼 들릴지도 모르지만, earthship은 지구상에 실제 존재하는 것입니다. 이 환경친화적인 구조물은 폐타이어로 만들어집니다. 약 2천 개의 타이어가 작은 집을 만드는 데 필요합니다. 타이어를 쌓아서 벽을 만들고, 열을 보존할 수 있게 진흙과 짚으로 채웁니다. 햇빛에 노출될 수 있도록 유리창이 집 앞쪽을 차지하고 있습니다. 집 벽면은 열을 아주 잘 가두기 때문에 햇빛 이외의 난방원이 필요하지 않습니다. earthship은 전기도 불필요하도록 태양열 판을 갖추고 있습니다. 이 주택은 전통적인 가옥보다 비용이 훨씬 적게 들기 때문에 환경주의자뿐만 아니라 저소득층 가정들에게도 아주 유용합니다.

10 ④

여: 난 스페인 여행을 위한 안내서가 필요해. 이거 어떤 것 같아?

남: 좋아보이는데, 출간된 연도를 확인해봐.

여: 오, 너무 오래전이네! 난 2015년이나 그 이후에 출간된 것을 사고 싶어.

남: 응. 그렇지 않으면 정보가 너무 예전 것일 거야.

여: 난 스페인 지도도 하나 얻어야 해, 그래야 길을 잃지 않으니까.

남: 동의해. 지도들은 아주 유용해. 이 안내서들 중 일부는 지도가 포함되어 있어.

여: 훌륭하네. 그리고 각 책의 제목 옆에 있는 별들은 뭐야?

남: 그건 웹사이트의 고객 평가야. 다섯 개의 별이 가장 높은 평가지.

여: 정말? 음, 나는 별점이 세 개보다 많은 책을 사고 싶어.

남: 그럼 이게 가장 나은 선택인 것 같다.

여: 응, 하지만 그건 재고가 없네. 난 곧 떠날 거라서 지금 바로 배송될 수 있는 걸 원해.

남: 오, 그렇다면 대신 이걸 주문해.

11 ②

남: 강변에 있으니 참 좋네. 물소리가 나를 편안하게 해줘.

여: 나도 그래. 난 이런 곳에서 산책하는 걸 좋아해.

남: 저기 봐! 저기에 대여점이 있어. 자전거를 타고 싶니?

여: 물론이지. 자전거 타고 멀리 가 보자.

문제풀이

① 응, 내 자전거를 네게 빌려줄 수 있어.

③ 응, 너와 쇼핑 가고 싶어.

④ 응, 강에서 하는 수영은 재미있을 것 같아.

⑤ 아니, 오토바이는 이 공원에서 허용되지 않아.

12 ①

여: 맥스, 수학 숙제하는 데 네 컴퓨터를 사용하고 있니?

남: 응, 답을 계산해야 하고 그래프를 만들어야 해.

여: 와, 복잡할 것 같아. 그걸 어떻게 할 거니?

남: 나는 특별한 프로그램을 쓸 거야.

문제풀이

② 칭찬해줘서 고마워.

③ 그 질문에 답하기가 힘들었어.

④ 난 이번 수학 시험에 통과할 거라고 확신해.

⑤ 이 숙제를 끝내는 데 난 일주일이 걸렸어.

13 ②

[휴대전화벨이 울린다.]

여: 여보세요? 데이비드! 네가 전화해 주니 기쁘구나.

남: 안녕하세요, 엄마. 죄송해요, 지난번에 전화한 이후로 너무 오랜만이죠. 시험공부를 하느라 바빴어요.

여: 괜찮아. 새 룸메이트는 어떠니?

남: 그 애는 친절해요. 전 그 애가 맘에 들어요. 그나저나, 엄마가 지나 이모와 여행 가는 게 이번 주말이에요?

여: 음, 우리는 샌디에이고로 차를 몰고 갈 예정이었어. 그런데 지나 이모가 몸이 안 좋아서 가지 않을 거야.

남: 오, 정말 안됐네요. 그러면 엄마를 보러 집에 가도 돼요? 중간고사가 마침내 끝났거든요.

여: 정말? 너를 다시 보게 될 거라니 정말 좋겠구나.

남: 같이 뭔가를 하는 게 어떨까요? 드라이브하거나 공원에 갈 수 있어요.

여: 좋은 생각이야. 그런데 나중에 이야기해도 될까? 난 지금 뛰어가야 해. 20분 후에 의사와 진료 예약이 있어.

남: 알았어요. 그러면 내일 전화할게요.

여: 그래, 그때 계획을 세울 수 있겠구나.

문제풀이

① 알았어, 우리 거기서 만나자.

③ 음, 너는 그녀에게 사과해야 해.

④ 일정을 변경해서 미안해.

⑤ 물론이야, 그녀가 그 소식을 들으면 기뻐할 거야.

14 ⑤

여: 당신 집들이에서 정말 즐거웠어요, 제이크.

남: 그랬다니 기뻐요. 음식은 어땠어요?

여: 음식은 맛있었어요. 당신이 대접한 그 샐러드에 들어있던 게 뭐죠?

남: 아, 그 샐러드요. 케이준 프라이드치킨으로 만든 거예요.

여: 곁들여진 드레싱이 맛있었어요. 이런 맛이 나는 건 전에 먹어보지 못했어요.

남: 모두들 그걸 좋아했던 것 같아요. 게눈 감추듯 먹어 치웠죠.

여: 그걸 어떻게 만드는지 나한테 가르쳐 줄 수 있어요?

남: 원하시면 요리법을 드릴게요.

여: 전 요리법을 잘 따라하지 못해요. 늘 뭔가를 망쳐버리거든요.

남: 그럼 다음에 오시면 그걸 어떻게 만드는지 보여줄게요.

문제풀이

① 닭고기 수프와 샐러드를 먹고 싶어요.

② 아, 그건 따라한 게 아니에요. 제가 독창적으로 만든 요리법이에요.

③ 집들이가 몇 시에 시작하죠?

④ 제게 요리법을 알려주면, 우리가 함께 만들 수 있어요.

15 ⑤

여: 에밀리와 그녀의 남동생은 부모님의 20주년 결혼기념일을 위해 뭔가 특별한 것을 하기로 했다. 오후에 그들은 부모님이 좋아하는 연극을 보러 극장에 가시게 했다. 부모님이 외출하신 동안, 에밀리와 남동생은 부모님이 집으로 돌아오실 시간에 맞춰 대접할 수 있도록 맛있는 식사를 준비했다. 그들은 아버지가 가장 좋아하는 와인도 샀다. 그러나 그들이 식사 준비를 마무리하기 전에, 갑자기 아버지가 지갑을 가지러 집에 돌아오셔서 그들이 만든 것을 보고 말았다. 이런 상황에서, 에밀리의 아버지가 에밀리와 그녀의 남동생에게 할 말로 가장 적절한 것은 무엇인가?

아버지: 정말 기특하구나. 네 엄마에게는 아무 말도 하지 않으마.

제] mess 엉망진창, 어질러 놓은 것

① 너희들이 얼마나 어질렀는지!
② 연극은 엉망이었어. 정말 끔찍한 날이야!
③ 네 녀석들이 저 음식을 다 먹었다니 믿을 수가 없구나.
④ 훌륭한 파티였어. 우릴 초대해줘서 고마워.

16 ④ 17 ⑤

남: 저는 김 박사이고, 치과의사입니다. 치아를 돌보기 위해 우리 모두가
해야 할 일들은 아주 많습니다. 이것은 우리가 무엇을 먹는지 지켜보
는 것도 포함되죠. 모두가 설탕이 치아에 나쁘다는 것을 압니다. 하지
만 문제를 일으키는 다른 놀라운 음식과 음료들이 있습니다. 심지어
얼음조차도 치과에 방문하도록 만들 수 있습니다. 얼음은 단순히 얼
어붙은 물이지만 그것을 깨무는 것은 치아를 손상시킬 수 있습니다.
알코올이 들어간 음료들은 또다른 문제입니다. 알코올은 입안을 마르
게 하고 충치는 마른 입안에서 더 쉽게 형성됩니다. 그리고 빵도 충치
를 생기게 한다는 것을 아시나요? 빵의 녹말은 침에 의해 당으로 바
뀌고 당은 충치를 만들죠. 감자 칩도 마찬가집니다. 그것들은 치아 사
이에 껴서, 그것들의 녹말이 점차 당이 됩니다. 이러한 음식과 음료를
최대한 피하는 것은 좋은 생각입니다. 그리고 물론, 매 식사 후 양치
하도록 하세요!

어휘
frozen 얼어붙은, 꽁꽁 언 bite 깨물다 cavity 충치 starch 녹말 saliva 침
gradually 점차 [문제] strengthen 강화하다

문제풀이
16 ① 당이 충치를 유발하는 이유
   ② 치아를 튼튼하게 하는 다양한 운동들
   ③ 충치를 때우는 가장 좋은 방법들
   ④ 치아를 상하게 하는 음식과 음료들
   ⑤ 치아를 효과적으로 깨끗하게 유지하는 법

DICTATION Answers

본문 p.116

01 feel lonely / very empty / require too much time
02 got my tire stuck / run-down parking lot / have those holes
    fixed
03 we still have any / get signed / monthly rental fee /
    security deposit
04 put on his skis / the vertical stripes / holding ski poles
05 is coming up / saving money / pay you back / writing a
    résumé
06 go on our picnic / get it fixed / That's a shame
07 a thicker one / The black ones / two orange notebooks
08 sounds like a kid's story / go with the story / it's considered
09 are piled up / allow sunlight exposure / great for low-
    income families
10 the year it was published / I don't get lost / it's not in stock
11 ride bicycles

12 need to calculate
13 I've been busy studying / feel well / have a doctor's
    appointment
14 housewarming party / you served / The dressing / give
    you the recipe
15 wedding anniversary / prepared a delicious meal /
    finishing touches
16~17 to take care of our teeth / a trip to the dentist's office / by
    your saliva / brush your teeth

20 영어듣기 모의고사
본문 ▲ p.120

01 ⑤	02 ①	03 ②	04 ④	05 ①	06 ⑤
07 ①	08 ④	09 ⑤	10 ②	11 ⑤	12 ①
13 ④	14 ④	15 ③	16 ①	17 ⑤	

01 ⑤

여: 주목해 주시겠습니까? 우리가 입장하기 전에 여러분께 몇 가지 규칙
을 설명하라는 요청을 받았습니다. 우선 공연 중에는 조용히 해주기
바랍니다. 오케스트라가 연주하고 있을 때는 잡담이 있어서는 안 됩
니다. 곡과 곡 사이에는 이야기를 할 수 있지만, 속삭여 주시기 바랍
니다. 어떤 경우에도 사진은 찍을 수 없습니다. 그리고 휴대전화는 지
금 모두 꺼 주십시오. 이 규칙들을 지키지 않는 학생은 객석에서 나가
달라는 요청을 받게 될 겁니다.

어휘
remain 계속[여전히] …이다 performance 공연, 연주 whisper 속삭이다
follow (규칙 등을) 따르다 auditorium *객석; 강당

02 ①

남: 우리 할머니 댁 근처 들판에 풍력 발전용 터빈이 세워지고 있어.
여: 요즘 점점 더 많은 터빈을 볼 수 있어.
남: 풍력 에너지는 공해를 일으키지 않기 때문에 많은 사람들이 풍력 에
너지를 지지하잖아.
여: 맞아. 그리고 에너지의 원천이 바람이니까 절대 고갈되지 않지.
남: 물론, 풍력 에너지를 사용해서 생기는 문제점들도 있어.
여: 난 생각나는 게 없는데.
남: 우선 그 지역 내의 바람의 양이 그렇게 일정하지 않아. 그게 얼마나
많은 양의 에너지를 생산하는지 추정하는 걸 어렵게 해.
여: 없는 것보다는 훨씬 나아. 또, 누구에게도 해를 입히지 않잖아.
남: 음, 사람에게는 아닐지 모르지만, 야생 동물에게는 어떨까? 일부 연
구에 따르면 새와 박쥐가 터빈에 의해 죽을 수 있다고 해.

여: 나는 그건 생각 못 했던 것 같아.

어휘
put up 세우다, 짓다 wind turbine 풍력 발전용 터빈 in favor of …을 찬성하여 pollution 오염, 공해 source 원천 run out 고갈되다 drawback 결점, 문제점 for starters 우선 첫째로 constant 변함없는, 일정한 estimate 추정하다 wildlife 야생 동물

03 ②

남: 실례합니다, 톰슨 선생님. 잠시 말씀 좀 나눌 수 있을까요?
여: 물론이죠. 사무실 안으로 들어오세요. 그리고 저를 바바라라고 불러주세요.
남: 고마워요, 바바라.
여: 이곳 Stratford 고등학교에서의 첫 한 주간 수업은 어떠셨어요?
남: 아주 순조로웠어요. 학생들도 훌륭하고, 다른 선생님들도 많이 도와주셨고요.
여: 다행이네요. 전 이 학교가 모두에게 즐거운 곳이 되도록 아주 열심히 일해 왔습니다.
남: 훌륭한 일을 하셨네요. 여기서 일하게 돼서 무척 기쁩니다.
여: 좋아요. 근데 무슨 문제라도 있나요? 걱정이 있어 보여요.
남: 음, 제 시간표 때문에요. 하루에 여섯 반을 가르치는 건 너무 많은 시간을 차지하는 것 같아서요.
여: 힘들다는 거 압니다. 하지만 시간이 지나면 수월해질 거예요.
남: 저도 그러길 바랍니다. 전 정말 제 학생들 모두에게 최선을 다하고 싶어요.

어휘
go well 잘되다 pleasant 즐거운 concerned 걱정스러운 take up (시간·장소 등을) 차지하다 go by (시간이) 지나가다 do one's best 최선을 다하다

04 ④

남: 우린 드디어 도착했어! 경치가 대단하지 않아?
여: 응. 힘든 등반이었지만 드디어 정상에 왔네. 저기 아래 고궁이 전통적인 건축 양식과 지붕 때문에 정말 아름답게 보여.
남: 그리고 그 뒤에 있는 건물들을 봐!
여: 저 고층 빌딩들이 여기 위에서 보니 멋지다.
남: 흠… 주요 도로가 거의 비어있네. 이거 놀라운걸.
여: 마라톤 때문에 폐쇄됐어.
남: 그렇구나. 아, 왼쪽에 있는 저 건물 보이지? 높은 아파트 건물? 독특한 건축 양식 때문에 매우 유명해.
여: 아주 멋지네. 그런데 경치 중에 내가 가장 좋아하는 부분이 뭔지 알아?
남: 뭔데?
여: 건물들 뒤로 산 정상에 있는 탑이야.
남: 응, 나도 그게 좋아.

어휘
make it 성공하다; 제시간에 가다; *도착하다 tough 힘든 palace 궁, 궁전 architecture 건축 양식; 건축(물) skyscraper 고층 건물 neat 깔끔한; *멋진, 훌륭한

05 ①

남: 릴리, 중요한 부탁 하나만 들어줄 수 있을까?
여: 글쎄. 무슨 일인데?
남: 내가 이번 금요일에 새 아파트로 이사하는데 도와줄 사람이 필요해.
여: 네가 옮겨야 할 짐이 많아?
남: 그렇게 많진 않아. 그리고 새 아파트가 지금 사는 아파트에서 몇 블록 밖에 안 떨어져 있어.
여: 새 아파트가 몇 층에 있는데?
남: 음… 4층이야. 게다가 엘리베이터도 없고 말이지.
여: 엘리베이터가 없어? 농담이지!
남: 나도 농담이면 좋겠다. 그래서 네 도움이 꼭 필요한 거야.
여: 알았어. 도와줄게, 하지만 보답으로 내 부탁도 들어줘야 해.
남: 물론이지, 뭐든지 말해.
여: 좋아. 내가 내일 밤에 영화를 볼 동안에 내 남동생 좀 봐줄래?

어휘
stuff 물건 in return 보답으로, 답례로 watch 돌보다

06 ⑤

여: 실례합니다, 경관님. 잠깐 말씀 좀 나눌 수 있을까요?
남: 물론이죠. 뭘 도와드릴까요?
여: 제 이름은 캐롤 해리스이고, 길모퉁이에 있는 꽃가게 주인입니다.
남: 1번가 꽃집이요? 저 거기 알아요. 밸런타인데이 때 거기서 아내에게 장미를 사줬어요.
여: 아. 어쩐지 낯이 익다 했어요. 그건 그렇고, 제가 문제가 좀 있어서요.
남: 무슨 문제죠?
여: 제 가게 앞 인도에 여러 명의 십 대들이 무리 지어 있어요.
남: 그 아이들이 말썽을 일으켰습니까?
여: 그렇진 않아요. 근데 그 아이들은 오후 내내 음악을 틀어놓고 시끄럽게 떠들고 있어요. 그 애들이 제 손님들 중 몇 명에게 겁을 줘서 쫓아버리는 것 같아요.
남: 알겠습니다. 제가 가서 그 아이들에게 얘기하겠습니다. 놀고 싶으면 공원으로 가야죠.
여: 정말 감사합니다, 경관님.

어휘
own 소유하다 avenue 거리, …가(街) familiar 낯익은 a bit of 약간의 sidewalk 인도 cause (문제 등을) 일으키다, 야기하다 scare … away …을 겁주어 쫓아버리다 hang out (어슬렁거리며) 시간을 보내다

07 ①

[전화벨이 울린다.]
여: Capitol Travel입니다. 무엇을 도와드릴까요?
남: 저는 인천에서 마드리드로 가는 왕복 비행기 표를 예약하고 싶어요.
여: 언제 출발하세요?
남: 1월 15일에 떠납니다. 그곳에서 열흘간 머물 거예요.
여: 알겠습니다. 직항으로 가는 일반석 표는 1,200달러에요. 비즈니스석은 500달러 더 들어요.
남: 저는 당연히 일반석으로 비행하고 싶은데, 여전히 비싸네요.
여: 학생이세요? 저희는 1월에 학생들을 위해서 직항을 10퍼센트 할인해 드리고 있어요.
남: 좋네요! 저는 대학생입니다. 하지만 여전히 제 예산을 벗어나네요.

여: 흠… 만약에 일정이 허락한다면, 이스탄불 경유를 포함하는 비행기로 바꾸는 게 어떠세요? 이렇게 하면 할인된 가격보다 300달러 더 적게 들 것이고, 경유도 겨우 3시간이에요.

남: 그게 더 나은 거래 같아요. 그 표를 예매하고 싶어요.

여: 알겠습니다. 여행자 보험도 구매하시겠어요? 50달러면 됩니다.

남: 아니요, 괜찮아요. 전 이미 보험에 들었어요.

어휘

round-trip 왕복 (여행)의 economy 일반석 permit 허락하다 stopover (비행기의) 중간 기착(지) layover 도중하차, 경유 insurance 보험 insure 보험에 들다

문제풀이

마드리드행 직항편 비행기 표는 1,200달러인데, 학생 할인 10퍼센트를 적용하면 1,080달러이다. 그런데 남자는 이보다 300달러 더 저렴한 이스탄불 경유 비행기를 선택했으므로, 남자가 지불할 금액은 780달러이다.

08 ④

여: Guitar Girls의 새 앨범을 샀어.

남: 그 노래 좋니?

여: 응. 가장 좋은 노래는 'Art World'야. 거의 10분 길이야.

남: 'Art World'라고? 지구가 얼마나 아름다운지에 관한 노래 말이니?

여: 응, 바로 그거야!

남: 그건 원곡이 아니야. 약 20년 전에 조지 킴이 녹음했어.

여: 그건 몰랐어. 그가 그 곡을 작곡했니?

남: 그가 곡은 썼는데 누가 가사를 썼는지는 모르겠어. 어쨌든, 그의 버전은 인기가 많지 않았어.

여: 그래? 음, 그 곡이 Guitar Girls에게는 대성공이 될 거야. 모두 그 얘기를 하고 있거든.

남: 이유가 궁금한걸.

여: 놀라울 건 없어. 환경에 관한 노래인데, 그건 요즘 사람들의 관심거리 잖아.

남: 맞는 말인 것 같아.

어휘

word 단어; *(노래의) 가사 version 버전, 형태 big hit 대성공, 큰 인기 make sense 타당하다, 말이 되다

09 ⑤

남: 겨울 방학이 빠르게 다가오고 있습니다. 그래서 우리 대학에서는 여러분이 경력을 쌓을 수 있게 도와주기 위해서 인턴십 박람회를 열 것입니다. 만약 여러분이 겨울 방학 기간에 약간의 직업 경험을 갖고 싶다면, 이 행사를 꼭 놓치지 않도록 하세요. 이 박람회는 1월 25일 오전 10시부터 오후 5시까지 Wilson 강당에서 개최될 것입니다. 많은 다양한 분야에 있는 회사의 대표들이 이 행사에 참여할 것입니다. 이 박람회는 여러분이 유용한 조언을 듣기 위해서 그들과 이야기할 수 있는 기회를 줄 것입니다. 입장료는 없지만, 참석하려면 1월 18일까지 온라인으로 등록해야 합니다. 학생들이 이 행사 이전에 직업 서비스 사무실에 이력서 견본을 제출하면, 경력 상담사가 이력서를 검토한 후에 도움이 되는 의견과 함께 돌려줄 것입니다.

어휘

approach 다가오다 internship 인턴 fair 박람회 break 방학, 휴가 representative 대표 field 분야 present 참석한 entrance fee 입장료

register 등록하다 submit 제출하다 résumé 이력서 review 검토하다 suggestion 제안, 의견

10 ②

여: 뭘 보고 있니, 제이콥? 저건 봉사 활동 센터에서 온 거니?

남: 응. 국제 봉사 활동 프로그램 목록이야. 하나 골라 보려고.

여: 그렇구나. 가고 싶은 특정 지역이 있어?

남: 딱히 없어. 그런데 아시아는 가고 싶지 않아. 너무 멀어.

여: 어떤 종류의 봉사 활동에 참여하고 싶니? 넌 교육에 관심 있잖아, 그렇지?

남: 응, 하지만 예술 말고는 아무거나 괜찮아. 난 예술적 소질은 전혀 없거든.

여: 그래. 이 프로그램 중 어떤 건 오랜 시간이 걸리네. 6개월은 너무 길어, 안 그래?

남: 아니. 일 년 미만의 어떤 것이든 괜찮아.

여: 흠, 그럼 두 개 중에서 고르면 되겠다. 이게 더 저렴하네.

남: 응, 그런데 4주는 충분한 시간이 아닌 것 같아. 그것보다 더 긴 게 좋아.

여: 그럼 이게 최선의 선택이네. 재미있어 보여, 그렇지?

남: 응! 멋진 경험이 될 거야.

어휘

volunteer service 자원봉사(단) region 지역 [문제] nurse 간호하다 medical 의료의 aid 원조, 지원

11 ⑤

여: 체스 하는 법을 가르쳐 줄래?

남: 물론이지. 체스는 아주 신나는 게임이야. 넌 분명히 잘하게 될 거야.

여: 하지만 난 기본 규칙조차 모르는걸!

남: 걱정하지 마. 게임의 규칙은 단순해.

어휘

[문제] piece (판 놀이에 쓰이는) 말, 알

문제풀이

① 난 다른 게임을 하고 싶어.

② 모든 사람이 가르치는 데 능숙한 건 아니야.

③ 체스 말은 검은색이거나 흰색이야.

④ 내 체스판과 말을 써도 돼.

12 ①

남: 우리 기숙사 2층에는 누가 머물고 있니?

여: 새로운 학생 무리가 이사 왔어. 왜?

남: 매일 밤 아주 시끄럽거든. 잠을 잘 수가 없어!

여: 가서 그들에게 말해야겠어.

어휘

dormitory 기숙사 get to sleep 잠이 들다 [문제] mop 대걸레로 닦다 had better …하는 게 좋을 것이다

문제풀이

② 그들은 지금 바닥을 대걸레로 닦고 있어.

③ 난 살 새집을 찾고 있어.

④ 방에서는 조용히 걷는 게 좋을 거야.

⑤ 잠을 잘 자는 게 기분을 더 좋게 해주지.

13 ④
남: 이런. 우리 오늘 오후에 가기로 한 소풍을 취소해야 할 것 같아.
여: 왜? 날씨가 이렇게 좋은데.
남: 그렇긴 한데 이따가 비가 온대.
여: 말도 안 돼. 창밖을 봐. 하늘에 구름 한 점 없잖아.
남: 음, 없는 것 같네. 그것참 이상하다.
여: 난 네가 소풍을 가고 싶어하는 줄 알았는데. 낭만적일 것 같다고 했었잖아.
남: 정말 낭만적일 거라고 생각해. 난 단지 비를 맞고 싶지 않을 뿐이야.
여: 이해가 안 되는데. 넌 왜 비가 올 거라고 생각하니?
남: 바로 여기 날씨 앱에 그렇게 쓰여 있어.
여: 글쎄, 일기 예보가 항상 정확하진 않잖아.

어휘
cancel 취소하다 romantic 낭만적인 get rained on 비를 맞다 [문제] hard 열심히; *심하게 forecast 예측, 예보 accurate 정확한

문제풀이
① 그 앱은 얼마였니?
② 비가 이렇게 심하게 오다니 믿기지가 않아!
③ 우린 대신에 바닷가에 가야 할 것 같아.
⑤ 너 신문 다 보고 나 좀 줄래?

14 ④
여: 넌 오늘 밤에 수학 시험공부할 거니?
남: 물론이지. 그 시험이 최종 성적의 절반을 차지하잖아. 난 잘해야 해.
여: 그러게 말이야. 난 이 과목에서 좋은 성적을 받지 못하면 졸업을 못할지도 몰라.
남: 난 네가 아주 잘할 거라고 생각해.
여: 그러고 싶어. 나 오늘 밤에 데릭을 만나서 같이 공부할 거야. 너도 우리랑 같이할래?
남: 그거 좋지. 어디서 만날 거야?
여: 캠퍼스 커피숍에서.
남: 오, 그렇다면 사양할래. 그냥 혼자서 공부하는 게 좋겠어.
여: 왜 마음이 바뀌었니?
남: 거긴 시끄러워서 집중할 수가 없거든.

어휘
final 최종의 grade 성적 Tell me about it. 그러게 말이야. graduate 졸업하다 on one's own 홀로, 혼자서 [문제] concentrate 집중하다 noise 소음

문제풀이
① 난 이번 시험은 공부할 필요가 없어.
② 난 도서관이 어디 있는지 몰라.
③ 데릭이 전화해서 우리 계획을 취소했어.
⑤ 난 친구들 몇 명을 만나서 같이 공부할 거야.

15 ③
여: 모스크바의 장기 출장에서 돌아오는 길에, 제이슨은 아기가 있는 어느 부부의 앞자리에 앉아 있다. 비행기가 이륙하자마자, 아기가 울기 시작하더니 울음을 그치질 않는다. 제이슨은 승무원에게 다른 자리

로 옮길 수 있는지 묻는다. 승무원은 빈자리가 없지만 다른 승객들에게 얘기를 해보고 자리를 바꿔 줄 사람이 있는지 알아보겠다고 말한다. 10분 후에 승무원이 돌아와서, 유감이지만 우는 아기 근처에 앉겠다는 사람이 없다고 말한다. 이런 상황에서, 제이슨이 승무원에게 할 말로 가장 적절한 것은 무엇인가?
제이슨: 어쨌든 고마워요. 신경 써 주셔서 감사합니다.

어휘
on one's way …하는 중에 take off 이륙하다 flight attendant 비행기 승무원 passenger 승객 [문제] appreciate 고마워하다 effort 노력, 수고

문제풀이
① 그거 잘됐네요. 정말 고마워요.
② 아니, 됐어요. 이 자리도 괜찮아요.
④ 죄송하지만, 빈자리가 하나도 없네요.
⑤ 괜찮아요. 우리 아기가 곧 울음을 그칠 거예요.

16 ① 17 ⑤
남: 주목해 주시겠습니까? 이미 아시는 바와 같이, 우리 연례 학교 축제가 다음 주 토요일에 마을 공원에서 열립니다. 이것은 매년 수백 명의 학생들과 그 가족이 참가하는 인기 있는 행사입니다. 그러나, 올해 축제에 입장하는 데 방문객에게 2달러를 내게 한 결정에 대한 일부 불만이 있었습니다. 과거에는 요금이 없었고, 사람들은 제게 이것이 바뀐 이유를 묻고 있습니다. 간단한 답변은 축제가 매년 커지고 있다는 것입니다. 이전엔 학교 축구장에서 열렸었지만, 충분한 공간이 없었습니다. 지금은 공원을 사용하기 위해 마을에 돈을 지불해야 합니다. 게다가 올해 축제에는 세 개의 공연과 많은 음식, 몇 가지의 오락 게임, 놀이기구가 있을 것입니다. 축제 참가자에게서 받는 돈 없이는 이것 중 어떤 것도 가능하지 않을 것입니다. 양해해 주셔서 감사드리며 모든 분들이 축제를 즐기시길 바랍니다!

어휘
annual 매년의, 연례의 attend *참석하다; (…에) 다니다 complaint 불만 charge 청구하다 fee 수수료, 요금 [문제] necessity 필요성 policy 정책, 방침 venue 장소 conflict 갈등 surround 둘러싸다 entrance 출입구, 문; *입장

문제풀이
16 ① 정책 변화의 필요성
　② 가족과 친구의 중요성
　③ 콘서트 취소를 막기 위한 노력
　④ 축제 장소 변경의 이유
　⑤ 입장료를 둘러싼 갈등

DICTATION Answers　본문 p.122

01　explain a few rules / may not take photographs / follow these rules

02　in favor of / drawbacks of using wind energy / killed by the turbines

03　went pretty well / make this school a pleasant place / take up too much time

04　look awesome / because of its unique architecture / on the top of

05 stuff to move / there's no elevator / a favor in return / watch my little brother

06 a bit of a problem / scaring some of my customers away / hang out

07 want to fly economy / out of my budget / That seems to be

08 10 minutes long / who wrote the words / that makes sense

09 build your career / some useful advice / prior to the event

10 international volunteer programs / any artistic talent / Anything less than a year

11 how to play chess

12 make a lot of noise

13 cancel our picnic / Don't be silly / get rained on

14 final grade / might not graduate / study on my own

15 On his way back / move to another seat / change seats with

16~17 had some complaints / has been growing each year / pay the town / attending the festival

21 영어듣기 모의고사

본문 ▲ p.126

01 ①	02 ①	03 ④	04 ⑤	05 ④	06 ⑤
07 ③	08 ⑤	09 ③	10 ②	11 ③	12 ③
13 ③	14 ②	15 ⑤	16 ②	17 ④	

01 ①

남: 나의 삶에서 내가 뭘 할 것인지를 전혀 알 수 없었던 때가 있었다. 만일 당신이 그렇게 느낀다면, 이것이 당신을 위한 책이다. 무엇이 정말 중요한지를 내가 알아가기 시작하도록 도와준 것이 바로 이 책이었다. 이것이 (미래를 점쳐 주는) 수정 구슬은 아니라는 것을 명심해야 한다. 하지만 이것은 당신이 누구인지에 대해 많은 유용한 정보를 제공한다. 지금 자신에 관한 몇 가지 해답을 찾고 있는 모든 사람들에게 이 책을 강력히 추천한다.

어휘
figure out 이해하다 keep in mind that …임을 명심하다 crystal ball (점치는 데 쓰는) 수정 구슬 currently 현재 search for …을 찾다

02 ①

여: 우리 새 아파트가 정말 마음에 들어요. 전에 옥상 테라스가 있는 곳에 살아본 적이 없거든요.

남: 네, 멋지지 않아요? 여기 앉아서 도시의 멋진 전망을 볼 수 있을 거예

요.

여: 사실은 그보다 더 쓸모가 있겠다고 생각 중이었어요.

남: 어 그래요? 당신 생각은 뭐였죠?

여: 음, 당근과 상추를 여기 위에다 심는 건 어때요? 어쩌면 완두콩도요.

남: 일이 많지 않을까요?

여: 별로 그렇지도 않아요. 주말에 그것을 할 수 있을 거예요. 재미있는 취미가 될 거고, 우린 먹을 신선한 채소도 얻게 되잖아요.

남: 알았어요. 좋은 생각이에요.

여: 좋아요! 오늘 오후에 씨앗을 사러 가는 게 어때요?

어휘
rooftop 옥상 patio 파티오(보통 집 뒤쪽에 만드는 테라스) lettuce 상추 pea 완두콩 seed 씨앗

03 ④

남: 자, 어떻게 생각하시나요?

여: 아름답네요. 최근에 주방을 개조한 건가요?

남: 네, 집주인이 고친지 일 년도 채 안 됐어요.

여: 알겠어요. 어떤 종류의 난방이 있는지 알려주시겠어요?

남: 천연가스를 사용해서 요금은 적절하게 나올 거예요.

여: 그거 잘됐네요. 그럼 그분들은 집값으로 얼마를 원하는 거죠?

남: 요구하는 가격은 현재로서는 22만 5천 달러입니다.

여: 잘 모르겠네요… 이 근처 동네에 비해 약간 비싼 것 같아요.

남: 음, 가격은 협상하는 동안 내려갈 것 같습니다.

여: 네… 생각 좀 해보고 오후에 전화 드릴게요.

남: 좋습니다. 여기 제 명함입니다. 아무 때나 전화 주셔도 됩니다.

어휘
renovate 개조[보수]하다 update 최신의 것으로 하다, 갱신하다 heating 난방 reasonable (가격이) 적정한, 너무 비싸지 않은 neighborhood 이웃, 인근 negotiation 협상, 협의 business card 명함

04 ⑤

남: 안녕하세요, 올리비아. 문제가 있나요?

여: 네, 시먼스 씨, 문제가 있어요. 분장실이 별로 깨끗하지 않아요.

남: 유감이군요. 잭슨 씨한테 옷걸이에 당신의 드레스를 걸어두라고 말했는데요.

여: 그렇게 하기 했는데, 빈 상자들이 바닥에 있네요.

남: 아, 그래요? 아마도 전에 그 분장실을 사용한 사람이 그것들을 치우는 걸 잊었나 봐요.

여: 그리고 바닥에는 부츠가 한 켤레 있어요.

남: 유감이네요. 다른 건 또 없어요?

여: 사실은, 있어요. 둥근 탁자 위에 오래된 꽃다발도 있어요.

남: 알겠어요. 사람을 시켜서 당장 방을 청소하도록 할게요.

여: 고마워요. 그리고 거울에 있는 사진들을 떼어주실 수 있나요? 그것 때문에 얼굴이 잘 보이지 않아요.

남: 알겠어요. 청소하는 사람더러 그걸 떼어내도록 할게요.

어휘
dressing room 분장실, 탈의실 hang 걸다 hanger 옷걸이 bunch 송이, 다발 get rid of …을 없애다 cleaner *청소부; 세제

05 ④

[전화벨이 울린다.]

남: AP 쇼핑의 필입니다. 무엇을 도와드릴까요?

여: 안녕하세요, 필. 저 휘트니예요.

남: 아, 안녕하세요. 당신이 자리에 없는 걸 알고 당신한테 오는 전화를 받고 있었어요.

여: 고마워요. 근데, 사장님이 휴대전화 번호를 바꾸셨나 봐요, 그렇죠?

남: 네, 지난주에요. 번호를 갖고 있지 않나요?

여: 아뇨, 그리고 전 사장님이 이번 주에 출장 가신 줄 알았어요. 오늘 출근하지 못한다고 말씀드려야 하는데.

남: 무슨 일 있어요?

여: 음, 전 아무 문제 없어요. 하지만 제 남편이 무릎을 다쳤어요.

남: 그것참 안됐군요. 통증이 심한가요?

여: 꽤 심해요. 그래서 제가 남편을 병원에 데려다줘야 해요. 사장님 번호를 제게 문자메시지로 보내주면 좋겠어요.

남: 알겠어요. 지금 바로 보내줄게요. 남편이 괜찮으면 좋겠네요.

여: 고마워요, 필.

어휘
on business 볼일이 있어, 업무로 **pain** 고통 **text message** (휴대전화로) 문자 메시지를 보내다

06 ⑤

남: 안녕하세요, 윌리엄스 교수님. 잠시 얘기 좀 나눌 수 있을까요?

여: 음, 사이먼, 내 업무 시간은 보통 오후란다.

남: 알지만, 급한 일이라서요.

여: 알겠다. 무슨 문제니?

남: 오늘 마감 시간까지 에세이를 제출할 수 없을 것 같아요.

여: 이해가 되지 않는구나. 끝내는 데 일주일이 넘는 시간이 있었잖니.

남: 그건 끝냈는데요, 윌리엄스 교수님, 제 노트북이 고장 났어요.

여: 음, 작업을 백업해 놨어야지.

남: 했는데요, 집에 USB 드라이브를 두고 왔어요. 2시간 연장을 해주시면 오늘 이따가 이메일로 보내드릴 수 있어요.

여: 알았다, 사이먼, 하지만 이런 일이 다신 없어야 해.

남: 없을 거예요. 감사합니다.

어휘
office hours 업무[집무] 시간 **emergency** 위급 상황 **submit** 제출하다 **deadline** 마감 시간 **back up** 백업하다 **drive** 드라이브, 구동 장치 **extension** 연장

07 ③

남: 저녁 식사 어떠셨어요?

여: 아주 좋았어요, 근데 계산서가 정확하지 않은 것 같아요.

남: 정말요? 제가 한번 보겠습니다.

여: 우린 스테이크 2인분과 와인 두 잔을 시켰어요. 그럼 28달러예요.

남: 네, 계산서에 그렇게 되어 있네요.

여: 하지만 보세요, 우리 계산서에 디저트에 대한 금액도 있어요.

남: 네, 5달러짜리 초콜릿 파이 한 조각이요. 그래서 총 33달러네요.

여: 하지만 우린 초콜릿 파이를 먹지 않았어요.

남: 아, 드시지 않으셨다고요? 죄송합니다. 그건 계산서에서 **빼도록** 하죠.

여: 고마워요.

어휘
charge 요금 **in total** 총, 통틀어 **take off** (금액 등을) 빼다, 공제하다

문제풀이
여자의 계산서에 33달러가 나왔으나, 그중 5달러짜리 초콜릿 파이는 먹지 않았으므로, 여자는 28달러만 내면 된다.

08 ⑤

여: 맞춰 봐! 나 방금 새 과학 선생님을 만났어. 아주 멋지시더라.

남: 어, 그래? 성함이 뭐니?

여: 피터슨 선생님이야. 선생님의 교실에는 커다란 아인슈타인 포스터가 있어.

남: 왜? 그가 선생님의 존경하는 인물이기 때문이니?

여: 맞아. 그는 차세대 아인슈타인을 발굴하기 위해 교사가 되었다고 하셨어.

남: 그게 너일지도 몰라.

여: 글쎄다. 어쨌든 선생님은 하버드를 졸업하셨으니 똑똑하실 거야.

남: 와, 하버드. 좋은 학교잖아. 항상 연구하면서 시간을 보내실 게 틀림없어.

여: 사실, 그분은 아주 활동적이신 것 같아. 등산과 테니스를 즐겨 하신다고 하셨어.

남: 흥미로운 분인 것 같아. 나도 그분을 만날 수 있으면 좋겠다.

여: 교장 선생님께서 내일 아침에 모두에게 선생님을 소개하실 거야.

남: 잘됐다. 그때 만날 수 있겠구나.

어휘
hero 영웅, 존경하는 사람 **graduate** 졸업하다 **principal** 교장

09 ③

여: 일광 절약 시간은 여름 달 동안에 현지 시각을 한 시간 앞으로 옮기는 관습입니다. 이 체제는 1908년에 캐나다의 도시인 Arthur 항구에서 처음으로 시행되었습니다. 이는 1차 세계대전 중에 많은 나라에서 인기를 얻었습니다. 그것은 에너지를 절약하는 방책으로 여겨졌지만 일부 전문가들은 이렇게 하는 것이 효과적이라고 믿지 않습니다. 흥미롭게도, 자동차 사고 수는 일광 절약 시간 중에 줄어듭니다. 밖이 환할 때 운전하는 것이 더 안전하기 때문에 이렇게 할 가능성이 가장 큽니다. 심장마비 비율은 시간이 바뀌는 첫 주에 증가합니다. 이는 시간이 앞으로 당겨지면서 한 시간의 수면 부족에 기인한 것입니다.

어휘
daylight saving time 일광 절약 시간, 서머타임 **practice** 관습 **local time** 현지 시각 **popularity** 인기 **expert** 전문가 **vehicle** 차량 **reduce** 줄이다 **heart attack** 심장마비 **be attributed to** …에 기인하다 **loss** 손실

10 ②

여: 안녕하세요, 손님! 텐트를 찾고 계시나요?

남: 네, 그렇습니다. 전 다음 주에 캠핑 여행에 쓸 하나가 필요해요.

여: 알겠습니다. 음, 저희는 여러 가지 선택할 수 있는 스타일들이 있는데요. 돔형 텐트가 아주 인기가 많습니다.

남: 그거 마음에 드네요. 터널형 텐트도 마음에 들고요.

여: 그런데 피라미드형 텐트는 싫으세요?

남: 싫어요. 좀 예쁘지 않아서요. 아무튼, 전 저와 제 친구 둘에게 충분히

큰 게 필요해요.

여: 알겠습니다. 초보자이시면, 저는 팝업 텐트를 사는 걸 추천해 드립니다.

남: 사실, 이게 저희 모두에게 첫 캠핑이 될 거라서요. 팝업 텐트는 설치가 더 쉬운가요?

여: 네, 그렇습니다. 그리고 야영지로 하이킹이나 운전을 해서 가실 건가요?

남: 하이킹이요. 숲으로 5킬로미터 정도 하이킹을 합니다.

여: 그렇다면 저는 더 가벼운 텐트를 고르실 것을 추천해 드립니다.

남: 좋은 생각입니다. 더 가벼울수록, 더 좋죠!

여: 그런 경우라면, 이것이 제가 구매하시라고 권해 드리는 텐트입니다. 실망하지 않으실 거예요!

pop-up tent 팝업 텐트(자동으로 튀어나오는 뼈대를 가진 둥그스름한 텐트) set up 설치하다, 세우다 campsite 야영지

11 ③

남: 네가 뉴욕에 나를 만나러 오다니 정말 신이 나!

여: 응, 타임스스퀘어를 보고 싶어. 거기 어떻게 갈 수 있니?

남: 버스 타고 갈 수 있어. 혹은 네가 좋다면 걸어갈 수도 있지.

여: 그냥 걸어가자. 날씨가 좋잖아.

어휘

[문제] ferry 연락선, 페리

문제풀이

① 거기가 어딘지 알 것 같아.
② 연락선은 얼마나 오래 타니?
④ 우리가 버스를 잘못 탄 것 같아.
⑤ 음, 거기에는 볼 게 없어.

12 ③

여: 실례합니다. 이 그림이 얼마인가요?

남: 그건 500달러입니다. 그 화가가 아주 유명해요.

여: 너무 비싼데요. 할인해주실 수 있을까요?

남: 죄송합니다. 그 가격은 이미 할인된 거예요.

어휘

discount *할인; 할인하다 [문제] purchase 구입한 것 for free 무료로

문제풀이

① 물론이죠. 더 저렴한 것을 보여드릴게요.
② 네. 구매한 것에 대해 10퍼센트 할인을 받았어요.
④ 와! 당신이 비싼 그림을 소장한 줄 몰랐네요.
⑤ 미술과 학생이면 무료로 둘러봐도 돼요.

13 ③

여: 시작하기 전에, 시간을 내주셔서 감사합니다.

남: 아니에요, 초대해 주셔서 감사합니다. 항상 당신의 프로그램을 즐겨 보았어요.

여: 감사합니다. 그럼 당신의 사업적 성공의 비밀을 얘기하면서 시작해볼까요?

남: 음, 전 정말 별 비밀이랄 게 없습니다. 그 누구와도 별다른 걸 하지 않

아요.

여: 하지만 당신은 대부분 사람보다 더 많은 걸 성취하고 수백만 달러의 재산을 벌었잖아요.

남: 제 생각엔 그건 제가 항상 최선을 다하고 절대 포기하지 않았기 때문인 것 같아요.

여: 그렇군요. 그럼 당신이 그저 운이 좋았다고는 생각하지 않나요?

남: 전혀 아니죠. 끈기와 노력이 열쇠이지, 운은 아니에요.

여: 그럼 그게 바로 요즘 젊은이들에게 해주고 싶은 조언인가요?

남: 네. 노력을 절대 멈추지 않으면 성공할 거예요.

어휘

secret 비밀, 비결 accomplish 성취하다 try[do] one's best 최선을 다하다

문제풀이

① 아니요, 나이가 중요하다고 생각하진 않아요.
② 맞아요. 어떤 사람들은 단지 매우 운이 좋죠.
④ 그래요. 당신은 사업의 비밀을 배울 필요가 있어요.
⑤ 잘 모르겠어요. 전 좋은 조언을 많이 받아왔어요.

14 ②

남: 안녕, 브렌다. 들어와.

여: 안녕, 폴. 나를 초대해 줘서 고마워. 와, 네가 여기서 일을 하다니 믿을 수가 없어.

남: 그래, 이건 굉장한 직업이야. 나는 항상 방송계에 관심이 있었어.

여: 이곳이 드라마를 촬영하는 곳이니?

남: 몇 개만 여기서 촬영해. 원한다면 *Golden Princess*를 촬영하는 곳을 보여줄 수 있어.

여: 그러면 정말 좋겠다! 내가 가장 좋아하는 프로그램 중 하나거든.

남: 그러면 날 따라와. 바로 이 복도 끝에 있어.

여: 우리가 유명한 배우들을 볼 수 있을 거로 생각하니?

남: 아니. 그들은 지금은 촬영하고 있지 않아. 아무도 거기에 없을 거야.

여: 괜찮아. 그래도 여전히 흥미로울 거야.

남: 조심만 해. 그들이 사용하는 장비는 매우 비싸거든.

여: 아무것도 만지지 않겠다고 약속할게!

어휘

broadcasting 방송계 film 촬영하다 equipment 장비

문제풀이

① 내가 그것들을 시험 삼아 써 봐도 괜찮겠니?
③ 쉿! 이제 촬영이 시작될 거야.
④ 내가 배우들한테 사인을 받아 줄게.
⑤ 이 장비를 얼마 주고 샀어?

15 ⑤

남: 스티븐은 시카고에서 자랐지만, 그는 언제나 다른 문화에 관심이 있었다. 그는 학교에서 초급과 중급 중국어 강좌를 수강해 왔다. 그는 정말로 중국어를 배우는 것을 즐겨 해서 교환 학생으로 중국에서 공부하기로 결심한다. 하지만 그가 비행기로 중국에 가서 첫 수업을 듣게 되자, 그는 교사가 말하는 것을 이해하려고 고군분투한다. 강의가 중국어로 진행되기 때문에 그가 다른 학생들을 따라잡는 것은 힘들다. 수업이 끝나자, 그는 숙제가 있다는 것을 이해하지만 정확하게 무엇인지 알지 못한다. 그래서 그는 옆에 있는 학생에게 숙제에 대해서

묻기로 결심한다. 이런 상황에서, 스티븐은 그 학생에게 뭐라고 말하겠는가?

스티븐: 다음 수업을 위해서 우리가 무엇을 해야 할지 말해주세요.

[어휘]

intermediate 중급의 exchange student 교환 학생 struggle to-v …하느라 고군분투하다 keep up with …을 따라잡다 assignment 숙제 [문제] translation 번역

[문제풀이]

① 수업 시간에 왜 필기를 하지 않았어요?
② 이 수업에 어떻게 등록할 수 있는지 알고 있어요?
③ 올해 여름 방학은 언제 시작해요?
④ 중국어 번역 프로그램을 갖고 있어요?

16 ② 17 ④

여: 안녕하세요, 여러분, 그리고 저희 투어에 오신 것을 환영합니다. 제 뒤를 보시면, 로마인들에 의해 만들어진 고대 수도교의 잔해를 보시게 될 겁니다. 최초의 수도교는 기원전 300년경에 등장했지만, 고대 로마인들은 수 세기 동안 많은 수도교를 만들었습니다. 로마에만 약 800킬로미터의 수도교가 있었습니다. 왜 로마인들은 이런 구조물들이 필요했을까요? 이 도시는 점점 더 많은 사람이 로마로 이동해 오면서 발전했습니다. 그래서, 더 깨끗한 물이 도시 전역에 걸쳐 운반되어야 했습니다. 따라서 로마인들은 수도교를 만들었고, 그 수도교는 물이 흐르게 하려고 중력을 사용했습니다. 이런 체계는 또한 로마인들이 그들의 도시를 깨끗하게 유지하고 강이나 호수에서 먼 장소에도 물을 운반할 수 있도록 했습니다. 그래서 로마인들은 어디에나 도시를 건설할 수 있었고 여전히 꾸준한 물 공급이 이루어지고 있습니다. 비록 대부분의 로마 수도교는 사실상 지하에 있었지만, 일부는 여러분이 오늘날 보고 계시는 것처럼 매우 높은 곳에 있었습니다.

[어휘]

ruin 《*pl.*》 잔해, 유적 aqueduct 수로, 수도교 structure 구조(물), 건축물 gravity 중력 supply 공급 underground 지하에 [문제] paved road 포장도로 architectural 건축학의

[문제풀이]

16 ① 고대 로마의 포장도로　② 고대 로마의 수도 체계
　③ 고대 로마 도시의 특징　④ 고대 로마의 건축 양식
　⑤ 식수를 찾는 효과적인 방법

DICTATION Answers

본문 p.128

01 figuring out / is currently searching for

02 enjoy a nice view / how about planting / fresh vegetables to eat

03 had it updated / be reasonable / during negotiations / give me a call

04 on a hanger / a pair of / a bunch of old flowers

05 answering your phone / I won't be in today / text message the boss's number

06 before the deadline / my laptop broke / email it to you

07 my bill is correct / charge for a dessert / have it taken off

08 graduated from / enjoys hiking and playing tennis / introduce him to everyone

09 gained popularity / during daylight saving time / This is attributed to

10 kind of ugly / hiking to the campsite / You won't be disappointed

11 take a bus

12 give me a discount

13 for having me / try my best / Absolutely not

14 been interested in broadcasting / one of my favorite shows / The equipment they use

15 has taken / keep up with / the student next to him

16~17 see the ruins of / over the centuries / make the water flow / have a steady water supply

22 영어듣기 모의고사

본문 ▲ p.132

01 ⑤	02 ⑤	03 ②	04 ⑤	07 ⑤	06 ④
07 ①	08 ④	09 ④	10 ③	11 ⑤	12 ③
13 ③	14 ③	15 ⑤	16 ①	17 ③

01 ⑤

여: 전 세계 많은 나라가 맥주와 담배 같은 물품에 추가적인 세금을 부과합니다. 그 나라들은 이런 물품들의 소비를 줄이기 위해 이렇게 하는 것입니다. 현재, 미국의 입법자들은 청량음료 세라는 새로운 종류의 세금을 검토 중입니다. 의사들은 수십 년 동안 청량음료와 관련한 건강상의 위험을 알고 있었습니다. 청량음료에는 엄청난 양의 설탕과 카페인이 있습니다. 이것들은 비만과 당뇨병까지 일으킬 수 있습니다. 청량음료에 추가 세금을 부과함으로써, 입법자들은 시민들이 콜라를 덜 마시고 그 결과 더 건강해지기를 바랍니다.

[어휘]

place a tax on …에 과세하다 extra 추가의 beer 맥주 cigarette 담배 consumption 소비 lawmaker 입법자 soft drink 청량음료 risk 위험 decade 10년 obesity 비만 diabetes 당뇨병

02 ⑤

여: 이따 스터디 모임에 먹을 간식을 사자.
남: 좋아, 그런데 우리가 사는 간식에 대해 주의해야 해. 대부분은 건강에 좋지 않은 첨가물이 들어 있어.
여: 첨가물? 소금과 설탕 같은 거 말이니?
남: 아니, 식품이 상하지 않게 하는 화학 물질이나 인공 색소 말이야.

여: 그게 뭐가 그렇게 나쁜데?

남: 오랫동안 먹으면 몸에 아주 해로울 수 있어.

여: 아, 어떤 식품 색소는 암을 일으킬 수 있다고 들은 게 기억나.

남: 음, 그건 100퍼센트 확실한 건 아니지만 위험할 수 있어.

여: 맞아, 근데 유감스럽게도 요즘 인공 첨가물이 없는 식품은 거의 없어.

어휘 additive 첨가물 chemical 화학 물질 go bad 상하다 artificial color 인공 색소 dye 염료

03 ②

남: 실례합니다. 오늘의 쇼에 대해서 질문 좀 드려도 될까요?

여: 물론이죠. 어느 잡지사 소속이시죠?

남: 전 잡지사 소속이 아닙니다. 지역 신문에 기고하고 있습니다.

여: 아, 그렇군요. 질문하세요.

남: 고맙습니다. 첫 번째 질문입니다. 선생님의 신작 발표회의 주제는 무엇입니까?

여: 주제는 '힘'입니다. 그게 제가 가죽과 같은 튼튼한 옷감을 사용한 이유이죠.

남: 알겠습니다. 그리고 올해의 최고 슈퍼모델이 참가하게 되어서 기대되시나요?

여: 프란체스카 로마노 말씀인가요? 네, 그녀가 제 발표회의 모델을 하게 돼서 무척 기쁩니다.

남: 그리고 끝으로, 이것이 선생님의 마지막 발표회라는 소문이 있던데요. 사실입니까?

여: 천만에요. 전 패션에 대해 열정을 갖고 있어요. 그래서 영원히 옷을 디자인할 겁니다.

남: 시간 내주셔서 감사합니다.

어휘 theme 주제 collection 수집물; *신작 (발표회) fabric 직물, 옷감 leather 가죽 supermodel 슈퍼모델 model 모델; *모델로 일하다 rumor 소문 passionate 열정적인

04 ⑤

남: 경비원으로 일하는 건 지루해요. 모든 게 매일 똑같아요. 안내소의 직원조차도 항상 똑같은 재킷과 모자를 써요.

여: 네, 하지만 사람들은 항상 다르잖아요. 저기 수하물을 끌고 가는 여자가 보이죠?

남: 네, 보여요.

여: 그녀의 강아지를 봐요. 안경을 쓰고 있어요. 그걸 매일 보진 않죠.

남: 음, 매일은 아니죠.

여: 또, 안내 데스크에 있는 저 남자는 매우 독특한 티셔츠를 입고 있네요. 해골이 그려져 있어요.

남: 당신 말의 의미를 알 것 같네요.

여: 그리고 그 사람 뒤로 말다툼을 하는 두 사람이 있어요.

남: 당신 말이 맞을지도 모르겠어요. 이 직업은 제가 생각했던 것보다 더 흥미롭네요.

어휘 security guard 경비원 luggage (여행용) 짐, 수하물 skull 두개골 argue 말다툼하다

05 ⑤

남: 안녕, 폴라. 뭘 하고 있니?

여: 안녕, 대니얼. 그냥 아르바이트를 찾고 있어.

남: 온라인에서 찾고 있니?

여: 물론이지. 고용주들이 일자리를 올려놓는 웹사이트들이 많거든.

남: 그렇구나. 그래서 어떤 종류의 일자리들이 있니?

여: 음, 대부분 식당에서 서빙하는 일들이야.

남: 그것 중의 어떤 자리에라도 지원했니?

여: 아니. 문제가 있어. 대부분의 고용주는 19세 이상인 사람을 원해.

남: 음, 안됐다. 얘, 잠깐만! 우리 삼촌이 다음 주에 쇼핑몰에 샌드위치 가게를 여실 거야. 어쩌면 삼촌이 도움이 필요할지도 모르겠어.

여: 그거 잘됐다! 삼촌께 전화해서 혹시 일할 사람을 찾고 계시는지 여쭤볼 수 있니?

남: 물론이지. 기꺼이 도와줄게.

어휘 employer 고용주 post (온라인에 글을) 게시하다. 올리다 vacancy 결원, 공석 wait tables 식당에서 서빙 일을 하다

06 ④

남: 아주 멋진 노트북이구나! 네 거니, 켈리?

여: 응, 생일 선물로 아버지께 받았어.

남: 정말 좋겠구나! 나도 하나 있으면 좋겠는데.

여: 네 생일이 다음 달에 다가오지 않니?

남: 응. 너 기억력이 좋구나.

여: 부모님께 생일 선물로 하나 사달라고 하지 그러니? 사주실지도 모르잖아.

남: 아닐걸. 이번 학기에 성적이 좋아야만 선물을 사주신다고 했지.

여: 그건 별문제 아니야. 그저 열심히 공부하고 숙제를 다 하렴.

남: 그럴게. 그리고 부모님의 기대를 충족시키면 노트북을 사 달라고 할 거야.

여: 행운을 빌게.

어휘 laptop 노트북 expectation 기대

07 ①

남: 무엇을 도와드릴까요?

여: 컴퓨터로 화상통화를 하고 싶은데, 필요한 장비가 없어서요.

남: 음, 우선 헤드셋이 필요하시겠네요.

여: 작은 마이크가 달린 헤드폰 세트를 말씀하시는 건가요?

남: 맞습니다.

여: 네. 저건 얼마인가요?

남: 저희 기본 모델은 20달러이고, 가장 좋은 모델은 55달러예요. 어떤 걸 원하시나요?

여: 기본 모델로 주세요. 또 뭐가 필요할까요?

남: 노트북에 내장된 웹캠이 있죠?

여: 아니요, 제 노트북은 정말 오래되어서요. 그래서 전 웹캠이 필요할 것 같네요. 그것도 파시나요?

남: 네. 이 단순한 15달러짜리부터 40달러짜리 모델까지 여러 가지가 있습니다.

여: 음, 그냥 채팅용이니까, 기본적인 걸로 살게요.

남: 알겠습니다. 다른 건요? USB 메모리 스틱이 단돈 10달러에 할인 판매 중이에요.

여: 아뇨, 그거면 돼요.

어휘
video call 화상통화 equipment 장비 headset 헤드셋 microphone 마이크 standard 표준의 built-in 내장된 webcam 웹캠(컴퓨터에 연결할 수 있는 비디오카메라)

문제풀이
여자는 화상통화용으로 헤드폰 기본 모델($20)과 기본적인 웹캠($15)을 구매한다고 하였으므로, 여자는 35달러를 지불할 것이다.

08 ④

[전화벨이 울린다.]

여: 분실물 센터입니다. 도와드릴까요?

남: 오늘 파란 지갑이 들어왔나요?

여: 아직까진 없었지만, 아직 이른 시간이니까요. 저희 버스 중 하나에 두고 내리셨나요?

남: 네, 오늘 아침에 43번 버스에요.

여: 알겠습니다. 큰 지갑인가요?

남: 아뇨, 평균 크기에요.

여: 알겠어요. 돈이 지갑 안에 많이 있었나요?

남: 별로요. 겨우 7달러에요. 하지만 그게 선물 받은 거라서 정말로 찾고 싶어요.

여: 이해합니다. 저희가 지갑을 식별하는 데 도움이 될 만한 게 안에 들어 있나요?

남: 네. 어머니 사진이 있어요. 그리고 제 도서관 카드도요.

여: 그렇군요. 이름과 전화번호를 알려 주세요. 누군가 지갑을 찾으면 전화를 드릴게요.

남: 좋아요. 제 이름은 도널드 존스이고 번호는 812-111-1520이에요.

어휘
lost and found 분실물 취급소 average-sized 평균 크기의 identify 알아보다

09 ④

남: 안녕하십니까. 제 전시회의 개막식에 오신 것을 환영합니다. 제 전시회의 제목은 *아이들의 눈을 통해서*이며, 앞으로 7일간 이곳에서 열릴 것입니다. 이 전시회의 주제는 집 없는 아이들입니다. 저는 여러분과 함께 그들의 이야기를 나누려는 희망을 가지고 사회적으로 혜택 받지 못하는 아이들의 사진을 찍으며 남미 전역을 여행했습니다. 여러분 중 일부는 사진을 찍을 때 제가 무엇을 포착하려고 애쓰는지를 물어보셨습니다. 그에 대한 대답은 '용기'입니다. 저는 제가 촬영하는 모든 아이들에게서 용기를 봅니다. 그런 이유로 이 전시회에서 제가 가장 좋아하는 사진은 '불'이라는 제목의 것입니다. 여러분이 보시다시피 어린 볼리비아 소녀의 눈은 용기의 불로 활활 타오르고 있습니다.

어휘
exhibition 전시(회) entitle 제목을 붙이다 collection 신작 (발표회) homeless 집 없는, 노숙자의 disadvantaged (사회적으로) 혜택 받지 못한 capture (사진 등으로) 포착하다 courage 용기

10 ③

남: Computer City에 오신 것을 환영합니다. 오늘 무엇을 도와드릴까요?

여: 노트북을 세일한다고 들었는데요. 제가 고르는 걸 도와주실래요?

남: 물론입니다. 저희는 세 가지 다른 크기의 화면을 가진 노트북들이 있습니다.

여: 저는 제일 작은 크기는 싫어요. 제가 종종 노트북으로 영화를 보거든요.

남: 알겠습니다. 저장 공간은 어떠세요? 256기가바이트는 충분한가요?

여: 네, 하지만 그것보다 적으면 안 됩니다. 저는 제 모든 프로그램을 위한 충분한 공간이 필요해요.

남: 알겠습니다. 그리고 자주 들고 다니실 건가요?

여: 네. 저는 업무상 출장 때 그걸 가지고 다닐 계획이라서, 무게가 2킬로그램보다 덜 나갔으면 해요.

남: 그건 문제없습니다. 이 두 개의 모델을 보시죠. 둘 다 아주 인기 있습니다.

여: 둘 다 좋아 보이지만, 저는 흰색이 더 마음에 드네요. 그게 제 사무실과 어울릴 거예요.

남: 탁월한 선택이세요. 오늘 엄청난 할인을 받게 되실 겁니다.

여: 그랬으면 좋겠네요!

어휘
laptop 노트북 storage 저장 weigh 무게가 나가다 match 어울리다 terrific 정말 멋진 bargain 할인

11 ⑤

여: 안녕, 톰. 휴일에 어디 갔었니?

남: 유럽으로 가족 여행을 갔어. 우리 가족은 거기로 이주할 계획이야.

여: 정말이니? 네가 곧 유럽에서 살 거라는 말이니?

남: 응, 하지만 이사 준비하는 데 시간이 좀 걸릴 거야.

어휘
emigrate (타국으로) 이주하다 [문제] scenery 경치 contact 연락하다

문제풀이
① 내년에 유럽을 여행할 거야.
② 삼촌이 독일에 살고 있어서 찾아갔었지.
③ 응, 경치가 예상한 만큼 아름다웠어.
④ 네가 거기로 이사하면 이메일로 연락할게.

12 ③

남: 내 컴퓨터가 너무 느리게 작동하고 있어. 파일을 여는 데 시간이 오래 걸려.

여: 바이러스 퇴치 프로그램을 돌려 봐. 내 컴퓨터도 똑같은 문제가 있었는데 그게 정말 도움이 됐어.

남: 좋은 생각이야! 그런데 어떻게 해야 해?

여: 인터넷에서 프로그램을 내려받을 수 있어.

어휘
run 작동하다, 작동시키다 anti-virus program 바이러스 퇴치 프로그램 [문제] up-to-date 최신의 copy 복사하다 material 자료

문제풀이
① 인터넷에서 컴퓨터를 살 수 있어.

② 요즘 대부분 컴퓨터에는 바이러스가 있어.
④ 내 컴퓨터는 네 것보다 더 최신이야.
⑤ 이 사이트는 자료를 복사하는 것을 허용하지 않아.

⑤ 그건 너희 어머니께서 네게 별도의 도움을 주라고 말씀하셨기 때문이지.

13 ③

[문 두드리는 소리]
남: 안녕하세요?
여: 안녕하세요. 전 멜리사예요. 아래층에 살아요.
남: 만나서 반가워요, 멜리사.
여: 음, 전 불만이 있어서 왔어요.
남: 아, 정말요? 무엇에 대해서요?
여: 당신 아파트로부터 우리 집 천장을 통해 나는 소음이 너무 심해요.
남: 죄송해요. 우리 가족이 작은 파티를 열어서 손님들을 좀 초대했거든요. 곧 끝날 거예요.
여: 최악인 부분은 내 침실 바로 위에서 아이들이 뛰어다니고 큰 소리로 웃는 게 들린다는 점이에요.
남: 그 점에 대해서는 정말 죄송해요. 애들을 조용히 시킬게요.
여: <u>고마워요. 전 자는 걸 방해받고 싶지 않거든요.</u>

어휘
downstairs 아래층에 **ceiling** 천장 [문제] **give one's regards to** …에게 안부를 전하다 **scold** 야단치다, 꾸짖다 **behavior** 행동 **disturb** 방해하다 **unreasonable** 불합리한

문제풀이
① 당신 가족에게 제 안부를 전해주세요.
② 제 아이들의 행동에 대해 아이들을 야단칠게요.
④ 정말 멋진 파티네요! 초대해주셔서 고마워요.
⑤ 이 밤 시간에 불평하시는 것은 불합리해요.

14 ③

남: 아, 미영아, 거기 있었구나. 내 사무실로 좀 들어올래?
여: 네, 오라일리 선생님.
남: 자리에 앉으렴. 자, 이제 네게 하고 싶은 얘기가 있단다.
여: 뭔데요, 선생님?
남: 음, 사실은 난 네게 매우 만족하고 있어. 이번 학기 들어 네 영어 점수가 많이 올랐어, 그렇지?
여: 네, 맞아요. 감사합니다.
남: 공부 습관이나 뭐 그런 걸 바꿨니?
여: 정확히 그런 건 아니에요. 저희 어머니가 제가 만약 영어 시험에서 A를 받으면 새로운 스케이트보드를 사주겠다고 하셨거든요.
남: 뭐라고? 그래서 영어에만 좀 더 노력을 많이 한 거니?
여: 음… 네. 전 정말 그게 갖고 싶어요.
남: <u>상을 받기 위해서가 아니라 너 자신을 위해서 공부해야지!</u>

어휘
mark 자국; *점수 **significantly** 상당히 **make an effort** 노력하다 [문제] **cheat** 속이다, (시험·경기 등에서) 부정행위를 하다 **reward** 보상[보답]하다; 보상 **impressed** 인상 깊은, 감명받은 **attitude** 태도

문제풀이
① 네가 부정행위를 했다는 것에 화가 난다.
② 난 학생들에게 선물로 보상하지는 않을 거야.
④ 학습에 대한 너의 새로운 태도에 난 무척 감명받았단다.

15 ⑤

남: 마틴은 월요일 오후에 엄마에게 학교에서 보낸 그의 하루에 대해 이야기하면서 집에 있다. 그는 아침에 기금 모금 행사가 있었다고 설명한다. 선생님들과 학생들이 아프리카의 어린이들을 돕기 위해서 유니세프를 위한 기부금을 모았다. 마틴은 대부분의 용돈을 주말에 써버리고 그날 주머니에 돈이 많지 않았기 때문에, 1달러밖에 기부하지 못했다. 지금 그는 더 많이 기부하지 않은 것을 후회한다. 그는 아프리카에 굶주리고 있는 어린이들이 있는데, 자신은 맛있는 간식을 먹는 것에 대해 가책을 느낀다고 말한다. 이런 상황에서, 마틴의 어머니가 마틴에게 할 말로 가장 적절한 것은 무엇인가?
마틴의 어머니: <u>네가 도울 기회가 또 있을 거야.</u>

어휘
fundraising 기금 모금 **donation** 기부 (*v.* donate) **allowance** 용돈 **regret** 후회하다 **guilty** 죄책감이 드는 **starve** 굶주리다 [문제] **crumb** 부스러기

문제풀이
① 그건 네가 생각하는 것만큼 중요하지 않단다.
② 제발 과자 부스러기 좀 치우렴.
③ 걱정하지 마. 다음번엔 더 잘할 수 있을 거야.
④ 왜 네 용돈을 전부 기부하지 않았니?

16 ① 17 ③

여: 반려동물이 훌륭한 동반자일지는 모르지만, 비행기 여행을 함께 하는 것은 어려울 수 있습니다. (항공사마다) 방침이 다르니, 여행 전에 반려동물과 관련된 항공사의 규정을 주의 깊게 확인하세요. 2달 미만의 반려동물은 일반적으로 허용되지 않습니다. 그리고 반려동물이 여행할 수 있을 만큼 건강하다는 수의사의 증명서가 필요할 수도 있습니다. 모든 동물은 반려동물 보관함에 넣어 수송되어야 하는데 그것은 일반적으로 수하물과 함께 비행기 아래쪽에 보관됩니다. 보관함에는 연락처 정보와 반려동물의 사진이 분명히 붙어 있어야 합니다. 바닥에는 수건을 깔아두어야 합니다. 보관함은 여행 중에 동물이 움직일 수 있도록 최소한 동물보다 세 배는 커야 합니다. 또한, 응급 상황 발생 시 접근이 가능하도록 반려동물 보관함은 잠그지 않은 채로 두는 것이 중요합니다. 국제 여행을 하는 분들께서는 반려동물이 도착지의 국가가 요구하는 적절한 예방 접종을 반드시 받게 해야 합니다. 사전 계획이 반려동물의 안전과 편안함을 보장할 수 있습니다.

어휘
companion 동반자 **regulation** 규정 **proof** 증명(서) **veterinarian** 수의사 **luggage** 짐, 수하물 **label** 라벨을 붙이다 **lined with** …로 안을 댄 **movement** 움직임 **unlocked** 잠그지 않은 **access** 접근하다 **emergency** 비상사태 **vaccination** 예방 접종 **destination** 목적지, 도착지 **ensure** 보장하다 [문제] **launch** 시작[개시]하다; *출시하다

문제풀이
16 ① 반려동물과의 비행
② 반려동물 주인들을 위한 최고의 상품들
③ 새롭게 출시된 항공사 서비스
④ 장기 여행을 위한 짐 싸기 기술

⑤ 반려동물 예방 접종의 중요성

DICTATION Answers

본문 p.134

01	place extra taxes on / health risks of / cause obesity
02	have unhealthy additives / keep food from going bad / cause cancer
03	write for / taking part / passionate about fashion
04	wearing the same / skull on it / arguing with each other
05	employers post vacancies / applied for / need some help
06	How lucky you are / ask your parents to / meet their expectations
07	need a headset / The standard one / on sale
08	one of our buses / it's average-sized / help us identify it
09	next seven days / homeless children / try to capture / the fire of courage
10	having a sale on laptops / watch movies on my laptop / carrying it around often
11	plans to emigrate
12	running very slowly
13	to complain / laughing loudly / keep them quiet
14	improved significantly / change your study habits / making an extra effort
15	a fundraising event / regrets not having donated more / who are starving
16~17	healthy enough to / three times as large as / leave pet carriers unlocked / proper vaccinations

23 영어듣기 모의고사

본문 p.138

01 ⑤	02 ⑤	03 ③	04 ④	05 ②	06 ①
07 ②	08 ②	09 ②	10 ①	11 ①	12 ②
13 ②	14 ③	15 ①	16 ③	17 ⑤	

01 ⑤

남: 주목해 주시겠습니까? 이것은 학생 여러분과 교직원 여러분 모두에게 중요한 공지입니다. 기상 예보에 의하면, 오늘 밤 일몰 직후에 기온이 영하로 떨어질 것으로 예상됩니다. 현재 우리가 겪고 있는 폭설은 곧 멈출 것으로 예상됩니다만, 이미 도로 위에 쌓인 눈이 두세 시간 안에 위험한 빙판으로 바뀔 것 같습니다. 도로 상황이 위험해질 가능성이 있기 때문에, 여러분 모두는 6시까지 하교하기 바랍니다. 6시

30분에 학교 건물이 잠길 것입니다. 학생들의 안전이 제일 중요하므로, 모든 선생님께서는 학생들의 귀가를 지도해 주실 것을 요청 드립니다. 여러분 모두 오늘 저녁에 집으로 무사히 돌아가기를 바랍니다. 들어주셔서 감사합니다.

어휘
announcement 공고 forecast (날씨) 예보 temperature 기온 sunset 일몰 hazardous 모험적인, 위험한 potential 잠재력[성] paramount 다른 무엇보다 중요한 assist in …을 거들다[돕다] guide 안내[인도]하다 journey 여행; *(단거리의) 이동

02 ⑤

여: 매튜, 왜 침울한 얼굴이니?
남: 방금 제일 친한 친구인 톰을 만났어. 그의 상사가 그를 마음에 들어하지 않는대.
여: 아, 그에게 어떤 조언을 해줬니?
남: 음, 난 아직 대학에 있으니 무슨 말을 할지 모르겠더라. 그냥 듣기만 했어.
여: 잘했어. 어차피 그는 어떤 조언을 원하진 않았을 거야.
남: 정말? 하지만 그는 불평을 아주 많이 했어.
여: 네가 제일 친한 친구니까 아마도 그는 너에게 자신의 문제를 털어놓고 싶었을 거야. 그는 너에게 말하면서 편안함을 느끼는 게 틀림없어.
남: 너는 그가 이제 기분이 더 나아지고 스트레스도 덜 받게 되었다고 생각하니?
여: 응. 때론 그냥 들어주는 게 사람들에게 큰 도움이 될 수 있어.
남: 그렇구나. 그게 아마도 그가 내게 고마워했던 이유인 것 같아.

어휘
long face 시무룩한 얼굴 pour out (감정 등을) 쏟아 놓다

03 ③

여: 오늘 시간을 내주셔서 감사합니다.
남: 오, 천만에요.
여: 첫 번째 질문입니다. 여기서 일하는 것이 즐거운가요?
남: 물론이죠. 저는 모든 동물을 좋아해요. 특히 제가 돌보는 침팬지와 원숭이를요.
여: 그들을 어떻게 관리하세요?
남: 우선, 저는 그들이 사람들과 같이 있을 때 편안함을 느끼도록 해 줍니다. 그들 중 몇몇은 야생에서 이곳으로 왔어요. 그래서 저는 그들이 새롭고 낯선 환경으로 스트레스를 받지 않도록 해야 해요.
여: 여기서 태어난 동물들은 어떤가요?
남: 그들은 다루기 훨씬 쉽죠. 그들은 사람들과 함께 있는 것에 익숙해요.
여: 이 일을 하기 위해서 특별한 학위를 받으셨나요?
남: 예. 여기서 일하려면 축산학, 동물학, 또는 관련 분야의 학위가 필요합니다.
여: 그렇군요. 답변 감사합니다. 당신처럼 되고 싶은 사람들에게 많은 도움이 될 것으로 생각해요.

어휘
make sure 반드시 (…하도록) 하다 unfamiliar 낯선, 익숙지 않은 animal science 축산학 zoology 동물학 field 분야

04 ④

여: 안녕, 여보! 여행은 어땠어요?
남: 매우 생산적이었어요. [코를 킁킁거리며] 이 좋은 냄새는 뭐죠?
여: 아, 요리하고 있었어요. 부엌으로 가요.
남: 그래요. [잠시 후] 흠… 왜 냉장고 문이 열려 있죠?
여: 아, 당신을 맞이하느라 문 닫는 걸 잊었어요.
남: 그렇군요. 싱크대는 접시로 가득하네요. 많은 음식을 만들고 있었나 보군요.
여: 아뇨, 사실 가스레인지에서 돼지갈비를 만들고 있어요.
남: 오! 돼지갈비 정말 좋아요!
여: 알아요. 부엌이 뭔가 달라진 걸 눈치챘어요?
남: 물방울무늬 커튼이 새롭네요. 멋져 보여요.
여: 네, 당신이 그걸 좋아한다니 기뻐요. 자, 내가 저녁 요리를 끝낼 동안 샤워하는 게 어때요?
남: 좋아요. 먼저 식탁에서 바나나를 가져갈게요.
여: 알았어요, 너무 늦지 말아요!
남: 곧 내려올게요. 그러고 나서 여행 이야기를 더 해줄게요.

어휘
sniff (코를) 킁킁거리다 pork chop 돼지 갈빗살 polka-dot 물방울무늬의

05 ②

여: 네 새해 결심이 뭐니?
남: 글쎄, 아직 결정하지 못했어. 작년에도 결심한 걸 달성하지 못해서 제대로 할 수 있을지 모르겠거든.
여: 결심이 뭐였는데?
남: 두 가지였어. 담배를 끊고 규칙적으로 운동하는 거.
여: 각각의 결심을 얼마나 지킬 수 있었는데?
남: 1월 중순까지였을 거야. 아주 창피한 얘기지만 지키기가 너무 어려웠어. 넌 어땠어?
여: 나? 내 결심은 일기 쓰는 거랑 적어도 한 달에 한 권씩 고전을 읽는 거였어.
남: 그래서?
여: 실제로 둘 다 잘 지켰어. 때때로 포기하고 싶었지만 계속했지. 그래서 마침내 결심했던 두 가지가 습관이 됐어.
남: 와, 고무적인데. 올해 내 결심이 무엇이 될지 알겠어. 다시 담배를 끊으려 노력하고, 일주일에 적어도 세 번은 운동할 거야.
여: 좋은 계획이야! 나도 너랑 같이 체육관에 다닐래.

어휘
resolution 결심 regularly 정기적으로 stick to …을 계속하다; *…을 지키다 ashamed 부끄러운, 창피한 keep a journal 일기를 쓰다 classic 고전, 명작 be tempted to-v …하고 싶어지다 inspiring 영감을 주는, 자극하는 Sounds like a plan. (제안에 동의할 때) 좋은 생각[계획]이야.

06 ①

[전화벨이 울린다.]
여: 여보세요?
남: 안녕, 줄리. 나 닉이야.
여: 어, 무슨 일이야?
남: 너 낚시하러 가는 걸 좋아한다고 했지, 맞니?
여: 응, 내 취미 중의 하나야.

남: 사실, 나 다음 주에 친구들과 낚시하러 갈 거야. 네가 나한테 낚싯대를 빌려줄 수 있는지 궁금해.
여: 물론이야, 내일 우리 집에 들러서 가져가.
남: 정말 고마워. 처음이라서 좀 걱정이야.
여: 낚시는 그렇게 어렵지 않아. 네가 해야 할 일은 낚싯대를 물속에 넣고 기다리는 거야. 금방 익숙해질 거야.
남: 어, 그렇구나. 갈 생각에 아주 신이 나.
여: 네가 잡은 물고기로 매운탕을 끓이는 거 잊지 마. 맛이 좋아!

어휘
fishing pole 낚싯대 come by 잠깐 들르다 get used to …에 익숙해지다

07 ②

여: 안녕하세요. 달걀을 팔고 계시네요?
남: 아, 네, 저희 달걀은 방목한 닭에서 나온 거라서 최고입니다!
여: 그래요? 닭도 파시나요?
남: 네, 닭은 한 마리에 5달러입니다.
여: 그럼 달걀은 얼마인가요?
남: 달걀 12개들이 한 판에 2달러이고 두 판에 3.5달러입니다.
여: 음, 달걀 2달러어치로 오믈렛을 몇 개 만들 수 있나요?
남: 오믈렛 하나에 달걀 두 개씩이니까, 오믈렛 여섯 개를 만들 수 있죠.
여: 좋아요, 그럼 달걀 두 판을 살게요. 5달러짜리 지폐인데 거스름돈이 있으신가요?
남: 물론이죠, 여기 있습니다. 오믈렛 맛있게 해 드세요.

어휘
for sale 팔려고 내놓은 free-range (닭 등을) 놓아 기르는, 방목의 dozen 12개짜리 한 묶음 omelet 오믈렛 worth 가치, 값어치 bill 지폐

문제풀이
여자는 달걀 두 판을 산다고 했으므로, 여자가 지불할 돈은 3.5달러이다.

08 ②

여: 국제 우주 정거장을 방문한 가장 최근의 우주선에 대해 들었니?
남: 아니. 그게 뭐가 특별한데?
여: 처음으로 민간 기업이 우주 비행에 자금을 지원했거든.
남: 아주 흥미로운데. 전에 우주 비행은 항상 정부가 자금을 지원하는 줄 알았어.
여: 맞아. 이 민간 우주선은 SpaceX Dragon이라고 불리는데, 계획대로 안전하게 우주 정거장에 착륙했어.
남: 어떻게 그들은 그걸로 돈을 벌지?
여: 나도 잘 몰라. 처음으로 할 수 있다는 걸 증명하고 싶었겠지.
남: 좋은 첫 시도인 것 같아.
여: 무슨 일이 있든, 우린 민간 우주여행의 시대에 살고 있어.
남: 이게 정말로 변화를 일으키겠는데. 이것이 어떻게 우주여행에 영향을 주는지를 보면 아주 흥미로울 거야.

어휘
spacecraft 우주선 International Space Station 국제 우주 정거장 fund 자금을 대다 government 정부 land 착륙하다 prove 증명하다 affect 영향을 주다

09 ②

여: 아무도 당신에게 Free Rice 웹사이트에 관해 이야기하지 않았다면, 지금이 그곳을 방문할 때입니다. 이 사이트는 굉장합니다. 여러분의 시간 중 단 몇 분 동안 여러분은 여러분의 어휘력을 향상하게 하는 게임을 하면서 동시에 전 세계의 굶주린 사람들에게 유엔의 도움으로 쌀을 보낼 수 있습니다. 게임은 간단합니다. 네 개의 선택지 중에서 한 단어의 올바른 정의를 고르면 됩니다. 정답을 맞힐 때마다 20알의 쌀알이 기아로 고통받고 있는 사람들에게 기부됩니다. 사이트 내 광고들은 (광고비를) 쌀알로 지급합니다. 그러면 유엔 세계 식량 프로그램은 쌀을 가난한 사람들에게 배급합니다. 이 사이트는 10월 7일에 출시되었습니다. 그날 1,000알도 안 되는 쌀알이 보내졌습니다. 그러나 10월 24일까지 매일 3천만 이상의 쌀알들이 기부되고 있었습니다.

어휘

pay ... a visit …에 방문하다 simultaneously 동시에 definition 정의, 뜻 grain 곡물; *낱알 hunger 기아, 굶주림 distribute 분배하다 launch 착수하다, 시장에 내다

10 ①

여: 케빈, 스포츠 센터의 일정표는 왜 보고 있는 거야?
남: 난 수영 강습을 듣고 싶어. 어떤 걸 골라야 좋을지 모르겠어.
여: 오, 난 네가 수영할 줄 안다고 생각했는데.
남: 사실, 할 줄은 알아. 전에 강습을 받았거든.
여: 그러면 넌 상급반을 들어야겠네.
남: 아냐. 그건 아주 오래전이라서. 기초반이나 중급반이면 될 것 같아.
여: 알겠어. 그럼 이 토요일 수업은 어때?
남: 일주일에 한 번은 충분하지 않아. 난 좀 빨리 배우고 싶거든. 일주일에 적어도 두 번은 모이는 수업이 필요해.
여: 음, 너에겐 몇 가지 선택권이 있어. 어느 시간대가 너한테 제일 좋아?
남: 난 9시부터 5시까지 일해. 오전 8시 전이나, 오후 7시 이후에 시작하는 수업만 들을 수 있어.
여: 그러면 넌 이 두 가지 중의 하나를 고를 수 있어.
남: 난 정원이 더 적은 수업으로 할래. 도와줘서 고마워, 마리아!

어휘

probably 아마도 advanced 앞선; *상급의 intermediate 중급의 at least 적어도

11 ①

남: 이건 누구야? 친숙하게 들리는데.
여: 내가 제일 좋아하는 밴드인 The Rock Stars야. 이건 그들의 최신 앨범이지.
남: 정말 마음에 들어. 어디서 앨범을 살 수 있니?
여: 온라인으로 주문할 수 있어.

어휘

copy (책·CD 등의) 한 부 [문제] order 주문하다 newspaper stand 신문 가판대 copy machine 복사기 out of order 고장 난

문제풀이

② 난 록 음악을 좋아했던 적이 없어.
③ 저쪽에 신문 가판대가 있어.
④ 이 콘서트 표들을 구해줘서 고마워.

⑤ 이 사무실의 복사기는 고장 났어.

12 ②

여: 실례합니다. 아이스크림 가게가 어디에 있는지 알려 주시겠어요?
남: 네, 쇼핑센터의 다른 편에 있어요. 앞으로 직진해서 왼쪽으로 도세요.
여: 고마워요. 이제 찾을 수 있을 것 같아요.
남: 서두르는 게 좋을 거예요. 곧 문을 닫거든요.

어휘

straight ahead 앞으로 똑바로 [문제] if it weren't for …이 아니라면[없다면] lend 빌려주다

문제풀이

① 그래서 아이스크림은 맛있게 드셨나요?
③ 별말씀을요. 당신이 그것을 찾았다는 소식을 들어서 기뻤어요.
④ 당신이 없다면 저도 그것을 찾을 수 없을 거예요.
⑤ 천만에요. 언제든 그걸 빌려드릴 수 있어요.

13 ②

여: 복권에 당첨됐다면서요. 축하해요!
남: 고마워요.
여: 그 돈으로 뭐 할 거예요? 차를 사실 건가요?
남: 아뇨, 차는 이미 있는 걸요. 내가 절대 갖지 못할 거로 생각했던 요트를 사고 싶어요.
여: 정말 멋진 생각이네요! 요트를 탈 줄 아세요?
남: 아뇨, 하지만 배울 거예요.
여: 아내에게 당신 계획에 관해 얘기했나요? 그녀가 기뻐할 거예요.
남: 네, 말했죠. 하지만 제 아내는 그 돈을 전부 자선 단체에 기부하길 원해요.
여: 오, 정말요? 당신 아내는 정말 관대하네요.
남: 네, 아내는 다른 사람들을 돕는 데 그 돈을 쓰고 싶어해요.

어휘

win the lottery 복권에 당첨되다 sailboat 요트 charity 자선 (단체) generous 관대한

문제풀이

① 어서 빨리 아내에게 복권에 당첨됐다고 말하고 싶어요.
③ 아내는 그 돈으로 새 스포츠카를 살 거예요.
④ 아내가 제 계획에 대해 너무 화내지 않길 바라요.
⑤ 전 그저 아내가 그렇게 많은 돈을 그만 좀 썼으면 좋겠어요.

14 ③

여: 기타 치신 지 얼마나 되셨어요?
남: 일 년 정도요.
여: 와, 훌륭하네요. 어디서 배우셨어요?
남: 교습을 받은 건 아니에요. 독학했어요.
여: 와, 저라면 그렇게 못했을 거 같아요.
남: 음, 어릴 때부터 곡을 연주해와서 그런지, 전 새로운 악기를 쉽게 배워요.
여: 또 어떤 악기를 연주하시는데요?
남: 어렸을 때는 피아노를 쳤고, 십 대 때는 바이올린을 연주했어요.
여: 와, 저 바이올린 정말 좋아해요. 아직도 연주하세요?

남: 아뇨. 그리고 싶지만 지금 제가 가진 악기는 기타뿐이에요.
여: 그거 유감이네요. 재능을 낭비하면 안 돼요.

어휘

teach oneself 독학하다 instrument 악기 teenager 십 대 [문제] talent 재능, 소질 go to waste 쓸모없게 되다

문제풀이

① 좋은 기타를 하나 사세요.
② 바이올린을 잘 관리해야 해요.
④ 하지만 저는 기타 치는 법을 가르쳐 드릴 수 없어요.
⑤ 그럼 당신의 아이들에게 독학하라고 하는 것은 어때요?

15 ①

여: 캐롤라인은 기분이 몹시 안 좋다. 그녀가 제일 좋아하는 밴드가 그녀가 살고 있는 도시에 오지만 표를 구하지 못했다. 캐롤라인의 어머니는 그녀가 안됐다고 생각한다. 캐롤라인은 올해 학교 성적이 정말 좋았기 때문에 그녀의 어머니는 그녀에게 뭔가 특별한 것을 해주고 싶어 한다. 어머니는 음반업계에 어떤 사람을 알고 있어서 캐롤라인을 위해 무대 뒤 출입증을 얻을 수 있었다. 이제 캐롤라인은 그녀가 제일 좋아하는 밴드를 만날 수 있다. 이런 상황에서, 캐롤라인이 그녀의 어머니에게 할 말로 가장 적절한 것은 무엇인가?
캐롤라인: 고마워요, 엄마! 엄마가 최고예요.

어휘

industry 산업; *업계 backstage pass 무대 뒤 출입증 [문제] must 꼭 봐야 할 것 front row 앞줄

문제풀이

② 이 콘서트는 꼭 봐야 한다는 걸 기억하세요.
③ 전 그 밴드를 싫어해요. 그 출입증은 조이에게 주세요.
④ 하지만 전 성적도 좋잖아요. 왜 가면 안 된다는 거죠?
⑤ 제가 앞줄 표를 원한다는 걸 어떻게 아셨어요?

16 ③ 17 ⑤

여: 학생 여러분, 좋은 아침입니다. 알다시피, 우리는 자연으로부터 많은 가치 있는 것들을 배울 수 있습니다. 오늘, 우리는 동물들에게서 영감을 받은 발명품들을 살펴볼 것입니다. 첫째로, 디자이너들이 수영복에 쓰일 새로운 소재를 개발했다는 것을 알고 있었습니까? 그것은 상어의 피부를 모델로 하여 만들어졌습니다. 상어는 바다에서 가장 빠른 생명체 중 하나입니다. 이 소재는 물속에서 사람들이 더 빠르게 움직이도록 도울 것입니다. 다음으로, 비행기들이 어떻게 날아가는지 생각해 보십시오. 비행기를 처음으로 만든 사람들은 아마도 새들에게서 아이디어를 얻었을 것입니다. 그다음으로, 표면에 거의 모든 것을 붙일 수 있는 작고 강한 패드가 있습니다. 발명가들은 어디서 그 아이디어를 얻었을까요? 그들의 아이디어는 게코(도마뱀붙이)의 발에서 왔습니다. 이 작은 도마뱀들의 발에는 벽을 기어오르는 것을 돕는 특별한 피부가 있습니다. 마지막으로, 윙슈트는 또 다른 훌륭한 예입니다. 사람들은 그들 스스로 이 아이디어를 생각해낸 것이 아닙니다. 그들은 그 디자인의 기반을 날다람쥐에 두었습니다. 이러한 발명품들이 인상적이지 않나요? 동물로부터 유래된 다른 발명품들이 있나요? 몇 분간 생각해 보시고, 여러분의 아이디어를 반 친구들 모두와 논의해 보도록 하겠습니다.

어휘

invention 발명(품) (v. invent) draw inspiration from …에서 영감을 얻다 material *직물, 천; 재료 be modeled after …을 본뜨다 creature 생명체, 생물 attach 붙이다 gecko 도마뱀붙이, 게코 lizard 도마뱀 come up with …을 떠올리다[찾아내다] base A on B A를 B에 근거[기반]를 두다 impressive 인상적인, 인상 깊은 [문제] fabric 직물, 천

문제풀이

16 ① 수영복의 진화
② 천연 섬유의 다양성
③ 자연에서 영감을 얻은 발명품들
④ 우연히 발명된 제품들
⑤ 제품 시험에 이용되는 동물들

DICTATION Answers   본문 p.140

01 drop below zero / turn into hazardous ice / a safe journey home

02 give him any advice / pour out his problems / feels better

03 enjoy working here / make sure that / get a special degree

04 forgot to close it / The polka-dot curtains / I'll be right down

05 achieve last year's resolution / stick to them / tempted to give up

06 lend me a fishing pole / get used to it / cook some fish stew

07 eggs for sale / each / make with $2 worth / take two dozen eggs

08 has funded a space flight / make money / how this affects

09 send rice to hungry people / people suffering from hunger / 1,000 grains of rice

10 how to swim / will be fine / at least twice a week

11 buy a copy of it

12 Go straight ahead

13 you won the lottery / how to sail / all the money to charity

14 playing the guitar / taught myself / I was a teenager / the only instrument

15 feels bad for / something special for her / get backstage passes

16~17 They're modeled after / how airplanes fly / can attach almost anything / did not come up with

01 ②	02 ④	03 ②	04 ③	05 ④	06 ①
07 ④	08 ④	09 ①	10 ③	11 ①	12 ③
13 ④	14 ⑤	15 ③	16 ③	17 ②	

01 ②

여: 안녕하세요, 여러분. 대림 아파트 경비실에서 안내 말씀드립니다. 오늘 밤은 몹시 추울 것으로 예상됩니다. 그러므로 파이프가 얼지 않도록, 밤새 수도꼭지에서 물이 약간씩 흐르도록 해 주시기를 여러분 모두에게 요청합니다. 이렇게 하시지 않으면, 파이프들이 얼어 터져서 대규모 수리가 필요하게 될지도 모릅니다. 이로 인해 몇 시간 동안 수돗물이 나오지 않거나 난방을 할 수 없게 될 것입니다. 그러므로, 여러분이 저희의 제안대로 행동해 주실 것을 강력하게 권고합니다. 감사드리며, 좋은 저녁 시간 되십시오.

어휘
security 안전, 경비(부) unusually 대단히, 몹시 protect 보호하다, 막다 pipe 관, 파이프 freeze 얼다 run (물 등이) 흐르다 faucet 수도꼭지 burst 터지다, 파열하다 require 필요로 하다, 요구하다 major 주요한 repair 수리 act on …에 따라 행동하다

02 ④

남: 식료품 목록에 내 걸 추가해 줄래, 제니?
여: 물론이죠. 뭔데요?
남: 탄산음료 한 병이야.
여: 탄산음료요? 근데 그런 식품이 매우 나쁘다고 항상 말씀하시잖아요. 그래서 우리가 그걸 전혀 안 사는 거고요.
남: 음, 그래. 거기엔 설탕이 잔뜩 들어있고 치아에 나쁘지. 하지만 토요일 파티를 위해 사려고 했어.
여: 음, 알았어요.
남: 걱정하지 마. 탄산음료를 마시는 건 습관이 되면 문제가 되는 거야.
여: 그럼 특별한 경우에는 괜찮아요?
남: 맞아. 그리고 네 언니의 고등학교 졸업식은 아주 특별한 행사지.

어휘
grocery 식료품(점) soda 탄산음료 graduation 졸업(식) special occasion 특별 행사

03 ②

남: 줄이 정말 기네요, 그렇죠?
여: 그러게요, 전 여기에 30분 동안 서 있었어요!
남: 왜 여기서 기다리고 계신 거예요?
여: 음, 어제 제가 이 스웨터를 샀는데, 보세요! 스웨터에 구멍이 있어요!
남: 아, 네, 그렇군요. 교환할 건가요, 아니면 환불받을 건가요?
여: 음, 색상과 스타일은 정말 마음에 들어서 그냥 교환하려고요.
남: 전 보통 이곳에서 사는 옷의 품질에 만족하고 있어요.
여: 저도 그래요. 근데 이번에는 좀 실망스럽네요. 이곳에서 품질이 낮은

상품을 팔려고 하는 게 아니라면 좋겠어요.
남: 그렇게 하진 않을 거예요. 제가 여기 주인을 아는데, 들여오는 물건에 대해 상당히 까다로워요.
여: 그러길 바라요.

어휘
hole 구멍 exchange 교환하다 rather 약간, 다소 low-quality 품질이 낮은 owner 주인 picky 까다로운

04 ③

남: 이 반려동물 가게에서 반려동물을 찾아보자. 저 흰 고양이가 귀엽다.
여: 음, 난 고양이를 갖고 싶진 않아.
남: 개는 어때? 저 짙은 색 점박이의 작은 개를 봐.
여: 사랑스럽지 않니? 하지만 난 개도 그렇게 원하지 않아.
남: 음, 넌 뭘 원하니?
여: 잘 모르겠어. 저 새들을 봐. 꼬리가 긴 저런 작은 새 중 하나를 살까 봐. 그 새들은 노래 부르는 걸 정말 좋아하잖아.
남: 응, 하지만 그게 널 짜증 나게 할 수도 있어. 물고기가 더 조용할 거야.
여: 하지만 선반에 있는 저 어항들은 비싸 보여.
남: 아, 점원이 먹이를 주고 있는 토끼는 어떨까?
여: 귀엽다. 이제 내가 뭘 원하는지 알겠어.

어휘
spot 점, 반점 adorable 사랑스러운 annoying 짜증스러운 fish bowl 어항 shelf 선반 feed 먹이를 주다

05 ④

[전화벨이 울린다.]
여: 여보세요?
남: 안녕, 여보. 마침내 집에 왔군요. 오늘 하루는 어땠어요?
여: 정말 고단한 하루였어요. 아침에 일어난 순간부터 계속 볼일 보러 돌아다닌 것 같아요.
남: 유감이네요. 오늘 저녁 7시쯤 집에 도착할 거라서 알려주려고 전화했어요.
여: 알았어요. 하지만 솔직히 말해서, 난 오늘 저녁에는 너무 피곤해서 요리를 할 수가 없어요.
남: 괜찮아요. 우리 그냥 외식하는 거 어때요?
여: 아! 당신 사무실 근처의 그 작은 식당에서 만두를 먹고 싶어요.
남: 좋아요. 거기서 6시 30분에 나랑 만나는 거 어때요?
여: 그럴 기운이 없어요. 내가 그 식당에 전화해서 사가지고 갈 걸 좀 주문하면 어떨까요?
남: 좋아요, 문제없어요. 내가 집에 가는 길에 들를게요.
여: 잘됐네요. 이따 봐요, 여보!

어휘
exhausting 지치게 하는 run an errand 볼일을 보다, 심부름하다 eat out 외식하다 dumpling (고기) 만두 stop by 잠시 들르다

06 ①

[휴대전화벨이 울린다.]
여: 여보세요?
남: 안녕, 엄마. 집에 계세요?

여: 응, 막 들어왔어. 별일 없지?

남: 그렇지 않아요. 저 농구를 하다가 발목을 삐었어요.

여: 저런. 심각하니?

남: 혼자 걸을 수 없어요. 엄마가 와서 절 데리고 가셔야 해요.

여: 알았어. 지금 어디에 있니?

남: 농구장 옆 동사무소 밖이에요.

여: 10분 뒤에 도착할 거다. 앉아서 아픈 발목을 편하게 두렴.

남: 고마워요 엄마. 얼음과 수건도 가져다주시겠어요? 발목 부은 게 가라 앉으면 좋겠어요.

여: 물론이지. 편히 있으면 내가 금방 거기로 가마.

어휘

twist one's ankle 발목을 삐다 serious 심각한 community center 동사무소, 지역센터 take the weight off (…을 쉬게) 앉다 swelling (살갗의) 부기 go down 내려가다, 낮아지다; *가라앉다 before you know it 순식간에, 눈 깜짝할 사이에

07 ④

여: 우리 엄마 생신이 내일인데, 난 아직 아무것도 사지 못했어.

남: 꽃을 드리는 게 어때? 잠깐만. [잠시 후] 여기 내가 제일 좋아하는 사이트야.

여: 와, 꽃이 엄청 많네! 난 이 꽃바구니가 마음에 들어.

남: 어떤 크기의 바구니를 원해? 작은 건 40달러, 중간은 50달러, 큰 건 60달러야.

여: 난 제일 큰 거로 할래. 그리고 꽃 말고도 뭔가를 추가하고 싶은데.

남: 이것 좀 봐. 초콜릿이나 비누 꽃을 바구니에 추가할 수 있어. 엄마한테 메시지도 쓸 수 있어.

여: 선택 사항이 너무 많다. 나는 비누 꽃이 마음에 들지만 하나당 5달러야. 그건 너무 비싸네.

남: 다른 선택 사항들도 비싸. 초콜릿은 30달러, 카드는 10달러야.

여: 흠… 네 말이 맞아. 난 그냥 비누 네 개를 추가하고 내가 직접 카드를 만들래.

남: 좋은 선택이야! 오, 언제 배송받아야 해?

여: 내일 오후 1시쯤.

남: 그 시간대 배송은 15달러야.

여: 알겠어.

어휘

add 더[추가]하다 in addition to …에 더하여 soap 비누 pricey 값비싼

문제풀이

여자는 가장 큰 크기의 꽃바구니($60)에 비누 꽃($5) 네 개를 추가했고 배송비로 15달러를 내야 하므로, 여자가 지불할 금액은 95달러이다.

08 ④

남: 안녕, 지현. 영국 여행은 어땠니?

여: 좋았어. 런던에 가서 빅 벤을 보게 되었어.

남: 빅 벤, 웨스트민스터 궁에 있는 유명한 시계탑 말이야?

여: 응. 원래의 궁이 1834년 화재로 파괴된 이후에 오거스터스 퓨진이 설계했어.

남: 사진을 본 적 있는 것 같아. 그거 아주 크지, 그렇지?

여: 응. 거의 100미터 높이이고, 각 시계 문자판이 폭이 7미터 정도야.

남: 하나 이상의 시계가 있어?

여: 응. 탑의 네 면에 각각 하나씩 있어.

남: 흥미롭다. 사진은 찍었니?

여: 응, 많이 찍었어! 내일 보여줄게.

어휘

palace 궁, 궁전 destroy 파괴하다 nearly 거의 clock face 시계 문자판

09 ①

남: Circle Cruise 일일 관광을 신청해주셔서 감사합니다. 배는 내일 오전 10시에 바로 출발하므로 늦지 마시기 바랍니다. 점심이 제공될 것이며 다과도 제공될 예정이지만, 여러분께서 드실 간식을 가지고 타셔도 좋습니다. 항해는 대략 3시간 정도 지속될 것이며, 이곳 Eastside 부두에서 출발해서 되돌아올 것입니다. 저희는 도시 주위를 항해하기 때문에, 여러분은 아침 내내 멋진 경치를 보게 되실 것입니다. 따라서 카메라를 가져오십시오. 또한 일기예보 상 해가 날 것으로 예상되므로, 반드시 자외선 차단제를 바르시기 바랍니다. 어린이를 동반하여 탑승하시는 부모님들은 자녀들을 잘 지켜보시길 바라며, 모든 분들은 항시 구명조끼를 착용하셔야 합니다. 다시 한번 신청해주신 것에 감사드리며, 아침에 뵙겠습니다.

어휘

sign up (for) (…을) 등록[신청]하다 promptly 바로, 즉시 refreshment 《pl.》 다과 approximately 대략 dock 부두 sail 항해하다 view 경치, 조망 forecast (일기) 예보 sunscreen 자외선 차단제 urge 강력히 권고하다 life jacket 구명조끼

10 ③

여: 얘! 그거 내가 다니는 스포츠 센터의 전단지니? 넌 다니고 싶지 않은 줄 알았는데.

남: 그랬지, 그런데 지금 여름 특별가를 제공하고 있어.

여: 어디 보자. [잠시 후] 패키지 E가 내가 이용하는 거야. 아주 좋은 가격이지.

남: 알아, 그런데 요가는 정확히는 내가 마음에 둔 게 아니야. 난 야외에서 하는 걸 더 좋아해.

여: 아, 그렇다면, 테니스나 골프를 추천하지만 그런 패키지들은 더 비싸.

남: 맞아, 그렇지만 괜찮아. 한 달에 100달러 미만이면, 좋아.

여: 그럼, 몇 가지 가능한 선택 사항들이 있네. 이건 어때?

남: 수업이 일주일에 한 번보다는 많았으면 하는데. 일주일에 한 번은 충분하지 않아.

여: 그래, 이해해. 나는 가능한 한 자주 요가 수업에 가려고 해.

남: 난 또한 수영장도 이용할 수 있었으면 좋겠어. 난 수영하는 걸 정말 좋아하니까 많이 이용할 거야.

여: 요구사항이 정말 많구나! 그래도 너의 요구사항을 모두 만족시키는 수업이 하나 있다.

남: 응, 이게 괜찮을 것 같네. 나 오늘 등록할 것 같아!

어휘

flyer (광고·안내용) 전단 deal 거래, 합의 pool 수영장 demand 요구 (사항) fulfill 수행[완료]하다; *만족시키다 [문제] special 특별한; *특별 상품 frequency 횟수 access 접근권, 이용할 권리

11 ①

여: 새 학교는 어떠니, 마이클?
남: 괜찮은데, 선생님들은 별로 마음에 들지 않아요.
여: 왜 그러니?
남: 그분들은 때때로 너무 엄격하신 것 같아요.

어휘
[문제] strict 엄격한

문제풀이
② 과학 선생님은 정말 친절하세요.
③ 새 학교는 길 아래에 있어요.
④ 그런 일이 생기면 전 화가 나요.
⑤ 선생님들이 보통 저에게 좋은 성적을 주세요.

12 ③

남: 벌써 수업에 등록했니?
여: 아니, 고르는 게 힘들어. 넌 뭘 추천해?
남: 난 고급 수학 과정을 들어. 나랑 같이 수강할래?
여: 물론이지, 그런데 난 공부하는 데 도움이 필요할지도 몰라.

어휘
sign up for …에 신청[등록]하다 yet (부정문에서) 아직; *(의문문에서) 이미, 벌써
have trouble (in) v-ing …에 곤란을 겪다 advanced 고급의

문제풀이
① 거기까지 내 차를 타고 가면 돼.
② 난 이미 그 수업에 등록했어.
④ 우리 반에는 학생이 많아.
⑤ 좋아, 고급 화학 과정을 수강할게.

13 ④

남: 미셸, 우리 이번 주말에 만날래?
여: 그래, 난 특별히 계획한 것도 없어. 넌 뭘 생각하고 있는데?
남: 음, 우리 여럿이서 여행을 가려고. 너도 우리랑 같이 갈래?
여: 좋지. 어디로 갈 건데?
남: 너 설악산 가본 적 있어? 연중 이맘때면 아름답거든.
여: 아니, 하지만 정말 설악산을 보고 싶어! 거기서 뭘 할 건데?
남: 글쎄, 등산하고 거기에 있는 오두막집에서 잘까 생각 중이야.
여: 뭘 먹을 거야?
남: 내가 가스버너를 가져갈 거고, 모두가 음식을 조금씩 가져올 거야.
여: 재미있을 것 같긴 한데, 난 등산을 그다지 좋아하지 않아. 길이 너무
   가파르면 겁이 나거든.
남: 걱정할 필요 없어. 내가 도와줄게!

어휘
get together 만나다 take a trip 여행하다 cabin 오두막집 scared 겁먹은
path 길 steep 가파른, 험준한 [문제] activity 활동

문제풀이
① 제발 나한테 화내지 마.
② 그냥 네가 좋아하는 음식을 좀 가져와.
③ 나도 위험한 활동을 하는 것을 좋아해!
⑤ 그렇다면 넌 스스로 요리를 해야 해.

14 ⑤

여: 얘, 테드, 새로 산 내 스케이트보드 좀 봐!
남: 와! 정말 멋있다! 그거 어디서 샀어?
여: 아빠가 내 생일 선물로 사주신 거야.
남: 정말 비싸 보여. 묘기를 부릴 수 있어?
여: 아니 아직, 계속 연습 중이야. 스케이트보드를 타면서 점프하는 걸 배
   우고 싶어.
남: 나도 그것처럼 생긴 스케이트보드가 있었는데, 도둑맞았어.
여: 정말? 어디서 도둑맞았는데?
남: 쇼핑센터에서. 밖에 뒀는데, 나와보니 없어졌어.
여: 자물쇠로 잠갔니?
남: 아니, 내 잘못이었어. 그냥 밖에 있던 자전거들 옆에 놔뒀었거든.
여: 참 안됐다. 슬펐겠다!

어휘
trick 묘기, 재주 practice 연습하다 lock up 자물쇠로 잠그다

문제풀이
① 정말? 네 묘기를 정말 보고 싶다.
② 넌 쇼핑센터에서 뭘 샀니?
③ 이건 내 스케이트보드야. 난 네 것을 훔치지 않았어.
④ 네 자전거를 확실히 자물쇠로 잠갔니?

15 ③

여: 제인은 직장에서 긴 하루를 보냈다. 그녀는 무척 바빴고 상관이 그녀
   에게 화를 냈다. 그녀는 녹초가 되었다. 집에 가서 저녁을 만드는 건
   그녀가 제일 원하지 않는 일이다. '피자나 시켜 먹을까 봐.' 그녀는 속
   으로 생각한다. 아파트 계단을 올라가 문을 열었는데, 불이 다 꺼져
   있고 그녀의 남편은 없다. "남편이 어디 있는 거지?" 그녀는 궁금해
   하면서 휴대전화를 꺼낸다. 그런데 그녀가 주방에 들어갔을 때, 그녀
   는 촛불이 켜져 있는 저녁 식탁에 꽃다발까지 놓여 있는 것을 발견한
   다. 남편이 따뜻한 포옹으로 그녀를 맞이하면서 "당신이 힘든 하루를
   보내서 마음이 아파요."라고 말한다. 이런 상황에서, 제인이 그녀의
   남편에게 할 말로 가장 적절한 것은 무엇인가?
제인: 여보, 정말 사려 깊네요.

어휘
exhausted 기진맥진한 the last … (that) 결코 ~할 것 같지 않은 … stair
《pl.》 계단 candlelit 촛불을 밝힌 complete with …을 완비한 greet 맞이
하다 [문제] thoughtful 생각에 잠긴; *배려심 있는, 친절한 promotion 승진

문제풀이
① 피자를 시켜요.
② 그건 당신 잘못이 아니에요.
④ 생각나게 해 줘서 고마워요.
⑤ 승진 축하해요.

16 ③   17 ②

남: 여러분은 짐을 꾸리고 호화로운 유람선에서의 꿈에 그리던 휴가를 보
   낼 준비가 되었습니다. 하지만 악몽은 배 위에 숨어 있을 수 있습니다.
   그것은 노로바이러스라고 하는데, 이는 여객선이나 사람들이 작은 공
   간에 모여있는 곳에서 흔히 발견됩니다. 이것은 감염자와 접촉함으
   로써 퍼집니다. 그것은 또한 그 바이러스에 오염된 음식과 물을 통해

서도 확산됩니다. 노로바이러스의 감염 증상은 일반적으로 심각하지 않습니다. 거기에는 배탈과 구토가 포함됩니다. 노로바이러스에 감염된 사람은 때때로 두통과 고열에 시달립니다. 증상은 일반적으로 며칠 동안만 지속되지만, 14일까지 전염성이 있을 수 있습니다. 다행히, 노로바이러스 감염의 예방은 쉽습니다. 규칙적으로 꼼꼼하게 손을 씻는 것이 질병에 걸릴 확률을 상당히 줄여줄 것입니다. 또한, 난간과 손잡이와 같이 자주 접촉되는 표면은 정기적으로 청소되어야 합니다.

어휘

luxurious 호화로운 cruise ship 유람선 nightmare 악몽 norovirus 노로바이러스(전염성 위장염을 일으키는 바이러스) spread 퍼뜨리다, 확산시키다 (spread-spread) contact 접촉 infected 감염된 contaminated 오염된 symptom 증상 severe 극심한, 심각한 upset stomach 배탈 vomit 토하다 contagious 전염성의 prevention 예방 thoroughly 완전히, 철저히 significantly 상당히 frequently 자주 surface 표면 handrail 난간 doorknob 손잡이 [문제] rash 발진

DICTATION Answers   본문 p.146

01	unusually cold weather / a little water running / freeze and burst / act on our suggestion
02	add something to / pick some up / make it a habit
03	a hole in it / get a refund / rather disappointed / quite picky
04	with dark spots / with a long tail / that the clerk is feeding
05	I've been running errands / too tired to cook / stop by on my way
06	twisted my ankle / take the weight off / before you know it
07	Here is my favorite site / I'd like to add something / the other options are pricey too
08	the original palace was destroyed / I've seen pictures of it / about seven meters wide
09	be served / your own snacks / great views / wear a life jacket
10	special summer deals / something outdoors / a few options available / a lot of demands
11	How do you like
12	signed up for
13	get together / taking a trip / hike up the mountain / the path is too steep
14	do any tricks / it got stolen / left it outside / lock it up
15	got mad at / lights are out / complete with flowers / with a loving hug
16~17	on a luxurious cruise ship / where people are gathered / become contaminated / reduce your chances of catching

25 영어듣기 모의고사
본문 ▲ p.150

01 ④	02 ⑤	03 ②	04 ⑤	05 ④	06 ⑤
07 ③	08 ①	09 ⑤	10 ②	11 ⑤	12 ⑤
13 ③	14 ②	15 ⑤	16 ⑤	17 ④	

01 ④

남: 오늘 새로운 Sosoft 로션을 써 보세요! 2주 후면 여러분의 피부톤과 피부색에서 뚜렷한 차이를 보시게 될 것을 보장합니다. 주근깨, 여드름, 그리고 주름이 사라지고, 여러분의 피부가 더욱더 부드럽고 매끄러워질 것입니다. Sosoft 로션은 건성, 중성, 지성 등 모든 피부 유형에 사용 가능합니다. 여러분의 피부에 맞는 로션을 선택하시고, 이 제품이 효과가 있음을 확인해보세요. 매일 저녁 발라 보시면 여러분은 반드시 만족하실 겁니다. 오늘 이 자리에서 열리는 대대적인 Sosoft 로션 판촉 행사를 놓치지 마십시오.

어휘

guarantee 약속하다, 장담하다 noticeable 눈에 띄는 freckle 주근깨 acne 여드름 wrinkle 주름 fade 사라지다 smooth 매끄러운 oily 기름기가 많은, 지성의 suit …에 적합하다 confident 자신감 있는; *확신하는 work 효과가 있다 apply 쓰다, 적용하다; *바르다 miss out (일·호기 등을) 놓치다 promotion 홍보, 판촉

02 ⑤

남: 점심으로 뭘 먹고 싶어?

여: 거리를 따라 내려가면 새로 생긴 햄버거 레스토랑이 있어. 거기 가보는 게 어때?

남: 난 햄버거를 안 먹거든. 난 채식주의자가 되기로 결심했어.

여: 그거 정말이야? 많은 맛있는 음식을 포기하는 걸 텐데.

남: 응, 하지만 채식주의 식단이 더 건강하다고 들었어.

여: 음. 난 절대 채식주의자가 되고 싶지 않아.

남: 왜? 넌 더 건강해지는 걸 원하지 않는 거야?

여: 채식을 하는 사람들은 상처가 낫는데 오래 걸리고, 쉽게 약해지고, 피곤해질 수 있다고 들었어.

남: 오, 그건 몰랐네. 채식주의자가 되는 걸 다시 생각해봐야겠다.

어휘

vegetarian 채식주의자; 채식주의의 heal 치유하다 ought to-v …해야 하다

03 ②

[전화벨이 울린다.]

남: 안녕하세요, 무엇을 도와드릴까요?

여: 안녕하세요. 전 204호에서 전화하는 건데요, 질문이 있어요.

남: 알겠습니다. 무엇에 관한 질문인지 알려주시면, 담당자에게 전화를 연결해 드릴게요.

여: 제가 방에서 국제 전화를 하려고 하는데, 걸 수가 없어요.

남: 아, 그런 경우라면 제가 바로 말씀드릴 수 있습니다.

여: 제가 9번을 누르고 나서 (전화)번호를 눌렀는데, 작동이 안 돼요.

남: 국제 전화를 하려면, 0번을 먼저 누르고 그다음에 9번을 누르셔야 합니다.

여: 아, 그렇군요. 그리고 제가 전화 카드를 사용해야 하나요?

남: 아뇨. 직접 거실 수 있는데, 그러면 체크아웃하실 때 그 요금이 청구서에 합산될 겁니다.

여: 잘됐네요. 도와주셔서 감사합니다.

어휘
direct 향하다; (길을) 안내하다; *보내다 appropriate 적절한 work 작동[기능]하다 charge 요금

04 ⑤

남: 나는 이 퀴즈 프로그램이 지겨워.

여: 그럼 자연 프로그램을 보자. 그건 8시에 시작해.

남: 지금 몇 시니? 아, 10분 뒤면 시작하는구나.

여: 좋아! [잠시 후] 어, 이 리모컨이 작동하지 않아.

남: 아마도 새 건전지가 필요한가 봐.

여: 내가 어제 새 건전지를 리모컨에 넣었어.

남: 건전지를 식물 옆에 있는 줄무늬 상자에서 가져왔니?

여: 응. 왜?

남: 그건 모두 다 쓴 거야. 나중에 재활용할 거거든. 새것은 TV 아래 가운데 서랍에 있어.

여: 아, 그래? 그럼 새 걸 가져올게.

남: 좋아. 그리고 오른쪽에 있는 서랍을 닫아 줄래? 내가 열어둔 것 같네.

어휘
tired of …이 싫증 난 remote control 리모컨 striped 줄무늬의 plant 식물, 작은 나무 dead 수명이 다 된 drawer 서랍

05 ④

여: 윌슨 씨, 신선한 블루베리 좀 남은 거 있어요?

남: 죄송해요, 카터 씨. 오늘 이미 블루베리가 다 팔렸네요.

여: 하지만 내일 더 들여오실 거죠?

남: 네. 아침에 따서 오후에 이곳으로 배달돼요.

여: 보통 몇 상자 들여오세요?

남: 아, 30상자 정도요. 그리고 그날 장사가 끝날 때쯤에 항상 다 팔아요.

여: 그게 걱정이에요! 여기서 블루베리를 정말 사고 싶은데, 제가 일찍 오질 못하니까요.

남: 그럼 제가 따로 좀 챙겨두면 어때요?

여: 그렇게 해주시겠어요? 제가 그것들을 가지러 8시쯤에 들를게요.

남: 좋아요. 얼마나 필요하세요?

여: 두 상자를 살게요. 블루베리 파이를 만들 거라서요.

남: 잊지 않도록 적어 둘게요.

어휘
pick 따다 sell out 다 팔다 set ... aside …을 따로 챙겨 두다 write ... down …을 적어 두다

06 ⑤

남: 조앤. 기분이 좋아 보이는 것 같네. 기말시험을 잘 봤니?

여: 아직 성적을 받지 않았어.

남: 아. 음, 3개월의 여름 방학이 시작되어서 좋은가보네.

여: 실은, 학교에 며칠 안 가니까 벌써 좀 지루해졌어.

남: 저런. 이번 여름에 즐겁게 보낼 특별한 계획이라도 있니?

여: 응, 도로 여행을 몇 번 갈 거야.

남: 도로 여행? 어떻게?

여: 어제 운전면허 시험에 합격했거든. 내가 원할 때는 엄마가 차를 언제든 빌려줄 수 있다고 하셨어.

남: 굉장한데! 그래서 네가 신나 보이는구나. 원하는 곳은 어디든 갈 수 있겠네.

어휘
be in a good mood 기분이 좋다 entertain 즐겁게 하다 awesome 엄청난, 굉장한

07 ③

[전화벨이 울린다.]

남: Home Chicken입니다, 주문하시겠습니까?

여: 네, 프라이드치킨 두 마리와 양념치킨 한 마리요.

남: 식사 중에 세트 메뉴로 원하시는 게 있나요?

여: 세트 메뉴에는 뭐가 있나요?

남: 9달러 50센트에 프라이드치킨 한 마리와 감자튀김, 콜라를 살 수 있어요. 치킨만 주문하면 9달러이고요.

여: 양념치킨은 얼마예요?

남: 보통 양념치킨은 10달러예요. 양념치킨 세트 메뉴도 50센트만 더 내시면 됩니다.

여: 알겠어요. 프라이드치킨 세트 두 개와 보통 양념치킨 한 마리로 할게요.

남: 그럼 가지러 오시겠습니까, 아니면 배달해 드릴까요?

여: 배달해 주세요.

남: 30분 내로 배달해 드리겠습니다. Home Chicken에 전화해 주셔서 감사합니다.

어휘
fried 기름에 튀긴 spicy 양념을 한 combo 콤보(여러 종류의 음식 세트) deal 거래 regular 보통의

문제풀이
여자는 프라이드치킨 두 세트($9.5×2)와 보통 양념치킨 한 마리($10)를 주문했으므로, 여자가 지불할 금액은 29달러이다.

08 ①

여: 아름다운 그림이네요. 판매하는 건가요?

남: 네. 500달러에 팔고 있어요.

여: 그렇군요. 유명한 화가가 그렸나요?

남: 네, 그녀의 이름은 애너벨라 롱고예요. 요즘 상당히 유명하죠.

여: 그녀에 대해 들어본 적이 있어요. 수채화 그림으로 유명하죠?

남: 맞아요, 이 그림은 캔버스 천에 그린 유화지만요.

여: 그렇군요. 그림에 있는 남자는 누구죠?

남: 2005년에 사망한 그녀의 남동생이에요. 그녀가 이 그림을 완성한 후 일 년 뒤의 일이었죠.

여: 정말 안됐군요. 근데 그 그림을 제 거실 벽에 걸면 보기 좋을 것 같아요.

남: 분명히 그럴 거예요. 가로 50센티미터, 세로 75센티미터예요.

여: 완벽한 크기네요. 이것을 사겠어요.

남: 좋습니다. 현금과 신용카드 모두 받습니다.

어휘
watercolor painting 수채화 oil 유화 물감 canvas 캔버스 천 complete 완성하다

09 ⑤

여: 모두 주목해 주시겠습니까? 금요일 오후 6시 30분부터 시작해서 건물 전체에 새로운 무선 시스템이 설치되는 동안 인터넷 접속이 불가능할 것입니다. 이것은 중요한 프로젝트로 일요일 오전이 되어야 완료될 것으로 예상됩니다. 이것은 물론 주말 동안 근무하려는 분들이라면 누구에게든 영향을 미치게 될 것입니다. 여러분께서는 또한 원격으로 인트라넷 사이트에 접속하실 수 없을 것이므로, 그에 맞춰 계획을 세워 주십시오. 저희는 여러분께 새로운 시스템에 접속하는 방법을 보여드리기 위해서 월요일에 무엇보다도 먼저 회의를 개최할 것입니다. 월요일부터 이 시스템은 지하층을 제외한 사무실의 모든 곳에서 이용 가능합니다. 만일 질문이 있다면 IT 부서의 데이브에게 휴대 전화로 연락하시기 바랍니다.

어휘
connection 접속, 연결 disable 무능하게 하다 wireless 무선의 install 설치하다 affect 영향을 미치다 intranet 인트라넷, 내부 전산망 remotely 멀리서, 원격으로 accordingly (상황에) 부응해서, 그에 맞춰 basement 지하층 department 부서

10 ②

남: 실례합니다. 이 러닝머신을 할인 판매하고 있나요?
여: 네, 그렇습니다. 이 다섯 개의 모델들이 지금 모두 20퍼센트 할인 중입니다.
남: 오, 훌륭하네요! 하지만 전 어느 것을 골라야 할지 모르겠어요.
여: 제가 도와드릴 수 있습니다. 찾고 계신 기능이 무엇인가요?
남: 음, 저는 최고 속도가 시간당 10마일보다 빠른 것을 원해요.
여: 그건 문제가 없습니다. 경로 설정이 몇 개나 되는지 생각하고 계신가요?
남: 네, 그렇습니다. 전 적어도 고를 수 있는 10개의 다른 경로가 필요해요.
여: 그렇군요. 그리고 벨트의 길이도 고려하시겠네요.
남: 전 꽤 키가 커서 55인치보다 더 길어야 할 것 같네요.
여: 알겠습니다. 요구하시는 또 다른 특징들이 있나요?
남: 네. 저는 그걸 제 침대 아래에 보관하고 싶어서요, 접을 수 있어야 해요.
여: 그렇다면, 제게 당신을 위한 완벽한 트레드밀이 있습니다.

어휘
treadmill 러닝머신, 트레드밀(걷거나 달리는 운동 기구) feature 특징 maximum 최대(의) store 저장[보관]하다 foldable 접을 수 있는 (v. fold)

11 ⑤

남: 안녕하세요, 올리비아 머레이 씨를 찾고 있어요.
여: 네, 전데요. 아, 막 이사 오신 새로운 이웃분이시군요.
남: 네, 이 우편물을 전해드리려고 들렀어요. 저희 우편함에 있더라고요.
여: 정말 고맙습니다. 착오가 있었나 봐요.

어휘
drop by 잠시 들르다 mailbox 우편함 [문제] lend 빌려주다

문제풀이
① 제게 이메일을 보내주실래요?
② 제게 그걸 빌려주셔서 감사해요.
③ 당신이 제게 편지를 보낸 줄 몰랐어요.
④ 당신은 좀 더 자주 우편함을 확인해야 할 것 같아요.

12 ⑤

여: 왼쪽에 저희 초콜릿 만드는 기계가 있어요.
남: 이곳 공기는 매우 좋은 냄새가 나네요!
여: 오른편은 배송 부서입니다. 둘러 보시겠어요?
남: 네. 초콜릿을 어떻게 포장하는지 궁금해요.

어휘
shipping 배송 department 부서 [문제] taste 맛을 보다 parcel 소포 curious 궁금한 package 포장하다

문제풀이
① 감사하지만 사양할게요. 이미 그것을 맛봤어요.
② 그럼 거기에서 저희가 배를 타고 가나요?
③ 네, 밸런타인데이를 위해 초콜릿을 좀 사려고요.
④ 아뇨, 제 소포는 우체국에서 이미 부쳤어요.

13 ③

남: 메건, 안색이 별로 안 좋아 보여.
여: 응, 어젯밤에 안 좋은 일이 있었거든.
남: 무슨 일이 있었는데?
여: 누가 내 배낭에서 지갑을 훔쳐갔어!
남: 정말? 넌 어디에 있었는데?
여: 방과 후에 공부하느라 공공 도서관에 있었어.
남: (돈을) 많이 잃어버렸어?
여: 아니, 지갑에 돈은 별로 없었어. 신용카드도 없었고.
남: 다행이네.
여: 근데 우리 아빠가 도서관에 가서 경비가 너무 소홀한 것에 대해 사서들한테 고함을 치셨다는 거야.
남: 이런, 네가 난처했겠구나.

어휘
terrible 끔찍한, 소름 끼치는 public 공공의 yell 고함치다 security 경비, 보안 [문제] embarrassed 난처한, 창피한 treat 대하다, 다루다

문제풀이
① 나도 거기서 공부해 본 적 없어.
② 맞아. 거긴 보안이 꽤 좋아.
④ 우리 아빠는 가끔 도서관에 가서.
⑤ 그 사람들이 너희 아버지를 그렇게 대했다니 믿기지 않아.

14 ②

남: 나 육상부 그만둘까 생각 중이야, 제시카.
여: 왜? 넌 육상부에서 가장 빠른 선수 중 하나잖아.
남: 영어와 과학 성적이 떨어지고 있어. 부모님은 내가 너무 많은 걸 하고 있다고 생각하셔.

여: 너 미술 수업도 받지, 그렇지?
남: 응, 일주일에 세 번 받아.
여: 육상부보다 그게 더 재미있어?
남: 대답하기 힘들어. 둘 다 좋거든.
여: 음… 어떻게 결정할 거야?
남: 우선 육상부 코치 선생님이랑 얘기해보고 이번 학기에 쉴 수 있는지 알아볼 생각이야.
여: 그래. 내년에 다시 할 수도 있잖아.

문제풀이

① 넌 이미 좋은 성적을 받고 있잖아.
③ 대신 배구팀을 고려해 봐.
④ 부상이 육상을 포기하게 할 순 없어.
⑤ 너희 부모님께서 허락하시지 않을 것 같은데.

15 ⑤

여: 리사는 Rocky Bay에 사는 유명한 화가이자 환경 운동가이다. 그녀는 시간과 돈을 환경 조직에 투자한다. 그녀는 예술을 통해, 재활용 프로그램에 참여하는 것과 같은, 사람들이 환경을 도울 수 있는 방법에 대한 의식을 높이려고 노력한다. 그림을 통해 그녀는 자연의 아름다움과, 인간의 활동이 어떻게 그 아름다움을 파괴할 수 있는지를 보여주고 싶어 한다. 그녀의 한 미술 전시회가 진행될 동안, Rocky Bay에 새로 이사 온 주민이 리사의 그림에 매우 감명을 받아 그림 두 점을 샀을 뿐만 아니라, 환경 단체에 많은 돈도 기부했다. 이런 상황에서, 리사가 그 남자에게 할 말로 가장 적절한 것은 무엇인가?
리사: 당신의 구매와 관대함에 감사드려요.

문제풀이

① 이제 당신은 모든 걸 다 갚으셨어요.
② 인생에서 이렇게 초조한 적은 없었어요.
③ 제 그림에 손대지 말아 주세요.
④ 원하시면 그에게 주세요. 전 상관없어요.

16 ⑤ 17 ④

남: 오늘 워크숍에서 저는 실용적인 것을 말씀드리고 싶네요. 우리 모두는 우리 인생의 어느 지점에서 새로운 집이나 아파트로 이사하게 됩니다. 그것은 스트레스를 줄 수 있는데, 특히 우리가 이사를 준비할 때에는 더욱 그렇습니다. 우리의 소유물들을 포장하는 것은 복잡한 과정이지만 그것을 현명하게 하는 것은 이사를 더 쉽게 만들어줍니다. 예를 들어 가전제품을 쌀 때, 반드시 설명서대로 같은 상자 안에 각 장치를 넣도록 하세요. 깨질 물건을 보호하는 것도 중요합니다. 주방 도구들은 신문으로 싸야 하며, 가구의 작은 부분들은 침대 시트로 감싸져야 합니다. 이것은 손상을 막는 데 도움이 될 것입니다. 또

한, 책처럼 무거운 물건들은 커다란 하나의 상자보다는 몇 개의 작은 상자에 담겨져야 합니다. 이것은 운반하기에 훨씬 쉽게 해줍니다. 기억하세요, 약간의 세심한 계획들이 스트레스 받는 이사를 순조롭고 스트레스가 없는 경험으로 바꿀 것입니다.

문제풀이

16 ① 새집을 현명하게 고르기
   ② 쉽게 가구를 옮기는 법
   ③ 당신의 집을 정리된 상태로 유지하기
   ④ 집안 꾸미기의 유용한 요령들
   ⑤ 가정용품 포장을 위한 조언

DICTATION Answers
본문 p.152

01 see a noticeable difference / wrinkles will fade / Apply it / great promotion

02 be a vegetarian / is healthier / weak and tired

03 direct your call / it's not working / use a calling card / when you check out

04 on in 10 minutes / those are dead / close the drawer

05 delivered here / sell out / set some aside / write it down

06 ahead of you / take a few road trips / passed my driving test

07 may I take your order / with any of your meals / picking it up / I'd like it delivered

08 Is it for sale / quite well known / completed this painting

09 a new wireless system / unable to connect / be available everywhere

10 any treadmills on sale / That's not a problem / it needs to be foldable

11 just moved in

12 smells so good

13 stole my wallet / at a public library / having such poor security

14 quitting the track team / take art lessons / take this semester off

15 environmental activist / raise awareness / destroy that beauty / gave a lot of money

16~17 some point in our lives / a complicated process / Protecting fragile items / help prevent damage

01 ②	02 ⑤	03 ③	04 ⑤	05 ④	06 ③
07 ③	08 ⑤	09 ①	10 ②	11 ④	12 ⑤
13 ①	14 ①	15 ②	16 ⑤	17 ②	

01 ②

여: 저희 백화점 내에서 쇼핑하고 계시는 고객님들께 간단한 안내 말씀 드리겠습니다. 저희는 고객님들에 대한 감사의 표시로 작은 행사를 준비했습니다. 지금으로부터 딱 1시간 후인 5시에 추첨을 통하여 고객님 중 열 분께 다양한 선물을 드리겠습니다. 오늘 구매를 하신 분들은 누구나 당첨이 될 수 있습니다. 영수증 하단의 응모권을 잘라서 8층에 위치한 상자에 넣기만 하십시오. 모든 분들이 이 행사에 참여하시길 바랍니다. 여러분의 쇼핑 경험이 최대한 유쾌한 것이 되게 하려고 늘 최선을 다하겠습니다. 대단히 감사합니다.

어휘
announcement 발표 token 표시 appreciation 감사 raffle off 추첨을 하다 various 다양한 purchase *구매; 구매하다 be eligible to-v …할 자격이 있다 entry 참가 등록 receipt 영수증 submit 제출하다 locate 위치시키다

02 ⑤

여: 요즘 컴퓨터 앞에서 많은 시간을 보내는 것 같더라, 제임스.
남: 응, 사실이야. 모든 업무 보고서를 제시간에 끝내려고 초과 근무를 많이 하고 있어.
여: 음, 키보드를 그렇게 많이 사용하고 있다면, 너는 조심해야 할 거야.
남: 왜 그렇지? 타자를 치다가 다치지는 않잖아, 그렇지?
여: 사실은, 그럴 수 있어. 같은 동작을 계속해서 반복하면 손목을 상하게 할 수 있어.
남: 그건 몰랐어.
여: 규칙적으로 휴식을 취하는 게 중요해. 단 몇 분만이라도 몸이 회복하는 데 도움이 될 수 있지.
남: 알겠어. 또 뭘 할 수 있을까?
여: 손목을 바로 해. 타자 칠 때 손목을 책상 위에 대지 마.
남: 조언해줘서 고마워.

어휘
overtime 초과근무 get hurt 상처를 입다 type (컴퓨터) 타자를 치다. 입력하다 motion 운동. 움직임 damage 손상을 주다 wrist 손목 break 휴식 recover 회복하다 straight 똑바른 rest 놓다. 기대다

03 ③

남: 안녕하세요, 혹시 이 침대를 쓸 수 있는지 아시나요?
여: 아마도 아닐 거예요. 그 위에 가방이 하나 있네요. 접수처에서 침대 번호 안 받으셨어요?
남: 네. 이게 제 침대여야 하거든요. 근데 보시다시피 위에 배낭이 있어서요.

여: 아, 누군가 잠시 그 위에 올려놓은 걸 거예요. 번호를 가지고 있다면, 그 침대가 당신 거예요.
남: 감사합니다. 그건 그렇고, 전 미국에서 온 마틴이라고 해요.
여: 만나서 반가워요. 전 크로아티아에서 온 소피예요.
남: 여기에 얼마나 계셨어요?
여: 약 3주 동안요. 런던은 처음이세요?
남: 네, 오늘이 첫날이에요. 추천해 줄 좋은 장소가 있나요?
여: 주말의 Camden 시장을 강력히 추천해요. 거기서 생동감 있는 분위기를 경험하실 수 있어요.

어휘
reception 접수처 lively 생동감 있는 atmosphere 분위기

04 ⑤

여: 여기 와 봐! 여기엔 풀을 먹는 동물들이 많이 있는 것 같아.
남: 맞아. 아! 양이 길에 걸어 다니고 있어.
여: 응, 아주 귀엽네. 저 작은 여자애 봐. 엄마 손을 잡고 양한테서 눈을 못 떼잖아.
남: 그 애는 양이 셔츠를 입고 있어서 궁금한 모양이야. 정말 웃겨.
여: 이봐, 저 사육사가 손에 새끼 토끼를 들고 있어.
남: 저 꼬마 남자애들이 토끼를 쓰다듬고 있네. 매우 부드러워서 놀란 것 같아.
여: 오 저기 봐! 저기에 얼룩말이 있어. 내가 제일 좋아하는 동물이야.
남: 더 가까이 가서 보자.
여: 그런데 저 남자가 얼룩말 앞에 있는 자기 아들의 사진을 다 찍을 때까지 기다려야겠어.

어휘
take one's eyes off of …에서 눈을 떼다 curious 궁금한 pet 쓰다듬다

05 ④

남: 안녕, 제니퍼. 너 내일 시간 있어?
여: 응, 있어. 왜?
남: 실은 내일 웨딩드레스를 사러 갈 건데 네 도움이 필요해.
여: 웨딩드레스? 너 결혼하니? 왜 나한테 말 안 했어!
남: 진정해. 다음 달에 내 여동생이 결혼하거든. 동생이 나한테 드레스를 고르는 걸 도와달라고 했어.
여: 그 애 약혼자는 어쩌고? 신부의 오빠가 드레스 고르는 걸 도와주는 게 좀 이상하지 않니?
남: 그 사람이 요즘 무척 바빠서 내가 같이 가주겠다고 했어. 아무튼, 난 네 조언이 필요해. 넌 패션 잡지사에서 일하니까 드레스 고르는 안목이 있을 거로 생각했거든.
여: 내가 도울 수 있다면 기쁠 거야. 재미있겠다.
남: 고마워. 내일 오후 3시 괜찮니? 내가 너희 집으로 데리러 갈게.
여: 좋아. 그때 보자.

어휘
gown *드레스; 가운 fiancé 약혼자 odd 이상한 bride 신부 have an eye for …을 보는 안목이 있다

문제풀이
① 그의 가족과 저녁 먹기
② 그의 여동생에게 웨딩드레스 빌려주기
③ 그의 여동생을 위해 웨딩드레스 디자인하기

④ 그의 여동생이 웨딩드레스 고르는 것을 도와주기
⑤ 그와 패션 잡지 일을 같이하기

06 ③

여: 안녕하세요? 방해해서 죄송합니다, 교수님.
남: 괜찮아, 에이미. 들어와 앉으렴. 무슨 일이니?
여: 제 프로젝트팀을 바꿀 수 있는지 알고 싶어서요.
남: 왜 그러지? 네 팀원들과 문제라도 있니?
여: 아뇨, 팀원들은 괜찮아요. 하지만 전 그들과 수업시간표가 달라요. 그래서 함께 모이기가 쉽지 않아요.
남: 알겠다. 어디 보자… 좋아. 크레이그와 매디슨의 팀으로 합류하는 건 어떠니? 피터가 수강 취소를 해서 그 애들은 한 사람 더 필요하거든.
여: 잘됐네요. 크레이그와 전 전공이 같고, 매디슨은 저랑 심리학 수업을 같이 듣거든요. 감사합니다, 교수님.
남: 천만에.

bother 방해하다 get together 모이다 drop 그만두다. 중단하다 major 전공 psychology 심리학

07 ③

여: 안녕하세요. 뭘 도와드릴까요?
남: 네. 어디 보죠. *The Fallen City*와 *Lost Paradise*를 빌릴게요.
여: 고객님께서 선택하신 두 영화 모두 신작이네요. 각각 5달러입니다.
남: 네. 아, *X-Men* DVD는 반납됐나요?
여: 확인해 볼게요. [타자치는 소리] 네, 있네요. 그건 신작이 아니라서 3달러입니다.
남: 좋아요. 그것도 빌릴게요.
여: 오늘은 그게 전부인가요? 신작 DVD 두 편과 신작이 아닌 것 한 편이 있으시네요.
남: 음, 이 최신 판타지 소설 두 권도요. 여기 있습니다.
여: 네, 그 책들은 권당 1달러입니다.
남: 좋아요. 오늘은 이게 다예요.

어휘
release 발매 rent (사용료를 내고) 빌리다 latest 최신의 volume 용량; *책 fantasy 판타지. 공상(의 세계)

문제풀이
남자는 5달러짜리 신작 DVD 두 편과 신작이 아닌 3달러짜리 DVD 한 편, 그리고 권당 1달러짜리 소설 두 권을 빌렸으므로, 남자가 지불할 금액은 15달러이다.

08 ⑤

여: 아주 특이한 건물이군! 이런 건 본 적이 없어.
남: 그래. 유명한 건축가인 조지아 존스가 설계했어.
여: 아, 그녀에 대해 들어봤어. 이 지역 출신이지?
남: 음, 그녀는 파리에서 태어났지만 여기서 생애 대부분을 살았지.
여: 그럼 더는 살아있지 않은 거니?
남: 응, 그녀는 2007년에 죽었어. 하지만 생전에 그녀는 거의 100개의 건물을 설계했어.
여: 그녀의 스타일은 정말 독특하다.

남: 그녀는 학생일 때 아시아에서 공부했고, 그래서 그녀의 작품은 동양과 서양의 스타일이 결합되어 있어.
여: 그렇구나. 그럼 이 건물이 그녀의 가장 유명한 작품이니?
남: 아니. 이것도 상당히 흥미롭긴 한데, 그녀의 다른 건물들만큼 유명하지는 않아.
여: 안에 들어갈 수 있을까? 내부를 보고 싶어.
남: 물론이지. 가자.

어휘
design 디자인[설계]하다 architect 건축가 combine 결합하다 well-known 유명한. 잘 알려진 interior 내부

09 ①

남: 왕립 박물관은 세계에서 가장 인상적인 박물관 중의 하나입니다. 그 박물관에는 다양한 문화권으로부터 온 매우 아름다운 예술품과 골동품이 소장되어 있으며, 그 결과 유럽 최고의 관광 명소 중의 하나가 되었습니다. 박물관이 제대로 기능하게 하려고 저희는 박물관 운영을 돕는 데 관심이 있는 사람들을 위한 지속적인 자원봉사 프로그램을 운영합니다. 일반적으로 80명가량인 자원봉사자들은 저희에게 매우 소중합니다. 그분들 중 대부분은 미술 전시관에 근무하시는데, 그곳에서 그분들은 방문객들의 박물관 경험을 향상시키기 위해서 방문객들에게 정보를 제공합니다. 일부 자원봉사자들은 방문자 만족도 설문조사를 곁에서 보조함으로써 박물관의 행정부서를 돕습니다. 다른 분들은 기념품 가게와 카페테리아에서 방문객들을 응대합니다. 저희는 정말 그분들 없이는 이토록 많은 것을 이루지 못했을 것입니다.

어휘
breathtaking (너무 아름답거나 놀라워서) 숨이 막히는 antique 골동품 tourist attraction 관광 명소 function 기능[작용]하다 ongoing 진행 중의 volunteer 자원봉사(자) number (총수가) …에 달하다 enhance 높이다. 향상시키다 administrative 관리상의. 행정의 satisfaction 만족 survey (설문) 조사

10 ②

여: 아, 이것 봐. 여기 새로운 일본어 수업 시간표가 있어.
남: 아, 맞다! 너 일본어 배우는 걸 시작하고 싶어하지, 그렇지? 음, 들을 수 있는 초보자 강좌가 많이 있는 것 같은데.
여: 음… 난 금요일은 불가능해. 매주 금요일에 독서 모임에 가거든.
남: 그럼 주말 수업은 어때?
여: 토요일에는 9시부터 5시까지 아르바이트를 해.
남: 음, 일요일에 하는 수업도 하나 있어.
여: 응, 그렇지만 그 수업은 중급자부터 상급자를 위한 것이야.
남: 아, 그러네. 그럼 평일 저녁 수업은 어때?
여: 음… 수업료가 120달러 미만이었으면 하는데. 이미 매달 내야 하는 돈이 많아서 말이야.
남: 그럼 평일 아침에는 갈 수 있겠니? 저렴한 아침 수업이 있어.
여: 응, 네 말이 맞아. 그 수업이라면 되겠다.
남: 잘됐네! 등록하러 가자.

어휘
beginner 초보자 intermediate 중급의 advanced 고급의 tuition 수업료 make it (모임 등에) 가다. 참석하다 sign up 등록하다

11 ④

여: TV에 무슨 문제가 있니, 제이크?
남: 확실하진 않지만, 작동이 안 돼. 심지어 켜지지도 않아!
여: 리모컨에 문제가 있을지도 몰라. 마지막으로 건전지를 바꾼 게 언제니?
남: 기억이 안 나. 새 건전지를 가져올게.

어휘
[문제] nearby 가까이에 grocery store 식료품점. 슈퍼마켓

문제풀이
① 가까이에서 수리점을 찾을 수가 없어.
② 한 달 전에 그 TV를 수리했어.
③ 너는 이 프로그램도 봐야 해.
⑤ 아래층의 슈퍼마켓에서 건전지를 팔아.

12 ⑤

남: 안녕, 클레어. 학교 현장 학습으로 어디에 갔었니?
여: 우리는 Whitewater 공원에 갔어. 거기 동물원이 아주 컸어.
남: 좋았겠네. 나도 작년에 거기 갔었어. 너는 거기서 뭐가 제일 좋았니?
여: 나에겐 자이언트 판다가 제일 흥미로웠어.

어휘
field trip 현장 학습. 견학 [문제] zookeeper 동물원 사육사 disappointed 실망한

문제풀이
① 넌 친구들이랑 거기에 가봐야 해.
② 우리 반 친구들은 그 공원에 가기로 정했어.
③ 난 크면 동물원 사육사가 되고 싶어.
④ 난 동물이 많지 않아서 실망했어.

13 ①

여: 와, 이 기사 좀 봐!
남: 무슨 내용인데?
여: 돌고래가 어떻게 거울에 비친 자기 모습을 알아볼 수 있는지에 관한 내용이야.
남: 그래? 난 인간과 침팬지만 그렇게 할 수 있다고 생각했었는데.
여: 믿기 힘들지만, 분명히 돌고래도 자기 모습을 알아볼 수 있대.
남: 맞아. 돌고래가 아이큐가 높다는 걸 과학 잡지에서 읽었어.
여: 그게 어느 잡지였니? 난 그것에 대해 더 읽고 싶어.
남: 이번 달에는 해양 포유동물에 관한 특집 기사도 실려 있어.
여: 재미있겠다.
남: 너랑 도서관에 가서 그 잡지를 보여줄 수 있어.
여: 오, 정말? 지금 당장 가자.

어휘
article (신문) 기사 recognize 인식하다 apparently 분명히 mammal 포유 동물 [문제] judge 판단하다

문제풀이
② 알아. 심지어 아기들도 거울을 보잖아.
③ 넌 그 과학 잡지를 읽어야 해.
④ 오, 사람들을 아이큐로 판단해서는 안 돼.
⑤ 괜찮아. 오늘 밤에 TV에서 돌고래들을 볼 수 있어.

14 ①

남: 피터슨 선생님, 제발 제 휴대전화를 돌려주세요.
여: 넌 규칙을 어겼어, 폴.
남: 정말 죄송해요. 무음 상태로 해 놓았다고 생각했는데, 그러지 못했어요. 그냥 단순한 실수였어요.
여: 네 휴대전화에서 나오는 소리는 수업에 정말 방해가 됐어. 왜 그렇게 크게 해 놓았니?
남: 죄송합니다. 집에서는 그 소리가 잘 들리지 않아서 그렇게 해 놓은 거예요.
여: 아무튼, 별로 부모님께 있었던 일을 말씀드릴 거다.
남: 제발 그러지 말아주세요… 이번 한 번만 용서해 주시면 안 될까요, 피터슨 선생님? 부모님이 아시면 제 휴대전화를 빼앗아가실 거예요. 이걸 사려고 한 달 내내 일했는데 산 지 겨우 사흘밖에 안 됐어요.
여: 음, 알겠다. 다시는 절대 교실에 그걸 가지고 오지 마라. 그렇지 않으면 다음에는 네 부모님께 전화를 드릴 거야.
남: 안 그럴게요. 약속해요.

어휘
innocent 결백한; *악의 없는 disturb 방해하다 punishment 벌 forgive 용서하다

문제풀이
② 선생님을 위해서 음악 소리를 키울게요.
③ 부모님이 제게 새 걸 사주실 거예요.
④ 저도 선생님 휴대전화가 마음에 들어요. 멋져요.
⑤ 운전 중에는 통화하면 안 되는 거 모르세요?

15 ②

여: 한 젊은이가 첫 직장을 얻고서 매우 신이 났다. 그런데 문제가 하나 있었다. 그는 아침 일찍 일어나는 것을 싫어했는데도 불구하고, 야간 근무가 아니라 주간 근무를 하게 되었다. 잠에서 깨려고 무슨 짓을 해봐도, 그는 항상 다시 잠이 들어버렸다. 그러나 그는 아주 열심히 노력해서 첫 주에는 매일 제시간에 출근할 수 있었다. 그러나 그 후부터 그는 지각하기 시작했다. 세 번째 지각하자 상사는 또 지각을 하면 그를 해고하겠다고 말했다. 그 젊은이는 상사에게 자신의 문제를 얘기했다. 상사는 그것에 대해 생각을 해 보고는 해결책이 있다고 그에게 말했다. 이런 상황에서, 상사가 그 젊은이에게 할 말로 가장 적절한 것은 무엇인가?
상사: 자네가 야간 근무를 하게 해주겠네.

어휘
day shift 주간 근무(조) (↔ night shift) fire 해고하다 solution 해결책
[문제] night duty 야간 근무 overtime 초과근무

문제풀이
① 해결책이 없네.
③ 내일은 아침 6시까지 출근하게.
④ 오늘은 추가 근무를 할 필요가 없네.
⑤ 지금부터는 반드시 제때 오도록 하게.

16 ⑤ 17 ②

남: 당신의 부모님은 아침마다 가장 먼저 커피 한 잔을 마시나요? 아니면 아마도 공부할 때 깨어있으려고 에너지 음료를 사는 친구가 있을 수도 있을지 모릅니다. 커피와 에너지 음료 모두 카페인이라는 것을 함

유하고 있어서 더 정신을 초롱초롱하게 해줍니다. 효과는 있지만 너무 많이 마시면 건강에 나쁠 수도 있습니다. 더군다나, 그것은 수면 사이클을 방해하고 정말 쉴 필요가 있을 때 휴식을 취하지 못하게 할 수도 있습니다. 그러나, 졸음을 막는 데 쓸 수 있는 자연스러운 방법들도 많이 있습니다. 가장 간단한 방법은 일어나서 돌아다니는 것입니다. 그렇게 하는 것은 뇌로 더 많은 산소를 가져다줘서, 당신을 깨어 있게 해줍니다. 과일이나 견과류와 같은 건강한 간식을 먹는 것도 도움이 됩니다. 이 식품들은 졸린 신체가 에너지를 얻는 데 필요한 단백질을 함유하고 있습니다. 마지막으로, 밝은 빛은 잠을 쫓아내는 쉬운 방법이 될 수 있습니다. 어두운 장소에 있으면 신체는 잘 시간이라고 생각할지도 모릅니다.

어휘

alert 정신이 초롱초롱한 contain 함유하다 interrupt 방해하다 oxygen 산소 protein 단백질 chase away 쫓아내다 [문제] insomnia 불면증 routine 일상의 일, 일과 sleepiness 졸음

문제풀이

16 ① 불면증을 치료하는 방법
② 왜 일상적인 일을 가지는 것이 중요한가
③ 다른 에너지 드링크는 어떻게 작용하는가
④ 어떤 음식이 수면의 질을 향상시키는가
⑤ 졸음과 싸우는 최고의 방법은 무엇인가

DICTATION Answers

본문 p.158

01	as a token of / has made a purchase / located on the eighth floor
02	a lot of overtime / get hurt / take regular breaks / Keep your wrists straight
03	this bed is available / How long have you been here / any good places you'd recommend
04	walking around on the path / how soft it is / in front of the zebra
05	You're getting married / Isn't it a bit odd / have an eye for
06	change my project team / not easy to get together / dropped the course
07	new releases / I'll rent that as well / the latest two volumes
08	was designed by / no longer alive / combines Eastern and Western styles
09	keep the museum functioning / in order to enhance / visitor satisfaction surveys
10	start learning Japanese / a weekend class / need the tuition
11	change the batteries
12	your class field trip
13	recognize themselves / about ocean mammals / go to the library
14	broke the rules / innocent mistake / Never bring it
15	work the day shift / made a great effort / would be fired / had a solution
16~17	feel more alert / get too much of it / keep yourself from getting sleepy / chase away sleep

27 영어듣기 모의고사

본문 ▲ p.162

01 ④	02 ④	03 ①	04 ④	05 ③	06 ⑤
07 ④	08 ③	09 ④	10 ①	11 ③	12 ④
13 ③	14 ⑤	15 ⑤	16 ①	17 ④	

01 ④

남: 이 메시지는 저희의 귀중한 모든 독자들을 위한 것입니다. 저희는 저희 웹사이트에 주시는 여러분의 헌신에 대해서 감사하게 여기며, 언제나 흥미 있고 유용한 콘텐츠를 제공하려고 최선을 다해 왔습니다. 그러나 화요일 오전 이래로, 저희 웹사이트로의 접속이 기술적인 문제로 인해 차단되었습니다. 그 결과, 웹사이트에 접속하려고 시도했던 사용자들이 대신에 오류 페이지를 받게 되었습니다. 모든 접속은 오늘 저녁때까지 복구될 겁니다. 저희는 이것이 일으켰을 모든 불편에 대해서 정말 유감스럽게 생각합니다. 저희는 이 문제가 저희 웹사이트에 대한 여러분의 충성심에 영향을 미치지 않기를 바라며, 반드시 이런 종류의 문제가 앞으로는 일어나지 않도록 할 것입니다. 여러분의 인내와 협조에 감사드립니다.

어휘

valued 귀중한 appreciate 고마워하다 dedication 헌신, 전념 content 【컴퓨터】 콘텐츠 access 접속; 접속하다 technical 기술적인 restore 복구하다 loyalty 충성심 occur 일어나다, 발생하다 patience 참을성, 인내심 cooperation 협조, 협동

02 ④

남: 오늘 아침 왜 늦었어요, 줄리? 교통이 막혔나요?
여: 아니요, 사실 전 지하철 타고 출근해요.
남: 그럼 무슨 일이 있었죠?
여: 지하철을 갈아타려고 했는데, 열차가 꽉 차서 탈 수가 없었어요. 다음 열차를 기다려야 했죠.
남: 스트레스가 되겠네요.
여: 정말 그랬어요. 출퇴근 시간에는 지하철을 이용하는 사람들이 항상 너무 많아요. 혼잡한 시간에는 열차가 더 자주 다녀야 해요.
남: 그럼 버스를 타보는 건 어때요?
여: 아, 그건 훨씬 더 심할 수도 있어요. 버스를 기다리는 시간이 더 길거든요.
남: 대도시에서 통근하기란 항상 힘든 것 같네요.

어휘

get stuck in traffic 교통이 막히다 transfer 갈아타다 rush hour (출퇴근) 혼잡 시간 wait 기다림; *기다리는 시간 commute 통근하다

03 ①

남: 우린 곧 라디오 방송국에 도착할 거예요. 인터뷰 준비됐어요?
여: 네. 제 새 영화에 대해서 가능한 한 자주 언급해야 한다고 알고 있어요.
남: 맞아요. 그리고 하나 더 있어요. Willis 서점에서 영화 포스터에 사인

을 하러 가는 일정이 막 잡혔어요.

여: Willis 서점이요? 알겠어요. 그럼 그에 대해서도 언급해야 하나요?

남: 맞아요. (개봉) 날짜가 빠르게 다가오고 있어서, 우리는 영화를 최대한 많이 언급해야 해요.

여: 알겠어요.

남: 좋아요. 우리가 지금 영화를 많이 홍보하면, 당신의 다음 배역은 더 클 수도 있어요.

여: 동의해요. 최선을 다할게요.

어휘

mention 언급하다 schedule 일정을 잡다, 예정하다 appearance 모습; *출연 approach 다가가다[오다] promote 촉진하다; *홍보하다

04 ④

여: 새로 태어난 제시카의 아기를 본 적 있어요?

남: 아직이요. 아기가 신생아실에 있어요?

여: 네, 와서 보세요.

남: 오, 다른 침대들은 비어 있으니까 아기는 창문 바로 옆에 있는 두 아기 중 한 명이겠네요.

여: 맞아요. 아기가 누구일지 맞춰봐요.

남: 평화롭게 자고 있는 아기인가요?

여: 아니요, 그건 제시카의 아기가 아니에요.

남: 그럼 모자를 쓰고 있는 아기가 틀림없네요. 어, 울고 있어요. 배고픈가 봐요.

여: 저기 있는 간호사가 우유를 먹이려고 젖병을 가져오고 있네요.

남: 오, 벌써 5시 되기 15분 전이에요. 수유 시간인가요?

여: 아마도요. 사실, 신생아들은 배고플 때마다 수유를 해줘야 하죠.

어휘

nursery 육아실 peacefully 평화롭게 bottle 젖병 feed (음식을) 먹이다 newborn 갓 난; *신생아

05 ③

[휴대전화벨이 울린다.]

남: 여보세요?

여: 안녕, 프랭크. 나 비벌리 스미스야.

남: 안녕, 비벌리. 무슨 일이야?

여: 음, 프랑스에 있는 회사에서 일하는 데 관심이 있어서 전화했어. 사실 곧 회사 한 곳에 지원할 거야.

남: 오, 잘됐다. 너도 알다시피, 내가 지금 5년간 파리에서 일하고 있잖아.

여: 그곳에 사는 거 즐겁니?

남: 당연하지. 파리는 멋진 도시야.

여: 그 말을 들으니 기뻐. 나도 역시 그곳에 살면서 일하고 싶어.

남: 여기 생활에 대한 정보를 좀 보내 줄까?

여: 나중에 해 줘. 사실 나는 다른 걸 부탁하려고 전화했어. 내가 회사에 이력서를 보내기 전에 내 이력서를 검토해 줄 수 있어?

남: 물론이지. 네가 언어 능력을 향상시킬 필요가 있다면, 좋은 프랑스어 개인 교사를 소개해 줄 수도 있어.

여: 그건 아직은 필요가 없어. 하지만 내가 그 일자리를 얻게 되면, 네게 우리가 연락하도록 주선해 달라고 할게.

어휘

apply for …에 지원하다 résumé 이력서 tutor 개인 교사

06 ⑤

[전화벨이 울린다.]

남: 여보세요.

여: 안녕, 폴. 나 러시아어 수업 같이 듣는 캐런이야.

남: 아, 안녕, 캐런. 무슨 용건이니?

여: 음, 프랑스로 수학여행을 가는 데 네가 아직 관심 있는지 궁금해서.

남: 응, 있긴 한데, 자리가 다 찬 줄 알았어.

여: 이젠 아냐. 크리스틴이 팔이 부러져서 못 가거든.

남: 그거 안됐구나.

여: 응, 나도 그녀가 안 됐어.

남: 어쨌든, 알려줘서 고마워.

여: 천만에. 넌 학교에 전화해서 등록해야 할 거야.

남: 알겠어. 지금 바로 할게.

어휘

class trip (학급) 수학여행 That's a shame. 그거 유감이다.

07 ④

남: 점심 괜찮았어요. 그렇지 않았나요?

여: 네, 빨리 나오고 가격도 저렴했어요.

남: 그럼 계산하고 회사로 다시 가죠.

여: 값이 얼마였죠?

남: 우린 중간 크기의 피자 한 판과 음료수 두 잔을 시켰어요. 피자 한 판은 15달러였고 음료수는 한 잔에 1달러 50센트였어요.

여: 난 피자가 10달러라고 생각했어요. 왜 지금은 15달러죠?

남: 하나에 2달러 50센트인 토핑 두 가지를 추가 주문했잖아요.

여: 아, 맞아요. 이상 없네요. 내가 얼마를 내면 되죠?

남: 내가 피자를 먹자고 했으니까 내가 낼게요.

여: 정말요? 나도 보탤 수 있는데요.

남: 아뇨, 내가 낼게요.

여: 고마워요.

어휘

medium 중간의 soda 탄산음료, 소다수 extra 추가의, 여분의 topping 토핑 (요리 위에 곁들인 것) owe 빚지다, 지불할 의무가 있다

문제풀이

두 사람은 피자($10)에 토핑 두 가지($2.5×2)를 추가로 얹고 음료수 두 잔($1.5×2)을 주문했는데, 남자가 전부 계산한다고 했으므로, 남자가 지불할 금액은 18달러이다.

08 ③

남: 안녕하세요, 손님. 무엇을 도와드릴까요?

여: 3층에 있는 싱글 룸에 묵고 있는데, 만족스럽지가 않아요.

남: 유감입니다. 어떤 문제가 있으신가요?

여: 객실이 너무 작아요. 옮길 수 있는 더 큰 객실이 있을까요?

남: 음, 더 큰 객실이 있는데 내일 전까지는 이용하실 수 없어요.

여: 괜찮아요. 바다 전망인가요?

남: 아니요, 하지만 산을 보실 수 있어요. 그리고 현재의 객실보다 두 배가 커요.

여: 그럼 더 비싸겠네요.

남: 네, 1박에 120달러입니다. 하지만 10퍼센트 할인해 드릴 수 있어요.

여: 좋은 것 같네요.

남: 잘됐네요. 아침에 모두 처리해 놓겠습니다. 그리고 문제가 생긴 것에 대해 사과드립니다.

여: 아, 괜찮아요. 도와주셔서 고마워요.

09 ④

여: Red Creek 캠프는 십 대들이 자신감을 향상하고 재미있는 시간을 보내면서 체력을 강화하도록 도와주기 위해 고안된 특별 하계 프로그램입니다. 저희 직원들은 국제적인 운동 전문가팀을 포함하고 있습니다. 그들은 확실히 모든 캠프 참가자들이 그들의 건강에 누가 되지 않으면서, 신속하게 강해지도록 해줄 것입니다. 이 프로그램은 11세에서 18세까지의 소년소녀들이 이용 가능하며, 8월의 첫 3주간 운영됩니다. 활동 프로그램에는 스포츠, 등산, 래프팅, 그리고 요리 교실이 있습니다. 등록하시기에 앞서, 모든 캠프 참가자들은 힘든 신체적 활동에 참가할 수 있을 만큼 건강하다는 것을 증명하는 의사의 확인서를 받아야 합니다. 더 많은 정보를 원하시면 저희 웹사이트 www. rccamp.com을 방문해 주십시오.

10 ①

여: 우린 뮤지컬 티켓을 예매해야 해요. 당신은 언제가 좋아요?

남: 나는 휴가 기간에 가고 싶어요. 내 휴가는 5월 24일부터 6월 1일까지예요.

여: 그 기간에 좋은 뮤지컬들이 많이 있어요.

남: 모든 뮤지컬들이 같은 가격인가요?

여: 아니요. 일부는 80달러이고, 다른 일부는 100달러에요.

남: 왜 가격에 차이가 있죠?

여: 더 비싼 뮤지컬들은 훨씬 더 평이 좋은 경향이 있어요. 그 뮤지컬들이 보통 훨씬 더 좋지요.

남: 그럼, 볼 만한 가치가 있다고 생각한다면, 좋은 뮤지컬을 봅시다. 나는 돈을 더 내도 상관없어요.

여: 좋아요! 동의해요. 아, 회원권이 있다는 게 방금 생각났네요!

남: 무슨 말이에요?

여: 일부 공연에 할인을 받을 수 있다는 의미예요.

남: 그거 잘됐네요! 그렇다면 회원 할인이 되는 뮤지컬로 선택합시다.

11 ③

남: 오늘은 야구 경기하기에 훌륭한 날씨야!

여: 정말 그래. Bears 팀이 오늘 이길 것 같아.

남: 왜 그런 말을 하니? 최근에 지고 있잖아.

여: 그래, 그런데 새로운 선수들 몇 명을 영입했거든.

12 ④

여: 저 상자 안에 있는 그 책들은 뭐야?

남: 그것들을 없애려고 해. 더 이상 안 읽거든.

여: 정말? 아직 상태가 좋은데.

남: 사실, 그것들을 지역 도서관에 기증할 거야.

13 ③

남: 질, 다이어트는 어떻게 되어가고 있니?

여: 아, 모르겠어. 정말 힘들어.

남: 음, 괜찮아 보이는데. 뭐가 문제야?

여: 3파운드를 뺐는데, 그게 내 최대치 같아. 아무리 운동을 많이 해도 체중은 그대로야.

남: 체중이 더는 줄지 않는다는 말이니?

여: 안 줄어. 조금도. 어떻게 해야 하지?

남: 음, 네가 운동을 하고 건강한 식사를 하기만 한다면, 나는 네가 괜찮을 거로 생각해.

여: 알아, 하지만 나는 여전히 날씬해지고 싶어.

남: 음, 운동은 보통 근육을 만들어. 그리고 근육은 지방보다 무겁지. 그래서 실제로는 네가 체중이 늘어도 더 날씬해지고 더 건강해질 수 있는 거야.

여: 그럼 나는 내 계획을 계속해야겠어.

14 ⑤

남: 안녕, 앨리스. 어제 어디에 있었어? 방과 후에 너를 보지 못했어.

여: 나는 오디션이 있었어. 학교 밴드가 새 가수를 찾고 있어.

남: 잘됐다. 프로 가수가 되는 것이 네 꿈이잖아, 그렇지?

여: 그래. 이건 정말 좋은 첫 단계가 될 거야.

남: 그들이 너를 선택할 거라고 확신해. 너는 노래를 잘하잖아.

여: 고마워, 그렇지만 오디션 내내 노래를 형편없이 부른 것 같아.

남: 그렇지 않을 거야. 왜 그렇게 말하니?

여: 밴드 멤버들이 감명받은 것처럼 보이지는 않았어. 그들은 날 뽑지 않을 거야.

남: 그건 모르는 거잖아. 그들은 언제 오디션 결과를 발표해?

여: 이번 주 금요일이나 되어야 할 거야. 그들이 날 선택하지 않으면 어떡하지?

남: 금요일까지 잊어버려. 지금은 네가 할 수 있는 게 없잖아.

어휘

doubt 의심하다 impressed 인상 깊게 생각하는

문제풀이

① 걱정하지 마. 널 응원하게 갈게.
② 나도 그렇게 생각해. 넌 계속 연습해야 해.
③ 난 신경 안 써. 네가 최선을 다할 거라고 믿거든.
④ 넌 이미 무대에서 공연을 잘 하잖아.

15 ⑤

남: 조셉은 최근 대학 졸업생이며 현재 첫 일자리를 찾고 있다. 어느 날, 호기심에 그는 인터넷에서 자기 이름을 검색하기로 결심한다. 그는 그의 사진을 많이 발견하고 놀란다. 그가 그 사진들을 클릭할 때, 그는 그 사진들이 모두 그의 친구인 멕의 블로그에서 나온다는 것을 발견한다. 그녀의 블로그는 대중에게 공개되어 있고 그녀가 그의 이름을 써놓아서, 누구든지 그 사진들을 찾을 수 있다. 조셉은 요즘 면접관들이 인터넷에서 입사 지원자들에 관한 정보를 찾는다고 들었다. 그는 면접관이 온라인에 게시된 사진에 근거해서 그에 대한 나쁜 인상을 받는 것을 원하지 않는다. 이런 상황에서, 조셉은 멕에게 뭐라고 말하겠는가?

조셉: 네 블로그에서 내 사진을 삭제해 주겠니?

어휘

graduate 졸업하다; *졸업생 currently 현재, 지금 out of curiosity 호기심에서 nowadays 요즘에는 interviewer 면접관 candidate 후보자; *지원자

문제풀이

① 내가 네 블로그에 접속할 수 있게 해 주겠니?
② 그 사진의 사본을 나에게 보내 주겠니?
③ 요즘 어떤 종류의 카메라를 사용하고 있니?
④ 블로그 만드는 것에 관해 언제 만나서 대화할 수 있을까?

16 ① 17 ④

여: 좋은 아침입니다, 학생 여러분. 오늘 저는 16세기와 17세기의 셰익스피어의 연극에 관해 이야기하려고 합니다. 추측하건대 여러분들은 셰익스피어의 연극을 보는 것은 오직 귀족들의 취미였을 거라고 생각할 것입니다. 그러나 17세기에는 하층 계급의 사람들조차 공연을 볼 수 있었습니다. 셰익스피어의 연극들은 종종 런던의 Globe 극장에서 공연되었는데, 그곳에서 사람들은 발코니의 좌석을 살 형편이 되지 않아도 연극을 볼 수 있었습니다. 대신에, 그들은 입장해서 바닥에 서 있는데 단지 1페니를 냈습니다. 그것이 바로 그들이 '바닥 관람객'이

라는 별명으로 불리게 된 이유입니다. 바닥 관람객들은 공연에 대해 매우 열정적이었던 것으로 알려져 있습니다. 그들이 좋아하지 않았던 등장인물들에게 야유를 하고는 했고 때로는 그들에게 무언가를 던지기도 했습니다. 저는 여러분들이 셰익스피어의 작품에 대해 바닥 관람객이 그랬던 것만큼 열광하시기를 희망합니다.

어휘

noble 귀족 afford (…을 살) 여유가 되다 balcony 발코니 nickname 별명; *별명을 붙이다 groundling 바닥 관람객(엘리자베스 왕조 시대에 극장 1층에서 적은 입장료를 내고 공연을 보았던 하층민들) enthusiastic 열렬한, 열광적인 [문제] nobleman 귀족, 상류층 inspiration 영감, 영감을 주는 것 sarcasm 비꼼, 빈정댐, 풍자

문제풀이

16 ① 누가 셰익스피어의 연극을 즐겼는가
② 셰익스피어는 왜 귀족들을 싫어했는가
③ 셰익스피어는 어떻게 영감을 얻었는가
④ 어디서 셰익스피어의 연극이 공연되었는가
⑤ 셰익스피어가 그의 연극에서 어떻게 풍자를 사용했는가

DICTATION Answers

본문 p.164

01 dedication to our website / As a result / may have caused

02 get stuck in traffic / run more often / The wait between buses

03 as often as possible / is approaching quickly / promote the film a lot

04 right next to the window / wearing a hat / to feed him

05 enjoyed living there / reviewing my résumé / put us in touch

06 going on the class trip / broke her arm / call the school to register

07 quick and cheap / two extra toppings / do I owe you

08 satisfied with it / twice as big as / apologize for the problem

09 help teenagers strengthen their bodies / without endangering / get a permission form / difficult physical activities

10 I'm on vacation / it's worth it / get a discount on

11 They've been losing

12 in good condition

13 your dieting going / that's my maximum / stays the same / are slimmer and healthier

14 I had an audition / What makes you say that / What should I do if

15 out of curiosity / Her blog is public / get a bad impression

16~17 a hobby just for nobles / even if they couldn't afford / for being very enthusiastic about / as the groundlings were

28 영어듣기 모의고사

본문 ▲ p.168

01 ②	02 ③	03 ④	04 ⑤	05 ⑤	06 ③
07 ④	08 ②	09 ③	10 ④	11 ①	12 ④
13 ⑤	14 ②	15 ④	16 ②	17 ②	

01 ②

여: 잠시만 주목해 주시겠습니까? 이렇게 비 오는 화요일 아침에 저희 매장에 방문하신 여러분 모두에게 감사드립니다. 물론, 여러분이 이곳에 오신 이유는 제 뒤의 테이블에 앉아계신 여성분 때문입니다. 샬롯 채프먼 씨는 문학 분야에서 성공하셨으며, 놀랄 만큼 젊은 나이에 그렇게 하셨습니다. 심지어 우리 나라의 슬럼프에 빠진 출판 산업을 혼자의 힘으로 살려낸 것이 그녀의 공이라고 믿는 분들도 있습니다. 그녀는 앞으로 한 시간 동안 자신의 최근 소설 작품에 사인을 해주실 것이므로, 여러분이 질서정연하게 줄을 서주신다면, 지금 당장 시작하도록 하겠습니다.

어휘
make a name for oneself 유명해지다 literary 문학의 amazingly 놀랍게도 credit (공로를) …에게 돌리다 revive 회복[부활]시키다 slump 급감[폭락]하다 publishing industry 출판 업계 copy 복사본; *(책·신문 등의) 한 부 orderly 질서 있는

02 ③

여: 엄마는 내가 아르바이트를 그만두길 바라서. 그게 학업을 방해한다고 생각하시거든.
남: 음, 그럴지도 모르지. 하지만 아르바이트는 네가 스스로 더 책임감이 있어야 한다는 걸 가르쳐주잖아.
여: 맞아. 나는 기한 전에 일을 끝내야 해.
남: 게다가 넌 돈을 벌잖아. 직장은 돈을 관리하는 법을 배울 기회를 주지.
여: 그래. 난 벌써 상당히 저축했어.
남: 엄마께 성인이 되었을 때를 위해 유용한 기술을 배우고 있다는 말씀도 드려.
여: 맞아. 그리고 내 성적은 아직 꽤 좋으니까.
남: 그래. 알겠니? 엄마를 설득하는데 네가 내세울 만한 점들이 많이 있잖아.
여: 네 말이 맞아. 조언해줘서 고마워.

어휘
distract 집중이 안 되게 하다 deadline 기한, 마감 시간 manage 관리하다 convince 납득시키다

03 ④

[전화벨이 울린다.]
남: RCI에 전화해 주셔서 감사합니다. 무엇을 도와드릴까요?
여: 귀사에서 TV에 광고하는 국제전화 프로그램에 대해서 여쭤볼 게 있어서요.

남: 네. 우선 현재 저희 회사에 계정을 갖고 계시는지 여쭤봐도 될까요?
여: 지금은 없어요. 예전에는 있었는데, 이사하면서 TeleSmart로 바꿨거든요.
남: 음, 사실 저희는 되돌아오시는 고객님들에게 특별 할인을 제공해 드립니다.
여: 그래요? 그건 몰랐어요.
남: 네, 초기 설치비에서 5퍼센트 할인을 받으실 거예요.
여: 알게 되어서 다행이네요. 이제 국제전화 요금제에 대해서 질문 좀 해도 될까요?
남: 담당자를 바꿔 드릴게요. 잠시만 기다려 주시겠어요?
여: 물론이죠, 문제없어요.

어휘
advertise 광고하다 account 계정 initial 처음의 installation 설치 transfer 옮기다 person in charge 책임자

04 ⑤

[전화벨이 울린다.]
여: 여보세요?
남: 안녕, 제니퍼. 네 오빠야. 뭐하고 있니?
여: 엄마 생신 저녁상을 차리고 있었어.
남: 아, 멋지구나. 어떤 탁자에서 먹을 거니?
여: 거실에 있는 둥근 탁자에서. 네 개의 의자를 놓고 한 사람당 하나씩 와인 잔을 두었어.
남: 테이블을 멋지게 꾸몄니?
여: 응, 각각의 접시 위에 새 모양으로 접힌 냅킨이 있어.
남: 멋지다. 내가 제안했던 대로 탁자 가운데에 꽃바구니를 하나 두었니?
여: 응. 좋은 생각이었어. 하지만 탁자 위에 양초를 두지는 않았어.
남: 왜? 엄마는 양초를 좋아하시잖아.
여: 그렇긴 하지만 양초들은 너무 많은 공간을 차지할 것 같아.

어휘
set 놓다; 상을 차리다 fold 접다 plate 접시, 그릇

05 ⑤

남: 안녕하세요, 그웬. 사무실로 복귀한 걸 환영해요. 일전에 있었던 사고는 유감이에요.
여: 고마워요. 다친 곳 없이 빠져 나와서 다행이죠.
남: 맞아요, 당신은 정말 운이 좋았어요. 제가 당신을 위해서 할 수 있는 일이 있으면 알려주세요.
여: 실은, 제 차가 2주 동안 정비소에 들어가 있을 거예요.
남: 알겠어요. 출퇴근할 때 차를 태워 드릴까요?
여: 음, 제가 화요일과 목요일에는 남편의 차를 이용할 수 있어요. 그날은 남편이 재택근무를 하는 날이거든요.
남: 와, 남편분이 부러워요.
여: 그래서 말인데, 화요일과 목요일에는 제가 태워 드리고, 나머지 요일에는 당신이 절 태워 주시면 어떨까요?
남: 네, 그러면 좋겠네요.
여: 네. 그렇게 해주시면 제 생활이 훨씬 더 편해질 거예요.

어휘
escape 탈출하다 uninjured 다치지 않은 envy 부러워하다

06 ③

여: 안녕하세요. 어디로 가시겠어요?

남: 가능한 한 빨리 서울역으로 가주세요.

여: 알겠습니다. 안전벨트를 매 주세요.

남: 거기 가는 데 얼마나 걸릴까요?

여: 교통이 막히지 않으면 20분 정도 걸릴 거예요.

남: 아, 빠듯하겠는데요. 30분 안에 도착하지 않으면 기차를 놓칠 거예요.

여: 최선을 다하겠지만, 제시간에 도착한다고 보장할 수는 없어요.

남: 괜찮아요. 제 잘못이죠. 친구와 여행을 갈 예정이거든요. 유감스럽게도, 저는 기차 시간표를 잘못 읽고 집에서 너무 늦게 나왔어요.

여: 교통이 너무 혼잡하지 않길 바랍시다.

어휘

stuck in traffic 교통이 막힌 tight 빠듯한 guarantee 보장하다 misread 잘못 읽다 (misread-misread)

07 ④

여: 그래서, 제 TV에 무슨 문제가 있는지 알아내셨나요?

남: 예, 손님. 유감스럽게도 화면이 손상되었네요. 화면을 교체해야겠어요.

여: 틀림없이 이사할 때 그런 것 같아요. 비용이 얼마나 들까요?

남: 음, 어디 봅시다… 50인치 텔레비전이면…

여: 죄송한데, 이건 47인치밖에 안 돼요.

남: 아, 그렇네요. 그러면 새 화면은 80달러의 비용이 듭니다. 그리고 배송 비용이 있어요.

여: 그게 얼마인가요?

남: 30달러예요. 그리고 20달러의 설치 비용이 추가될 겁니다.

여: 흠… 제가 예상한 것보다 더 비싸네요. 현금으로 내면 할인 받을 수 있나요?

남: 네. 총액에서 10퍼센트 할인을 받습니다.

여: 좋네요. 오래 걸릴까요?

남: 화면은 도착하는 데 사흘이 걸려요.

여: 알겠습니다. 기다릴 수 있을 것 같네요.

어휘

figure out 알아내다 replace 교체하다 installation 설치 (v. install)

문제풀이

남자는 TV를 수리하는 데 47인치 화면($80)을 교체하기로 하고, 여기에 배송비($30)와 설치비($20)를 추가로 내는데, 현금으로 계산하는 조건으로 총액에서 10퍼센트 할인을 받는다. 이에 남자가 지불할 금액은 117달러이다.

08 ②

여: 안녕하세요. 새 차를 찾고 있어요. 비싸지 않은 게 필요해요.

남: 음, 특별 저가 모델이 있어요. 저희 것들 중에 가장 저렴한 것입니다.

여: 이름이 뭐죠?

남: 그 모델은 Extra라고 해요. 지난달에 막 출시됐어요.

여: 그렇군요. 얼마인가요?

남: 정상 가격은 12,000달러예요. 그런데 할부 판매도 있습니다.

여: 그것에 대해 알려 주시겠어요?

남: 지금 1,000달러를 지불하시면 앞으로 5년 동안 한 달에 200달러를 내실 거예요.

여: 적정한 것 같군요. 시승을 해봐도 될까요?

남: 네. 운전이 아주 잘 된다고 느끼실 거예요.

여: 그러길 바라요. 자동차는 어디 있어요?

남: 저기에 있는 빨간 차예요. 우선 고객님의 운전면허증을 봐야겠네요.

어휘

budget 예산; *저가의, 저렴한 release 출시하다 installment plan 할부 판매 reasonable 타당한; *(가격이) 적정한 test drive 시승 driver's license 운전면허증

09 ③

남: 여러분의 여름 방학을 가장 유용하게 보내고 싶으신가요? 그렇다면 North Valley 영어 캠프로 오시는 건 어떤가요? 아름다운 산악 호숫가에 위치한 저희 캠프는 모두 원어민으로 구성된 강사진이 특징입니다. 프로그램은 7월에 시작해서 2주 동안 지속됩니다. 참가자들은 저녁에는 재미있고 유익한 수업을 들을 것이고, 낮에는 스포츠, 공예, 보트 타기 등을 포함한 재미있는 야외 활동에 참여하게 되는데, 이 모두는 전적으로 영어로 진행됩니다. 무엇보다도 North Valley 영어 캠프는 비용도 적정한데, 2주 프로그램이 단 700달러입니다. 등록 신청서와 추가 정보는 저희 웹사이트 www.nvecamp.net에서 찾아보실 수 있습니다.

어휘

make the most of …을 최대한 활용하다 feature 특징으로 삼다. 포함하다 entirely 전적으로, 전부 informative 유익한 craft 공예 conduct (특정한 활동을) 하다 affordable (가격이) 알맞은 registration 등록 additional 추가의

10 ④

남: 우린 곧 브리즈번에 있는 호스텔의 방을 예약해야 해.

여: 맞아. 우린 2주 뒤에 떠나니까. 우리 예산이 얼마지?

남: 우린 하룻밤에 50달러까지 쓸 수 있어.

여: 그렇구나. 여기 인기 있는 호스텔 목록이 있어. 우린 그것들 중 대부분을 감당할 수 있어.

남: 오, 좋다. 근데 한여름이니까, 수영장이 없는 곳에서 묵지는 말자.

여: 동의해. 바비큐는 어때? 그게 중요할까?

남: 응! 바비큐 파티는 다른 여행자들을 만나는 멋진 방법이거든.

여: 알겠어. 조식은 어때? 조식이 포함된 곳으로 얻어야 할까?

남: 그러자. 난 근사하고 성대한 식사로 하루를 시작하는 걸 좋아해.

여: 알겠어. 그러면 이곳에 묵자.

남: 좋아. 내가 전화해서 지금 바로 방을 예약할게.

어휘

afford …할 여유가 있다 pool 수영장 reserve 예약하다

11 ①

여: 이봐, 제임스! 린다의 생일 파티가 내일이야, 맞지?

남: 응. 정말로 신날 것 같아. 그 애 선물은 샀어?

여: 아직. 나랑 같이 가서 괜찮은 걸 찾아보겠니?

남: 좋아. 선물 가게에 가자.

남: 지선아, 간단한 질문 좀 해도 될까?

여: 물론이지, 어서 해.

남: 네가 복사기를 거의 다 썼는지 궁금해서 말이야. 내가 거의 10분째 기다리고 있거든.

여: 아, 아니. 아직 멀었는데. 난 아직도 이 프로젝트를 100부 가까이 복사해야 해. 아마 30분 정도 더 걸릴 거야.

남: 30분이라고? 네가 복사기를 그렇게 오래 사용할 거면 다른 사람들에게 알려줘야지.

여: 사실, 난 모두에게 4층이나 5층에 있는 복사기를 사용할 것을 제안하는 이메일을 보냈어.

남: 그랬니? 미안해. 오늘 아침에 이메일을 확인하지 않았거든.

여: 괜찮아. 이 일이 모두에게 불편을 끼친다는 것을 알긴 하지만, 어쩔 수 없는 일이었어.

남: 내가 그냥 다른 층으로 가면 될 것 같아.

어휘
copy machine 복사기 copy 복사본. 부 unavoidable 불가피한 [문제]
notify 알리다

문제풀이

① 음, 다음에는 나한테 반드시 알려줘.

③ 이해해줘서 고마워.

④ 네가 마침내 다 끝냈다니 다행이야.

⑤ 부디 이메일 서버를 곧 고쳤으면 좋겠어.

15 ④

여: 윌리스 씨는 연세가 지긋한 남성이다. 젊은 시절에 그는 아파트 건물을 몇 채 샀다. 세월이 지나서 그 지역이 발전함에 따라 임대료가 점점 더 올랐고, 이제 윌리스 씨는 매우 부유하다. 수십 년간 그는 열심히 일해 왔고, 그의 건물에 거주하는 사람들의 모든 문제를 직접 처리해왔다. 그렇지만 이제 그는 스페인으로 이사해서 안락하고 편안한 삶을 살기로 했다. 그는 앞으로 그의 건물들의 모든 문제를 처리할 건물 관리인을 고용했다. 그가 떠나기로 되어 있는 날 며칠 전에, 윌리스 씨의 세입자들 중의 한 사람이 그에게 전화해서 배관이 새고 있다고 말한다. 이런 상황에서, 윌리스 씨가 그 세입자에게 할 말로 가장 적절한 것은 무엇인가?

윌리스 씨: 이제부터는 건물 관리인에게 전화해야 할 거예요.

어휘
elderly 연세가 드신 rent 집세 decade 10년 relaxation 휴식 hire 고용하다 tenant 세입자 leak 새다 [문제] no longer 더 이상 ⋯ 아닌 fix 고치다
appreciate 진가를 알아보다

문제풀이

① 난 더 이상 그 건물의 소유주가 아니에요.

② 걱정하지 마세요, 제가 지금 당장 고치러 가겠어요.

③ 휴가가 끝난 후에 모든 것을 처리할게요.

⑤ 고맙습니다. 제 노고를 알아주다니 기쁘네요.

16 ② **17** ②

남: 어떤 직장에 지원할 때, 이력서는 잠재적인 고용주 여러분의 이력에 관해 알 수 있는 한 가지 방법입니다. 그러나, 많은 사람들이 커버

어휘
[문제] pick up ⋯을 태우러 가다

문제풀이

② 물론이지, 내일 널 태우러 올 수 있어.

③ 미안해, 너에게 줄 선물을 이미 샀어.

④ 린다가 저녁에 오라고 말했어.

⑤ 내일이면 난 파티에 갈 준비가 될 거야.

12 ④

남: 안녕, 클로이. 뭘 읽고 있니?

여: 전자책에 관한 잡지 기사야. 전자책은 아주 편리하고 다른 여러 장점이 있다고 쓰여 있어.

남: 전자책을 사용하는 것의 부정적인 면도 언급하니?

여: 응. 전자책이 네 시력을 손상시킬 수도 있다고 쓰여 있어.

어휘
handy 편리한 advantage 이점 mention 언급하다 downside 부정적인 면 [문제] focus on ⋯에 집중하다 damage 손상을 주다 eyesight 시력
replace 대체하다

문제풀이

① 아니. 난 종이책 읽는 걸 좋아해.

② 아니. 이 잡지는 내 것이 아니야.

③ 응. 우리는 그것의 장점에 더 집중해야 해.

⑤ 응. 전자책이 가까운 미래에 종이책을 대체할 거라고 쓰여 있어.

13 ⑤

여: 안녕하세요. 이게 다인가요?

남: 네, 이 인스턴트 커피 세 상자가 전부입니다.

여: 알겠습니다. 할인 카드 있으신가요?

남: 네, 여기요.

여: 감사합니다. 총 24달러입니다.

남: 16달러여야 하지 않나요? 저는 이게 '두 개 사면, 한 개가 무료'라고 생각했어요.

여: 아, 그렇군요. 이 한 상자는 진한 커피인데, 이건 더 비싸요. 부드러운 커피만 할인 중이에요.

남: 아, 저는 세 개가 모두 부드러운 커피인 줄 알았어요. 제가 실수로 하나 잘못 집어 온 거군요.

여: 괜찮습니다. 그걸 취소하고 부드러운 커피로 한 상자 더 계산할게요.

남: 감사합니다. 진열대에 가서 부드러운 커피를 하나 더 가져와도 될까요?

여: 네, 그리고 이 진한 커피 상자는 다시 가져다 놓아 주세요.

어휘
grab 붙잡다. 움켜잡다

문제풀이

① 괜찮아요, 제가 거스름돈을 드릴 수 있어요.

② 좋습니다. 화면 위에 서명하세요.

③ 고맙습니다, 도와주셔서 정말 감사합니다.

④ 아무래도 커피를 그렇게 많이 드시면 안 되겠네요.

레터 또한 (채용) 절차의 중요한 부분임을 잊습니다. 커버레터는 자신에 대한 소개이며, 일반적으로 고용주가 여러분에 관해 접하는 첫 번째 정보입니다. 커버레터는 당신이 어떤 직종에 지원하고 싶은지를 설명해야 합니다. 그런 다음 왜 당신이 그 자리에 적합한지에 대한 언급을 포함해야 합니다. 예를 들어, 당신은 당신이 가진 회사의 업무 환경에 유용한 기술이나 성격적인 특징을 설명할 수 있습니다. 회사 표창과 같은 특별한 업적이 있으면 그것도 언급돼야 합니다. 같은 이력서를 다수의 회사에 보내도 괜찮지만, 커버레터는 그렇게 하면 안됩니다. 대신, 여러분이 관심 있는 각 회사의 구체적인 특징을 강조하도록 (커버)레터를 맞추십시오. 고용주는 여러분이 지원서를 제출하기에 앞서 회사를 조사하는 데 시간을 들였는지 알고 싶어합니다.

어휘

résumé 이력서 potential 잠재적인 well-suited 적절한 personality 성격 trait 특성 achievement 업적, 성과 multiple 다수의 personalize (개인의 필요에) 맞추다 highlight 강조하다 specific 구체적인 prior to …에 앞서 application 지원서

DICTATION Answers
본문 p.170

01 made a name for / be signing copies / form an orderly line

02 distracting me from my schoolwork / how to manage / plenty of points

03 you advertise on TV / offer a special discount / the person in charge

04 a wine glass / in the center of / take up too much space

05 have escaped uninjured / you can drive / my life much easier

06 Put on your seat belt / get stuck in traffic / get there on time

07 must have happened / there's a delivery fee / pay in cash

08 released last month / sounds reasonable / see your driver's license

09 native speakers / last for two weeks / fun outdoor activities

10 leave in a couple of weeks / to meet other travelers / with a nice, big meal

11 get a present

12 mention any downsides

13 have a discount card / buy two, get one free / grabbed the wrong one

14 the copy machine / sent an email / inconvenient for everyone

15 the residents of his buildings / hired a building manager / a pipe is leaking

16~17 potential employers / well-suited for the position / such as company awards / prior to submitting an application

29 영어듣기 모의고사
본문 ▶ p.174

01 ①	02 ④	03 ②	04 ③	05 ②	06 ③
07 ④	08 ③	09 ②	10 ③	11 ④	12 ④
13 ①	14 ⑤	15 ⑤	16 ③	17 ⑤	

01 ①

[자동응답기가 울린다.]

남: 안녕하세요, 이 메시지는 론 워터맨 씨에게 드리는 것입니다. 여기는 온라인 서점 Book World입니다. 고객님과 연락이 닿지 않아서 메시지를 남깁니다. 지난주 서평자 모집에 응모해주셔서 감사합니다. 고객님께서 *The Tipping Point*에 대한 서평을 작성하도록 선정되셨다는 것을 알려 드리려고 전화했습니다. 6월 24일까지 저희 웹사이트에 서평을 작성해주세요. 그 책에 관한 고객님의 생각을 자유롭게 써주시면 됩니다. 서평은 적어도 200단어 길이는 되어야 한다는 점을 유념해주세요. 질문이 있으시면 080-262-5544로 저희에게 전화해주세요. 감사합니다.

어휘

submit 제출하다 entry 참가 등록 book reviewer 서평가 express (의견을) 나타내다, 표현하다 note 주목[주의]하다

02 ④

남: 어젯밤에 놀라운 기사를 읽었어. 그 기사에 따르면 과학자들은 지구에 추락한 외계 우주선을 발견했다고 하던데.

여: 그걸 정말 믿니?

남: 음, 당연히 처음엔 말도 안 되는 것처럼 보였지. 하지만 내가 인터넷에서 다른 기사들을 봤을 때, 진짜 UFO 목격에 관한 기사들이 많은 걸 발견했지.

여: 난 잘 모르겠다. 내 생각엔 그 사이트들 좀 위험한 것 같은데.

남: 정말? 난 재밌다고 생각했는데!

여: 너한텐 재밌겠지만 어떤 사람들은 그런 가짜 뉴스를 정말 믿는다고. 그건 우리 사회에 해로울 수 있어.

남: 네가 무슨 말 하는지 알겠다. 그리고 내 생각엔 그런 사이트들은 단지 광고에서 수입을 얻고 싶어하는 것 같아.

여: 맞아. 우리는 우리가 찾은 모든 뉴스 사이트들을 너무 빨리 믿어서는 안돼.

어휘

alien 외계의; 외계인 spaceship 우주선 crash (차량·운전자 등이) 충돌하다; *(항공기가) 추락하다 sighting (순간·찰나의) 목격 entertaining 재미있는 fake 가짜의, 거짓된 advertising *광고(하기); 광고업

03 ②

여: 안녕하세요. 대니얼 로한 씨인가요?

남: 네. 도와 드릴까요?

여: 이 사진에 있는 사람을 알아보시겠어요?

남: 아뇨, 본 적 없는 여잔데요.

여: 이 여자분이 당신이 자기 핸드백을 훔쳤다고 하는데요.
남: 전 그런 짓을 한 적이 없어요.
여: 어제 오후 2시에 시내에 계셨어요?
남: 아뇨, 여기 집에서 TV를 보고 있었어요.
여: 그걸 증명해 줄 사람이 있나요?
남: 글쎄요, 아내는 외출 중이었어요… 하지만 전 여기에 있었다고요!
여: 그 여자분은 당신이라고 확신하고 있습니다. 유감스럽지만 몇 가지 질문을 더 드리려면 당신을 데리고 가야 합니다. 저랑 같이 가시죠.
남: 제 변호사를 만나고 싶어요.

어휘
recognize 알아보다 claim 주장하다 verify 증명하다 positively 분명히
identify 확인하다, 알아보다

04 ③

남: 생물학 섹션에서 뭘 찾고 있니?
여: 세포 생물학에 관한 책을 찾고 있어. 사실 내가 찾는 책은 선반 꼭대기에 있어. 그걸 내리는 걸 도와줄 수 있겠니?
남: 그럼. 여기 있어.
여: 고마워. 있잖아, 웃기는 일이네. 맨 위 선반은 꽉 차 있는데 그 맨 위에서 두 번째 선반은 비어 있어.
남: 맞아. 책이 골고루 꽂혀야 할 것 같아.
여: 맞아. 그런데 넌 뭘 찾고 있어?
남: 유명한 생물학자에 관한 정보를 찾고 있는데 어디에서 찾아야 할지 잘 모르겠어.
여: 아마도 사서가 도와줄 거야. 그녀가 컴퓨터에서 작업하고 있지 않니?
남: 아니, 그녀는 큰 배낭을 멘 남자와 얘기하고 있어. 그도 질문할 게 좀 있나 봐.
여: 음, 너는 그녀가 그를 도와주는 일을 끝내면 물어보면 되겠구나.

어휘
biology 생물학 cellular 세포의 packed 꽉 들어찬 spread out 넓은 공간을 쓰다[차지하다] biologist 생물학자 librarian 도서관 사서

05 ②

여: 나 너무 스트레스받아.
남: 무슨 일이니?
여: 인터넷으로 전자 제품들을 찾고 있는데, 너무 어려워.
남: 왜 지금 그것들을 찾고 있어?
여: 이번 주말에 이사하는데, 내가 사용하는 가전 제품들이 전부 내 룸메이트 것이라서.
남: 용산에 있는 전자 상가에 가보는 게 어때? 거기 가격이 대체로 싸거든.
여: 거기 이미 가봤는데, 내가 기대했던 것보다 가격이 비쌌어.
남: 그럼, 중고 제품을 사는 건 어때?
여: 좋은 생각이긴 한데, 어디서 사?
남: 내가 좋은 가게를 알아. 내일 널 데려가 줄 수 있어.
여: 아, 고마워.

어휘
stressed out 스트레스가 쌓인 electronic 전자의 belong to …에 속하다
secondhand 중고의

06 ③

여: 살사 댄스 수업은 잘 받고 있니?
남: 사실은 그 수업을 그만 들을까 생각 중이야.
여: 뭐? 너 그걸 매우 좋아한다고 했고, 그래서 스페인어까지 배우기 시작했잖아!
남: 응, 여전히 살사를 배우는 건 좋은데, 내가 새로 온 살사 선생님인 마리아나에 대해서 말했던가?
여: 응, 너희 수업에 새로운 선생님이 오셨다고 했잖아. 선생님은 어때?
남: 너도 알다시피 지난번 선생님이었던 소피 선생님은 상당히 세심하고 배려가 있으셨거든. 안타깝게도 마리아나 선생님은 그렇지 않아.
여: 난 그분이 재능 있는 무용가이고 대회에서 우승도 많이 했다고 들었는데.
남: 응, 춤 실력은 아주 뛰어나. 하지만 선생님으로서는, 학생들을 별로 고려하지 않는 것 같아.
여: 무슨 뜻이야?
남: 선생님은 다른 학생들 앞에서 내 실수를 바로 지적해. 그건 너무 창피해. 그분은 인내심이 없는데, 사실 그건 선생님에게 정말 필요한 거잖아.
여: 아, 그건 정말 좋지 않구나. 다른 수업을 찾아보는 건 어때?
남: 그래야 할 것 같아.

어휘
quit 그만두다 attentive 세심한 considerate 사려 깊은 talented 재능이 있는 competition 경기, 시합 point out …을 지적하다 embarrassing 난처한, 쑥스러운 patience 인내심

07 ④

여: 실례합니다, 저는 세탁기를 찾고 있어요.
남: 음, 선택할 수 있는 다양한 엄선된 모델들이 있습니다. 마음에 두고 있는 특정한 것이 있으세요?
여: 저는 알레르기 케어 기능이 있는 것을 찾고 있어요.
남: 이것은 어떤가요? 이것이 인기 있는 알레르기 케어 기능을 가진 대용량의 드럼식 모델입니다.
여: 흠… 이건 2천 달러네요. 제가 내려고 한 비용보다 더 비싸요.
남: 음, 손님이 저희 매장의 신용카드를 신청하시면 25퍼센트 할인을 받으실 수 있습니다.
여: 나쁘지 않군요. 제가 또 여름 세일 10퍼센트 할인을 받을 수 있나요?
남: 아니요, 만약에 신용카드 할인을 선택하시면, 여름 세일 할인은 받으실 수 없습니다.
여: 알겠어요. 그래도 어쨌든 그것을 구매할게요. 신용카드는 어떻게 신청하면 되죠?
남: 이 양식만 작성하시면 됩니다.

어휘
a wide selection of 다양하게 엄선된 … specific 특정한 cycle 순환; 회전 capacity 용량 sign up for …을 신청하다 apply 적용하다 fill out …을 작성하다

문제풀이
여자는 2,000달러의 세탁기를 구매하려고 하는데 매장의 신용카드를 신청하여 25퍼센트의 할인을 받으므로, 여자가 지불할 금액은 1,500달러이다.

08 ③

남: 실례합니다. 저 목걸이에 관해 몇 가지 질문을 해도 될까요?
여: 물론이죠. 뭘 알고 싶으세요?
남: 음, 우선, 그건 어디에서 온 것이죠? 이집트 것인가요?
여: 네, 그래요. 한 탐험가가 이집트 피라미드에서 발견해서 박물관에 기증했어요.
남: 그렇군요. 금으로 만들어진 것 같은데요.
여: 그래요. 몇백만 달러의 가치가 있다고 여겨지죠.
남: 놀랍군요. 이집트 왕의 것이었나요?
여: 사실은 유명한 여왕인 클레오파트라의 것이었어요.
남: 그렇게 고가의 가치가 있다는 게 놀랄 일이 아니군요. 특별한 의미가 있나요?
여: 음, 그것은 특별한 종교 행사에만 착용 됐어요.
남: 아주 흥미롭네요. 질문에 답해주셔서 고맙습니다.
여: 천만에요. 나머지 전시도 즐겁게 보세요.

어휘
explorer 탐험가 donate 기증하다 made out of …으로 만들어진 no wonder …인 게 놀랍지 않다, 당연하다 ceremony 의식 exhibit 전시(회)

09 ②

여: 미군은 원격으로 비행 가능한 무인 항공기인 드론의 사용을 늘리고 있습니다. 전문가들은 지금으로부터 10년 후면 3만 대의 드론이 사용될 것으로 예측합니다. 비록 전쟁 지역에서 자주 볼 수 있지만, 모든 드론이 상대편 표적을 공격하는 것은 아닙니다. 다수는 드론에 부착된 비디오카메라로 정보를 수집하는 데 사용됩니다. 실제로, 아주 많은 정보가 나와서 미국 공군은 자료를 분석하는 데 약 7만 명을 고용하고 있습니다. 드론은 민간용으로도 사용될 수 있는지 점검을 받고 있습니다. 이것은 경찰관과 소방관, 응급의료 구조원들이 자신의 활동에 이 기술을 사용할 수도 있다는 것을 의미합니다. 좋든 싫든, 드론은 우리 생활의 일부인 것처럼 보입니다.

어휘
drone 드론 unmanned 무인의 aircraft 항공기 remotely 멀리서, 원격으로 opposition 반대; *반대측, 상대편 attached to …에 부착된 analyze 분석하다 civilian 민간인의 emergency medical responder 응급의료 구조원 operation 활동, 작전

10 ③

여: 그게 뭐야?
남: 자동차 회사의 팸플릿이야. 회사 월급이 올라서 새 차를 한 대 사려고.
여: 잘됐다. 어떤 종류의 차를 살 거야?
남: 음, 신형 Falconi로 결정했어.
여: 와, Falconi는 좋은 자동차지. 근데 뭐가 문제야?
남: 다섯 가지의 다른 모델이 있는데, 어떤 거로 할지 결정을 못 했어.
여: 오픈카를 원하니?
남: 음, 아니. 근데 에어컨은 있어야 해.
여: 알았어. 가죽 시트는?
남: 응, 난 가죽 시트가 좋아. 그리고 DVD 플레이어 말고 CD 플레이어가 있었으면 좋겠어.
여: 음, 그럼 저 모델을 사는 게 좋겠다.

남: 그렇네. 도와줘서 고마워.

어휘
raise 올리다; *(임금) 인상 decide on …로 결정하다 convertible 컨버터블(지붕을 열고 닫을 수 있는 차)

11 ④

남: 스마트폰을 사지그래, 루시?
여: 사실은, 곧 하나 사려고 해. 좋은 모델을 이미 찾아뒀어.
남: 정말? 어떤 건데?
여: 별로 비싸지 않고 화면이 커.

어휘
[문제] on one's way 가는 길에

문제풀이
① 전화기를 고치러 가는 길이야.
② 난 지금 그걸 쓰고 있는데 아주 만족스러워.
③ 난 괜찮아. 난 원할 때 언제든 그 전화기를 살 수 있어.
⑤ 난 인터넷을 검색했어. 너도 그것을 찾아봐야 해.

12 ④

여: 안녕, 존. 뭘 하고 있니?
남: 아, 상을 차리는 중이야. 우리 가족은 오늘 밤 정원에서 저녁 파티를 열 거야.
여: 멋진데. 파티에 누가 오니?
남: 삼촌의 가족이 올 거야. 우린 한동안 만나지 못했어.

어휘
set the table 식탁을 차리다 host (파티를) 주최하다, 열다 [문제] decorate 장식하다

문제풀이
① 내가 너희 정원을 장식해 줄게.
② 난 파티에 갈 수 있어. 거기에서 만나자!
③ 네가 도와주면 좋겠어. 이 탁자는 아주 무거워.
⑤ 아버지가 음식을 좀 사 오실 거야. 그가 요리도 하실 거야.

13 ①

여: 실례합니다. 우리 전에 만난 적 없나요?
남: 없는 것 같은데요.
여: 당신이 무척 낯익어요. 분명히 만난 적이 있는 것 같은데. 이름이 톰이죠, 그렇지 않나요?
남: 아뇨, 제 이름은 데이비드예요.
여: 아, 당신을 보니까 Greenville에서 만났던 톰이라는 남자가 생각나요.
남: 전 Greenville에 사는 남동생이 있어요. 근데 그 애 이름은 팀이죠.
여: 맞아요! 팀! 그는 전기 회사에 다녔어요.
남: 네. 그리고 커다란 검은색 개를 키웠죠?
여: 네, 커다랗고 털이 많은 검은색 개요.
남: 그 애가 바로 제 남동생이에요.
여: 이제야 이해가 되네요. 실수해서 죄송해요.
남: 늘 있는 일이에요. 우린 많이 닮았거든요.

문제풀이
② 언제 그에게 당신을 소개해 드릴게요.

③ 괜찮아요. 저도 당신을 기억하지 못했잖아요.

④ 괜찮아요. 제가 요즘 많이 달라 보이거든요.

⑤ 죄송하지만, 무슨 말씀을 하시는 건지 모르겠어요.

14 ⑤

남: 캐리, 나 큰일을 저질렀어!

여: 무슨 일인데?

남: 학교에서 노트북 컴퓨터를 빌렸는데 그걸 버스에 두고 내렸어!

여: 오, 저런! 가능한 한 빨리 그걸 찾아봐야겠네.

남: 알지. 책을 읽느라 좌석 밑에 노트북을 두었거든. 그러고 나서 갑자기
내가 내려야 할 정류장이라는 걸 깨닫고는 급하게 내려 버렸어!

여: 그리고 노트북에 대해선 잊어버렸구나.

남: 난 이제 어떻게 하지?

여: 봐. 여기 내 휴대전화가 있어. 버스 회사에 전화해. 그들에게 무슨 일
이 일어났는지를 말하고 분실물 보관소를 확인해 달라고 요청해 봐.

남: 만약 없다고 하면 어떡하지?

여: <u>그럼 (학교) 사무실에 가서 설명해야지.</u>

어휘
laptop 휴대용 컴퓨터 stop 정거장 lost and found 분실물 센터 [문제]
cheer up 힘내다 catch the bus 버스를 잡아타다

문제풀이
① 힘내. 다음엔 더 잘할 거야.

② 넌 몇 시에 버스를 타야 하니?

③ 그럼 넌 서둘러 공부해야겠구나.

④ 걱정하지 마. 내가 네 고장 난 컴퓨터를 고칠 수 있어.

15 ⑤

여: 조지는 반려묘 두 마리를 키운다. 그는 바쁜 일정에도 반려묘를 돌보
는 데 최선을 다한다. 출근하기 전과 퇴근한 후로 그는 그들에게 먹이
를 주고 그들과 놀아준다. 하지만 그는 고양이들이 더 많은 관심과 주
의가 필요하다는 것을 안다. 어느 날, 조지가 직장에서 집에 돌아왔을
때, 그의 고양이 중 한 마리가 없어졌다. 매우 바쁜 주였고 그는 매
우 피곤하다. 그는 어떻게 해야 할지 망설인다. 바로 그때, 문을 두드
리는 소리가 들린다. 조지가 문을 열자, 작은 소녀가 보인다. 그녀는 옆
집에 사는데 그의 고양이를 찾았다고 말한다. 그 소녀는 그가 집에 올
때까지 그 고양이와 놀아 주었다고 설명한다. 조지는 소녀에게 고마
워하지만, 그는 그녀가 그 고양이를 정말 좋아한다는 것을 알 수 있었
다. 이런 상황에서, 조지가 소녀에게 뭐라고 말하겠는가?

조지: <u>내가 집에 없을 때 내 고양이를 돌봐주겠니?</u>

어휘
take care of …을 돌보다 despite …에도 불구하고 feed 밥[우유]을 먹이다;
먹이를 주다 care 돌봄 knock 노크하다; *노크 (소리)

문제풀이
① 와, 넌 고양이들을 정말 좋아하는구나.

② 주소를 잘못 찾아온 것 같은데.

16 ③   17 ⑤

남: 주목해 주십시오, 여러분. 이 Cyrus 주민센터에 있는 초급 회화 수업
에 이렇게 많은 분들께서 신청해 주셔서 기쁩니다. 이 강좌는 여러분
이 초보자들을 위한 회화 기법을 배울 수 있게 해 드립니다. 알맞은
색과 붓 고르기부터 사실적인 그림자 만들기까지, 이 재미있고 긴장
을 풀어주는 취미를 시작하는 데 여러분이 아셔야 할 모든 것을 배울
것입니다. 수업은 석 달 동안 진행됩니다. 저희는 매주 화요일과 목요
일 오후 6시 30분부터 8시 30분까지 두 시간 동안 만날 예정입니다.
매 수업은 여기 203호에서 열립니다. 수업 이후에 더 남아서 자유로
이 그림 작업을 해도 좋습니다. 그러나, 오후 10시에 주민센터가 문을
닫는다는 점을 유념하셔서 그때까지는 건물을 나가셔야 합니다. 지불
하신 수강료에는 수업에 필요한 모든 물품의 가격이 포함되어 있습니
다. 그러니, 여러분이 가지고 오셔야 할 유일한 것은 상상력입니다. 오
늘 저녁 우리가 시작하기 전에, 여러분 모두 반 수강생들에게 이름과
등록하신 이유를 말씀해 주셨으면 합니다.

어휘
sign up (for) (…을) 신청하다 community center 주민센터, 동사무소
brush 붓 realistic 현실적인 shadow 그림자 note 주목[주의]하다
supply 《pl.》 보급품, 물품

DICTATION Answers   본문 p.176

01	submitting an entry / express your thoughts / feel free
02	an alien spaceship that crashed / it seemed crazy / what you mean
03	stole her purse / verify that / positively identified
04	on the top shelf / packed / working on the computer
05	stressed out / belong to my roommate / secondhand items / take you there
06	quitting the class / attentive and considerate / won many competitions / points out my faults
07	a wide selection of / sign up for / fill out this form
08	donated it to the museum / worth several millions dollars / during special religious ceremonies
09	could be in use / are used for gathering information / analyze the data
10	got a raise / decided on / five different models / air-conditioning
11	found a nice model
12	setting the table
13	look very familiar / remind me of / That explains it then
14	borrowed a laptop / got off the bus / the lost and found
15	take care of / feeds them / until he got home
16~17	run for three months / stay longer / exit the building / why you signed up

01 ④	02 ①	03 ②	04 ④	05 ③	06 ⑤
07 ③	08 ⑤	09 ⑤	10 ④	11 ④	12 ①
13 ①	14 ④	15 ①	16 ④	17 ①	

01 ④

여: 주목해 주십시오, 여러분. 여러분도 아시다시피 저희는 운동회 날을 맞아 여러 가지 행사를 마련했습니다. 그러나 시작하기 전에 경기의 안전에 관한 말씀을 드리고자 합니다. 팀을 나눈 후에, 혹시 있을지 모르는 부상을 방지하기 위해 제일 먼저 준비 운동을 함께 할 것입니다. 그리고 어떠한 경기도 거칠게 하면 위험할 수 있습니다. 다른 선수들을 발로 차거나 밀거나, 혹은 치지 마십시오. 우리는 재미있고 편안한 하루가 되어야 하는 날에 그 어떤 부상도 발생하지 않기를 바랍니다. 자, 1팀 학생들을 호명하겠습니다.

어휘
organize (행사 등을) 계획[준비]하다 safety 안전 divide A into B A를 B로 나누다 injury 부상 go through …을 겪다[경험하다] warm-up (경기 전의) 준비 운동 rough 거칠게, 난폭하게

02 ①

남: 저기에 있는 저 흰색 SUV 보여? 저게 내 꿈의 차야.
여: 농담이겠지.
남: 왜 그런 말을 해? 아주 멋진 차잖아. 난 아마 살 여유가 안 될 것 같지만.
여: 하지만 그게 너무 낭비라서 그래, 에릭. SUV는 크기가 너무 커서 일반 차량보다 훨씬 더 많은 휘발유가 필요해. 매우 쓸데없잖아.
남: 나는 동의하지 않아. SUV는 일반 차가 할 수 없는 일을 할 수 있어.
여: 비포장도로를 가거나 화물을 나르는 것 같은?
남: 그래, 바로 그거야.
여: 그런 용도로 쓰이는 SUV를 언제 봤니? 사람들은 SUV를 도시에서 운전하는 데 쓰잖아.
남: 음, 난 그러지 않을 거야. 난 외딴곳에서 캠핑을 하거나 거친 시골에서 운전할 용도로 내 차를 가지고 싶은 거야.

어휘
wasteful 낭비하는, 낭비적인 vehicle 차량 pointless 무의미한, 쓸모없는 off-road (포장된) 도로에서가 아닌 cargo 화물 remote 외딴 countryside 시골 지역, 전원 지대

03 ②

[문 두드리는 소리]
남: 안녕하세요, 여기 주문하신 것 있습니다. 다 있는지 확인해 보십시오.
여: 고맙습니다, 맛있겠는데요. 얼마를 드려야 하죠?
남: 3만 원입니다.
여: 여기 있습니다. 한 가지 말씀드리고 싶은데, 오늘 저녁에 주문할 때 당황스러웠어요.

남: 오, 왜 그러셨나요?
여: 주문을 받는 사람이 프랑스어밖에 못하더라고요. 잠시 제가 전화를 잘못 걸었나 했거든요.
남: 아 네. 전화 받을 사람이 없으면 그분이 전화를 받아요.
여: 이해해요. 제가 전화를 끊을 뻔했다는 것뿐이에요.
남: 제가 주인한테 얘기할게요.
여: 아뇨, 이제 어떻게 된 일인지 알게 되었으니까 괜찮아요.
남: 네, 하지만 다른 손님들도 당황하실 수 있으니까요.

어휘
owe 빚을 지다 come to (총계가) …이 되다 confused 당황한 hang up 전화를 끊다 mention 언급하다 now that …이니까, …이므로

04 ④

여: 이 영화 포스터들 멋지다!
남: 그래! 이게 다 1990년대의 유명한 영화야.
여: 그것들 중에 본 거 있어?
남: 정장을 입은 개미가 있는 걸 봤어. 넌?
여: 투수에 관한 이 영화가 눈에 익어. 그는 공을 던지려고 와인드업을 하고 있어.
남: 나도 그 영화를 봤어. 오, 이 영화가 상을 받았다고 들었어. 두 남자가 만리장성을 응시하고 있네.
여: 그 영화의 줄거리를 아니?
남: 아니, 하지만 보고 싶어. 넌 어떤 걸 보고 싶니?
여: 자신의 얼굴 앞에 손으로 하트를 만들고 있는 여자애가 있는 게 좋을 것 같아.
남: 난 우주선을 타고 날아가는 저 세 명의 외계인에 관한 걸 보고 싶어.

어휘
familiar 익숙한, 친숙한 pitcher 투수 wind up (투수가) 와인드업을 하다 award 상 stare at …을 응시하다 the Great Wall of China 만리장성 alien 외계인 spaceship 우주선

05 ③

여: 화요일 오전에 일찍 봉투를 가지고 내려와 주셔서 감사합니다.
남: 당연히 그래야죠. 정해진 날짜이기도 하고 전 그것들을 일찍 내다버리고 싶었거든요.
여: 단 한 가지 문제가 있다면, 병, 캔, 플라스틱류, 그리고 판지를 모두 한 봉지에 담으셨다는 거예요.
남: 그렇게 하면 문제가 되나요?
여: 음, 저희는 주민들께 그것들을 서로 다른 봉투에 분리해 달라고 부탁드리고 있습니다.
남: 알겠습니다. 제가 여기로 이사 온 지 얼마 안 돼서 몰랐어요. 죄송합니다.
여: 괜찮아요. 하지만 그것들을 분리해야 각자 적합한 공장으로 보내지거든요. 말하자면 종이들은 재생지 공장에, 플라스틱 병들은 플라스틱 공장으로 보내지는 거죠.
남: 다음번엔 꼭 기억할게요.
여: 좋아요. 재활용 쓰레기통은 받으셨나요?
남: 아뇨. 어떤 종류의 재활용 쓰레기통인가요?
여: 재활용품을 모으는 용도예요. 실내 혹은 실외에서 사용할 수 있죠. 하나 갖다 드릴게요.

남: 좋아요. 감사합니다.

어휘
assign (날짜·시간 등을) 지정하다 cardboard 판지, 보드지 resident 거주자, 주민 separate 분리하다 proper 적절한 plant 공장 recycling 재활용 bin 통, 쓰레기통 recyclable 재활용 가능한; *재활용품

06 ⑤

여: 실례합니다, 윌슨 선생님.
남: 그래, 리아? 오늘은 무슨 일 때문이지?
여: 내일 수학 시험이 있는데요…
남: 그래, 너 요즘 수학 때문에 애를 먹고 있더구나. 개인 지도가 필요할 것 같니?
여: 그건 생각해 봐야겠어요. 하지만 지금은 시험을 나중에 치러도 되는지 여쭤보려고 왔어요.
남: 아주 흔한 요청은 아니구나. 무슨 일이지?
여: 오빠가 내일 대학을 졸업하는데, 부모님께서 제가 꼭 졸업식에 참석하길 원하세요.
남: 음, 그건 아주 중요한 행사인 것 같구나. 그렇다면 좋아, 그럼 시험은 수요일 점심시간에 여기서 보는 거로 하자.
여: 고맙습니다, 윌슨 선생님.
남: 그래, 하지만 공부 열심히 해라. 쉽지 않을 거야.
여: 열심히 할게요. 오빠에게 도와 달라고 할 거예요!

어휘
tutor 개인 지도를 하다 for now 지금 당장은 common 흔한, 보통의 request 요청 attend 참석하다 ceremony 의식 fairly 상당히, 꽤

07 ③

남: 그 접시들이 마음에 드십니까?
여: 네, 접시 몇 개를 살까 생각 중인데, 아직 어떤 것을 사야 할지 결정하기 못했어요.
남: 저희가 폐업을 해서 많이 할인된 가격에 팔고 있어요.
여: 어머, 세상에. 이 접시들은 얼마인가요?
남: 세트당 300달러였어요. 그런데 이제 한 세트에 100달러밖에 안 해요. 한 세트에는 5개의 접시가 들어있어요.
여: 좋네요. 제가 원하는 색을 고를 수 있나요?
남: 네, 파란색만 빼고요. 파란색 접시는 세 개밖에 안 남아서, 세트로 팔 수가 없네요.
여: 아, 그런데 저는 그 접시들이 정말 좋아요. 개별적으로 사면 얼마죠?
남: 오늘은 개당 30달러입니다.
여: 음… 흰색 접시 세트를 100달러에 사고 싶어요. 그리고 아마도 파란색 접시도 사야겠네요.
남: 파란색 접시 세 개를 모두 사신다면, 개당 5달러씩 할인해 드릴게요.
여: 훌륭해요! 그럼 모두 살게요. 정말 고맙습니다.

어휘
plate 접시 consider 고려[생각]하다 close down (가게를) 폐쇄하다 individually 개별적으로, 각자 따로

문제풀이
여자는 흰색 접시 세트를 100달러에 샀고, 개당 30달러인 파란색 접시 세 개를 각각 5달러가 할인된 금액($25×3)에 샀으므로, 여자가 지불할 금액은 175달러이다.

08 ⑤

여: 내일 동아리 모임에 참석할 거지, 사이먼?
남: 잘 모르겠어. 오후 6시에 바이올린 강습이 있거든.
여: 그건 문제가 안 돼. 우린 늦어도 5시 30분이면 끝날 거야.
남: 좋아. 네가 이번 주 모임을 진행할 거지?
여: 응, 내 차례야. 의논할 게 몇 가지 있어.
남: 그래. 올해 축제 날짜를 정해야 해.
여: 그리고 새 회원을 끌어올 방법도 생각해야 해.
남: 현재 회원 12명 모두 거기 오니?
여: 사라와 토드 빼고 모두. 그들은 현장 학습에 가서 없을 거야.
남: 그렇구나. 그럼 평소처럼 211호에서 하는 거지?
여: 맞아. 반드시 오도록 해.
남: 걱정하지 마. 내일 보자!

어휘
at the latest (아무리) 늦어도 attract 끌어들이다, 유인하다 current 현재의 field trip 현장 학습 make sure to-v 반드시 …하도록 하다

09 ⑤

남: 만약 집에 뜰이 있다면, red thread disease에 주의하시는 게 좋겠습니다. 이 질병은 지금 전국에서 발견될 수 있습니다. 이것은 잔디를 공격하며, 잔디밭의 표면이 젖어 있고 잔디의 뿌리가 건조할 때 발생합니다. 너무 자주 물을 주는 것 또한 이 질병을 악화시킵니다. red thread disease의 첫 번째 증상은 흰색을 띠는 부분입니다. 이 부분들은 2인치에서 3피트 넓이까지 이를 수 있습니다. 시간이 흐르면서 이들이 이어져서 넓은 면적의 손상된 잔디를 형성할 수 있습니다. 손상된 잔디의 풀잎을 자세히 살펴보면, 여러분은 잔디의 잎을 가로지르는 작은 빨간 줄을 볼 수 있습니다. 만약 여러분의 잔디가 red thread disease를 갖고 있다고 생각한다면, 최고의 치료법은 질소를 함유하는 비료를 사용하는 것입니다. 그것이 여러분의 잔디가 회복하고 장래에 그 질병을 예방하는 데 도움이 될 것입니다.

어휘
thread 실 lawn 잔디, 잔디밭 symptom 증상 patch 부분, 조각 blade (칼·도구 등의) 날; *(한 가닥의) 풀잎 treatment 치료(법) fertilizer 비료 nitrogen 질소

10 ④

여: 이 웹사이트에 있는 반지 너무 예쁘다. 하나 주문할까 하는데.
남: 응, 그것들 멋진데. 금으로 주문할 거야, 아니면 은으로 할 거야?
여: 난 은을 정말 좋아해, 하지만 피부가 좀 가렵더라고. 그래서 금으로 살 거야.
남: 장밋빛 금이랑 노란빛 금 중에 어느 것을 선호해?
여: 난 둘 다 좋아. 하지만 금은 14캐럿이어야 해.
남: 왜 그래?
여: 난 그게 10캐럿보다 더 밝고 빛난다고 생각해.
남: 정말? 난 정말 잘 못 알아차리겠더라.
여: 진짜 그래. 그리고 난 반지 안쪽에 내 이름을 새기고 싶어.
남: 이 스타일들 중에 몇 개가 그 서비스를 제공하네.
여: 맞아. 그래서 난 이 두 개 중에서 결정해야 할 것 같아.
남: 너 빨리 사고 싶구나? 이게 더 빨리 배송된다.
여: 그러네! 그걸로 주문해야겠다.

어휘

itch 가렵게 하다 karat 캐럿(순금이 포함된 정도) notice 알아차리다 engrave (돌·쇠붙이 등에) 새기다 ship 수송하다 [문제] fashionable 유행하는

11 ④

여: 제이크, 내가 준 그 파일을 끝냈어요?
남: 아니요, 정말 미안해요, 브리아나. 아직 그것을 할 시간이 없었어요.
여: 오늘 오후 2시까지 끝내줘요.
남: <u>좋아요. 가능한 한 빨리 할게요.</u>

어휘

be done with …을 다 처리하다 [문제] cabinet 캐비닛, 보관장

문제풀이

① 좋아요, 그럼 파일을 내게 줘요.
② 당신의 파일 캐비닛은 이쪽에 있어요.
③ 이미 당신에게 그것을 보낸 것 같은데요.
⑤ 걱정하지 마세요. 그건 오늘 오후에 수리될 거예요.

12 ①

남: 이 피클이 아주 맛있네. 어디서 사니?
여: 특산품 매장에서 사. 프랑스에서 수입되는 거야.
남: 와, 언제 너랑 거기 같이 가고 싶어.
여: <u>좋아, 다음 주말에 널 거기 데리고 갈게.</u>

어휘

pickle 피클 specialty 특산품, 특제품 import 수입하다 [문제] take an order 주문을 받다 share 공유하다 recipe 레시피, 요리법 someday 언젠가, 훗날

문제풀이

② 음, 거기는 온라인으로 주문을 받지 않아.
③ 아, 네가 그 물품을 수입했는지 몰랐어.
④ 미안한데, 그 요리법을 공유할 순 없어.
⑤ 언젠가 프랑스로 여행 가고 싶어!

13 ①

여: 우리 올해 휴가는 어디로 갈까?
남: 난 베트남이나 괌 같은 열대 지방으로 가고 싶어.
여: 베트남에는 할 게 많다. 여러 곳들을 다 둘러보느라 바쁠 거야.
남: 사실 난 관광은 별로 하고 싶지 않아. 그냥 쉬고 싶어.
여: 그럼 괌이 더 좋겠네.
남: 너는 그것에 대해 어떻게 생각하는데?
여: 솔직히 말하면, 난 너무 덥지 않은 곳을 생각하고 있었어.
남: 그래, 네가 더위를 얼마나 못 참는지 내가 알지.
여: 하지만 괌은 작은 섬이니까 바다에서 상쾌한 바람이 불어올 거야.
남: 괜찮아. 더 시원한 곳으로 가자.
여: 아냐, 괌으로 가자. 난 네가 얼마나 바닷가에서 쉬고 싶어 했는지 알아.
남: <u>고마워! 네게 신세를 졌구나.</u>

어휘

site 현장, 장소 refreshing 심신을 상쾌하게 하는, 기운이 나게 하는 [문제] I owe you one. 고마워. 신세 졌어. know better than …할 만큼 어리석지 않

다 do not yield a single point 조금도 양보하지 않다 in the first place 애당초, 처음부터

문제풀이

② 그 정도는 알아.
③ 비용을 감당할 수 없을 것 같아.
④ 넌 조금도 양보하는 법이 없구나.
⑤ 난 처음부터 거기 가기 싫었어.

14 ④

여: 안녕하세요, 점보 칠리 치즈 핫도그 두 개 주세요.
남: 네, 몇 분이면 됩니다.
여: 얼마죠?
남: 총 5달러예요.
여: 여기 있습니다. 거기에 겨자와 케첩도 넣어 주시겠어요?
남: 물론이죠. 다른 건요?
여: 양파도 넣어주실 수 있을까요?
남: 그럼요. [잠시 후] 여기 있습니다. 칠리 치즈 핫도그 두 개입니다.
여: 고맙습니다. 오, 메뉴판에 quesadoga라고 쓰여 있는 걸 막 발견했는데요. 그게 정확히 뭔가요?
남: 오, 그건 저희 가게의 특별한 스타일의 핫도그 중 하나입니다. 매운 멕시칸 치즈와 매운 소스가 함께 뿌려져 있고 토르티야로 싼 핫도그죠.
여: <u>와! 다음번에 한 번 먹어봐야겠네요.</u>

어휘

jumbo 아주 큰 mustard 겨자 as well …도 역시 onion 양파 wrap 싸다, 둘러싸다 tortilla 토르티야

문제풀이

① 샌드위치도 하나 주실래요?
② 잠시만요, 전 핫도그 두 개만 주문했는데요.
③ 실례지만, 제 감자튀김을 잊으신 것 같은데요.
⑤ 제가 핫도그 위에 있는 치즈를 좋아할 것 같진 않아요.

15 ①

남: 매년 크리스마스가 되면 레이첼의 어머니는 크리스마스 만찬을 준비하고 식탁을 차리며, 식사 후에 뒷정리를 하신다. 레이첼과 그녀의 남동생, 그리고 그녀의 아버지는 가끔 치우기는 하지만, 대개 요리는 돕지 않는다. 올해는 레이첼의 어머니가 매우 편찮으셔서 한 번에 몇 분 정도밖에 서 있지 못하신다. 어머니는 모든 준비를 다 하려면 심한 스트레스를 받을 것이다. 레이첼은 올해는 자신이 크리스마스 만찬을 준비하기로 해서 그녀는 남동생에게도 도와달라고 부탁할 생각이다. 이런 상황에서, 이 사실을 알고 레이첼의 어머니가 레이첼에게 할 말로 가장 적절한 것은 무엇인가?
레이첼의 어머니: <u>우리 애들이 세상에서 최고야.</u>

어휘

set the table 식탁을 차리다 at a time 한 번에, 동시에

문제풀이

② 이번에는 아주 멋진 휴가가 될 거야.
③ 올해 크리스마스 만찬은 엉망이었어.
④ 너는 어떻게 도와줄 수 없다고 말할 수 있니?
⑤ 네가 작년처럼 요리를 도와준다는 이야기를 들으니 기분이 좋구나.

16 ④ 17 ①

여: 컴퓨터에 갑자기 전기가 끊기면 시스템에 해롭고, 또 당신은 작업하고 있던 중요한 파일을 잃게 될 수도 있습니다. 제너레이터-X가 당신을 위한 해결책입니다! 제너레이터-X는 당신의 컴퓨터에 15분 동안 추가 전력을 제공하는 배터리입니다. 이것은 당신에게 파일을 저장하고 컴퓨터를 제대로 끌 수 있는 시간을 줍니다. 그리고 제너레이터-X는 사용하기가 쉽습니다. 그것을 일반 콘센트에 꽂고 당신의 컴퓨터를 그 기기에 연결하세요. 정전 시 제너레이터-X는 배터리가 사용 중이라는 것을 나타내기 위해 삐 소리를 낼 것입니다. 그것이 사용자에게 문제를 알립니다. 전력이 복구되면 배터리는 자동적으로 충전됩니다. 이 놀라운 기기는 보통 100달러가 넘게 판매되지만, 저희는 단 74.99달러의 특가로 제공하고 있습니다. 저희 웹사이트인 www.generatorxsales.com으로 주문하러 오세요. 전국 백화점에서도 구하실 수 있습니다. 오늘 제너레이터-X를 사시고 다시는 컴퓨터 파일을 잃지 마세요!

어휘

shut down 끄다 properly 제대로 plug in 플러그를 꽂다 outlet 콘센트
device 장치 power outage 정전 beep 삐 소리를 내다 indicate 표시하다
alert 알리다, 경보를 발하다 restore 복구하다 recharge (재)충전하다

DICTATION Answers

본문 ● p.182

01	discuss sports safety / avoid possible injury / fun and relaxing
02	Because they're so large / carry cargo / camping in remote places
03	I owe you / placing the order / was about to hang up
04	winding up / making a heart / flying in a spaceship
05	the assigned day / residents to separate them / use it indoors or out
06	need some tutoring / at a later date / fairly important event
07	at greatly discounted prices / five plates in a set
08	at the latest / decide on a date / attract some new members
09	look out for / Watering too often / look closely at
10	makes my skin itch / I've never really noticed / get it quickly
11	finish it by 2(2:00)
12	They're imported from
13	go somewhere tropical / not quite so hot / refreshing wind
14	a couple of minutes / as well / What exactly is that
15	sets the table / be very stressful / cook Christmas dinner
16~17	shut down your computer / During a power outage / is automatically recharged / never lose a computer file

---

31 영어듣기 모의고사
본문 ● p.186

01 ①	02 ⑤	03 ③	04 ⑤	05 ②	06 ④
07 ③	08 ③	09 ①	10 ④	11 ④	12 ①
13 ②	14 ⑤	15 ④	16 ⑤	17 ③	

01 ①

남: 역사상 최초로 Glenmore 저수지 댐이 넘쳤던 수요일 저녁에 캘거리 인근 세 지역의 60명의 주민들이 대피를 준비했습니다. Elbow 강 하류에 밀려 들어온 물은 많은 집과 아파트를 위협했습니다. 남아 있던 주민들은 지하실이 진흙과 물로 차 있는 것을 보았습니다. 재산 피해는 수백만 달러가 될 수도 있습니다. 기상 예보관들은 최악의 사태가 더 올 것이라고 경고하고 있습니다. 천둥과 번개를 동반한 폭우와 강한 바람이 앞으로 며칠에 걸쳐 예상됩니다.

어휘

flee 도망치다 reservoir 저수지 overflow 범람하다 surge 큰 파도, 밀려듦
threaten 위협하다 face 직면[직시]하다 mud 진흙 basement 지하층
property 재산 thunderstorm 뇌우(천둥과 번개를 동반한 비)

02 ⑤

여: 광고하고 있는 이 비디오 게임을 봐.
남: 와… 표지에 피투성이 이미지가 있잖아.
여: 이런 게임이 아이들에게 팔리고 있다는 걸 믿을 수 있니?
남: 뭐? 나이 제한 같은 게 있어야 하잖아.
여: 그게 합리적인 것 같은데, 우리 나라에서는 비디오 게임 판매에 관한 규정이 없어.
남: 그건 몰랐어.
여: 이 게임의 표지를 보고 나서, 지역 정치인에게 편지를 쓸까 매우 고심 중이야.
남: 그래야 해. 나에게 아이가 있다면 이런 콘텐츠가 아이들에게 광고되는 걸 바라지 않을 거야.
여: 동의해. 그건 청소년들에게 너무 폭력적이야. 우린 그것에 대해 뭔가 해야만 해.

어휘

bloody 피투성이의 restriction 제한 sensible 합리적인 regulation 규정
politician 정치인 market 광고하다 violent 폭력적인 youngster 청소년

03 ③

남: 저희 결혼식을 계획하시는 데 비용이 얼마나 들까요?
여: 가격은 고객님이 필요한 서비스에 따라 달라집니다.
남: 기본 패키지에 어떤 게 포함되어 있죠?
여: 기본 패키지는 드레스를 선택하시는 걸 도와드리고, 훌륭한 음식조달 업체를 찾아 드리고, 그리고 결혼식장에 대한 모든 책임을 지는 것입니다.
남: 음, 솔직히 말해서 저희는 결혼식에 전념하기 위해 직장에서 휴가를 낼 여유가 없어요.

여: 그런 경우라면 기본 패키지에 추가해서 제가 꽃을 고르고, 장식을 하고, 그날 손님 접대를 도맡는 것을 포함하겠습니다.

남: 아, 좋습니다. 저희 결혼식을 모두 담당해주세요. 그럼 저희는 그것에 대해선 걱정할 필요가 없겠네요.

여: 그렇죠.

남: 알겠습니다. 그럼 모든 것이 다 포함된 패키지를 제공해주시고, 그날을 완벽하게 만들어주세요.

여: 물론이죠. 두 분이 만족하게 하도록 제가 최선을 다하겠습니다.

어휘
organize 준비[조직]하다 vary 다양하다 package 일괄계약 catering 음식 조달(업) take responsibility for …을 책임지다 in addition to …에 더하여 decoration 장식 handle 다루다 take charge of …의 책임을 지다 satisfy 만족시키다

04 ⑤

여: 정말 멋진 사진이구나!

남: 응, 마당에 있는 우리 가족이야.

여: 아버지가 아주 심각해 보이네. 왜 팔짱을 끼고 계시니?

남: 아, 말씀하실 때 항상 그러셔.

여: 그렇구나. 그럼 아버지 옆에서 다리를 꼰 채 앉아 있는 여성 분이 어머니야?

남: 아니, 그건 우리 큰누나야. 나이가 들어가면서 엄마를 똑 닮아가긴 하지. 엄마는 집으로 가는 길에 서 계셔. 꽃무늬 앞치마를 입고 있는 분이야.

여: 아, 어머니 미소가 좋으시다. 그리고 담장을 따라 줄지어 있는 저 꽃들이 아주 아름다워!

남: 동감이야. 그리고 문 옆에 둥근 화분 속 커다란 식물이 보이니?

여: 응, 보여.

남: 조부모님께서 특별 선물로 우리에게 그걸 주셨어.

어휘
yard 뜰, 마당 serious 심각한 have one's arms folded[crossed] 팔짱을 끼다 with one's legs crossed 다리를 꼰 채로 path 길 apron 앞치마 row 줄 fence 울타리, 담장 pot 화분

05 ②

남: 당신 이번 주말은 부모님 댁에서 보낼 거죠?

여: 왜요?

남: 당신 아버님 생신이잖아요, 아니에요?

여: 그래서요? 우린 이번 주말을 같이 보내기로 했잖아요.

남: 난 당신이 적어도 아버님 생신 저녁 식사에 갈 거로 생각했어요.

여: 아뇨, 난 잠깐 들러서 아버지 선물만 드릴 거예요.

남: 그리고 나서 우린 잭의 바비큐 파티에 가는 건가요?

여: 물론이죠. 하지만 먼저 내가 아버지 선물 사는 걸 도와줘요.

남: 그래요. 쇼핑몰에 가요. 지금 할인을 많이 하고 있어요.

여: 정말요? 잘됐네요! 지금 가요.

어휘
stop by …에 잠깐 들르다 drop off …을 내려놓다 mall 쇼핑몰

06 ④

여: 안녕, 브루스. 공원 갈 준비 됐어?

남: 아니, 준비 안 됐어.

여: 무슨 일이야? 너 지쳐 보여. 어젯밤에 늦게까지 안 잤어?

남: 아니, 11시에 잤어. 근데 새벽 5시에 깼어.

여: 토요일 아침 새벽 5시에? 네가 늦게까지 잘 줄 알았는데.

남: 그럴 계획이었지.

여: 자명종 꺼놓는 걸 잊어버렸어?

남: 아니, 자명종 때문에 깬 게 아니라, 이웃들 때문에 깼어. 새벽 내내 큰 소리로 싸우더라고.

여: 끔찍하다. 가서 조용히 해달라고 하지 그랬어?

남: 너무 피곤했거든. 그 대신에 그냥 새벽 내내 침대에 누워서 천장만 바라봤어.

여: 집에 가서 낮잠을 푹 자는 게 좋겠다. 공원은 내일 갈 수 있잖아.

어휘
turn off …을 끄다 alarm clock 자명종 argue 언쟁하다 go over 건너가다 lie 눕다, 누워 있다 (lay-lain) stare at …을 응시하다 ceiling 천장 take a nap 낮잠 자다

07 ③

남: 안녕하세요, 주문하시겠어요?

여: 안녕하세요. 가족용 식사를 주문하고 싶은데요.

남: 4인 가족용이요, 아니면 6인 가족용이요?

여: 차이점이 뭔가요?

남: 4인 가족용 식사는 10달러입니다. 6인 가족용 식사는 15달러입니다.

여: 음, 전 아이 세 명에 남편이 있어요.

남: 그럼, 4인 가족용 식사에 개당 60센트짜리 비스킷 두 개를 추가로 주문하시는 걸 추천합니다. 아이들이 대개 저희 비스킷을 좋아하거든요.

여: 음료는 어떻게 되나요? 4인 가족용 식사에 음료는 몇 개가 나오죠?

남: 4개가 나오니까 추가로 1달러를 내시고 하나 더 주문하세요.

여: 알았어요, 그렇게 할게요. 고마워요.

어휘
difference 차이 suggest 제안하다 extra 여분의, 추가의 biscuit 비스킷

문제풀이
여자는 4인 가족용 식사($10)와 60센트짜리 비스킷 두 개($0.6×2), 그리고 음료수 한 개($1)를 추가로 주문했으므로, 여자가 지불할 금액은 12.20달러이다.

08 ③

[전화벨이 울린다.]

남: 여보세요. 크리스입니다.

여: 안녕하세요, 크리스. 전 철인 3종 경기 클럽의 제인입니다.

남: 네, 안녕하세요.

여: 내일 있을 철인 3종 경기에 대해 알려 드리려고 전화했어요.

남: 당신 전화를 기다리고 있었어요. 이번에 처음으로 경기에 참여하거든요. 궁금해서 그러는데, 얼마나 많은 사람이 참여하나요?

여: 100명 정도 참여할 거예요. 음, 시작 시각이 오전 6시니까, 늦어도 5시 40분까지는 월드컵 경기장에 도착하셔야 해요.

남: 알겠습니다. 경기가 바로 시작되나요?

여: 스트레칭을 하는 시간이 잠깐 있지만, 네, 6시 정각에 시작될 거예요.

남: 그리고 수영이 먼저 시작되고요, 맞죠?

여: 네, 맞아요. 그 후에 자전거 타기와 마라톤이 있어요.

남: 알겠습니다. 정보 감사해요.

여: 천만에요.

어휘

triathlon 철인 3종 경기 inform 알리다 no later than 늦어도 …까지는
right away 곧바로 set aside for (…에 대비해) 챙겨두다 on the dot 정각
에 be followed by …가 이어지다[따라오다]

09 ①

여: 여러분 안녕하십니까, New City 동물원에 오신 것을 환영합니다. 이
곳은 주에서 가장 큰 동물원이며, 나라에서는 두 번째로 큰 동물원입
니다. 이곳은 500종이 넘는 수천 마리의 동물들을 보유하고 있습니
다. 동물원을 돌아다니는 세 가지 일반적인 방법이 있습니다. 보행자
용 길을 따라 걷거나, 안내를 받는 버스 투어를 하시거나, 머리 위 높
이에 위치한 곤돌라 중의 하나를 타시면 됩니다. 이런 모든 선택에 대
한 더 많은 정보는 바로 저쪽 식당 옆에 위치한 안내소에서 이용하실
수 있습니다. 동물원을 대표해서, 여러분께 몇 가지 기본적인 주의사
항들을 지켜주실 것을 요청합니다. 첫째, 미리 허가를 받지 않았다면
동물들에게 먹이를 주지 마십시오. 둘째, 플래시를 터뜨리며 사진을
찍지 마십시오. 그리고 셋째, 쓰레기는 항상 휴지통에 버려주십시오.

어휘

feature 특색으로 삼다 species (분류상의) 종(種) pedestrian 보행자
pathway 좁은 길 gondola 곤돌라 overhead 머리 위에 on behalf of
…을 대표해서 guideline 지침 permission 허가, 허락 trash can 휴지통

10 ④

남: 이 광고 봤어? 새 학교에서 전화나 실시간 영상 채팅으로 영어 수업을
제공한대.

여: 진짜 편할 것 같아! 하나 신청하려고?

남: 응, 하지만 어느 것이 나에게 가장 좋을지 잘 모르겠어.

여: 음, 기초반은 너무 쉬울 거야. 중급반이나 고급반이 너에게 맞을 거
야.

남: 동의해. 그리고 난 이미 비즈니스 영어 수업을 들은 적이 있어.

여: 그래? 음, 다른 수업들도 모두 좋아 보여.

남: 응, 그래 보여. 하지만 난 실시간 영상 채팅을 그다지 사용하고 싶지는
않아. 내 인터넷 연결이 충분히 빠르지 않거든.

여: 그렇구나. 그리고 일주일에 몇 번이나 수업을 듣고 싶어? 난 세 번이
가장 좋을 거 같은데.

남: 사실, 난 그건 너무 많은 거 같아. 요즘 일정이 꽤 바쁘거든.

여: 알겠어. 그러면 이 수업이 가장 나은 선택 같아 보여.

남: 응, 그러네. 빨리 시작하고 싶다!

어휘

advertisement 광고 offer 제공하다 convenient 편리한

11 ④

남: 실례합니다. 여기가 박물관으로 가는 입구인가요?

여: 아니요, 입구는 저쪽이에요.

남: 아, 고마워요. 그리고 몇 시인가요?

여: 4시 되기 몇 분 전이에요.

어휘

entrance 입구 have the time 시간이 있다; *몇 시인지 알다 [문제] awfully
정말, 몹시 plenty of 많은

문제풀이

① 네, 그 소식을 들으니 유감이네요.

② 유감스럽게도 오늘 제가 몹시 바빠요.

③ 물론이죠, 지금 시간이 많아요.

⑤ 오늘 그 입구가 개방되지 않는 것 같네요.

12 ①

여: 유학 가는 게 흥미로울 거로 생각하니?

남: 응, 그것에 대해 많이 생각해봤어. 그런데 그게 나를 위한 것인지는
모르겠어.

여: 왜 그렇게 생각하니?

남: 향수병이 너무 심할까 봐.

어휘

study abroad 유학하다 [문제] homesick 향수병을 앓는

문제풀이

② 부모님들은 유학을 다녀오셨어.

③ 응, 함께 공부하려면 오늘 밤에 만나자.

④ 미안하지만, 난 도서관에서 혼자 공부하는 걸 선호해.

⑤ 사실, 그걸 생각해 볼 기회가 없었어.

13 ②

남: 당신 팀장님한테 휴가 가는 것에 관해서 물어봤나요?

여: 아니요. 우린 아직 언제 휴가를 갈지도 안 정했잖아요, 그렇죠?

남: 아직 안 정한 것 같네요. 너무 바빠서 생각할 수가 없었어요.

여: 저도 마찬가지예요. 이게 절 스트레스받게 하는 참이에요. 그냥 지금
정하는 게 어때요?

남: 좋아요. 5월은 어때요?

여: 안돼요, 너무 일러요. 6월 15일 이후여야 해요. 6월 중순까지 바쁠 예
정이거든요.

남: 알겠어요. 그럼 7월 첫째 주 어때요?

여: 전 좋아요.

남: 훌륭해요. 그 주에 휴가 쓰는 것에 대해서 우리 둘 다 팀장님한테 물
어봅시다. 그러면 우린 항공권이랑 호텔을 예약할 수 있을 거예요.

여: 좋아요. 기분이 한결 낫네요.

남: 네, 우리가 드디어 계획해서 기쁘네요.

어휘

stress … out …을 스트레스받게 하다 [문제] overtime 초과근무, 잔업 take
time off 휴가를 내다

문제풀이

① 휴가를 갈 수 있게 해줘서 고마워요.

③ 미안하지만, 당신은 초과 근무를 해야 돼요.

④ 저도요. 정말 편안한 휴가네요.

⑤ 그럼 당신은 휴가를 낼 필요가 없겠네요.

14 ⑤

남: 여보세요. 수잔이에요?
여: 네, 제가 수잔인데요.
남: 수잔, 전 찰스 킴입니다. 정말 미안한데 오늘 모임에 못 나갈 거 같아요.
여: 왜요? 무슨 일 있나요?
남: 음, 아내가 지금 병원에서 진통 중이에요.
여: 정말요? 당신 부인이 임신 중이었다는 것조차 몰랐네요.
남: 네, 임신 중이에요. 그래서 우리 모임을 다시 조정할 수 있을까요?
여: 물론이죠. 당신은 부인과 함께 있어야죠.
남: 고마워요. 그리고 오늘 정말 미안해요.
여: 괜찮아요. 가족보다 중요한 건 없죠.

어휘

make it (모임 등에) 가다, 참석하다 in labor 진통 중인 pregnant 임신 중인 reschedule 일정을 변경하다 [문제] spoil 망치다

문제풀이

① 제가 당신의 하루를 망친 게 아니었으면 좋겠네요.
② 저도 모든 것에 대해 죄송해요.
③ 난 당신이 임신한 걸 이미 알고 있어요.
④ 당신이 지금 내게 이렇게 하다니 믿을 수가 없군요.

15 ④

남: 안드레아가 새 아파트로 이사 온 지 9개월이 되었다. 그녀의 옆집 이웃은 매일 밤 쓰레기를 복도에 내놓는다. 그는 젊은 대학생인데 청소부가 그것을 치울 거로 생각하는 것 같다. 안드레아는 그의 행동을 보고 충격을 받지만, 참으려고 노력한다. 하지만 날이 갈수록 안드레아는 점점 더 화가 난다. 그녀는 젊은이가 그렇게 하는 것을 그만두기를 바랐지만, 그는 지금까지 그만두지 않는다. 마침내 그녀는 그에게 대응하기로 한다. 이런 상황에서, 안드레아는 그녀의 이웃에게 뭐라고 말할 것인가?
안드레아: 복도에 쓰레기를 내놓아서는 안 돼요.

어휘

hallway 복도 behavior 행동 patient 참을성이 있는 day by day 날이 갈수록 confront 직면[대면]하다 [문제] garbage 쓰레기 hallway 복도 treat 대접하다

문제풀이

① 우리가 이제 좋은 친구가 되어서 기뻐요.
② 당신 말이 맞는 것 같아요. 다시는 그러지 않을게요.
③ 산에서 당신이 버린 쓰레기를 치워야 해요.
⑤ 언제든 들러 주세요. 제가 멋진 저녁을 대접할게요.

16 ⑤ 17 ③

여: 안녕하세요, 학생 여러분. 이번 주에는, 앤톤 반 레벤후크에 관한 보고서를 쓸 것인데, 그는 1676년에 박테리아를 발견한 사람입니다. 그래서 보고서를 위해, 여러분들은 박테리아가 무엇인지 알아야 할 필요가 있을 것입니다. 여러분 중 다수가 박테리아는 바이러스와 같다고 생각할 거라고 여겨지는데요. 그러나 사실은 그렇지 않습니다. 바이러스는 세포가 전혀 없는 반면에, 박테리아는 (세포가) 두 개로 분열하여 번식하는 단세포 생물입니다. 바이러스가 박테리아보다 훨씬

작고 단순해서, 반 레벤후크는 현미경을 통해 바이러스가 아닌 오직 박테리아만 볼 수 있었습니다. 박테리아가 나쁘게 들릴지 모르겠지만, 일부 박테리아는 사실 우리에게 이롭습니다. 박테리아는 두 개로 분열하여 번식하는데, 그것은 그들이 스스로 번식할 수 있기 때문에 어떤 해도 끼칠 필요가 없다는 것을 의미합니다. 반면에, 바이러스는 세포에 침입하여 더 많은 그들의 DNA를 만드는 데 그것들을 사용하며, 그것이 대부분의 바이러스들이 해로운 이유입니다. 보고서를 쓰기 전에 박테리아와 바이러스가 무엇인지 이해하는 것을 확실히 해두세요. 앤톤 반 레벤후크에 대해 조사함으로써 여러분들은 전반적인 생물학뿐만 아니라 박테리아에 대해 많은 것을 배울 거라 확신합니다.

어휘

bacteria 박테리아 suspect 의심하다; *…라고 여기다 virus 바이러스 single-celled 단세포의 organism 유기체, 생물(체) reproduce 번식하다 split 나뉘다, 분열하다 microscope 현미경 harm 피해, 손해 invade 침입하다 biology 생물학 in general 보통; *전반적으로 [문제] spread 펼치다; *퍼지다, 확산시키다

문제풀이

16 ① 최초의 박테리아
② 어떻게 바이러스가 확산하는가
③ 일부 바이러스의 이점
④ 박테리아가 인간에 미친 영향
⑤ 바이러스와 박테리아의 차이점

DICTATION Answers 본문 p.188

01	prepared to flee / reservoir dam overflowed / faced with mud / still to come
02	is being sold to kids / any regulations on / strongly considering writing
03	organize our wedding / take full responsibility / be concerned with it / to satisfy you
04	have his arms folded / with her legs crossed / along the fence
05	at your parents' place / drop off his present / a big sale going on
06	woke up / turn off / arguing loudly / take a long nap
07	take your order / What's the difference / come with / an extra dollar
08	inform you / taking part in it / set aside for stretching
09	more than 500 species / take a guided bus tour / take photographs with a flash
10	sign up for one / internet connection isn't fast / have a pretty busy schedule
11	entrance to the museum
12	to study abroad
13	haven't even decided / too busy to focus / ask our bosses for
14	can't make it / in labor / reschedule the meeting
15	moved into / leaves trash out / stop doing it

surrounding (*pl.*) 주위(의 상황) figure out 파악하다 fit 맞다 get away 도망치다 mood 기분

32 영어듣기 모의고사

본문 ▲ p.192

01 ③	02 ②	03 ②	04 ③	05 ②	06 ⑤
07 ③	08 ②	09 ④	10 ②	11 ①	12 ③
13 ③	14 ②	15 ③	16 ④	17 ④	

01 ③

여: 이 소설은 우리에게 19세기 미국인들의 생활 모습을 잘 보여 줍니다. 제가 이 책에서 정말 마음에 들었던 부분은 작가가 등장인물들을 전개해 나가는 방식이었습니다. 주인공의 삶이 제 삶과 전혀 달랐지만 저는 그에게 정말로 동화될 수 있었습니다. 이 책의 주제는 보편적이어서 나이, 문화, 혹은 배경과 관계없이 누구에게나 재미있는 읽을거리가 될 것입니다. 서로 다른 이야기들을 왔다 갔다해서 가끔은 줄거리의 구조가 약간 실망스럽기도 했습니다. 하지만 이야기 전체가 결말에 가서 연관되었을 때 모든 것이 다 의미가 있었습니다.

어휘
give insight into …에 대한 통찰력을 주다 author 저자 leading 주요한 identify with …와 동일시하다 theme 주제 universal 일반적인, 보편적인 regardless of …에 관계없이 background 배경 structure 구조 plot 줄거리 frustrating 좌절감을 주는, 실망시키는 at times 가끔은, 때로는 back and forth 앞뒤로; *이리저리 tie together 연관시키다 worthwhile 가치 있는

02 ②

여: 네 고양이의 수염이 아주 사랑스러워, 토드.
남: 입 주변의 털 말이니? 아주 쓸모 있기도 해.
여: 수염이 실제로 어떤 데 쓰인다는 거니?
남: 고양이는 주변 환경에서 길을 찾는 데 그것들을 활용해.
여: 아, 그럼 수염을 통해 주위의 상황에 관한 정보를 얻는 거야?
남: 맞아. 그리고, 수염이 고양이 몸체만큼 넓은 거 보이니?
여: 응. 그 이유는 뭔데?
남: 고양이가 어떤 장소를 지나갈 수 있는지 아닌지를 파악하기 쉽게 해 주거든. 급하게 도망치려고 할 때 유용하지.
여: 그 밖에 또 어떤 데 사용되니?
남: 고양이의 기분을 나타내주지. 기분이 좋으면 수염은 안정되어 있어. 하지만, 봐봐. 지금 수염을 뒤로 당기고 있어. 네가 고양이를 화나게 했나 봐!

어휘
whisker (고양이 · 쥐 등의) 수염 adorable 사랑스러운 navigate 길을 찾다

03 ②

[전화벨이 울린다.]
남: Great Home Supplies입니다. 무엇을 도와드릴까요?
여: 음, 제가 며칠 전에 그쪽 웹사이트에서 그릇을 두 개 주문했는데 어제 도착했어요. 근데 제가 포장을 풀어보니 그중 하나가 깨져 있더라고요.
남: 아, 정말 죄송합니다. 아마 배송 중에 문제가 있었던 것 같네요. 제가 즉시 처리해 드리겠습니다. 사용자명이 어떻게 되시죠?
여: 'smash1'이에요.
남: [타자치는 소리] 네, 포트메리온 그릇을 두 개 주문하셨네요.
여: 맞아요. 그럼 제가 어떻게 하면 되죠?
남: 저희 배달원이 내일 새 제품을 가지고 댁에 방문할 겁니다. 깨진 것은 그분에게 주세요.
여: 알았어요. 아, 제가 배송비를 또 내야 하나요?
남: 아뇨, 내지 않으셔도 됩니다. 그리고 저희 웹사이트의 '나의 페이지' 섹션에서 배송 상태를 확인하실 수 있어요.
여: 알겠어요. 감사합니다.

어휘
unpack 풀다, 끄르다 shipment 선적; 수송 handle 다루다, 처리하다 delivery fee 배송비 status 상태

04 ③

여: 안녕, 마이크. 재미있게 놀고 있어요?
남: 네, 여기가 정말 마음에 들어요.
여: 새라는 어디 있어요?
남: 그 애는 꽃무늬 수영복을 입고 수건 위에서 쉬고 있어요. 아주 귀여워 보여요, 그렇죠?
여: 네, 맞아요. 크리스와 댄은 어때요?
남: 그 애들은 저기 물속에서 공을 갖고 놀고 있어요.
여: 오. 몇몇 사람들이 오리배를 타고 있는 게 보이네요.
남: 그러네요. 저 배들은 재미있어요. 하나 타 보세요. 계곡 건너편 나무의 경치가 숨 막히게 아름다워요.
여: 그러네요. 하지만 저기 물에 왜 밧줄이 떠다니나요?
남: 수영하는 사람들을 얕은 물에 있게 하려는 거예요. 밧줄을 뜨게 하는 공도 멀리서 보일 만큼 커요.
여: 안전을 위한 좋은 생각이네요.

어휘
swimsuit 수영복 valley 계곡, 골짜기 breathtaking (너무 아름다워서) 숨이 막히는 float 뜨다, 떠다니다 shallow 얕은 from a distance 멀리서

05 ②

남: 아, 큰일 났네! 어떡하지?
여: 무슨 일이야?
남: 오늘 저녁에 여자친구를 영화관에 데려간다고 약속했는데, 아직 이 일을 다 못 끝냈어!
여: 지금 그녀한테 전화해서 못 갈 거라고 말해.

남: 일 때문에 또 취소하면 그녀는 화를 낼 거야! 이번이 세 번째거든. 네가 도와줄 수 있을까?

여: 글쎄, 엄마를 만나기 전에 널 도와줄 시간이 조금 있어. 내가 뭘 하면 돼?

남: 이 보고서들 좀 복사해 줄래?

여: 그래, 그건 쉽네. 다른 건?

남: 없어. 네가 복사해 주면, 내가 내일 아침에 사장님이 오시기 전에 스테이플러로 철하면 돼.

여: 좋아, 할 수 있어. 시간 좀 봐! 넌 지금 출발하지 않으면 늦을 거야!

남: 도와줘서 고마워! 다음 주에 점심 살게!

어휘
make a photocopy 복사하다 staple (종이·서류 등을) 철하다 [문제] copy 복사하다; 복사본 bind 묶다, 철하다

문제풀이
① 그와 함께 영화 보러 가기　② 보고서 복사하기
③ 영화표 예매하기　④ 복사물을 철하기
⑤ 그에게 점심 대접하기

06 ⑤

남: 여행 어땠어, 캐서린?

여: 정말 좋았어! 이집트 피라미드를 구경하고, 낙타를 타고 사막을 여행했어.

남: 와! 정말 좋았겠다!

여: 좋았어, 근데 거기에 있는 관광 명소에 약간 실망했어.

남: 무슨 말이니?

여: 글쎄, 너 피라미드 옆에 쇼핑몰이 있다는 사실을 알고 있었어?

남: 아니, 정말이야? 난 역사 유적지는 그 자체의 가치를 위해서라도 보존되어야 한다고 생각해.

여: 응! 그리고 지역 음식보다 패스트푸드 체인점이 더 많이 있었어.

남: 저런, 너무하는군!

어휘
fabulous 멋진, 굉장한 on camelback 낙타를 타고 tourist attraction 관광 명소[명물] preserve 보존하다 awful 심한, 끔찍한

07 ③

남: CD를 다 파시려고요?

여: 네, 별로 자주 듣지 않아서요.

남: 전 새로운 음악을 듣고 싶은데요. 클래식 기타 음악이 있나요?

여: 네, 저쪽을 한번 보세요.

남: 아, 여기 있네요. 이건 얼마죠?

여: 그건 한 개에 6달러인데, 클래식 음악 CD를 전부 다 사시면 한 개당 4달러 50센트에 드릴게요.

남: 클래식 CD가 모두 몇 개인데요?

여: 7개요.

남: 좋아요. 전부 다 살게요. 전 클래식 음악을 정말 좋아하거든요! 여기 제 신용카드요.

여: 현금으로 지불하시면 개당 50센트 할인해 드릴게요.

남: 안타깝지만, 현금이 하나도 없어요. 카드로 결제해야겠네요.

여: 알겠어요.

어휘
classical 고전의; 클래식의 all together 모두 다 in cash 현금으로

문제풀이
남자는 한 개당 4.5달러에 클래식 음악 CD 7개를 모두 사겠다고 했고, 현금 결제 시 할인을 받을 수 있지만 현금이 없어서 신용카드로 결제해야 하므로, 남자가 지불할 금액은 31.50달러($4.5×7)이다.

08 ②

남: 이 일이 실제로 일어났다는 게 믿어지지 않아.

여: 뭐에 대해서 이야기하는 거니?

남: 1937년에 스페인 내전 동안 발생한 부당한 공격에 대한 기사야. 독일 폭격기들이 스페인의 한 마을을 세 시간 동안 공격했어.

여: 세 시간! 정말 끔찍했겠다.

남: 끔찍했어. 거의 1,700명이 사망했어. 전쟁사에서 매우 중요한 사건이야.

여: 정말? 왜 그렇지?

남: 그건 공군이 얼마나 강력할 수 있는지를 보여줬거든. 폭격기는 그 마을을 쉽게 파괴했어. 그 후, 많은 나라가 미래에 자신들의 마을을 지킬 수 있도록 공군을 개혁하려 노력했지.

여: 정말 중요하구나.

남: 가장 충격적인 것은 표적이었어. 그건 한 국가가 군인들이 없는 도시를 공격한 최초의 사건이었어.

여: 그렇구나. 사람들이 틀림없이 정말 화가 났겠다.

남: 응, 대중들은 분노로 반응했고, 오늘날에도 사람들은 여전히 그 공격에 분노하고 있어.

어휘
article 기사 unfair 부당한 attack 공격 civil war 내전 bomber 폭격기 significant 중요한, 의미심장한 air force 공군 destroy 파괴하다 afterwards 그 뒤에 defend 방어하다 target 표적

09 ④

남: 다음 주 금요일에 시작해서 6월 내내 매일 밤 공연되는 클래식 뮤지컬 *미녀와 야수*가 시 예술 센터에 옵니다. 공연은 매일 밤 7시로 예정되어 있으며, 주말에는 오후 1시에 시작하는 추가 오후 공연이 있습니다. 입장권은 성인 50달러, 학생은 30달러이며, 예술 센터의 웹사이트에서 온라인으로 예매할 수 있습니다. 매 공연이 시작하기 전에 CD, 셔츠, 기타 관련 품목들이 로비에서 판매될 것입니다. 공연은 오리지널 출연진들과 다양한 국제 유명인사들이 출연하며, 우리 지역 출신의 최고 탤런트 몇 명이 함께 합니다. 출연진들에 대한 최신 정보를 원하신다면, 저희 사이트를 주목하십시오.

어휘
nightly 밤마다 하는 be scheduled for …로 예정되어 있다 additional 추가적인 related 관계있는 cast 등장인물, 출연진 celebrity 유명인사 keep an eye on …을 주목[주시]하다

10 ②

여: 안녕하세요. 전 당신의 꽃꽂이 수업 중 하나를 등록하고 싶은데요.

남: 물론입니다. 저희 여름 일정이 바로 여기 있어요. 올해는 특별 일일 수업도 제공합니다.

여: 그건 제가 찾고 있는 게 아니에요. 저는 적어도 2주는 지속되는 수업을 듣고 싶어요.

남: 좋습니다. 저희는 더 오랫동안 진행되는 수업도 있어요. 그것들은 일주일에 1회, 2회, 혹은 3회 모입니다.

여: 일주일에 3회요? 그건 너무 잦아요. 1회나 2회가 좋습니다.

남: 알겠습니다. 그럼, 가장 관심이 가는 꽃꽂이 종류가 있으신가요?

여: 저는 모든 것에 다 관심이 있어요. 오, 잠깐만요. 저 화환 수업은 이미 수강했었어요.

남: 그러면 그건 다시 하고 싶지 않으신가요?

여: 그렇습니다. 그리고 이 가격들은 뭔가요? 저는 당신의 수업들이 무료라고 생각했거든요.

남: 맞습니다만, 꽃값은 내셔야 해요. 어떤 수업들은 더 비싼 꽃들을 사용합니다.

여: 오, 저는 비싼 꽃을 써보고 싶어요. 그것들이 더 매력적이라고 생각해서요.

남: 맞습니다, 보통은 그렇죠. 제 생각엔 이 수업이 당신에게 딱 맞을 것 같네요.

flower arranging 꽃꽂이 (= flower arrangement) last 지속하다 free 무료의 attractive 매력적인 [문제] wreath 화환, 화관

11 ①

여: 생일 깜짝 선물 받을 준비 됐니?

남: 와! 강아지다! 난 항상 강아지를 원했어.

여: 그를 뭐라고 부르고 싶니?

남: 럭키라고 부르자.

어휘
name 이름; 이름을 지어주다

문제풀이
② 내 친구는 멋진 개를 가지고 있어.
③ 아직 그의 이름을 몰라.
④ 내일 그에게 전화할 거야.
⑤ 내 생일에 새 자전거를 원해.

12 ③

남: TV에서 뭘 보고 있니?

여: 이건 큰 축구 시합이야. 우리 팀이 이기길 바라.

남: 그들이 이길 수 있다고 생각해? 지고 있는 것 같은데.

여: 음, 아직 시간이 좀 남아 있어.

어휘
lose 지다, 패하다 (↔ win) [문제] rather 다소, 좀

문제풀이
① 곧 새 TV를 사고 싶어.
② 하지만 난 축구를 상당히 잘해.
④ 축구는 나에겐 좀 지루한 것 같아.
⑤ 큰 야구 시합은 잠시 뒤에 시작해.

13 ③

남: 무슨 일이야, 수잔?

여: 내 컴퓨터가 작동하지 않는데 할 일이 많아!

남: 내가 고칠 수 있을지도 몰라. 컴퓨터에 무슨 문제가 있는 거니?

여: 모르겠어. 이 프로그램을 열 때마다 컴퓨터가 꺼져.

남: 사용하고 있는 메모리 용량이 얼마인지 알아봤어?

여: 그건 어떻게 하면 되는데?

남: 음, 여길 대고 마우스 오른쪽 단추를 클릭해.

여: 그래, 큰 원이 있고 절반이 파란색으로 되어 있어.

남: 어, 그럼 메모리에는 문제가 없는데.

여: 내 컴퓨터에서 뭐가 문제인지 찾을 방법이 또 없을까?

남: 미안하지만 난 도울 수가 없구나. 그 컴퓨터 회사의 전화 상담 서비스에 전화해 보는 게 어때?

여: 좋은 생각이야. 전화번호를 찾아봐야겠다.

어휘
shut down 문을 닫다; *(기계가) 멈추다 help line 전화 상담 서비스 [문제] be infected with …에 감염되다

문제풀이
① 걱정하지 마. 난 괜찮아.
② 컴퓨터를 끄는 게 좋겠다.
④ 내 컴퓨터가 바이러스에 감염됐어.
⑤ 왜냐하면, 내 컴퓨터에 문제가 있어서.

14 ②

여: 아시아에 관한 다큐멘터리 괜찮았니?

남: 응, 정말 좋았어.

여: 넌 아시아 여행을 생각해 본 적 있어?

남: 응. 난 정말 태국에 가고 싶어.

여: 그건 좀 어려울 거야. 넌 태국어를 못하잖아.

남: 태국 사람들은 영어 공부를 아주 열심히 해! 난 거기서 영어로 말할 거야.

여: 그거 좋은 생각이다! 같이 가자!

남: 좋아. 태국 사람들과 쉽게 대화할 수 있도록 태국어 강의도 미리 듣자.

여: 가까운 언어 교육원에서 새로운 태국어 강의가 있다고 들었어.

남: 전화해서 정보를 좀 얻어보자.

어휘
documentary 다큐멘터리 Thailand 태국 Thai 태국어; 태국의

문제풀이
① 나도 태국 음식을 좋아해.
③ 넌 어제 그들에게 전화했어야 해.
④ 외국인들과 소통하는 건 어려워.
⑤ 영어 강좌를 듣는 게 어때?

15 ③

여: 미셸은 그녀의 동료 몇 명과 점심을 먹고 있다. 돈을 내야 할 때, 그녀는 지갑에 돈이 없다는 것을 알게 된다. 그녀는 매우 당황스럽고 돈을 좀 빌려야 한다. 그녀는 돈이 있었다고 확신해서 무슨 일이 있었는지를 알아내려고 노력한다. 그녀는 자기의 십 대 남동생이 그 전날 돈을 빌려달라고 부탁했지만, 그녀가 안 된다고 했던 게 갑자기 생각난다. 그녀는 화가 나서 동생에게 전화하고 동생이 그녀의 지갑에서 돈을 가져갔다고 비난한다. 그가 부인하지만 그녀는 그를 믿지 않고 그에게

소리를 지른다. 그런데 그녀는 사무실로 돌아오자 동료의 선물을 사느라고 돈을 쓴 것이 기억난다. 그녀는 전화기를 들고 사과를 하기 위해 그녀의 남동생에게 전화한다. 이런 상황에서, 미셸은 그녀의 남동생에게 뭐라고 말을 할까?

미셸: 미안해. 내가 엄청난 실수를 했어.

어휘

coworker 동료 embarrassed 당황스러운 accuse A of B A를 B로 비난[고소]하다 colleague 동료

문제풀이

① 나에게 돈을 빌려줘서 고마워.

② 내일 나에게 돈을 갚아도 돼.

④ 넌 좀 더 신중해야 했어.

⑤ 거짓말하지 말라고 내가 몇 번이나 말했니?

16 ④ 17 ④

남: 저는 유명한 사진을 하나 여러분들께 보여드리고 그것이 보여주는 역사에 대해 좀 이야기하려고 합니다. 이것은 뗏장집 앞에 있는 가족의 사진입니다. 정착민들이 북미의 대평원에 도착했을 때, 그들은 집을 짓기에 충분한 건축 자재를 가지고 있지 않았습니다. 그래서 대신에 그들은 잔디와 흙을 사용했습니다. 이런 집들은 '떼를 쌓아 만든 집' 혹은 '뗏장집'이라고 알려져 있습니다. 그들은 1900년대 초반에 처음 나타났습니다. 그 당시에 북미에 백만 채 이상의 뗏장으로 만든 집들이 지어졌던 것으로 추정됩니다. 뗏장집은 겨울에는 따뜻하고 여름에는 시원했으나, 몇몇 단점도 있었습니다. 뗏장집은 종종 매우 습했습니다. 그리고 사진을 보시면, 지붕 위에 소가 한 마리 있는 것을 눈치채실 겁니다. 정말로, 뗏장집들이 잔디로 만들어졌기 때문에, 배고픈 소들은 때때로 잔디를 먹기 위해 뗏장집에 올라갔습니다. 소들만 뗏장집에 끌리는 유일한 동물은 아니었습니다. 방울뱀도 종종 벽에 집을 만들었습니다. 1900년대 초반 뗏장집에서의 삶은 꽤 불편했을 것입니다.

어휘

soddie 뗏장집 (= sod house) settler 정착민, 개척자 the Great Plains 대평원 soil 흙 estimate 추정하다 disadvantage 불리한 점, 약점 damp 축축한, 눅눅한 rattlesnake 방울뱀 [문제] futuristic 시대를 앞서는, 초현대적인 eco-friendly 친환경적인 immigrant 이민자, 이주민

문제풀이

16 ① 소를 위해 지어진 집
② 초현대적인 친환경 주택
③ 북미 이주민의 역사
④ 북미 정착민이 지은 주거 형태
⑤ 1990년대 초 지하에 거주한 사람들

DICTATION Answers

본문 p.194

01 gives us insight / developed the characters / regardless of age

02 navigate through their environment / as wide as / fit through a certain space

03 was broken / a problem during shipment / pay the delivery

fee

04 in her flower-pattern swimsuit / in swan boats / from a distance

05 haven't finished this work / cancel again / make photocopies / staple them together

06 traveled through the desert / tourist attractions / next to the pyramids / more than local foods

07 look over there / for $4.50 each / take all of them

08 an unfair attack / update their air forces / responded in anger

09 throughout the month of June / are scheduled for / before each show / our best local talent

10 what I'm looking for / you're most interested in / use more expensive flowers

11 your birthday surprise

12 my team wins

13 isn't working / shuts down / help line

14 the documentary / study English hard / communicate easily

15 it's time to pay / accuses him of taking / spent the money on

16~17 enough building material to / were known as / were made of grass / be quite uncomfortable

33 영어듣기 모의고사

본문 ▲ p.198

01 ⑤	02 ①	03 ⑤	04 ②	05 ②	06 ④
07 ③	08 ③	09 ⑤	10 ④	11 ②	12 ①
13 ④	14 ⑤	15 ④	16 ③	17 ②	

01 ⑤

남: 수업이 끝나기 전에 중요한 주제에 대해 잠시 한 말씀 드리겠습니다. 여러분 모두 알다시피, 이번 겨울에 퍼진 굉장히 위험한 형태의 독감이 돌고 있습니다. 그것에 걸리거나 남에게 전염시키는 것을 피하는 가장 좋은 방법 중의 하나는 하루에 여러 번 비누로 손을 씻는 것입니다. 또한 입과 코 위로 마스크를 착용하는 것도 좋은 방법입니다. 모든 사람들은 독감에 걸릴 위험이 있지만, 우리가 많은 다른 사람들과 온종일 긴밀히 접촉하는 장소인 붐비는 학교와 같은 곳에서는, 이 위험이 더욱 커집니다. 그러므로 우리 자신과 다른 사람들을 보호하기 위해서 우리 모두 최선을 다합시다.

어휘

address 연설하다; *(…에게 직접) 말을 하다 flu 독감 go around 돌다; *(병 등

이) 퍼지다 at risk of …의 위험이 있는 magnify 확대하다

02 ①

남: 내가 Smith 예술 대학에 입학 허가가 났다는 걸 방금 알았어.
여: 축하해! 거긴 정말 일류 예술 대학이잖아.
남: 내가 최고로 바랐던 대학교야. 하지만 입학 허가를 받고 나니, 다시 생각하게 돼.
여: 무슨 말이니?
남: 그게 단지… Smith 대학에서는 다른 분야의 공부를 계속할 수 있는 선택권이 없을 거야.
여: 그렇지. 그런데 예술을 공부하고 싶으면—그리고 네가 그렇다고 알고 있는데—더 좋은 곳은 없잖아.
남: 알아.
여: 꿈꾸던 학교를 포기한다면 나중에 네 선택을 후회하게 될 거야.
남: 네 말이 맞는 것 같아.
여: 게다가, Smith 대학에 가서 마음에 들지 않으면, 언제든 다른 학교로 옮길 수 있어.

어휘
prestigious 명망 있는, 일류의 pursue 추구하다; *계속하다 regret 후회하다
transfer 옮기다

03 ⑤

남: 네, 다시 돌아왔습니다. 쇼의 이 시점에서 전화를 받아야겠네요. 여보세요?
여: 네, 안녕하세요, 라이언.
남: 안녕하세요! 전화해 주셔서 감사합니다.
여: 전 매일 아침 운전해서 사무실에 출근하는 동안 당신의 프로그램을 듣는답니다. 정말 멋져요.
남: 감사합니다. 하지만 운전 중에 전화하시는 게 아니었으면 좋겠군요! 위험하니까요!
여: 아뇨, 아녜요. 전 지금 주차장에 있어요. 하지만 노래를 신청하고 싶어요.
남: 물론이죠. 특별한 누군가에게 보내고 싶으신 건가요?
여: 네. 전 제인 존스의 'You're the Best'를 듣고 싶어요. 그리고 그 노래를 당신에게 바칩니다.
남: 저요? 왜죠?
여: 왜냐하면 당신이 매일 아침 제 얼굴에 미소를 띠게 해주니까요.
남: 그렇게 말씀해주시니 정말 좋네요!

어휘
parking lot 주차장 request 요청하다

04 ②

여: 너 도넛 먹을래? 맛있어 보인다.
남: 어, 그러네. 그중에 가루 설탕이 뿌려진 초콜릿 도넛 하나를 먹을까 봐.
여: 저 남자의 쟁반에 있는 것과 같은 별 모양 도넛도 두 개를 사자.
남: 그것들도 좋아 보이네. 우린 쟁반이 필요할 것 같아.
여: 맞아. 어디 있는데?
남: 판매대의 오른쪽 끝에 있어.

여: 좋아. 오! 표지판 좀 봐! 도넛을 하나 더 가져가야겠는걸.
남: 네 말이 맞아. 세 개 가격으로 네 개의 도넛을 살 수 있어.
여: 잘됐네. 그럼 저 여자가 가리키고 있는 것들 중 한 개를 먹는 게 어때?
남: 소매 없는 웃옷을 입은 여자? 그래.

어휘
sprinkle (쿠키 등에) 잘게 뿌려진 초콜릿·설탕 tray 쟁반 counter 계산대; *판매대 point at …을 가리키다 sleeveless 소매 없는 top 상의

05 ②

남: 난 이 지역을 잘 모르겠어.
여: 아까 거기서 좌회전했어야 해.
남: 어디에서?
여: 2킬로미터 전쯤에서 말이야. 내가 운전하게 했어야지.
남: 왜 아무 말도 안 했어? 이제 파티에 늦겠다.
여: 네가 다른 길을 아는 줄 알았어.
남: 음, 우린 길을 잃은 것 같아.
여: 길을 물어볼 만한 사람이 안 보이네. 그냥 차를 돌려서 반대 방향으로 가.
남: 하지만 아까부터 방향을 대여섯 번은 바꿨잖아. 난 돌아가는 길이 정확하게 생각이 안 나.
여: 걱정하지 마. 내가 길을 기억하니까.
남: 다행이다! 고마워.

어휘
recognize 알아보다 area 지역 turn around 방향을 바꾸다 the other way 반대로, 거꾸로 turn 회전, (차량의) 방향 전환

06 ④

여: 너 왔구나, 마이클. 왜 이렇게 늦었니?
남: 정말로 미안해, 나탈리. 얼마나 오래 기다리고 있었니?
여: 30분 정도야.
남: 오늘 정말로 힘든 날이었어. 그리고 전화기를 집에 두고 와서 네게 전화할 수가 없었어.
여: 그걸 가지러 집에 갔니?
남: 아니. 그럼 훨씬 더 늦었을 거야.
여: 오는 길에 무슨 일이 있었니?
남: 응. 자동차 추돌 사고를 목격했어.
여: 그거 끔찍하군.
남: 응. 그리고 목격자 진술을 하려고 경찰이 오기를 기다려야 했어. 막 네게 전화를 하려던 참에 전화기를 집에 두고 온 걸 알았지.
여: 그렇구나. 신경 쓰지 마.

어휘
witness 목격자 statement 진술

07 ③

남: 안녕하세요. Sayana 호텔에 오신 것을 환영합니다. 예약하셨나요?
여: 아뇨, 예약은 안 했습니다. 가격이 어떻게 되는지 궁금했어요.
남: 어떤 종류의 방을 찾고 계십니까?
여: 글쎄요, 저 혼자뿐이에요. 그래서 싱글룸이면 됩니다.
남: 알겠습니다. 기본 싱글룸은 일 박당 20달러입니다. 디럭스룸은 일 박

당 50달러이고요.

여: 기본 싱글룸과 디럭스룸 사이의 차이점이 무엇이죠?

남: 디럭스룸은 거품 욕조와 작은 주방을 포함하고 있어요.

여: 저는 그것들 모두 필요가 없을 것 같네요. 그냥 기본 방으로 할게요.

남: 알겠습니다. 그리고 얼마나 묵으실 겁니까?

여: 나흘 밤을 머무를 거예요. 아, 조식이 포함되어 있나요?

남: 아니요. 하루 10달러에 조식을 추가하실 수 있습니다.

여: 알겠습니다. 조식을 추가할게요, 하지만 내일 하루만요. 다른 날에는 계획이 있거든요.

어휘
rate 속도; 비율; *요금 deluxe 고급의 Jacuzzi 자쿠지(물에서 기포가 생기게 만든 욕조) stick with …을 계속[고수]하다

문제풀이
여자는 일 박당 20달러인 기본 싱글룸에 나흘간 머물 예정이고, 일 박당 10달러인 조식은 하루만 추가하기로 했으므로 여자가 지불할 금액은 90달러이다.

08 ③

남: 질문이 있어, 미나. 우주로 간 최초의 한국인이 누구였지?

여: 그녀의 이름은 이소연이야. 1978년에 태어난 과학자야.

남: 아! 여성이었어? 그건 몰랐어.

여: 응. 그녀는 2008년에 국제 우주 정거장에 갔지.

남: 흥미로운데. 한국이 고유의 우주선을 갖고 있니?

여: 아직 없어. 그녀는 러시아 우주선의 승객이었어.

남: 그렇구나. 우주에는 얼마나 오래 있었니?

여: 그녀는 다양한 실험을 하면서 우주 정거장에서 열흘을 보냈어.

남: 어떤 종류의 실험을 했니?

여: 무엇보다도, 낮은 중력이 식물과 곤충에 미치는 영향을 연구했어.

남: 와, 그녀는 대단하구나.

어휘
space station 우주 정거장 spacecraft 우주선 passenger 승객 conduct 실시[수행]하다 experiment 실험 effect 영향, 효과 gravity 중력

09 ⑤

여: 무지개 공원의 버스 관광에 오신 것을 환영합니다. 무지개 공원은 휴고 스미스 씨에 의해 설립된 유서 깊은 공원입니다. 휴고 스미스 씨가 처음 이곳에 왔을 때 많은 무지개를 봤기 때문에 이곳을 무지개 공원이라고 이름 지었습니다. 여러분께서 표를 구입하실 때 나눠 드린 지도를 보시면 이 지역에서 가볼 만한 명소들의 목록이 나와 있습니다. 그리고 안내소에서 얻을 수 있는 소책자에는 유용한 쿠폰들이 들어 있습니다. 이 투어는 약 한 시간 반이 소요될 것입니다. 우리는 약 30분 동안 차를 타고 공원 주위를 돌아볼 것입니다. 그리고 나서 사진도 찍으시고 쿠폰을 사용할 수 있는 장소들을 방문하실 수 있도록 여러분께 한 시간 동안의 자유 시간을 드릴 것입니다. 즐거운 관광이 되시길 바랍니다.

어휘
historical 역사적인 hand out …을 나누어 주다 attraction 명소 booklet 소책자

10 ④

여: 얘, 데이비드. 여기서 뭐 하고 있어?

남: 안녕, 제시카. 나 좀 도와줄래? 조카한테 줄 게임팩을 사고 싶은데 컴퓨터 게임에 대해서는 전혀 모르거든.

여: 컴퓨터 게임? 그 방면으로는 내가 거의 전문가지. 어디 보자… [잠시 후] 축구 게임은 어때? 요즘 대부분의 아이들이 축구를 좋아하거든.

남: 괜찮을 것 같은데, 사실 걔는 차에 푹 빠져 있어.

여: 그러면 이 자동차 경주 게임을 좋아하겠네. 그건 심지어 조이스틱도 줘.

남: 어디 보자. 와, 50달러가 넘네. 비싸다… 50달러 안쪽으로 샀으면 했는데.

여: 좋아. 이건 어때? 40달러야.

남: 그런데 그것의 등급을 봐. 내 조카는 겨우 열 살이야.

여: 네 조카가 그렇게 어린 줄은 몰랐네. 다른 걸 골라줄게… 이거야! 분명히 네 조카가 마음에 들어 할 거야. 그런데 그건 조이스틱이 없어.

남: 괜찮아. 아마 조이스틱은 내가 나중에 사주면 될 거야.

어휘
nephew 조카 expert 전문가 crazy 열중한, 열광적인 rate 등급을 매기다

11 ②

남: 시험은 금요일에 있을 거예요. 책 전체를 다루게 될 거예요.

여: 질문이 있어요. 시험에 논술 문제도 포함되나요?

남: 네, 빈칸 채우기 문제와 몇 개의 논술 문제가 있을 거예요.

여: 각각 몇 개가 될까요?

어휘
cover 다루다, 포함하다 entire 전체의 blank 빈; *빈칸 essay question 논문식 문제, 논술형 문제

문제풀이
① 이 빈칸들을 채워 주실 수 있나요?
③ 시험 결과가 언제 나왔나요?
④ 제 시험 성적에 만족하지 않아요.
⑤ 축하해요! 논술에서 A를 받았어요.

12 ①

여: 어디 가니, 브랜든?

남: 봉사활동 동아리 모임에 가는 중이야.

여: 오, 네가 그 동아리에 가입했는지 몰랐네. 무슨 일을 하니?

남: 노인들과 아이들을 도우려고 일해.

어휘
on the way to …에 가는 중인 [문제] senior 고령자, 노인

문제풀이
② 모임은 내일 밤 7시에 있을 거야.
③ 난 동아리 회원들과 다른 반에 있어.
④ 난 너와 같은 반에 있어. 날 못 봤니?
⑤ 사실 사진 동아리에 들까 생각 중이야.

13 ④

남: 안녕, 마리아. 공부하러 우리 집에 와줘서 고마워.

여: 날 불러줘서 고마워.

남: 여기로 와, 우린 거실에서 공부해도 돼. 엄마가 간식거리를 좀 만들어주셨어.

여: 좋아. *[잠시 후]* 와! 이 오래된 앨범들 좀 봐.

남: 응, 그 앨범들은 우리 아빠 거야. 아빠가 70년대의 클래식 앨범을 수집하시거든.

여: 난 옛날 록 음악에 푹 빠져있는데. 넌 어때?

남: 나도 좋아해. 우리 집안 내력인 것 같아.

여: 믿을 수가 없어. 이것 좀 봐. 이건 비틀즈의 최고 앨범이잖아.

남: 나도 그렇게 생각해. 공부하는 동안 그걸 들을까?

여: 그럼 주의가 산만해질 것 같아. 하지만 내가 나중에 그걸 빌릴 수 있을까?

남: 아마 그럴 거야. 하지만 아빠한테 허락을 받아야 해.

어휘

be into …을 좋아하다, …에 관심이 많다 run in one's family …의 집안 내력이다 distracting 정신을 산란[산만]하게 하는 [문제] permission 허가, 허락 release 발매, 출시

문제풀이

① 난 공부할 때 음악 듣는 걸 좋아해.

② 응, 난 그걸 저 길가의 음반 가게에서 샀어.

③ 고맙지만 괜찮아. 난 이미 비틀즈 앨범을 모두 다 갖고 있거든.

⑤ 물론이지, 하지만 새로 발매된 것에 대해서는 특별 연체료가 있어.

14 ⑤

[문 두드리는 소리]

여: 네? 무슨 일이시죠?

남: 안녕하세요, 부인. 13B호에 피자 배달 왔는데요.

여: 뭔가 착오가 있나 보군요. 전 피자를 주문하지 않았어요.

남: 음. 여기가 13B호인가요?

여: 네, 맞아요. 하지만 전 분명히 피자를 시키지 않았어요.

남: 이상하군요. 메모를 확인해 보고 누가 주문 전화를 했는지 볼게요.

여: 그러세요.

남: 에릭 피터슨이라는 분이 주문하셨어요. 남편분 성함이 에릭인가요?

여: 아뇨. 근데 에릭이라는 이름의 열두 살짜리 아들은 있어요.

남: 이런. 누가 피자를 시켰는지 알 것 같군요.

여: 죄송해요. 그 애는 항상 말썽을 일으키죠.

어휘

[문제] bother 괴롭히다 get into trouble 말썽을 부리다

문제풀이

① 바보 같으니, 제가 깜박했나 봐요.

② 절 좀 그만 괴롭히고 가세요.

③ 괜찮아요. 누구나 실수하는 법이니까요.

④ 남편이 배가 고팠나 봐요.

15 ④

남: 존은 초등학교 학생이다. 그는 아주 예의 바르고 공부도 열심히 하지만, 욱하는 성미가 있다. 어느 날 존은 필통 때문에 한 학생과 말다툼을 했다. 그 학생은 그 필통이 자기 것이라고 말했다. 존은 그것이 자신의 것이라고 주장했다. 그들은 잠시 소리를 질러댔고, 그러다가 존이 그 학생을 모욕했다. 교장 선생님은 존을 교무실로 데리고 가서 어

떻게 된 일인지 물었다. 존은 설명하면서 교장 선생님께 필통 안쪽에 쓰여 있는 자신의 이름을 보여드렸다. 이런 상황에서, 교장 선생님이 존에게 할 말로 가장 적절한 것은 무엇인가?

교장 선생님: 그 애가 잘못하긴 했지만, 문제를 싸움으로 해결해선 안 되지.

어휘

elementary 초등학교의 polite 예의 바른 bad temper 욱하는 성질 argument 말다툼 belong to …의 것이다 insist 주장하다 yell 소리치다 for a while 잠시 동안 insult 모욕하다 [문제] steal 훔치다 solve 해결하다

문제풀이

① 왜 그 안에 네 이름이 안 쓰여 있니?

② 모두 네 잘못이라는 걸 인정한다니 기쁘구나.

③ 다른 학생의 필통을 훔치는 건 나빠.

⑤ 왜 네 친구를 교무실로 데리고 왔는지 설명해주겠니?

16 ③ 17 ②

여: 오늘, 우리는 잔지바르에 대해 배워 볼 것입니다. 잔지바르는 인도양의 섬들의 무리입니다. 그것은 아프리카의 동쪽 해안에 위치한 나라인 탄자니아의 일부입니다. 잔지바르에 대해 들어보셨다면 그곳을 스쿠버 다이버들에게 훌륭한 여행지로 만드는 아름다운 해변과 멋진 산호초들에 대해 당신은 아마도 알고 있을 겁니다. 그러나 그곳을 매우 특별한 곳으로 만드는 것은 그것뿐만이 아닙니다. 게다가, 잔지바르는 흥미로운 역사를 가지고 있습니다. 19세기에 잔지바르는 정향을 생산하기 시작하여 이윽고 세계에서 가장 큰 정향 생산국이 되었습니다. 그들은 또한 생강, 계피, 그리고 후추를 재배하기 시작했습니다. 이러한 이유로 잔지바르는 '향신료의 섬'으로 불리곤 했습니다. 19세기에는 향신료가 매우 가치가 있었기 때문에, 잔지바르는 꽤 부유해졌습니다. 그러나, 다른 나라들이 그들만의 향신료 산업을 개발하기 시작하자, 잔지바르의 시장점유율이 하락했습니다. 그러나 그곳은 방문할 만한 멋진 장소로 남아있습니다. 그러니 여러분들이 그곳을 여행할 정도로 운이 좋다면, 그곳의 해변을 즐길 동안 향신료 산업의 중심지로서의 그곳의 과거를 기억해주세요.

어휘

gorgeous 아주 멋진, 아름다운 coral reef 산호초 clove 정향 ginger 생강 cinnamon 계피 pepper 후추 spice 향신료 share 몫, 지분 decline 감소[하락]하다 [문제] geographic 지리학적인 feature 특징 crisis 위기, 고비 neighboring 이웃의, 근처의

문제풀이

16 ① 잔지바르의 관광 명소들

② 잔지바르의 지리학적 특징

③ 잔지바르의 향신료 사업의 역사

④ 잔지바르의 경제적 위기의 원인

⑤ 주변국과 잔지바르의 관계

DICTATION Answers

 본문 p.200

01 washing your hands with soap / at risk of getting / protect ourselves and others

02 got accepted / give up on / transfer to another school

03	listen to your show / request a song / is for you
04	two star-shaped ones / for the price of three / pointing at
05	should have turned left / we're lost / turn around / exact way back
06	What took you so long / left my phone at home / make a witness statement
07	was wondering / stick with / is breakfast included
08	travel into space / conducting a variety of experiments / effects of low gravity on
09	was founded by / a list of attractions / an hour of free time
10	help me out / sort of an expert / just ten years old
11	include essay questions
12	I'm on my way
13	they're my dad's / it runs in my family / borrow it sometime
14	must be some mistake / check my notes / The order was placed
15	bad temper / belonged to / insulted the other boy / his name written inside
16~17	a group of islands / tourist destination / the largest clove producer / lucky enough to travel

34 영어듣기 모의고사

본문 ▲ p.204

01 ③	02 ④	03 ①	04 ③	05 ④	06 ⑤
07 ③	08 ④	09 ②	10 ④	11 ⑤	12 ③
13 ④	14 ⑤	15 ⑤	16 ③	17 ⑤	

01 ③

여: 날씨 좋은 여름날에 야외에서 시간을 보내는 건 즐겁습니다. 하지만 여름 날씨는 빠르게 변할 수 있기 때문에 주의해야 합니다. 한 시간도 안 되어 밝은 햇빛이 갑자기 위험한 뇌우로 바뀔 수 있습니다. 여러분이 바깥에 있을 때 이런 일이 발생하면, 가장 가까운 건물로 서둘러 들어가십시오. 만약 그럴 수 없다면, 낮은 지대로 뛰어가서 폭풍우가 지나가기를 기다리십시오. 하지만 반드시 전신주 근처에 서 있지 않도록 하십시오. 한 지역에서 가장 높은 것이 번개에 맞을 가능성이 가장 크기 때문에 탁 트인 공간 역시 피해야 합니다. 이 조언들을 염두에 두면 반드시 안전한 여름을 보낼 수 있을 겁니다.

어휘
thunderstorm 뇌우(雷雨) utility pole 전신주

02 ④

남: 이 공원이 폐쇄된다니 슬퍼.
여: 그래. 하지만 우리 주(州)는 더 이상 공원을 개방할 여유가 없대.
남: 알아. 하지만 이 장소는 나에게 좋은 추억이 있어. 가족들과 여기 와서 재미있게 보냈거든. 다른 선택안이 있어야 할 것 같은데.
여: 아이디어가 있어?
남: 음, 하나 있어. 그런데 그게 얼마나 현실성이 있는지는 모르겠어.
여: 들어보자.
남: 주 정부가 일부 공원에서 주차비를 청구해서 돈을 벌 수 있을 거야.
여: 괜찮은 아이디어야. 아마도 그 공원을 개방해 둘 만큼 충분한 돈을 모으게 될 거야.
남: 그걸 공원 공무원에게 얘기해야겠어.

어휘
can't afford to-v …할 여유가 없다 option 선택 (사항) realistic 현실성 있는, 실현 가능한 earn 벌다 charge 청구하다 raise 모으다 official 공식적인; *공무원

03 ①

남: 오늘 와주셔서 감사합니다.
여: 결정을 내리기 전에 당신을 한 번 더 만나 뵙고 싶었습니다. 시작하기 전에 저에게 질문이 있으십니까?
남: 사실, 있습니다. 절 고용하게 되면, 전 언제부터 일을 시작하나요?
여: 이번 주말까지 결정을 내리니 월요일부터 시작하겠군요.
남: 좋습니다. 그럼 전 누구와 일하게 되죠?
여: 경험이 많은 전문가들로 구성된 팀과 일하게 될 겁니다.
남: 그렇군요. 회사 측에서 제시하는 급여는 어떻게 됩니까?
여: 5년 계약을 하게 될 것입니다. 급여는 연봉 4만 달러에서 시작되며 정기적으로 인상될 겁니다.
남: 정말 감사합니다. 여기까지가 제 질문입니다.
여: 좋습니다. 이젠 제가 질문하도록 하겠습니다.

어휘
hire 고용하다 experienced 경험 있는, 노련한 professional 전문가 contract 계약 be eligible for …할 자격이 있다 periodic 주기적인 raise 임금 인상

04 ③

남: 시작할 준비 됐나요?
여: 네, 그런데 좀 긴장되네요. 당근 케이크를 만드는 게 처음이라서요.
남: 걱정하지 마세요. 간단해요.
여: 알았어요. 먼저 뭘 해야 하나요?
남: 먼저, 모든 재료가 탁자 위에 있는지 확인해야 해요.
여: 모두 여기 있는 것 같네요.
남: 달걀이 바구니에 있고, 당근이 조각으로 썰려 있는지 확인하세요.
여: 모든 것이 완벽해요. 그런데 오븐은 어디 있죠?
남: 탁자 바로 뒤에 있어요.
여: 아, 이제 보이네요. 그런데 토끼가 그려진 상자 안에는 뭐가 있나요?
남: 비어 있어요. 끝나고 나서 그 안에 케이크를 넣을 거예요. 집에 가져가셔도 돼요!

어휘
nervous 초조한 ingredient 재료 cut into pieces 토막 내다 empty 비

어 있는

05 ④

남: 어디 가니, 앤?

여: 엄마 심부름을 해야 해요. 엄마가 제게 몇 가지 일을 하라고 시키셨어요.

남: 아, 그래? 무슨 심부름이니?

여: 우선 전화 요금을 내야 해요.

남: 돈과 청구서는 받았니?

여: 네, 아빠. 그리고 세탁소에 엄마 옷을 맡기라고 하셨어요.

남: 도서관 옆에 있는 세탁소에 갈 거니?

여: 네. 실은, 그다음에 도서관에 갈 참이었어요. 숙제로 수필을 읽고 보고서를 써야 해서요.

남: 나한테 연체 도서가 좀 있거든. 네가 반납 좀 해줄 수 있겠니?

여: 그럼요. 책은 어디에 있어요?

남: 커피 테이블 위에 있단다. 내가 가져다줄게. 여기서 잠깐만 기다려라.

여: 네.

어휘
run an errand 심부름을 하다 bill 청구서 dry cleaner's 세탁소 overdue 연체된

06 ⑤

[휴대전화벨이 울린다.]

여: 안녕, 아빠. 무슨 일이세요?

남: 안녕, 얘야. 너 어디니?

여: 친구와 커피숍에서 공부하고 있어요.

남: 아. 나는 네가 지금쯤이면 저녁을 먹으러 집에 와 있겠다고 생각했단다.

여: 음, 학교에서 뭘 먹었어요.

남: 그렇구나. 엄마는 어디 있는지 아니?

여: 저녁을 집에서 안 먹는다고 알려 드리려고 아까 엄마한테 전화했어요. 아마 잠시 나가셨을 거예요.

남: 아, 이제 오네. 엄마가 포장 음식을 사 온 것 같구나.

여: 잘됐네요. 두어 시간 뒤에 봬요.

남: 알았어, 얘야. 이따 보자.

어휘
take-out food 포장 음식 a couple of 두서너 개의, 몇 개의

07 ③

여: 이 웹사이트에 네가 찾고 있던 책이 있어. 여러 가지 인쇄판이 있네.

남: 나는 네 번째로 펴낸 책이 필요해. 그게 최신판일 거야, 그렇지? 얼마야?

여: 응, 맞아. 종이책 버전은 25달러야. 그리고 전자책도 있어! 그건 단지 15달러야.

남: 나는 종이책을 사고 싶어. 배송비는 얼마야?

여: 일반 배송은 4달러야.

남: 그럼 여기 도착하는 데 얼마나 걸리지?

여: 음, 4달러짜리는, 여기까지 3일에서 5일이 걸려. 그러나 10달러를 내면 내일 받을 수 있어.

남: 아니야, 그건 괜찮아. 서두르지 않아도 되거든.

여: 알겠어. 그럼 일반 배송으로 할게. 다른 건?

남: 응. 내 친구의 것으로 한 권 더 살래. 그 애 생일이 다음 주거든.

여: 아, 그럼, 45달러 넘게 사면 배송비는 무료야.

남: 잘됐다! 그럼 종이책 두 권을 주문하는 거로 하자.

어휘
edition (책 · 간행물 등의) 판, 호 latest 최근의, 최신의 paperback 페이퍼백 (표지가 종이 한 장으로 된 책) shipping 선박; *배송 in a hurry 서둘러, 급히 copy 복사본; *(책 · 신문 등의) 한 부

문제풀이
남자는 권당 25달러인 종이책을 두 권 산다고 했고, 45달러 이상 사면 배송비는 무료라고 했으므로 남자가 지불할 금액은 50달러이다.

08 ④

여: 존, 미국에서 방울뱀을 본 적이 있니?

남: 응, 사막에서.

여: 꼬리 끝에 방울이 진짜로 달려 있니?

남: 응, 그래. 까마귀와 코요테, 인간과 같은 포식자에게 경고하려고 꼬리를 흔들어.

여: 방울뱀이 그런 소리를 내는 걸 들어본 적 있어?

남: 아니. 하지만 모든 방울뱀이 서로 다른 달랑거리는 소리를 내.

여: 왜 그런 거지?

남: 방울뱀이 클수록 소리가 더 커져. 그리고 더 높은 체온은 소리를 더 크게 만들지.

여: 아, 흥미로운걸. 방울뱀은 먹잇감을 물어서 죽이니?

남: 응. 그들이 무는 것은 독성이 강해서 북미에서는 80퍼센트가 넘는 뱀에 물린 사망자가 발생해.

여: 와! 그들은 아주 크니?

남: 응, 2미터 넘게 자랄 수 있어.

어휘
rattlesnake 방울뱀 rattle 방울, 딸랑이; 덜걱덜걱 소리 내다 warn 경고하다 predator 포식자 bite 물다; 물기 prey 먹이 fatality 사망자

09 ②

남: 안녕하십니까, 여러분. 오늘 Old Mine History Tour를 예약해주셔서 감사합니다. 관광의 첫 번째 부분은 폐광된 지 오래된 터널을 전동차를 타고 통과하는 것입니다. 이것은 여러분을 500미터가 넘는 지하에 모셔다드릴 흥미진진한 여정입니다! 그 이후에는 이 지역의 역사학자가 방문객을 위해 개조된 광산의 특별한 곳을 도보로 안내할 것입니다. 그리고 나서 우리는 지상으로 돌아오게 되며, 아름다운 박물관 정원에서 점심 식사가 제공될 것입니다. 점심 식사 후에는 금광 박물관을 둘러보고 기념품점을 방문할 시간이 주어질 것입니다. 기념품점에서는 이 지역에서 생산된 금으로 만든 수공예 보석류를 포함한 다양한 기념품을 판매합니다.

어휘
mine 광산 disused 사용되지 않는, 폐기된 underground 지하에 historian 역사가 on foot 도보로 convert 개조하다, 전환시키다 wander around 돌아다니다 souvenir 기념품 handmade 손으로 만든 jewelry 귀금속, 보석류

10 ④

여: 윌리엄, 너 내년에 교내에서 살 거지?
남: 아냐, 형이랑 지낼 교외 아파트를 찾아봤어. 넌?
여: 난 교내에서 살고 싶은데, 생활관 결정을 못 했어.
남: 그것들은 꽤 비싸. 교내에서 살 여유가 있는 건 확실해?
여: 응, 월 천 달러 미만이면 어떤 곳이라도 좋아.
남: 오, 좋네. 그러면 난 Grey 관을 추천할게. 작년에 내가 거기 살아봤는데, 정말 마음에 들었어.
여: 거기 방들이 좋긴 한데, 욕실을 같이 써야 해.
남: 그렇긴 하지. 그러면 너는 개인 욕실을 선호하는 거야?
여: 응, 맞아. 그리고 나는 급식이 포함된 곳을 원해.
남: 알겠어. 넌 6개월짜리 계약을 원해, 아니면 12개월짜리 계약을 원해?
여: 난 더 긴 게 좋아. 내내 이사 다니는 건 지긋지긋해.
남: 어떤 기분인지 알아. 아무튼, 이 생활관이 너한테 가장 좋은 선택인 것 같다.

어휘
off-campus 교외의 residence hall 생활관 private 개인의 meal plan 식권; *급식 contract 계약

11 ⑤

여: 안녕, 잭. 기분이 안 좋아 보이는데. 무슨 일이야?
남: 내 컴퓨터가 작동을 멈췄어. 바이러스에 걸린 것 같아.
여: 오, 이런. 지금 당장 끝내야 할 게 있니?
남: 응. 오늘 저녁에 보고서를 이메일로 보내야 해.

어휘
[문제] replace 대신[교체]하다

문제풀이
① 어떤 것이든지 큰 도움이 될 거야.
② 응. 난 컴퓨터 수리점에 전화할 거야.
③ 아니. 난 컴퓨터를 잘 다루지 못해.
④ 난 그것을 교체할 만한 충분한 돈이 없어.

12 ③

남: 안녕하세요. 제 디지털카메라를 가지러 왔는데요.
여: 알겠습니다. 성함을 말씀해주시겠습니까?
남: 제 이름은 토마스 부처예요. 사흘 전에 맡겼어요.
여: 죄송하지만, 손님의 카메라는 아직 수리가 안 됐습니다.

어휘
drop ... off …을 갖다 주다[맡기다] [문제] deliver 배송하다 refund 환불 technician 기술자

문제풀이
① 손님의 회사로 배송되는 것을 원하십니까?
② 여기를 보니 파란색을 주문하셨군요.
④ 죄송하지만, 이 상점에서는 환불을 해드리지 않습니다.
⑤ 지금 바로 기술자를 댁으로 보내드릴 수 있습니다.

13 ④

남: 그 상자 안에 든 게 뭐야? 너 쇼핑했어?

여: 응, 오늘 TV 한 대 새로 샀어.
남: 새 TV? 쓰던 게 뭐 잘못됐어?
여: 아니, 그냥 나한테 선물하고 싶었어.
남: 그걸 어디서 샀니?
여: 모퉁이에 있는 전자 상가에서.
남: 아, 나 거기 알아. 싸게 샀어?
여: 응, 천 달러밖에 안 하던데. 그런 좋은 가격에 그걸 사지 않을 수가 없었어.
남: '천 달러밖에'라고? 너무 비싼 거 같은데.
여: 음, 화면이 커.
남: 난 화면이 큰 TV가 뭐가 좋은지 정말 모르겠던데.
여: 오늘 밤에 놀러 와. 같이 DVD를 보면 너도 알게 될 거야.

어휘
treat 한턱내다 electronics 전자 제품 deal 거래 can't help but …할 수밖에 없다 screen 화면 appeal 매력

문제풀이
① 원한다면 가져가. 난 괜찮아.
② 우린 TV를 너무 많이 봐. 카드 게임이나 하자.
③ 난 전자 상가에서 하나 살 거야.
⑤ 팝콘 있어? 난 팝콘 없이는 영화를 못 봐.

14 ⑤

여: 아, 여기 계산서가 있어. 가서 계산하자.
남: 이런. 깜박하고 지갑을 안 가져온 것 같아.
여: 괜찮아. 내가 전부 낼 수 있어.
남: 너무 당황스러워.
여: 당황할 거 없어. 별일 아니잖아.
남: 내가 꼭 갚을게.
여: 무슨 소리야. 지난번에 네가 저녁을 샀잖아.
남: 응, 하지만 그땐 내가 먹자고 한 거였잖아.
여: 그럼 다음 주에는 내가 저녁을 먹자고 할 테니까 그때 네가 갚으면 돼.
남: 그래, 그러면 되겠다.
여: 계산하고 나서 우리 극장에 가면 되겠어.
남: 알겠어, 근데 먼저 우리 집에 들러서 내 지갑부터 챙기자.

어휘
check 계산서 forget 잊어버리다 (forgot-forgotten) wallet 지갑 embarrassing 당황스럽게 하는 embarrassed 당황스러운, 당황한 deal 사항, 중요한 일 pay ... back (빚 등을) 갚다 [문제] available 시간이 있는 insist 주장하다 stop by …에 들르다

문제풀이
① 글쎄, 난 이번 주에는 안 돼.
② 좋아. 나 지금 돈 많아.
③ 시간 좀 봐! 우리 회사에 늦겠어.
④ 전혀 미안해할 필요 없어.

15 ⑤

여: 타일러와 애비는 학교의 생물 수업 과제 파트너이다. 과제를 하려면 두 사람이 모여야 하는데, 애비는 늘 바빠 보인다. 마감일이 월요일 아침으로 다가와서, 타일러는 결국 혼자서 과제를 전부 다 해야 할 것 같아 걱정된다. 그는 과제를 하는 건 괜찮지만, 애비가 그것으로 점수

를 받아서는 안 된다고 생각한다. 애비는 방금 타일러에게 일요일 저녁이 될 때까지는 만날 수가 없다고 말했다. 이런 상황에서, 타일러가 애비에게 할 말로 가장 적절한 것은 무엇인가?

타일러: 점수를 받고 싶다면 이 과제에 참여해야 해.

어휘
biology 생물학 **get together** 모이다 **deadline** 마감 시간 **come up** 다가오다, 가까이 오다 **end up v-ing** 결국 …하다 **by oneself** 혼자 **credit** 학점

문제풀이
① 우리 월요일에 생물 수업이 있니?
② 일요일 오전에 도서관에서 봐.
③ 미안한데, 나 이번 주말에 정말 바빠.
④ 너와 파트너로 이 과제를 하게 돼서 기뻐.

16 ③ 17 ⑤

남: 안녕하세요. 오늘 원예 수업에서는, 약에 관해 이야기할 것입니다. 약국에서 여러분이 사는 약의 종류가 아니라, 정원에서 여러분이 재배할 수 있는 종류를 말합니다. 개인적으로 저는 제 약초 정원에 항상 많은 바질을 심습니다. 홀리 바질이라고 알려진 특정 종류의 바질은, 감기, 기침, 고열을 치료하는 데 아주 좋습니다. 그리고 복통을 겪고 있다면 카모마일을 꼭 심도록 하세요. 그것은 여러분이 섭취한 음식을 몸이 소화하도록 돕습니다. 생강은 또 다른 소화 보조제입니다. 그것은 여러분의 면역 체계를 강화하며, 여러분의 몸이 질병과 맞서 싸울 수 있게 해줍니다. 그리고 마지막으로, 혈압이 걱정되시면, 정원에 마늘을 위한 공간을 좀 만드세요. 맛있다는 것 외에도 그것은 여러분의 심장을 건강하게 유지하고 혈압을 낮게 하는 데 도움이 됩니다. 이 모든 식물들은 비싸지도 않고 기르기도 쉬우니, 어떤 정원에도 훌륭한 새 식구가 될 것입니다.

어휘
gardening 원예 **pharmacy** 약국 **cure** 치료하다, 고치다 **cough** 기침 **digestive aid** 소화 보조제 **strengthen** 강화하다 **immune system** 면역 체계 **make an addition** 첨가하다

문제풀이
16 ① 천연 소화 보조제
② 완벽한 정원 만들기
③ 식물의 건강상의 이점들
④ 가정 치료법의 위험성
⑤ 소화에 미치는 약초의 영향

DICTATION Answers
 본문 p.206

01 In less than / run to / have a safe summer
02 can't afford to keep it open / has to be another option / charging parking fees
03 making a decision / hire me / experienced professionals
04 all the ingredients / cut into pieces / put the cake in it
05 run some errands / pay the phone bill / read an essay / some overdue books
06 studying with my friend / went out / in a couple of hours
07 you've been looking for / How much is shipping / get a

copy
08 to warn predators / every single rattlesnake / bite their prey
09 an electric train / guide you on foot / time to wander around
10 They're all pretty expensive / have to share a bathroom / I'm tired of moving
11 need to get done
12 dropped it off
13 treat myself / get a good deal / see the appeal
14 have forgotten my wallet / no big deal / pay the bill
15 a school project / deadline is coming up / end up doing / get credit for it
16~17 you buy at the pharmacy / suffer from stomach problems / another digestive aid / make an excellent addition to

35 영어듣기 모의고사
본문 ▲ p.210

01 ⑤	02 ①	03 ⑤	04 ④	05 ④	06 ①
07 ④	08 ②	09 ③	10 ⑤	11 ①	12 ⑤
13 ④	14 ②	15 ③	16 ④	17 ④	

01 ⑤

남: 미국의 가장 중요한 천연자원 중 하나는 플로리다의 에버글레이즈입니다. 불행히도, 이 지역 대부분이 토지 개발 때문에 파괴되었습니다. 이 연약한 생태계는 보호되어야 합니다. 그곳은 흑곰과 인디고 뱀, 거북을 포함해 수천 가지 동물과 식물 종의 서식지입니다. 또한, 플로리다 환경의 중요한 일부이기도 합니다. 이 습지를 떠난 깨끗한 물은 플로리다 해안 근처의 작은 섬과 산호층에 중요합니다.

어휘
resource 자원 **destroy** 파괴하다 **due to** …에 기인하는, … 때문에 **delicate** 연약한, 섬세한 **ecosystem** 생태계 **wetland** 습지(대) **coral** 산호 **formation** 형성(물) **coast** 해안

02 ①

남: 이번 주말에 특별한 걸 했니, 클라라?
여: 친구랑 난 우리 집 근처에 4D 영화를 보러 갔어.
남: 아, 특수 효과가 있는 그런 극장이었니?
여: 맞아. 섬광과 바람, 분무, 심지어 향도 나왔어. 그리고 화면에서 일어나는 일에 맞게 의자가 움직였어.
남: 재미있는 경험인 것 같아.

여: 그랬어. 너도 언젠가 반드시 가보는 게 좋을 거야.

남: 음, 표가 일반 표 값의 세 배라서 보려고도 하지 않았어.

여: 날 믿어. 실망하지 않을 거야. 사실, 곧 또 다른 4D 영화를 보러 가고 싶어!

남: 어쩌면 우리 같이 갈 수도 있겠다.

어휘

special effect 특수 효과　flashing light 섬광　water spray 분무 definitely 분명히, 틀림없이

03 ⑤

남: 주문하시겠어요, 아니면 메뉴를 더 보시겠어요?

여: 전 아직 특별 요리들을 살펴보는 중이에요.

남: 제가 그것들을 설명해 드릴까요?

여: 네. 그래 주시면 정말 좋겠네요.

남: 오늘의 생선 요리는 신선한 연어입니다. 아홉 가지 허브로 양념한 것입니다.

여: 전 점심으로 생선을 먹었어요. 스테이크는 어떤가요?

남: 스테이크도 아주 좋은 선택입니다. 립아이 스테이크인데, 육즙이 풍부하죠. 세 가지 사이드 요리가 곁들여 나옵니다.

여: 저는 스테이크로 해야겠어요.

남: 어떻게 요리해 드릴까요?

여: 미디엄 레어로 해주세요.

어휘

look over …을 훑어보다　special (식당의) 특별 메뉴　salmon 연어　season 양념하다　juicy 즙[수분]이 많은　come with …에 부수되다, …이 딸려 나오다 side 곁들여 나오는 요리　medium-rare 약간 덜 익힌

04 ④

남: 내 캠핑 여행 사진을 좀 봐.

여: 어, 사람이 두 명밖에 없네. 너와 네 친구니?

남: 응. 우린 물고기를 몇 마리 잡았어.

여: 그렇구나. 바구니에 물고기 몇 마리가 있는 것 같네.

남: 맞아.

여: 호수에서는 수영할 수 없었던 것 같구나.

남: 응, 저기서 수영하는 것을 금지한다고 쓰인 표지판이 있었어. 하지만 우린 텐트를 가지고 가서 그 안에서 잤어.

여: 그래, 그건 멋진 돔형 텐트네.

남: 그냥 지난달에 하나 샀어.

여: 그렇구나. 그런데 넌 왜 그렇게 거기서 낚시하는 걸 좋아하니?

남: 난 그냥 자연 속에 있는 게 좋아. 아주 평화롭고, 그 호수 주변의 숲을 바라보는 게 좋아.

어휘

dome 돔형(의), 반구형 지붕(의)　stare into …을 응시하다

05 ④

[문 두드리는 소리]

여: 네? 누구세요?

남: 옆집 사는 크리스예요.

여: 오, 안녕하세요, 크리스. 무슨 일이시죠?

남: 우리가 서로를 그다지 잘 알지 못한다는 건 알지만, 제가 부탁을 하나 드려도 될까 해서요.

여: 그러세요.

남: 제가 내일 아침에 갑작스럽게 출장을 가야 해요. 일주일 동안 떠나 있을 거라서 저 대신 스파키를 돌봐주실 수 있을까 해서요.

여: 죄송하지만, 크리스, 못 도와 드릴 것 같네요. 전 개를 아주 무서워 하거든요.

남: 오, 아니에요.

여: 어릴 때부터 전 개를 무서워했어요.

남: 제 말은 그게 아니고요, 스파키는 개가 아니에요. 제 개 이름은 조조고 우리 부모님과 함께 지낼 거예요.

여: 그럼 스파키는 누구예요?

남: 스파키는 제 금붕어예요.

여: 아! 그렇다면 문제없어요.

어휘

unexpected 예기치 않은　nervous 불안해[두려워]하는　goldfish 금붕어

06 ①

[전화벨이 울린다.]

남: 안녕하세요. Ace Powercom입니다.

여: 여보세요. 고객 서비스 센터죠?

남: 네, 무엇을 도와드릴까요?

여: 인터넷 연결에 문제가 있어요.

남: 알겠습니다. 어떤 유형의 서비스를 이용하고 계세요?

여: 무선 인터넷 서비스인데, 와이파이 신호가 잡히지 않아요. 지난 2시간 동안 작동하지 않고 있어요.

남: 알겠습니다. 우선, 와이파이 공유기를 끄신 다음 5분 뒤에 다시 켜세요. 보통 그렇게 하면 문제가 해결되거든요.

여: 알겠어요. 그런데 만약 효과가 없으면요?

남: 그럼, 다시 전화를 주시면 댁에 방문할 기사를 주선해 드릴게요.

여: 그렇게 할게요. 도와주셔서 감사합니다.

어휘

customer service 고객 서비스 센터　connection 연결　signal 신호 router 라우터(중계 장치), 공유기　fix 해결하다　arrange 마련[주선]하다

07 ④

여: 안녕하세요. 무엇을 도와드릴까요?

남: 체리와 블루베리를 사고 싶은데요. 얼마죠?

여: 체리는 파운드당 2달러고, 블루베리는 파운드당 4달러예요.

남: 그럼 각각 2파운드씩 살게요.

여: 네. 다른 필요한 것이 있으신가요?

남: 아니요. 저는 단지 체리와 블루베리만 사려고 했어요.

여: 정말요? 저는 복숭아를 매우 추천해요. 올해 복숭아가 정말 맛있어요.

남: 음… 얼만데요?

여: 한 봉지에 단지 8달러예요. 한 봉지에는 5개의 복숭아가 들어있고요. 한 조각 먹어보실래요?

남: 네, 주세요. [잠시 후] 음… 정말 맛있네요. 그런데 저는 이미 체리와 블루베리를 사서요. 돈이 많이 들 것 같은데요.

여: 복숭아 한 봉지를 사시면, 전체 금액의 10퍼센트를 할인해 드릴게요.

남: 음, 괜찮은 제안이네요. 알겠어요. 복숭아도 한 봉지 살게요.

어휘
highly 매우, 대단히 total 총, 전체의

문제풀이
남자는 체리($2)와 블루베리($4)를 각각 2파운드씩 샀고 복숭아($8)도 한 봉지 구입하여 총 가격은 20달러인데, 총액에서 10퍼센트 할인을 받으므로 남자가 지불할 금액은 18달러이다.

08 ②

남: 그게 비행기 표인가요? 어디 가세요?
여: 네. 네덜란드에 갈 거예요.
남: 출장으로 가시나요?
여: 아니요. 남동생이 암스테르담에서 산업공학을 공부하고 있어서 보러 가는 거예요.
남: 암스테르담이요? 그게 수도인가요?
여: 네. 그 나라에서 가장 큰 도시이기도 해요. 빨리 가고 싶어요.
남: 음… 유럽 지도를 떠올려 보고 있는데요. 독일 근처죠, 그렇죠?
여: 네. 동쪽으로는 독일을 접하고 남쪽으로는 벨기에를 접하고 있죠.
남: 맞아요. 제 기억이 맞는다면, 상당히 작은 나라죠.
여: 네. 인구가 단 1,700만 명 정도예요.
남: 그럼 독일어를 쓰나요?
여: 아니요, 그렇지 않아요. 네덜란드어를 써요. 가기 전에 네덜란드어를 공부할 거예요.

어휘
industrial engineering 산업공학 capital 수도 border 국경, 경계; *(국경을) 접하다 population 인구 Dutch 네덜란드어

09 ③

여: 우선 학생회에 참가하겠다고 신청해주신 모든 분께 감사드립니다. 우리는 학기 내내 주간 회의를 하게 될 것입니다. 이 회의는 휴일 일정과 맞물리는 경우를 제외하고는 매주 수요일에 열릴 것입니다. 우리가 첫 번째로 해야 할 일은 회장과 부회장을 선출하는 것입니다. 일단 우리가 이 임원들을 선출하고 나면, 그들이 회의 안건을 정하는 것을 맡게 될 것입니다. 우리는 또한 몇몇 다양한 특별 위원회를 갖게 될 것이며, 모든 사람이 그중 하나에 자원해야 할 것입니다. 그리고 물론, 이것은 방과 후 활동이므로 여러분은 부모님께서 서명하신 허가서를 제출해야 합니다. 그럼, 질문 있습니까?

어휘
sign up 등록[신청]하다 student council 학생회 conflict with …와 충돌[상충]하다 order of business (처리해야 할) 문제, 과제 elect 선출하다 vice-president 부회장 officer 임원, 간부 in charge of …을 맡아서[담당하여] agenda 의제, 안건 committee 위원회 permission 허가

10 ⑤

남: 난 방금 다가올 사업 학회에 우리를 등록했어요. 이제 호텔을 골라야 해요.
여: 그래요. 학회 측에서 추천 장소 목록을 제공했어요?
남: 네. 그것은 심지어 회의장에서 도보 거리까지 포함하고 있어요.
여: 좋네요! 우리가 10분 이내에 걸어갈 수 있는 곳으로 골라요.

남: 그래요. 우리는 또한 무료 인터넷 연결이 되는 곳을 선택해야 해요.
여: 동의해요. 우리는 학회 후에도 일해야 할 테니까요.
남: 난 저녁에 운동도 하고 싶어요. 그게 내가 스트레스를 해소하는 방법이거든요.
여: 피트니스 센터가 있는 호텔을 찾을 수 있을 거라 확신해요.
남: 네, 목록에 몇 개가 있네요. 당신은 요구 사항들이 있어요? 이곳에는 쇼핑 센터도 있는데.
여: 아뇨, 난 쇼핑하러 갈 필요는 없어요. 하지만 난 스파에서 좀 쉬는 게 괜찮을 것 같아요.
남: 알겠어요. 그러면 우리의 모든 요구를 충족하는 호텔이 하나 있는 것 같네요.
여: 좋아요. 팀장님께 거기에 방 2개를 예약해 달라고 요청합시다.

어휘
upcoming 다가오는 work out 운동하다 relieve 덜다, 해소하다 requirement 요구 사항 relax 쉬다, 이완하다 meet 만족시키다 [문제] additional 추가의

11 ①

남: 샐리, 저녁 먹을 시간이야!
여: 몇 분만 더요, 아빠! 이 비디오 게임에서 이기고 있거든요.
남: 안 돼, 음식이 식고 있어. 지금 와서 먹으렴.
여: 알았어요. 금방 갈게요.

어휘
[문제] in the mood for …할 기분이 나서

문제풀이
② 저는 비디오 게임이 정말 싫어요.
③ 저녁 식사로 뭘 만들까요?
④ 아빠가 이 게임을 좋아하실 줄 알았어요.
⑤ 죄송해요, 저는 점심을 먹을 기분이 아니에요.

12 ⑤

여: 실례합니다, NY 백화점에 가려면 어떤 지하철 노선을 타야 하나요?
남: 2호선을 타세요.
여: 고마워요. 어디서 내려야 하는지도 말씀해 주시겠어요?
남: 시청역에서 내리세요. 여기서 세 정거장이에요.

어휘
get off 내리다, 하차하다 [문제] transfer 옮기다; *갈아타다

문제풀이
① 20분 정도 걸릴 거예요.
② 3호선으로 갈아타셔야 할 것 같아요.
③ 맞아요! 당신은 지하철을 타셔야 해요.
④ 서둘러야 해요. 지하철이 아주 붐비거든요.

13 ④

여: 얘, 패트릭!
남: 아, 안녕, 앨리스. 무슨 일이야?
여: 내 테니스 라켓을 갖고 오는 걸 깜박했어. 체육 시간에 그게 필요하거든. 너 갖고 있니?
남: 음, 갖고 있긴 한데, 그건 내 사물함에 있어.

여: 그것 좀 빌릴 수 있을까? 너도 알다시피 김 선생님은 수업 준비에 굉장히 엄격하시잖아.

남: 네가 지난번에 그걸 빌려 갔을 때 너는 나한테 돌려주지 않았어. 그래서 내가 너희 교실까지 가지러 가야 했어, 기억해?

여: 아, 그랬어? 미안해. 이번에는 수업이 끝나는 대로 바로 돌려줄게. 잊어버리지 않을 거야. (라켓을 빌려주면) 정말 날 크게 도와주는 거야.

남: 좋아, 하지만 다 쓰고 나서 그걸 바로 갖다 줘.

physical education 체육 locker 사물함 strict 엄격한 huge favor 어려운 부탁 [문제] deadline 기한, 마감 일자 late fee 연체료

문제풀이
① 사실 난 테니스를 그렇게 좋아하지는 않아.
② 우리가 같이 김 선생님의 수업을 듣게 되어 기뻐.
③ 고마워. 네가 날 도와준 걸 절대 잊지 않을게.
⑤ 기한은 어제까지였으니 연체료를 내야 해.

14 ②

남: 샐리, 당장 이리 와 봐!

여: 무슨 일이니?

남: 무슨 일이냐고? 네가 두 시에 날 데리러 왔어야 했잖아.

여: 그랬나? 난 그런 약속을 한 기억이 없는데.

남: 음, 그랬거든.

여: 난 정말 그런 말을 한 기억이 없어. 내가 왜 너한테 거짓말을 하겠어?

남: 네가 거짓말을 한다는 게 아니야. 단지 네가 오겠다고 약속했다는 거지.

여: 내가 그 약속을 언제 했는데?

남: 오늘 아침에 네가 날 내려줬을 때.

여: 이제 생각난다! 정말 미안해.

남: 괜찮아. 이제라도 생각이 났으니까.

여: 이걸 어떻게 보상해야 하지?

get over 넘다, 건너다; *가다 be supposed to-v …하기로 되어 있다 commitment 약속 drop off (차로 가는 도중에) 내려주다 [문제] make it up …에게 변상하다[물어주다] liar 거짓말쟁이

문제풀이
① 내게 진실을 말해줘.
③ 정말 금방 돌아온다고 약속할게.
④ 아니, 네가 날 데리러 오기로 되어 있었잖아.
⑤ 네가 날 거짓말쟁이라고 부르다니 믿을 수가 없어.

15 ③

남: 케이트는 법정 변호사이다. 그녀는 매일 법정에서 그녀의 의뢰인을 변호한다. 케이트는 자기 일에 대한 자부심이 대단하며 더 훌륭한 변호사가 되기 위해 더 열심히 노력한다. 케이트의 옆집에 사는 이웃인 재닛에게는 케이트와 그녀의 직업을 존경하는 어린 딸이 있다. 그 소녀는 겨우 3학년이지만 커서 케이트처럼 성공한 법정 변호사가 되기를 바란다. 케이트가 어린이들을 매우 좋아하고 청소년 센터에서 자원봉사도 한다는 것을 알고는, 재닛이 케이트에게 다가간다. 재닛은 케이트에게 자기 딸이 케이트의 일을 하루 동안 따라 다니면서 법정을 볼 수 있는지 물어본다. 이런 상황에서, 케이트가 재닛에게 할 말로 가장

적절한 것은 무엇인가?

케이트: 괜찮아요. 문제없어요.

trial lawyer 법정 변호사 court 법정 defend 방어하다; *변호하다 client 의뢰인, 고객 take pride in …을 자랑하다 admire 존경하다 adore 아주 좋아하다 approach 다가가다 tag along 따라붙다, 쫓아다니다 [문제] dare 감히 …하다

문제풀이
① 전 다음 주에 이사 갈 거예요.
② 절 위해 일해 주시겠어요?
④ 어떻게 감히 그들이 저한테 그런 걸 부탁하죠?
⑤ 당신은 당신 딸을 위해 파티를 계획했어야 했어요.

16 ④ 17 ④

여: 주목해 주시겠습니까? 이번 주말 일기 예보에 따르면 이상 고온 현상이 지속한다고 합니다. 대부분 지역은 현재 폭염 주의보로부터 안전합니다. 하지만 바람 부족과 고온이 함께 작용하여 이번 주말에 공기 상태를 위험할 정도로 나쁘게 만들겠습니다. 피할 수 없는 게 아니라면, 도시 거주민은 앞으로 이틀 동안 야외에서 시간을 보내시면 안됩니다. 이는 특히 어린이들과 고령자의 경우에 그렇습니다. 알레르기를 앓고 있거나 공기 오염에 민감한 사람도 야외에서 보내는 시간의 양을 제한하기 위해 노력해야 합니다. 그뿐만 아니라, 상황이 나빠지는 것을 막기 위해, 모든 분들께 운전을 가능한 한 최소로 하고 정원 쓰레기 태우기를 자제할 것을 요청합니다. 월요일 오후에는 한랭전선이 이 지역을 통과하며 많은 비를 뿌릴 예정입니다. 이로 인해 상황은 상당히 개선될 것입니다. 오늘 오후 최고 기온은 섭씨 30도로 예상됩니다. 새롭게 추가되는 정보를 들으시려면 주말 내내 저희 방송국에 채널을 고정해 주세요. 감사합니다.

call for 요구하다; *(날씨를) 예보하다 temperature 온도, 기온 heat 열; *더위 combine 결합하다 unavoidable 불가피한 the elderly 어르신들 sensitive 예민한, 민감한 refrain from v-ing …하기를 자제하다 cold front 한랭전선 significant 중요한, 의미 있는 improvement 향상, 개선 stay tuned 채널을 고정하다

DICTATION Answers
본문 p.212

01 due to land development / thousands of animals / is important for the small islands

02 with special effects / what was happening / tickets are three times

03 looking over your specials / seasoned with nine herbs / comes with three sides

04 caught some fish / weren't allowed to swim / stare into the forest

05 ask you a favor / unexpected business trip / I've been afraid of

06 having a problem with / turn it back on / arrange for an engineer

07 two pounds of each / highly recommend / discount on your total

08 Is that the capital / borders Germany to the east / has a population

09 take part in / be held on Wednesdays / in charge of / a permission form

10 includes the walking distance / how I relieve stress / I wouldn't mind relaxing

11 a few more minutes

12 where I should get off

13 strict about being prepared / give it back / a huge favor

14 get over here / such a commitment / you are lying / dropped me off

15 takes great pride / grow up to become / tag along to

16~17 combined with a lack of wind / spend time outdoors / make an effort to limit / refrain from burning / stay tuned

MEMO

만만한 수능영어

수능만만

영어듣기
35회

1. 고2, 3 학생들을 위한 수능듣기 실전 난이도를 반영한 훈련서

2. 최신 수능 및 평가원 모의고사를 철저히 분석, 출제 가능성 높은 문항 선별 수록

3. 어려운 연음, 핵심 내용, 관용 표현 등을 중심으로 한 Dictation 수록

4. 수능 듣기 유형 및 유형별 필수 어휘와 표현을 완벽 정리한 부록 MINI BOOK 제공